METHODS IN MOLECULAR BIOLOGY™

Series Editor
John M. Walker
School of Life Sciences
University of Hertfordshire
Hatfield, Hertfordshire, AL10 9AB, UK

For further volumes:
http://www.springer.com/series/7651

Unnatural Amino Acids

Methods and Protocols

Edited by

Loredano Pollegioni

Università degli studi dell'Insubria, Varese, Italy

Stefano Servi

Politecnico di Milano, Milano, Italy

Humana Press

Editors
Loredano Pollegioni
Dipartimento di Biotecnologie e Scienze Molecolari
Università degli studi dell'Insubria
Varese, Italy
loredano.pollegioni@uninsubria.it

Stefano Servi
Politecnico di Milano
Milano, Italy
stefano.servi@polimi.it

ISSN 1064-3745 e-ISSN 1940-6029
ISBN 978-1-4939-5888-7 ISBN 978-1-61779-331-8 (eBook)
DOI 10.1007/978-1-61779-331-8
Springer New York Dordrecht Heidelberg London

Printed on acid-free paper

Humana Press is part of Springer Science+Business Media (www.springer.com)

Preface

Even though nonproteinogenic amino acids are present in nature, they are usually defined as unnatural or nonnatural, meaning that they are other than the 22 natural L-α-amino acids (L-α-AAs) constituting natural peptides and proteins. The definition usually adopted is not specific for defined structural classes since among the group there are L-α-AAs differing in their chemical structure, α-AAs of D-configuration, non-α-AAs of both absolute configurations. Besides their structural diversity, interest in these compounds is due to a number of reasons such as

their occurrence in nature
their biological properties
their biosynthetic pathway (often related to posttranslational modification of natural AAs) and the enzymes acting on them
the chemical and enzymatic methods for their production
the analytical aspects
their use as probes
their incorporation into peptides and proteins (and the related biotechnological uses).

The volume of *Methods in Molecular Biology* series entitled "Unnatural Amino Acids" with the well-established format of detailed experimental procedures in the step-by-step protocols approach (an introductory overview, a list of the materials and reagents needed to complete the experiment, followed by a detailed procedure supported with a helpful notes section offering tips and tricks of the trade as well as troubleshooting advices) addresses these topics, assembling the material into five sections (parts). Part I (*Synthesis of Unnatural Amino Acids*) deals with methods where enzyme catalysis is used to produce nonnatural amino acids. Attention is focused mainly on the obtainment of those enzymes and also on the practical aspects of their application in biotransformations (target compounds are required for the study and preparation of drug candidates of increasing interest and complexity). The second section (Part II, *Applications of Unnatural Amino Acids*) deals with aspects concerning the presence of unnatural AAs in peptides with antibiotic properties. This includes the application of unnatural amino acids to the *de novo* design of selective antibiotic peptides, the site-specific incorporation of unnatural amino acids, and the related conformational changes observed in proteins, as well as the use of unnatural amino acids to probe structure–activity relationships in antimicrobial peptides.

Part III, *Use of Unnatural Amino Acids in Protein Synthesis*, deals with applications of specific techniques allowing unnatural amino acids to be genetically incorporated into proteins both in yeast and mammalian cells. In both cases, orthogonal tRNA/aminoacyl-tRNA synthetase pair is used taking advantage of suitable techniques and allowing high incorporation of the nonnatural AA. Most aspects concerning AAs of D-configuration are grouped in

Parts IV and V (12 chapters), namely, *Analysis and Applications*, and *Enzymes Active on D-Amino Acids*. The analytical studies address two main aspects:

- the determination of D-AAs in food – where their presence is associated with bio-availability and impairing of food nutritional value, thus becoming indicators of food quality
- their determination in biological fluids – e.g., investigations of different neurological and psychiatric disorders correlating the appearance/susceptibility to these disorders with the concentration of D-serine in brain.

the importance of enzymes acting on (D-specific oxidases, proteases) or producing (racemases, isomerases) D-AAs is also emphasized in chapters dedicated to the study of these specific biocatalysts.

We believe that the future will likely see many more researchers focusing on the investigation and utilization of unnatural amino acids: these molecules promise of becoming an argument with enormous potential. Comprehensive but convenient, this manual, giving detailed prescription in Molecular Biology, will contribute to drive the attention to the many fields of growing scientific interest in nonnatural amino acids.

Varese, Italy *Loredano Pollegioni*
Milano, Italy *Stefano Servi*

Contents

Part III Use of Unnatural Amino Acids in Protein Synthesis

Part IV D-amino Acids: Analysis and Applications

Part V Enzymes Active on d-amino Acids

Contributors

CLARA AICART-RAMOS • *Departamento de Bioquímica y Biología Molecular, Facultad de Ciencias Químicas, Universidad Complutense, Madrid, Spain*
YASUHISA ASANO • *Biotechnology Research Center and Department of Biotechnology, Toyama Prefectural University, Toyama, Japan*
ZEINAB ASSAF • *Lab SEESIB, Clermont Université, Université Blaise Pascal, Aubière, France*
FLORIAN BAUMGART • *Departamento de Bioquímica y Biología Molecular, Facultad de Ciencias Químicas, Universidad Complutense, Madrid, Spain*
VERENA BÖHRSCH • *Institut für Chemie und Biochemie, Freie Universität Berlin, Berlin, and Labor fur Biochemie, Beuth-Hochschule für Technik, Berlin, Germany*
STEPHANIE G. BURTON • *Cape Peninsula University of Technology, Bellville, and Research and Postgraduate Studies, University of Pretoria, South Africa*
LAURA CALDINELLI • *Dipartimento di Biotecnologie e Scienze Molecolari, Università degli Studi dell'Insubria, Varese, and "The Protein Factory", Politecnico di Milano and Università degli Studi dell'Insubria, Varese, Italy*
FRANCA CASTIGLIONE • *Dipartimento di Chimica, Materiali e Ingegneria Chimica "G. Natta", Politecnico di Milano, Milano, Italy*
PERE CLAPÉS • *Department of Chemical Biology and Molecular Modeling, Instituto de Química Avanzada de Cataluña IQAC-CSIC, Barcelona, Spain*
JOSEFA MARÍA CLEMENTE-JIMÉNEZ • *Departamento de Química-Física, Bioquímica y Química Inorgánica, Universidad de Almería, Almería, Spain*
T. ASHTON CROPP • *Department of Chemistry and Biochemistry, University of Maryland, College Park, MD, USA*
GLENN DALLÉRAC • *NeuroCentre Magendie and Université de Bordeaux, Bordeaux, France; Centre National de la Recherche Scientifique, Université de la Méditerranée, Marseille, France*
PAOLA D'ARRIGO • *Dipartimento di Chimica, Materiali ed Ingegneria Chimica "Giulio Natta" Politecnico di Milano, and "The Protein Factory", Politecnico di Milano and Università degli Studi dell'Insubria, Milano, Italy*
ROSEMARY A. DORRINGTON • *Department of Biochemistry, Microbiology and Biotechnology, Rhodes University, Grahanstown, South Africa*
GEORGE H. FISHER • *Department of Chemistry, Barry University, Miami Shores, FL, USA*
MENDEL FRIEDMAN • *Western Regional Research Center, Agricultural Research Service, United States Department of Agriculture, Albany, CA, USA*
NORIHIKO FUJII • *Research Reactor Institute, Kyoto University, Osaka, Japan*
NORIKO FUJII • *Research Reactor Institute, Kyoto University, Osaka, Japan*

THIERRY GEFFLAUT • *Lab SEESIB, Clermont Université, Université Blaise Pascal – CNRS, Aubière, France*
RENATO GENNARO • *Department of Life Sciences, University of Trieste, Trieste, Italy*
MICHAEL GERRITS • *RiNA GmbH, Berlin, Germany*
MARIANA GUTIÉRREZ • *Department of Chemical Biology and Molecular Modeling, Instituto de Química Avanzada de Cataluña-CSIC, Barcelona, Spain*
CHRISTIAN P.R. HACKENBERGER • *Institut für Chemie und Biochemie, Freie Universität Berlin, Berlin, Germany*
RICKEY P. HICKS • *Department of Chemistry, East Carolina University, Greenville, NC, USA*
NOBUMASA HINO • *RIKEN Systems and Structural Biology Center, Tsurumi, Yokohama, Japan*
TAKAHIRO HOHSAKA • *School of Materials Science, Japan Advanced Institute of Science and Technology, Nomi, Ishikawa, Japan*
HIROSHI HOMMA • *Department of Pharmaceutical Life Sciences, Kitasato University, Minatoku, Tokyo, Japan*
FLORIN-DAN IRIMIE • *Department of Biochemistry and Biochemical Engineering, Babes-Bolyai University, Cluj-Napoca, Romania*
JESÚS JOGLAR • *Department of Chemical Biology and Molecular Modeling, Instituto de Química Avanzada de Cataluña-CSIC, Barcelona, Spain*
MASUMI KATANE • *Department of Pharmaceutical Life Sciences, Kitasato University, Minatoku, Tokyo, Japan*
KLAUDIA KOVÁCS • *Department of Organic Chemistry and Technology, Budapest University of Technology and Economics and Institute of Enzymology, Hungarian Academy of Sciences, Budapest, Hungary*
FRANCISCO JAVIER LAS HERAS-VÁZQUEZ • *Departamento de Química-Física, Bioquímica y Química Inorgánica, Universidad de Almería, Almería, Spain*
CAROL E. LEVIN • *Western Regional Research Center, Agricultural Research Service, United States Department of Agriculture, Albany, CA, USA*
JIA LIU • *Department of Chemistry and Biochemistry, University of Maryland, College Park, MD, USA*
PAUL MAJKUT • *Institut für Chemie und Biochemie, Freie Universität Berlin, Berlin, Germany*
SERGIO MARTÍNEZ-RODRÍGUEZ • *Departamento de Química-Física, Bioquímica y Química Inorgánica, Universidad de Almería, Almería, Spain*
GWYNNETH F. MATCHER • *Department of Biochemistry, Microbiology and Biotechnology, Rhodes University, Grahanstown, South Africa*
RYAN A. MEHL • *Department of Chemistry, Franklin & Marshall College, Lancaster, PA, USA*
GIANLUCA MOLLA • *Dipartimento di Biotecnologie e Scienze Molecolari, Università degli Studi dell'Insubria, Varese, and "The Protein Factory", Politecnico di Milano and Università degli Studi dell'Insubria, Varese, Italy*
JEAN-PIERRE MOTHET • *NeuroCentre Magendie, INSERM U862, and Université de Bordeaux, Bordeaux, France; Centre de Recherche en Neurobiologie et Neurophysiologie de Marseille, Université de la Méditerranée, Marseille, France*

ARMANDO NEGRI • *Dipartimento di Patologia Animale, Igiene e Sanità Pubblica Veterinaria – sez. Biochimica, Università degli Studi di Milano, Milano, Italy*

SUSUMU OHTANI • *Institute for Frontier Oral Science and Department of Forensic Medicine, Kanagawa Dental College, Yokosuka, Japan*

CSABA PAIZS • *Department of Biochemistry and Biochemical Engineering, Babes-Bolyai University, Cluj-Napoca, Romania*

JENNIFER C. PEELER • *Department of Chemistry, Franklin & Marshall College, Lancaster, PA, USA*

MIRELLA S. PILONE • *Dipartimento di Biotecnologie e Scienze Molecolari, Università degli Studi dell'Insubria, Varese, and "The Protein Factory", Politecnico di Milano and Università degli Studi dell'Insubria, Varese, Italy*

LUCIANO PIUBELLI • *Dipartimento di Biotecnologie e Scienze Molecolari, Università degli Studi dell'Insubria, Varese, and "The Protein Factory", Politecnico di Milano and Università degli Studi dell'Insubria, Varese, Italy*

LOREDANO POLLEGIONI • *Dipartimento di Biotecnologie e Scienze Molecolari, Università degli Studi dell'Insubria, Varese, and "The Protein Factory", Politecnico di Milano and Università degli Studi dell'Insubria, Varese, Italy*

LÁSZLÓ POPPE • *Department of Organic Chemistry and Technology, Budapest University of Technology and Economics, Budapest, Hungary*

INNA RADZISHEVSKY • *Department of Biochemistry, Technion, Israel Institute of Technology, Haifa, Israel*

IGNACIO RODRIGUEZ-CRESPO • *Departamento de Bioquímica y Biología Molecular, Facultad de Ciencias Químicas, Universidad Complutense, Madrid, Spain*

FELIPE RODRÍGUEZ-VICO • *Departamento de Química-Física, Bioquímica y Química Inorgánica, Universidad de Almería, Almería, Spain*

ELENA ROSINI • *Dipartimento di Biotecnologie e Scienze Molecolari, Università degli Studi dell'Insubria, Varese, and "The Protein Factory", Politecnico di Milano and Università degli Studi dell'Insubria, Varese, Italy*

AMANDA L. RUSSELL • *Department of Chemistry, East Carolina University, Greenville, NC, USA*

SILVIA SACCHI • *Dipartimento di Biotecnologie e Scienze Molecolari, Università degli Studi dell'Insubria, Varese, and "The Protein Factory", Politecnico di Milano and Università degli Studi dell'Insubria, Varese, Italy*

KENSAKU SAKAMOTO • *RIKEN Systems and Structural Biology Center, Tsurumi, Yokohama, Japan*

MARTINE SANCELME • *Lab SEESIB, Clermont Université, Université Blaise Pascal, Aubière, France*

MARCO SCOCCHI • *Department of Life Sciences, University of Trieste, Trieste, Italy*

MASAE SEKINE • *Department of Pharmaceutical Life Sciences, Kitasato University, Minatoku, Tokyo, Japan*

REMIGIUSZ SERWA • *Institut für Chemie und Biochemie, Freie Universität Berlin, Berlin, Germany*

GABRIELLA TEDESCHI • *Dipartimento di Patologia Animale, Igiene e Sanità Pubblica Veterinaria – sez. Biochimica, Università degli Studi di Milano, Milano, Italy*

DAVIDE TESSARO • *Dipartimento di Chimica, Materiali ed Ingegneria Chimica "Giulio Natta" Politecnico di Milano, and "The Protein Factory", Politecnico di Milano and Università degli Studi dell'Insubria, Milano, Italy*
MARA TSESARSKAIA • *Department of Chemistry, Barry University, Miami Shores, FL, USA*
ALESSANDRO TOSSI • *Department of Life Sciences, University of Trieste, Trieste, Italy*
FABRICE TURPIN • *NeuroCentre Magendie, INSERM U862, and Université de Bordeaux, Bordeaux, France*
SHIGEYUKI YOKOYAMA • *RIKEN Systems and Structural Biology Center, Tsurumi, Yokohama, and Department of Biophysics and Biochemistry, The University of Tokyo, Hongo, Bunkyo-ku, Tokyo, Japan*
TOSHIHARU YAMAMOTO • *Department of Human Biology, Kanagawa Dental College, Yokosuka, Japan*
BEÁTA VÉRTESSY • *Institute of Enzymology, Hungarian Academy of Sciences, Budapest, Hungary*
FEDERICA VOLONTÈ • *Dipartimento di Biotecnologie e Scienze Molecolari, Università degli Studi dell'Insubria, Varese, and "The Protein Factory", Politecnico di Milano and Università degli Studi dell'Insubria, Varese, Italy*
LEI WANG • *The Jack H. Skirball Center for Chemical Biology & Proteomics, The Salk Institute for Biological Studies, La Jolla, CA, USA*
QIAN WANG • *Tianjin Institute of Industrial Biotechnology, Chinese Academy of Sciences, Tianjin, China*
HERMAN WOLOSKER • *Department of Biochemistry, Technion, Israel Institute of Technology, Haifa, Israel*
SOTIR ZAHARIEV • *International Centre for Genetic Engineering and Biotechnology, Trieste, Italy*

Part I

Synthesis of Unnatural Amino Acids

Chapter 1

Preparation of Unnatural Amino Acids with Ammonia-Lyases and 2,3-Aminomutases

László Poppe, Csaba Paizs, Klaudia Kovács, Florin-Dan Irimie, and Beáta Vértessy

Abstract

Ammonia-lyases catalyze a wide range of processes leading to α,β-unsaturated compounds by elimination of ammonia. In this chapter, ammonia-lyases are reviewed with major emphasis on their synthetic applications in stereoselective preparation of unnatural amino acids. Besides the synthesis of various unnatural α-amino acids with the aid of phenylalanine ammonia-lyases (PALs) utilizing the 3,5-dihydro-5-methylidene-4H-imidazol-4-one (MIO) prosthetic groups, the biotransformations leading to various unnatural β-amino acids with phenylalanine 2,3-aminomutases using the same catalytic MIO prosthetic group are discussed. Cloning, production, purification, and biotransformation protocols for PAL are described in detail.

Key words: Phenylalanine ammonia-lyase, Phenylalanine 2,3-aminomutase, Recombinant enzyme, Biotransformation, Unnatural amino acid

1. Introduction

The lyases are enzymes cleaving C–C, C–O, C–N, and other bonds by elimination, leaving double bonds or rings, or conversely adding groups to double bonds (1). Ammonia-lyases, acting on C–N bonds, catalyze the formation of α,β-unsaturated bonds by elimination of ammonia from their substrates (2). There are a number of ammonia-lyases which catalyze the reversible deamination of amino acids (Fig. 1) (2).

The synthesis of natural and unnatural amino acids in homochiral form is an important challenge of preparative chemistry (3). Ammonia-lyase catalyzed reactions can be used for the stereoselective production of amino acids by two alternative ways (2).

Loredano Pollegioni and Stefano Servi (eds.), *Unnatural Amino Acids: Methods and Protocols*,
Methods in Molecular Biology, vol. 794, DOI 10.1007/978-1-61779-331-8_1,

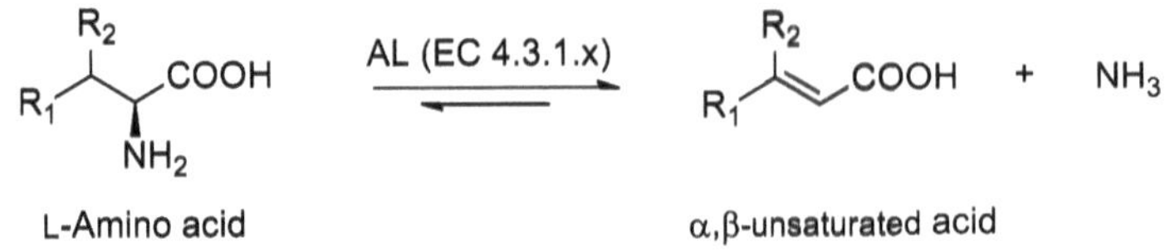

Fig. 1. Reversible deamination of L-amino acids by ammonia-lyases.

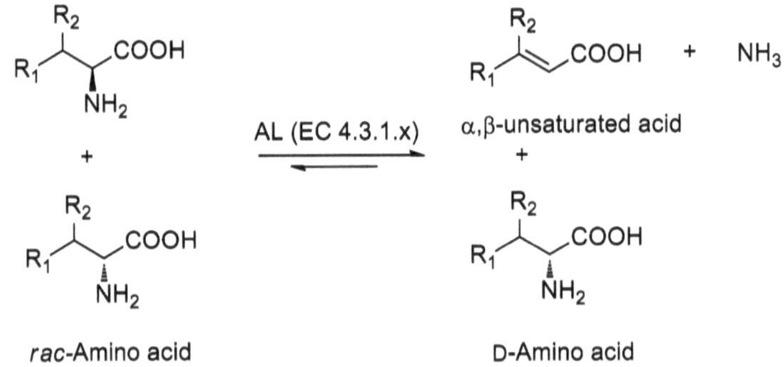

Fig. 2. Stereodestructive production of D-amino acids by ammonia-lyases.

One possibility is to utilize the stereoconstructive reverse reaction of ammonia-lyases (Fig. 1). Thus, addition of ammonia to unsaturated achiral precursors catalyzed by ammonia-lyases can be used for the preparation of L-amino acids.

On the other hand, the stereodestructive nature of the ammonia-lyase reactions can be exploited also for the production of D-enantiomers by enantiomer selective destruction of the L-enantiomers from their racemates (Fig. 2).

Although a number of ammonia-lyases can act on amino acids or amino acid derivatives, only a few can be applied for the stereoselective production of unnatural amino acids (2).

The enzyme L-aspartate ammonia-lyase (aspartase, EC 4.3.1.1) is one of the most specific enzymes known accepting only L-aspartic acid (R_1 = COOH, R_2 = H) as substrate (4). Therefore, the synthetic use of aspartase is limited to commercial production of its natural substrate from fumaric acid and ammonia with aspartase-containing immobilized cells of *Escherichia coli* in continuous, fixed-bed operation (5, 6). Recombinant production of aspartase enabled the modification of the catalytic properties of the biocatalyst by directed evolution (7, 8).

Due to its broader substrate specificity, the 3-methylaspartate ammonia-lyase (MAL, EC 4.3.1.2; acting on 3-methylaspartic acid, R_1 = COOH, R_2 = CH_3) has been used for broader synthetic processes. MAL-catalyzed addition of ammonia to chloro- and bromofumaric acid resulted in the formation of (2*R*,3*S*)-3-chloro- and (2*R*,3*S*)-3-bromoaspartic acid, respectively (9). MAL was also used for the synthesis of L-aspartic acids containing 3-alkyl-substituents in the (*S*)-configuration (10) and for producing the corresponding C-3 deuterated isotopomers (10, 11). Moreover,

N-substituted aspartic acids were obtained via enantiospecific addition of *N*-nucleophiles – such as hydrazine or methylamine – to fumaric acids using MAL (12).

The histidine ammonia-lyase (HAL, EC 4.3.1.3) eliminates the α-amino group from L-histidine (Fig. 1: R_1 = 1*H*-imidazol-4-yl, R_2 = H) and thus results in the formation of (*E*)-urocanate. Interaction of Zn^{2+} with the imidazole rings of the substrate and of His 83 of HAL during the catalysis can explain (13, 14) why HAL accepts only L-histidine, L-4-fluorohistidine (15), or L-4-nitrohistidine (16, 17) as substrates. In spite of the thermostability of HAL, this narrow substrate tolerance limits the use of this enzyme as biocatalyst.

HAL belongs to the 3,5-dihydro-5-methylidene-4*H*-imidazol-4-one (MIO)-containing ammonia-lyase family together with tyrosine ammonia-lyase (TAL, EC 4.3.1.23) and phenylalanine ammonia-lyase (PAL, EC 4.3.1.24) catalyzing the deamination of the corresponding L-amino acids (TAL: R_1 = 4-hydroxyphenyl, R_2 = H; PAL: R_1 = phenyl, R_2 = H) (18). The nonoxidative deamination catalyzed by HAL (19), PAL (20, 21), and TAL (22) requires the presence of the uncommon electrophilic MIO prosthetic group (23, 24) within the enzyme.

Although the substrate specificity of TAL is limited practically to its natural substrate L-tyrosine (22, 25–29), PAL accepts a wide range of L-arylalanines as substrate (2, 18).

PAL of genuses *Rhodosporium* and *Rhodotorula* as whole cell biocatalyst (30–34) or PAL of parsley as isolated enzyme (35–39) can be applied for the synthesis of a wide range of L-arylalanines by ammonia addition to their corresponding unsaturated precursors (reverse reaction in Fig. 1: R_1 = aryl, heteroaryl, R_2 = H). In this way, enantiopure L-arylalanines containing differently substituted phenyl rings (30–36, 38), polycyclic aromatic rings (31, 34), or even heterocycles (31, 34–39) were synthesized. A sequential chemoenzymatic process including an esterase from porcine liver and PAL from parsley was developed to produce the enantiopure L-arylalanines from the corresponding aromatic aldehydes (38) (Fig. 3).

The PAL of parsley was applied to produce the D-enantiomers of arylalanines in high enantiomeric purity by enantiomer selective destruction of the L-enantiomers from their racemates (37, 39) (Fig. 2: R_1 = aryl, heteroaryl, R_2 = H).

The strong similarity of L-phenylalanine and L-tyrosine 2,3-aminomutases (PAM (40–43) and TAM (44, 45), respectively) to the ammonia-lyase family is indicated by the presence of unusual catalytic MIO in these aminomutases and also by the fact that both PAM (46) and TAM (47) have ammonia-lyase activity.

Similarly to TAL, the substrate specificity of TAM is limited to convert (*S*)-α-tyrosine to either (*R*)- (48, 49) or (*S*)-β-tyrosine (44, 47).

PAM of plant origin can convert a wide range of (*S*)-α-arylalanines and even (*S*)-α-styrylalanine to the corresponding

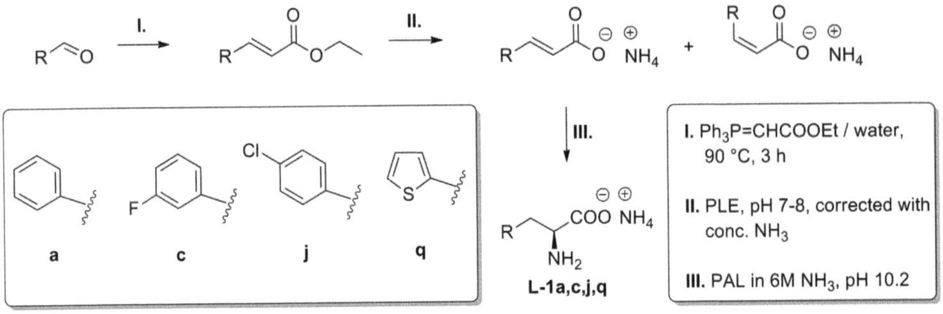

Fig. 3. The one-pot synthesis of L-arylalanines (38).

(*R*)-β-amino acids (50). Due to its PAL activity, PAM can be also applied to produce a mixture of (*S*)-α- and (*R*)-β-arylalanines (51). The combined use of PAM and PAL allows the preparation of (*S*)-β-arylalanines from their racemates (52). In this process, PAM isomerizes the (*R*)-β-arylalanine and PAL selectively degrades the forming (*S*)-α-amino acid.

Besides their biocatalytic usefulness, MIO enzymes have medical aspects like better understanding of the genetics of HAL regulation for the treatment of histidinaemia (53) or use PAL as a therapeutic agent for the treatment of phenylketonuria (54, 55).

In this chapter, the methods for cloning, production, purification, and applications for biotransformations of ammonia-lyases are described in detail using PAL of bacterial and parsley origin as examples.

2. Materials

2.1. Transformation

1. The gene encoding the *Photorhabdus luminescens* PAL (Phl6) was amplified by PCR from genomic DNA and it was cloned into a pUC18 cloning vector (56). The insert was excised from the resulting plasmid via *Eco*RI and *Xba*I restriction digestion. The gene fragment was separated from the vector DNA using agarose gel electrophoresis. The purified insert was then directionally ligated into the pBAD-24 expression vector. The resulting vector was confirmed by sequencing using the pBAD promoter forward and reverse primers (5′-CCTGACGCTTT TTATCGCAACTC-3′ and 5′-GAGGCATCCTGGTACCCC AG-3′, respectively).
2. A *rec*A, *end*A, *ara*BADC-, and *ara*EFGH+ TOP10 *E. coli* strain, which is able to transport L-arabinose without metabolizing it (see Note 1), was used for cloning and transformation.
3. TFB I solution, pH 5.8 (pH adjusted with 10% acetic acid): 100 mM RbCl, 50 mM $MnCl_2$, 30 mM potassium acetate, 10 mM $CaCl_2$, 15% glycerol. Store at 4°C.

4. TFB II solution, pH 6.8 (pH adjusted with 1 M KOH): 10 mM MOPS, 10 mM RbCl, 75 mM $CaCl_2$, 15% glycerol.

2.2. Expression of PAL from *Photorhabdus luminescens*

1. Low-salt LB (Luria–Bertani) broth medium: dH_2O (800 mL), Bacto tryptone (10 g), yeast extract (5 g), and NaCl (5 g) were added. pH was adjusted to 7.2 with NaOH. If LB agar medium was required, then agar (17 g) was also added. The volume was adjusted to 1 L with dH_2O. The media were sterilized by autoclaving.
2. Antibiotic stock solutions:

   Ampicillin stock solution: 20 mg/mL, in medium: 100 μg/mL.

   Kanamycin stock solution: 10 mg/mL, in medium: 30 μg/mL.
3. Lysis buffer: 150 mM NaCl, 50 mM Tris–HCl, pH 8.0, 10 mM 2-mercaptoethanol (BME), Complete Protease inhibitor cocktail (Roche), 2 mM PMSF, 5 mM benzamidine (BA).

2.3. Activity Measurement of PAL from *Photorhabdus luminescens*

1. Phenylalanine-Tris–HCl substrate solution: 0.033 g L-phenylalanine in 0.1 M Tris–HCl, pH 8.8 buffer (20 mL).
2. 0.1 M Tris–HCl, pH 8.8: Tris base (12.1 g) dissolved in dH_2O (900 mL) titrated with conc. HCl to adjust the desired pH. The volume of buffer was brought to 1,000 mL by adding dH_2O.

2.4. SDS Polyacrylamide Gel

1. 40% Acrylamide solution: BioRad.
2. 10% (w/v) SDS: SDS (10 g) dissolved in dH_2O (90 mL), then corrected with dH_2O to 100 mL total volume.
3. Resolving gel buffer: 1.5 M Tris–HCl, pH 8.8. Dissolve 27.23 g of Tris base to 80 mL dH_2O, bring to the desired pH with conc. HCl solution, add dH_2O to a total volume of 150 mM. Store at 4°C.
4. Stacking buffer: 0.5 M Tris–HCl, pH 6.8. Dissolve Tris base (6 g) in dH_2O (60 mL), bring to the desired pH with conc. HCl solution, add dH_2O to a total volume of 100 mL. Store at 4°C.
5. 10% APS (ammonium-persulfate), APS (100 mg) in dH_2O (1 mL).
6. TEMED (BioRad).
7. Sample electrophoresis buffer: dH_2O (3.55 mL), 0.5 M Tris–HCl buffer, pH 6.8 (1.25 mL), glycerol (2.5 mL), 10% (w/v) SDS (2 mL), 0.5 (w/v) bromophenol blue (0.2 mL), 50 mM DTT.
8. Running buffer stock solution (10×), pH 8.3: 30.3 g Tris base, 144 g glycine, 10 g SDS in 1 L dH_2O. Do not adjust pH.

2.5. Growth of Mutant *E. coli* Strains

1. Medium for microbiological trials (LB medium containing ampicillin and kanamycin): Bacto tryptone (10 g), NaCl (10 g),

and Bactoyeast (5 g) with bidistilled H_2O to 1 L, pH 7.5. Ampicillin sodium salt (85 mg) and kanamycin monosulfate (25 mg) were added after sterilization when the medium was cooled to 50–60°C.

2. LB-ampicillin–kanamycin agar plates: LB medium (1 L), agar (18 g), ampicillin sodium salt (85 mg), and kanamycin monosulfate (25 mg). The antibiotics were added to the LB medium containing agar after it was cooled to 50–60°C.

2.6. Isolation and Purification of wt-PAL from Parsley

2.6.1. Buffers for Protein Isolation

1. Sonication buffer: 20 mM K_2HPO_4/KH_2PO_4 buffer, pH 7.5, 25 mL.
2. Dialysis buffer: 20 mM K_2HPO_4/KH_2PO_4 buffer, pH 7.5, 5 L.

2.6.2. Stock Solutions

1. Benzamidine solution (0.5 mM): benzamidine (75 mg) in bidistilled H_2O (1 mM); stored at 4°C.
2. IPTG solution (100 mM): IPTG (23.8 mg) in bidistilled H_2O (1 mL); stored at −20°C.
3. Pefabloc solution (20 mM): Pefabloc (4.8 mg) in bidistilled H_2O (1 mL); stored at 4°C.
4. L-Phenylalanine solution (20 mM): L-phenylalanine (165 mg) with 0.1 M Tris buffer, pH 8.8, filled up to 50 mL.

2.6.3. Buffers for the Chromatographic Purification of the Protein

1. Buffers for ion exchange chromatography on a HiLoad 16/10 Q-Sepharose column: Buffer A: 20 mM K_2HPO_4/KH_2PO_4 buffer, pH 7.5, and Buffer B: 20 mM K_2HPO_4/KH_2PO_4 buffer, pH 7.5 with 1 M KCl.
2. Buffer for gel filtration chromatography on a HiLoad 26/60 Superdex 200 column: 150 mM K_2HPO_4/KH_2PO_4 buffer, pH 7.5, with 600 mM NaCl and 3% maltose.

2.6.4. Storage of Columns

1. Ion exchange column: bidistilled H_2O with 20% ethanol.
2. Gel filtration column: bidistilled H_2O with 0.5% NaN_3.

3. Methods

3.1. Transformation

The $CaCl_2/MgCl_2$ transformation protocol (56) was used for transforming *E. coli* TOP ten strain with plasmid pBAD24-Phl6. For stages 1 and 2, buffers TFB I and II, respectively, were used.

3.2. Expression of PAL from Photorhabdus luminescens

1. Sterile LB medium (50 mL) containing ampicillin (100 μg/mL) was inoculated with the transformed cells.

2. The culture was shaken at 220 rpm at 37°C until OD_{600} ~1–2 (approximately 16 h).
3. Subsequently, the inoculum (0.8 mL) was transferred into sterile LB medium (50 mL) containing ampicillin (100 μg/mL).
4. The culture was shaken at 220 rpm at 37°C until OD_{600} ~0.6.
5. Then the temperature was decreased to 30°C and the cells were induced by the addition of 0.02% L-arabinose.
6. The culture was shaken at 220 rpm at 20°C for additional 12 h.
7. The cells were harvested by centrifugation of the cell suspension at 2,500×*g*.
8. All of the subsequent procedures were carried out on ice-bath.
9. The pellets were resuspended in lysis buffer (5 mL) and the cell suspension was sonicated (3×45 s) at amplitude 40% and pulsation 60% using a Bandelin Sonopuls HD 2070 instrument. The sonication was performed until the viscosity of the suspension was significantly lowered.
10. The extract was centrifuged at 5,000×*g* for 30 min and the activity of PAL from the supernatant was determined. The progress of the reaction was monitored by UV detection at 290 nm of the produced (*E*)-cinnamate ($\varepsilon_{290} = 10^4$/M/cm at room temperature): 100 μL supernatant was added to 1,000 μL of 0.1 M Tris–HCl, pH 8.8, containing 20 mM L-phenylalanine.

3.3. SDS–PAGE

1. The PAL content of the crude extract, supernatant, and pellet from previous steps was analyzed by SDS–PAGE.
2. Sample preparation: mix the sample protein solution with 2× volume of sample buffer and boil at 90–100°C for 2–5 min. Run the SDS–PAGE using the buffers described above.

3.4. Growth of Mutant E. coli Strains Expressing wt-PAL from Parsley

1. A BL21(DE3) *E. coli* strain transformed with plasmids pT7-7(PAL) and pREP4-groESL was grown on an agar-plate, by overnight incubation at 37°C until the thickness of the bacteria culture became approximately 2 mm.
2. A single culture from the agar-plate was removed with a sterile loop and introduced in LB medium (50 mL) which contains ampicillin (85 mg/L) and kanamycin (25 mg/L).
3. The inoculum was incubated at 37°C and shaken (230 rpm) until the optical density (OD_{600}) is ~1.0.
4. Subsequently, the inoculum (20 mL) was transferred to a sterile LB medium (1 L), containing ampicillin (85 mg/L) and kanamycin (25 mg/L).
5. The cell culture was incubated and shaken at 37°C until OD_{600} ~0.6–0.7, when IPTG (it triggers the transcription of plasmid) (113 mg) was added.

6. The culture was shaken for further 20 h, followed by the isolation of the cells by centrifugation (2,500×*g* for 50 min), resulting in 4–5 g of wet cells.

3.5. Isolation and Purification of wt-PAL from Parsley

1. The isolated cells were resuspended in phosphate buffer (25 mL, 20 mM, pH 7.2) which contains Pefabloc (250 μL) and benzamidine (250 μL) as protease inhibitors, and benzonase (5 μL) as an endonuclease for DNA/RNA.
2. The ice-cooled cell suspension was sonicated for 5×2 min, followed by centrifugation at 5,000×*g* for 1 h at 4°C.
3. The supernatant was separated and dialyzed overnight in phosphate buffer (5 L, 20 mM, pH 7.2) which contains Pefabloc (100 μL) and benzamidine (100 μL).
4. The sterile-filtered solution was purified by FPLC on a HiLoad 16/10 Q-Sepharose HP anion-exchange column (Fig. 4) at a 2 mL/min flow rate using a gradient program visualized in Table 1 (20 mM phosphate buffer, pH 7.5, buffer A; 20 mM K_2HPO_4/KH_2PO_4, pH 7.5, with 1 M NaCl, buffer B). PAL was eluted with the mixture of buffers A and B (75–25%).
5. The fraction was concentrated to 5–6 mL volume, by centrifugation of the solution at 2,500×*g* using Centricon filter,

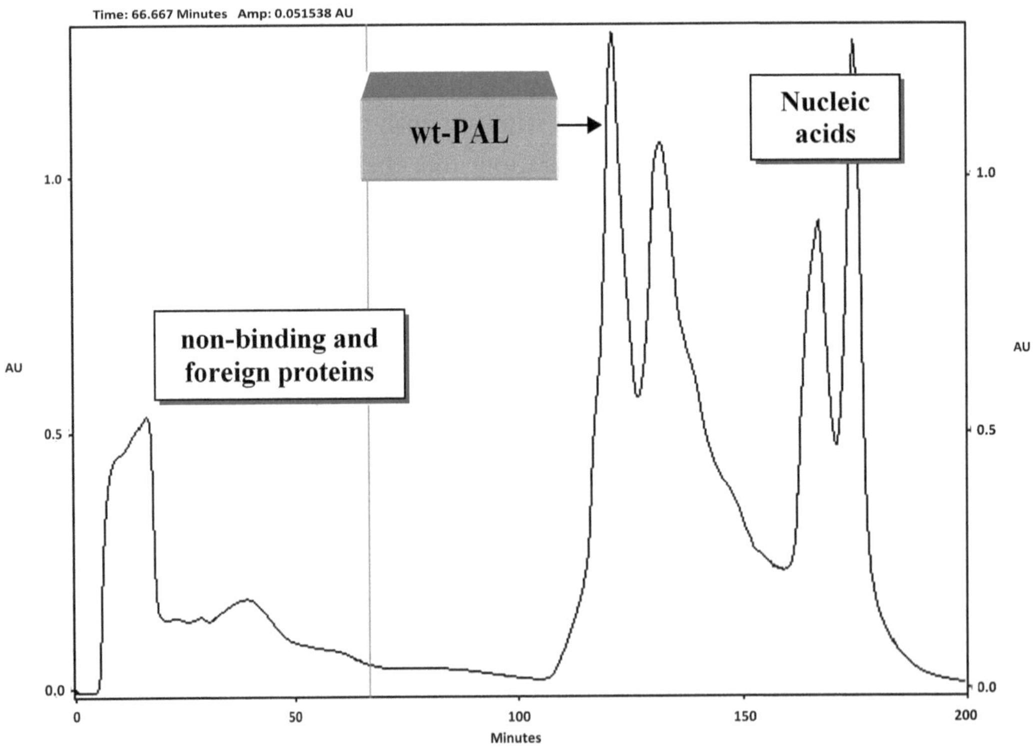

Fig. 4. Elution diagram of the protein–nucleic acid mixture containing wt-PAL on anion-exchange column.

Table 1
The gradient program for the anion-exchange chromatography

Time (min)	Buffer A (%)	Buffer B (%)
0	100	0
90	100	0
120	75	25
150	70	30
170	0	100
190	0	100

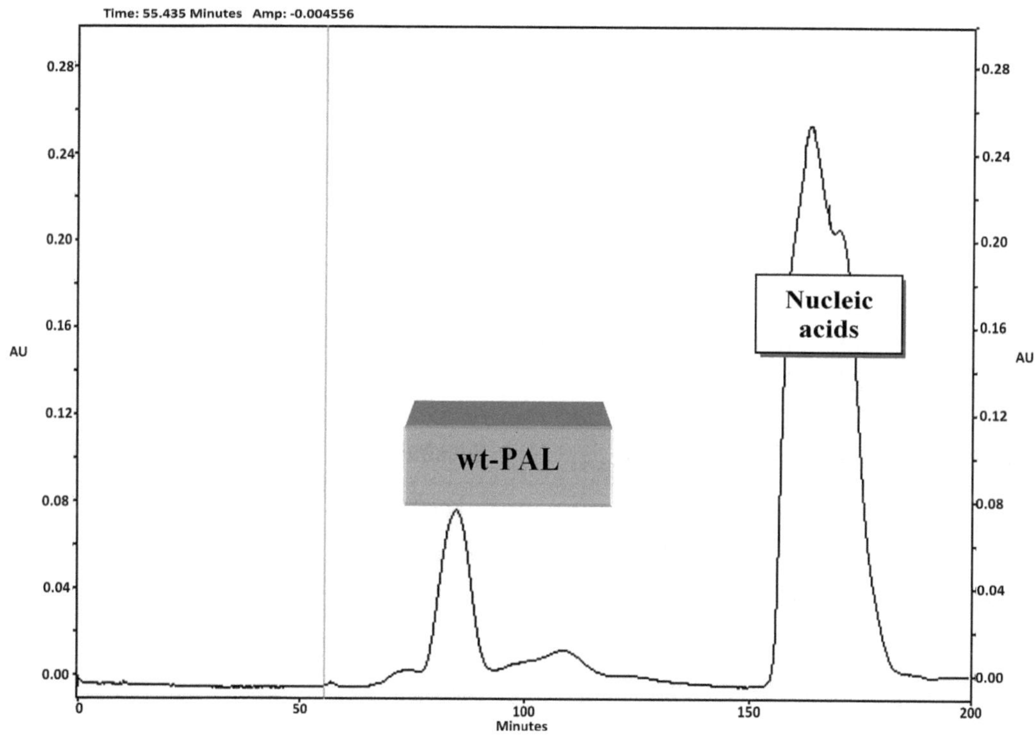

Fig. 5. Elution diagram of wt-PAL on a gel filtration chromatography.

followed by the next purification step by gel filtration on a HiLoad 26/60 Superdex™ 200 column, using phosphate buffer (150 mM, pH 7.5 with 600 mM of NaCl and 3% maltose) as eluent in isocratic manner at a 1.75 mL/min flow rate.

6. The fraction containing PAL was collected between 75 and 100 min (Fig. 5).

7. The glycerol–water solution of the enzyme can be kept at −80°C for 6 weeks without significant loss of activity.

3.6. Determination of the Protein Concentration

1. For the determination of protein concentration of the fractions containing pure PAL, the Warburg and Christian method was used by measuring the extinction at 280 and 260 nm. At 280 nm, absorb both aromatic amino acids and nucleic acids, but at 260 nm, only the nucleic acids show absorption.
2. The protein concentration can be determined according to the following formula: protein concentration (mg/mL) = $1.45E_{280nm} - 0.74E_{260nm}$.

3.7. Kinetic Parameters of wt-PAL from Parsley

1. Determination of V_{max} and K_m of the substrate amino acids in the PAL reaction: the kinetic constants were determined using a 1 mL cuvette, by recording for 5 min in an interval of 1 min the UV absorbance of the produced acrylates at wavelengths where the corresponding amino acids do not (see Note 2).
2. After incubation of the enzyme (25 μg) in 0.1 M Tris–HCl buffer at pH 8.8 and 30°C for 5 min, amino acids were added at various substrate concentrations of the enzymatic assays between 0.5 and 2.5 mM (see Note 3). The K_m and V_{max} values are shown in Table 2.

Table 2
Kinetic parameters of the PAL-catalyzed reactions with various (hetero) arylalanines (35, 36)

Substrate	K_m (μM)	$V_{max}/V_{max\ Phe}$	Substrate	K_m (μM)	$V_{max}/V_{max\ Phe}$
L-**1a**	32	1	L-**1m**	378	1.8
L-**1b**	65	1.1	*rac*-**1n**	4.2	0.8
L-**1c**	79	2.1	*rac*-**1o**	723	0.0058
L-**1d**	10	0.56	*rac*-**1p**	76.4	0.34
L-**1e**	85	0.85	*rac*-**1q**	135	1.01
L-**1f**	159	2.7	*rac*-**1r**	18.2	0.16
L-**1g**	76	0.16	*rac*-**1s**	48.4	0.49
L-**1h**	50	1.0	*rac*-**1t**	117	0.14
L-**1i**	94	2.0	*rac*-**1u**	2.3	0.17
L-**1j**	45	0.82	*rac*-**1v**	21.4	0.46
L-**1k**	704	0.80	*rac*-**1w**	13.4	0.35
L-**1l**	1,331	2.4	*rac*-**1x**	9.7	0.21

3. More than 20 L- or *rac*-(hetero)arylalanines were found to be substrates of wt-PAL from parsley (Fig. 6) (35–39).

3.8. Synthesis of Enantiopure Amino Acids with wt-PAL of Parsley

3.8.1. Kinetic Resolution of Racemic Amino Acids Catalyzed by PAL (37, 39)

1. PAL was added to the solution of the racemic amino acids (0.5 mmol) in Tris–HCl buffer (0.1 M, pH 8.8, 40 mL), and the reaction mixture was stirred under argon at 30°C.
2. The progress of the reaction was monitored by HPLC (see Note 4).
3. When the L-enantiomer was completely consumed (Fig. 7), the pH of the solution was adjusted to 1.5 with 5% HCl (see Note 5), then the solution was heated to 90°C for 10 min, cooled to room temperature, filtered, and applied to a Dowex 50 × 8 cation exchange resin column.
4. The elution of the pure D-enantiomer of amino acids was performed with ammonia solution (2 M). Reaction times, quantity of the enzyme in each case, and the yields for the products are presented in Table 3.

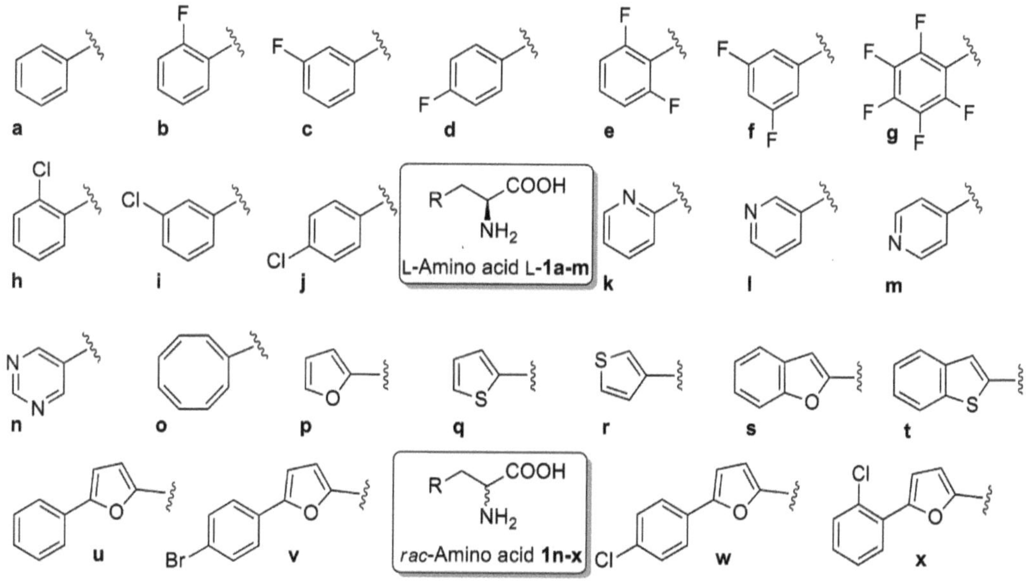

Fig. 6. Substrates of wt-PAL from parsley (35–39).

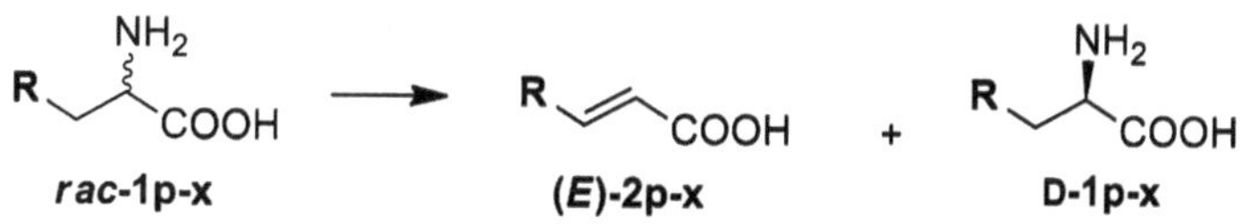

Fig. 7. PAL-catalyzed preparation of D-heteroarylalanines (37, 39).

Table 3
Preparative scale synthesis of D-heteroarylalanines catalyzed by PAL (37, 39)

Substrate	Product[a]	Yield[b] (%)	PAL (IU)	Time (days)
rac-**1p**	D-**1p**	44	2.5	4
rac-**1q**	D-**1q**	45	2	2
rac-**1r**	D-**1r**	45	7.5	7
rac-**1s**	D-**1s**	43	6	3
rac-**1t**	D-**1t**	44	6	3
rac-**1u**	D-**1u**	44	8	7
rac-**1v**	D-**1v**	45	12	8
rac-**1w**	D-**1w**	45	8	7
rac-**1x**	D-**1x**	43	12	8

[a] e.e. >98% for all the compounds
[b] Yields are given for the isolated compounds

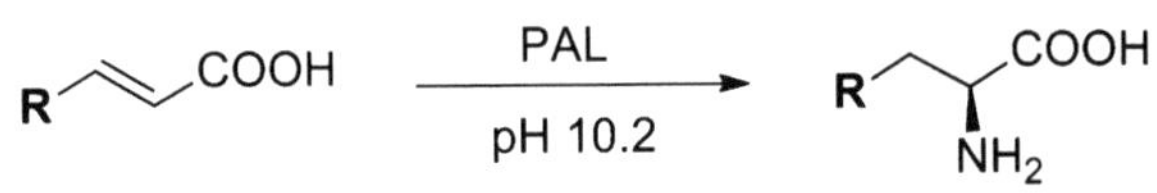

(*E*)-**2a-m, p-x** L-**1a-m, p-x**

Fig. 8. PAL-catalyzed preparation of L-(hetero)arylalanines (35–37, 39).

3.8.2. Synthesis of L-Amino Acids from (Hetero) Arylacrylates by PAL-Mediated Enantioselective Ammonia Addition (35–37, 39)

1. To the (hetero)arylacrylate (0.675 mmol) solution in half-concentrated ammonia solution (20 mL), CO_2 was bubbled until the pH of the solution was adjusted to 10.2.
2. After the addition of PAL, the reaction mixture was stirred under argon at 30°C (Fig. 8).
3. The progress of the reaction was monitored by HPLC.
4. When the formation of the L-amino acids was stopped, the reaction mixture was degassed under reduced pressure, the pH was adjusted to 1.5 with 5% HCl.
5. The resulting mixture was then filtered, heated to 90°C for 10 min, cooled to room temperature, and filtered again.
6. The filtrate was applied to a Dowex 50×8 cation exchange resin column.
7. Elution of the pure L-enantiomer of the amino acids was performed with 2 M ammonia solution.
8. Reaction times, quantity of the enzyme, and the yields of the products are presented in Table 4.

Table 4
Enantioselective synthesis of L-(hetero)arylalanines catalyzed by PAL (35–37, 39)

Substrate	Product[a]	Yield[b] (%)	PAL (IU)	Time (days)	Substrate	Product[a]	Yield[b] (%)	PAL (IU)	Time (days)
(*E*)-**2a**	L-**1a**	52	1–2	0.5	(*E*)-**2l**	L-**1l**	59	1	0.75
(*E*)-**2b**	L-**1b**	50	1–2	0.5	(*E*)-**2m**	L-**1m**	75	1	0.75
(*E*)-**2c**	L-**1c**	59	1–2	0.5	(*E*)-**2p**	L-**1p**	66	6	3
(*E*)-**2d**	L-**1d**	70	1–2	0.5	(*E*)-**2q**	L-**1q**	89	6	2
(*E*)-**2e**	L-**1e**	88	1–2	0.5	(*E*)-**2r**	L-**1r**	54	6	4
(*E*)-**2f**	L-**1f**	69	1–2	0.5	(*E*)-**2s**	L-**1s**	49	6	4
(*E*)-**2g**	L-**1g**	51	1–2	0.5	(*E*)-**2t**	L-**1t**	66	10	3
(*E*)-**2h**	L-**1h**	37	1–2	0.5	(*E*)-**2u**	L-**1u**	89	15	4
(*E*)-**2i**	L-**1i**	99	1–2	0.5	(*E*)-**2v**	L-**1v**	54	10	3
(*E*)-**2j**	L-**1j**	59	1–2	0.5	(*E*)-**2w**	L-**1w**	49	15	4
(*E*)-**2k**	L-**1k**	63	1	0.75	(*E*)-**2x**	L-**1x**	66	10	3

[a]e.e. >98% for all the compounds
[b]Yields are given for the isolated compounds

3.8.3. Synthesis of L-Amino Acids from the Corresponding Aldehydes by a One-Pot, Three-Step Reaction (38)

1. The mixture of one of the aldehydes as shown in Fig. 3 (5 mmol) and triphenyl-λ^5-phosphanilidene acetic acid ethyl ester (1.91 g, 5.5 mmol) in 15 mL water was vigorously stirred at 90°C until the entire quantity of aldehydes was completely transformed into (*E*)-cinnamates (approximately in 3 h, checked by HPLC). In each case, 3–5% (*Z*)-isomer was also formed.
2. After cooling to room temperature, the pH of the solution was adjusted to 8 with conc. ammonia solution and PLE (2 mg, 28 IU) was added.
3. The suspension was stirred and sonicated at 25–30°C while by the addition of conc. ammonia the pH of the solution was kept between 7 and 9.
4. After the completion of the hydrolysis of the cinnamic acid esters (approximately 6 h, checked by HPLC), water and conc. ammonia were added to reach a 6 M ammonia solution and a 2 mM cinnamate concentration.
5. CO_2 was bubbled until the pH of the solution was adjusted to 10.2.
6. After the addition of PAL (1 IU), the reaction mixture was stirred under argon at 30°C.
7. The progress of the reaction was monitored by HPLC.

Table 5
Yields and reaction times for the one-pot synthesis of L-arylalanines (38)

Product[a]	Yield (%)[b]	Time (days)
L-**10a**	78	2
L-**10c**	91	4
L-**10j**	78	2
L-**10q**	88	2

[a]e.e. >98% for all the compounds
[b]Yields are given for the isolated compounds

8. When the formation of the L-amino acids ceased, the reaction mixture was degassed under reduced pressure, the pH was adjusted to 1.5 with 5% HCl, filtered, then the solution was heated to 90°C for 10 min, cooled to room temperature, filtered, and applied to a Dowex 50×8 cation exchange resin column.
9. Elution of the pure L-enantiomer of the amino acids was performed with 2 M ammonia solution.
10. The reaction conditions and yields of the final products are shown in Fig. 3 and Table 5.

4. Notes

1. In the presence of L-arabinose, expression from pBAD is turned on while the absence of L-arabinose produces very low transcription levels from pBAD. Non-induced levels are repressed even further by growth in the presence of glucose. By varying the concentration of L-arabinose, protein expression levels can be optimized to ensure maximum expression of soluble protein. In addition, the tight regulation of pBAD by AraC is useful for the expression of potentially toxic or essential genes.
2. It was important to record separately the UV spectra of the L-(hetero)arylalanines and (hetero)arylacrylates. By comparing the two spectra, the wavelength at which only the (hetero) arylacrylate showed maximal absorption was chosen. The molar extinction coefficient (ε) of the (hetero)arylacrylates was determined at the chosen wavelength.
3. In case of (hetero)arylacrylates with high molar extinction coefficient (ε), the enzymatic assay was carried out at lower concentration of L-(hetero)arylalanines. During the kinetic

measurements, it was important to keep the absorbance of the producing (hetero)arylacrylates lower than 1.5.

4. HPLC analyses were performed using either the Astec Chirobiotic-Tag, Chirobiotic-T_2, or Chirobiotic-T columns (4.6×250 mm). Methanol:TEAA buffer (pH 4.1 or 6) 80:20 to 50:50 (v/v) was used as eluent in enantiomeric separations of *rac*-(hetero)arylalanines (retention times of the fully separated enantiomers were between 4 and 16 min). The retention times of (hetero)arylacrylates were around 2 min.
5. After adjusting the pH to 1.5, some of the (hetero)arylacrylates precipitated. In this case, the suspension was filtered and the filtrate was worked up as described in the main text.

Acknowledgments

This research was supported by the Hungarian National Office for Research and Technology (NKFP-07-A2 FLOWREAC), by the Hungarian National Science Foundation (OTKA, K68229, and CK78646), and by Howard Hughes Medical Institutes #55000342. This work is also related to the scientific program of "Development of quality-oriented and harmonized R+D+I strategy and functional model at BME" project (TÁMOP-4.2.1/B-09/1/KMR-2010-0002), supported by the New Hungary Plan. The financial support to Csaba Paizs from the Romanian Ministry of Education and Research (CNCSIS No. 527/2461) is gratefully acknowledged. We thank János Rétey (Institute of Organic Chemistry, Karlsruhe Institute of Technology, Germany), G. E. Schulz, and M. Baedeker (University of Freiburg, Germany) for providing us the *Petroselinum crispum* PAL expression system, András Holczinger (Department of Applied Biotechnology and Food Science, Budapest University of Technology and Economics, Budapest, Hungary) for providing us the gene of *Photorhabdus luminescens* PAL in pBAD-24 vector, and also M. Stieger (Hoffmann-La Roche, Basel, Switzerland) for vector pREP4-GroESL carrying the HSP-60 system.

References

1. Mahdi JG, and Kelly DR (2000) Chapter 2 – Lyases. In: Kelly DR (ed) Biotechnology, 2nd edn. VCH, Weinheim-New York, vol. VIIIb, pp. 41–171.
2. Poppe L, and Rétey J (2003) Properties and synthetic applications of ammonia-lyases. *Curr Org Chem* 7, 1297–1315.
3. Hughes AB (ed) (2009) Amino acids, peptides and proteins in organic chemistry: Volume 1 - Origins and Synthesis of Amino Acids. Wiley-VCH, Weinheim-New York.
4. Viola RE (2000) L-Aspartase: new tricks from an old enzyme. *Adv Enzymol Relat Areas Mol Biol* 74, 295–341.
5. Chibata, I., Tosa, T., and Sato, T. (1976) Production of L-aspartic acid by microbial cells entrapped in polyacrylamide gels. *Methods Enzymol* 44, 739–746.

6. Sato T, Tosa T (1993) Production of L-aspartic acid. *Bioprocess Technol* 16, 15–24.
7. Wang LJ, Kong XD, Zhang HY et al. (2000) Enhancement of the activity of L-aspartase from *Escherichia coli* W by directed evolution. *Biochem Biophys Res Commun* 276, 346–349.
8. Asano Y, Kira I, and Yokozeki K (2005) Alteration of substrate specificity of aspartase by directed evolution. *Biomol Eng* 22, 95–101.
9. Akhtar M, Cohen MA, and Gani G (1987) Stereochemical course of the enzymic amination of chloro- and bromo acid by 3-methylaspartate ammonia-lyase. *Tetrahedron Lett* 28, 2413–2416.
10. Akhtar M, Botting NP, Cohen MA et al. (1987) Enantiospecific synthesis of 3-substituted aspartic acids via enzymic amination of substituted fumaric acids. *Tetrahedron* 43, 5899–5908.
11. Archer CH, Thomas NR, and Gani D. (1993) Syntheses of (2*S*,3*R*)- and (2*S*,3*R*)(3-^{2}H)-3-methylaspartic acid: slow substrates for a *syn*-elimination reaction catalysed by methylaspartase. *Tetrahedron:Asymmetry* 4, 1141–1152.
12. Gulzar MS, Akhtar M, and Gani D (1997) Preparation of N-substituted aspartic acids via enantiospecific conjugate addition of N-nucleophiles to fumaric acids using methylaspartase: synthetic utility and mechanistic implications. *J Chem Soc Perkin Trans I* 649–655.
13. Klee CB (1972) Metal activation of histidine ammonia-lyase. Metal ion-sulfhydryl group relationship. *J Biol Chem* 247, 1398–1406.
14. Seff AL, Pilbák S, Silaghi-Dumitrescu I et al. (2011) Computational investigation of the histidine ammonia-lyase reaction: a modified loop conformation and the role of the zinc(II) ion. *J Mol* Mod 17, 1551–1563.
15. Klee CB, Kirk KL, Cohen LA et al. (1975) Histidine ammonia-lyase. The use of 4-fluorohistidine in identification of the rate determining step. *J Biol Chem* 250, 5033–5040.
16. Klee CB, Kirk KL, and Cohen LA (1979) 4-Nitro-L-histidine as a substrate for histidine ammonia-lyase: the role of β-hydrogen acidity in the rate-limiting step. *Biochem Biophys Res Commun* 87, 343–348.
17. Langer M, Pauling A, and Rétey J (1995) The role of dehydroalanine in catalysis by histidine ammonia-lyase. *Angew Chem Int Ed* 34, 1464–1465.
18. Poppe L, and Rétey J (2005) Friedel-Crafts type mechanism for the enzymatic elimination of ammonia from histidine and phenylalanine. *Angew Chem Int Ed* 44, 3668–3688.
19. Schwede TF, Rétey J, and Schulz GE (1999) Crystal structure of histidine ammonia-lyase revealing a novel polypeptide modification as the catalytic electrophile. *Biochemistry* 38, 5355–5361
20. Calabrese JC, Jordan DB, Boodhoo A et al. (2004) Crystal structure of phenylalanine ammonia-lyase: multiple helix dipoles, implicated in catalysis. *Biochemistry* 43, 11403–11416.
21. Ritter H, and Schulz GE (2004) Structural basis for the entrance into the phenylpropanoid metabolism catalyzed by phenylalanine ammonia-lyase. *Plant Cell* 16, 3426–3436.
22. Louie GV, Bowman ME, Moffitt MC et al. (2006) Structural determinants and modulation of substrate specificity in phenylalanine-tyrosine ammonia-lyases. *Chem Biol* 13, 1327–1338.
23. Poppe L (2001) Methylidene-imidazolone: a novel electrophile for substrate activation. *Curr Opin Chem Biol* 5, 512–524.
24. Rétey J (2003) Discovery and role of methylidene imidazolone, a highly electrophilic prosthetic group. *Biochim Biophys Acta Proteins Proteomics* 1647, 179–184.
25. Kyndt JA, Meyer TE, Cusanovich MA et al. (2002) Characterization of a bacterial tyrosine ammonia lyase, a biosynthetic enzyme for the photoactive yellow protein. *FEBS Lett* 512, 240–244.
26. Berner M, Krug D, Bihlmaier C et al. (2006) Genes and enzymes involved in caffeic acid biosynthesis in the actinomycete *Saccharothrix espanaensis*. *J Bacteriol* 188, 2666–2673.
27. Watts KT, Mijts BN, Lee PC et al. (2006) Discovery of a substrate selectivity switch in tyrosine ammonia-lyase, a member of the aromatic amino acid lyase family. *Chem Biol* 13, 1317–1326.
28. Xue Z, McCluskey M, Cantera K et al. (2007) Improved production of p-hydroxycinnamic acid from tyrosine using a novel thermostable phenylalanine/tyrosine ammonia lyase enzyme. *Enzyme Microb Technol* 42, 58–64.
29. Schroeder AC, Kumaran S, Hicks LM et al. (2008) Contributions of conserved serine and tyrosine residues to catalysis, ligand binding, and cofactor processing in the active site of tyrosine ammonia lyase. *Phytochemistry* 69, 1496–1506.
30. Tanaka H, Oudou T, and Uchida K (1988) Production of fluorine-containing phenylalanine derivative from fluorine-containing cinnamic acid. *Jap Pat* 63148992.
31. Renard G, Guilleux JC, Bore C et al. (1992) Synthesis of L-phenylalanine analogs by *Rhodotorula glutinis*. Bioconversion of cinnamic acids derivatives. *Biotechnol Lett* 14, 673–678.
32. Zhao JS, and Yang SK (1995) Red yeast catalyzed amination of olefinic bonds and synthesis

of optically pure *S*-amino acids. *Chin J Chem* 13, 241–245.

33. Zhao JS, and Yang SK (1997) A new asymmetric synthesis of L-alpha-amino acid via microbial transformation. *Acta Chim Sinica* 55, 196–201.
34. Liu W (1999) Synthesis of optically active phenylalanine analogs using *Rhodotorula graminis*. *US Pat* 5981239
35. Gloge A, Langer B, Poppe L et al. (1998) The Behavior of substrate analogues and secondary deuterium isotope effects in the phenylalanine ammonia-lyase reaction. *Arch Biochem Biophys* 359, 1–7.
36. Gloge A, Zoń J, Kővári Á et al. (2000) Phenylalanine ammonia-lyase: the use of its broad substrate specificity for mechanistic investigations and biocatalysis. Synthesis of L-arylalanines. *Chem Eur J* 6, 3386–3390.
37. Paizs C, Katona A, and Rétey J (2006) The interaction of heteroaryl-acrylates and alanines with phenylalanine ammonia-lyase from Parsley. *Chem Eur J* 12, 2739–2744.
38. Paizs C, Katona A, and Rétey J (2006) Chemoenzymatic one-pot synthesis of enantiopure L-arylalanines from arylaldehydes. *Eur J Org Chem* 1113–1116.
39. Paizs C, Toşa MI, Bencze LC et al. (2010) 2-Amino-3-(5-phenylfuran-2-yl) proprionic acids–phenylfuran-2-yl acrylic acids are novel substrates of phenylalanine ammonia lyase. *Heterocycles* doi:10.3987/COM-10S(E)60.
40. Walker KD, Klettke K, Akiyama T et al. (2004) Cloning, heterologous expression, and characterization of a phenylalanie aminomutase involved in taxol biosynthesis. *J Biol Chem* 279, 53947–53954.
41. Steele CL, Chen Y, Dougherty BA et al. (2005) Purification, cloning, and functional expression of phenylalanine aminomutase: the first committed step in taxol side-chain biosynthesis. *Arch Biochem Biophys* 438, 1–10.
42. Mutatu W, Klettke KL, Foster C et al. (2007) Unusual mechanism for an aminomutase rearrangement: Retention of configuration at the migration termini. *Biochemistry* 46, 9785–9794.
43. Wu B, Szymański W, Wijma HJ et al. (2010) Engineering of an enantioselective tyrosine aminomutase by mutation of a single active site residue in phenylalanine aminomutase. *Chem Commun* 46, 8157–8159.
44. Christenson SD, Liu W, Toney MD et al. (2003) A novel 4-methylideneimidazole-5-one-containing tyrosine aminomutase in enediyne antitumor antibiotic C-1027 biosynthesis. *J Am Chem Soc* 125, 6062–6063.
45. Christianson CV, Montavon TJ, Van Lanen SG et al. (2007) The structure of L-tyrosine 2,3-aminomutase from the C-1027 enediyne antitumor antibiotic biosynthetic pathway. *Biochemistry* 46, 7205–7214.
46. Walker KD, Klettke KL, Akiyama T et al. (2004) Cloning, heterologous expression, and characterization of a phenylalanine aminomutase involved in taxol biosynthesis. *J Biol Chem* 279, 53947–53954.
47. Christenson SD, Wu W, Spies MA et al. (2003) Kinetic analysis of the 4-methylideneimidazole-5-one-containing tyrosine aminomutase in enediyne antitumor antibiotic C-1027 biosynthesis. *Biochemistry* 42, 12708–12718.
48. Rachid S, Krug D, Weissman KJ et al. (2007) Biosynthesis of (*R*)-β-tyrosine and its incorporation into the highly cytotoxic chondramides produced by *Chondromyces crocatus*. *J Biol Chem* 282, 21810–21817.
49. Krug D, and Müller R (2009) Discovery of additional members of the tyrosine aminomutase enzyme family and the mutational analysis of CmdF. *ChemBioChem* 10, 741–750.
50. Klettke KL, Sanyal S, Mutatu W et al. (2007) β-Styryl- and β-aryl-β-alanine products of phenylalanine aminomutase catalysis. *J Am Chem Soc* 129, 6988–6989.
51. Wu B, Szymański W, Wietzes P et al. (2009) Enzymatic synthesis of enantiopure α- and β-amino acids by phenylalanine aminomutase-catalysed amination of cinnamic acid derivatives. *ChemBioChem* 10, 338–344.
52. Wu B, Szymański W, Wildeman S et al. (2010) Efficient tandem biocatalytic process for the kinetic resolution of aromatic β-amino Acids. *Adv Synth Catal* 352, 1409–1412.
53. Eckhart L, Schmidt M, Mildner M et al. (2008) Histidase expression in human epidermal keratinocytes: Regulation by differentiation status and all-trans retinoic acid. *J Dermatol Sci* 50, 209–215.
54. Gámez A, and Sarkissian CN (2005) Phenylalanine ammonia lyase, enzyme substitution therapy for phenylketonuria, where are we now? *Mol Genet Metabol* 86(Suppl 1), 22–26.
55. Wang L, Gámez A, Archer H et al. (2008) Structural and biochemical characterization of the therapeutic *Anabaena variabilis* phenylalanine ammonia lyase. *J Mol Biol* 380, 623–635.
56. Sambrook J, and Russell DW (2006) The condensed protocols from – Molecular cloning: a laboratory manual; Cold Spring Harbor Laboratory Press, New York, s.58., protocol 1.25.

Chapter 2

Multistep Enzyme Catalyzed Reactions for Unnatural Amino Acids

Paola D'Arrigo and Davide Tessaro

Abstract

The use of unnatural amino acids, particularly synthetic α-amino acids, for modern drug discovery research requires the availability of enantiomerically pure isomers. Starting from a racemate, one single enantiomer can be obtained using a deracemization process. The two more common strategies of deracemization are those obtained by stereoinversion and by dynamic kinetic resolution. Both techniques will be here described using as a substrate the D,L-3-(2-naphthyl)-alanine, a non-natural amino acid: the first one employing a multi-enzymatic redox system, the second one combining an hydrolytic enzyme together with a base-catalyzed substrate racemization. In both cases, the final product, L-3-(2-naphthyl)alanine, is recovered with good yield and excellent enantiomeric excess.

Key words: D-Amino acid oxidase, Amino transferase, Non-natural amino acid, Multi-enzyme reaction, Deracemization, Dynamic kinetic resolution

1. Introduction

Unnatural amino acids, the non-genetically-coded amino acids that either occur naturally or are chemically synthesized, and particularly synthetic α-amino acids, are a growing group of compounds required for a large number of biotechnological applications (from the pharmaceutical and cosmetics to the agrochemicals field) (1–4). Due to their structural diversity and functional versatility, they are widely used as chiral building blocks and molecular scaffolds in constructing chemical combinatorial libraries. Furthermore, they have played a significant role in the area of peptide research (5). Unnatural amino acids are not only of great value for the de novo design of peptides and proteins with a high propensity to fold with predetermined secondary or tertiary structure (6–8), but are also in their own valuable pharmaceuticals (e.g., L-DOPA), as well as

Loredano Pollegioni and Stefano Servi (eds.), *Unnatural Amino Acids: Methods and Protocols*,
Methods in Molecular Biology, vol. 794, DOI 10.1007/978-1-61779-331-8_2, © Springer Science+Business Media, LLC 2012

components of numerous therapeutically relevant compounds (e.g., D-3-(2-naphthyl)-alanine is found in the peptide drug Nafarelin) (9). It is thus evident that unnatural amino acids are becoming indispensable tools in drug discovery efforts.

The worldwide increase of the market for enantiopure raw materials largely requires the development of new products by new chiral technologies and improved enantioselective processes. In particular, the use of biocatalysts is largely investigated. In recent times, the term deracemization has been used to describe techniques becoming increasingly important for the preparation of chiral compounds as single enantiomers. The term refers to a process in which starting from a racemate, one single enantiomer is obtained. Deracemization methods are particularly suited for non-racemic amino acid preparation since the synthesis of the racemic forms are well established. Non-natural amino acids are particularly required in preparative scale and high enantiomeric excess for the preparation of peptidomimetics.

Here we present two alternative methodologies for obtaining the same compound, L-3-(2-naphthyl)-alanine (L-2-NpAla), by deracemization: the first method is the deracemization by stereo-inversion and the second one is by dynamic kinetic resolution (DKR). The substrate employed in the two methods has been D,L-3-(2-naphthyl)-alanine (D,L-2-NpAla), a non-natural amino acid of interest for the preparation of peptidomimetics and drugs candidates.

2. Materials

2.1. Deracemization by Stereo-Inversion

1. All reagents and solvents are purchased from Sigma-Aldrich.
2. ^{1}H NMR Spectra are recorded on Bruker ARX 400 instrument operating at the ^{1}H resonance frequency of 400 MHz. Chemical shifts (δ, ppm) are reported relative to tetramethylsilane (TMS) as an internal standard. All spectra are recorded in DMSO-d_6 (when not differently indicated) at 305 K.
3. Silica gel 60 F_{254} plates (Merck) are used for analytical TLC. Detection is achieved with UV light followed by I_2, ninhydrin or potassium permanganate staining.
4. HPLC analyses are performed on a Merck-Hitachi apparatus L6200 intelligent pump fitted with a UV detector.
5. The bioconversion is followed on a Zorbax SB-Aq column (150 mm×4.6 mm, 5 μm; Agilent Technologies) using trifluoroacetic acid (TFA) 0.1%/acetonitrile (70/30), flow rate at 1 mL/min, with benzoic acid as internal standard, detector at 254 nm.

6. The enantiomeric excess of L-2-NpAla is determined with a Crownpak CR+ (250 mm, 4 μm, Chiral Technologies), using aqueous perchloric acid (pH 1.5): methanol (9:1), flow rate at 1 mL/min, detector at 254 nm.
7. Optical rotations are determined with a Propol Digital Polarimeter Dr Kenchen, and $[\alpha]_D$ values are given in units of deg cm^2/g at 25°C.
8. Oxygen consumption assays are performed with a Hansatech oxygen electrode (Hansatech Instruments Ltd, Pentney, UK).
9. UV measurements are carried out with a UV/Vis Jasco V-560 spectrophotometer.

2.2. Deracemization by Dynamic Kinetic Resolution

1. The subtilisin in CLEA® form is marketed by CLEA Technologies and sold under the name Alcalase CLEA-ST.
2. The analytical equipments, unless otherwise specified, are the same used in Subheading 2.1.
3. The solvents used for HPLC analysis and sample preparation were purchased from Sigma-Aldrich and were all of HPLC grade.
4. The Chiralcel-OD and the Crownpak-CR+ column were bought from Daicel Chemical Industries, Ltd.
5. Chromatographic analysis for substrate conversion: Chiralcel-OD 250 mm, at 25°C, using hexane/isopropanol 997:3 at 1.2 mL/min, detector at 210 nm. Peaks: 6 min (naphthalene), 44.5 min (D-boc-2-NpAla-SEt), 48.5 min (L-boc-2-NpAla-SEt).
6. Chromatographic analysis for final product chiral evaluation: Crownpak-CR+ 250 mm, at 25°C, using [$HClO_4$(aq.) pH 1.5]/methanol 9:1 at 1.1 mL/min, detector at 210 nm. Peaks: 29.9 min (D-2-NpAla), 38.9 min (L-2-NpAla).
7. The sample taken from the reaction is dissolved in 1 mL of methanol and filtered through a 40-μm PPE membrane using a plastic syringe. The resulting solution is directly injected in the HPLC system.

3. Methods

3.1. Deracemization by Stereo-Inversion

Deracemization by stereo-inversion is a multistep chemo-enzymatic transformation which starts with a racemic amino acid derivative. While one of the enantiomers (R_f) is not affected by the enzymatic transformation, the other enantiomer (S_f) is transformed into a compound (A_i) which can, in turn, be transformed into the starting compound of opposite configuration (R_f) or into a racemate. In the first case, the convergent process gives only the enantiomer (R_f); in the second one, in successive cycles the enantiomeric excess increases at each cycle, eventually reaching complete conversion into one single enantiomer. At the end of the process, the enantiomer (S_f) is completely transformed into (R_f) (10). This methodology allows the complete recovery of a single enantiomer from a racemic mixture.

Here we present the deracemization of the amino acid D, L-3-(2-naphthyl)-alanine by stereo-inversion based on a combination of an amino acid oxidase of D-selectivity with an amino transferase with L-specificity.

D-Amino acid oxidase (EC 1.4.3.3, DAAO) is a highly stereoselective flavoenzyme which catalyses the dehydrogenation of the D-isomer of amino acids to give the corresponding α-imino acids and, after subsequent hydrolysis, α-keto acids and ammonia (11). Oxygen, the terminal redox acceptor, reoxidizes the reduced FAD cofactor to give hydrogen peroxide. In particular, DAAO from the yeast *Rhodotorula gracilis* exhibits a very high turnover number, tight binding with the coenzyme FAD, a broad substrate specificity and has an active site large enough to accommodate even substrates of considerable size (12, 13).

Amino transferases (ATs) catalyse the reversible transfer of an amino to a keto group in a reaction between an amino acid (as an amino donor) and a keto acid (as amino acceptor). Many bacterial ATs accept a wide array of keto acids as amino acceptors and are useful as biocatalysts in the preparation of AAs. Among the ATs used in biotransformations, aspartate amino transferase (EC 2.6.1.1, AAT) shows a wide substrate specificity both towards the amino donor and the acceptor. Problems arising from the application of this system for the production of L-amino acids from the corresponding α-keto acid concern the equilibrium constant of the reaction which, being closed to 1, requires a system to shift the equilibrium towards the desired side. Several multiple enzymatic systems have been applied to solve the problem, both with whole cell catalysts and with coupled enzyme strategies. The problem of equilibrium of the transamination reaction has been solved by using L-cysteine sulfinic acid (L-CSA) as the amino donor which furnishes an unstable β-keto sulfinic acid, which is readily decomposed into pyruvate and sulfur dioxide without the need of catalysis (14, 15). Moreover, because of the high acidity of L-CSA, the purification of the amino acid product by ion exchange chromatography is highly simplified.

We report the application of DAAO from *R. gracilis* to generate the 2-naphthylpyruvic acid (2-NpPA) from the racemic mixture of 3-(2-naphthyl)-alanine and the in situ transformation of

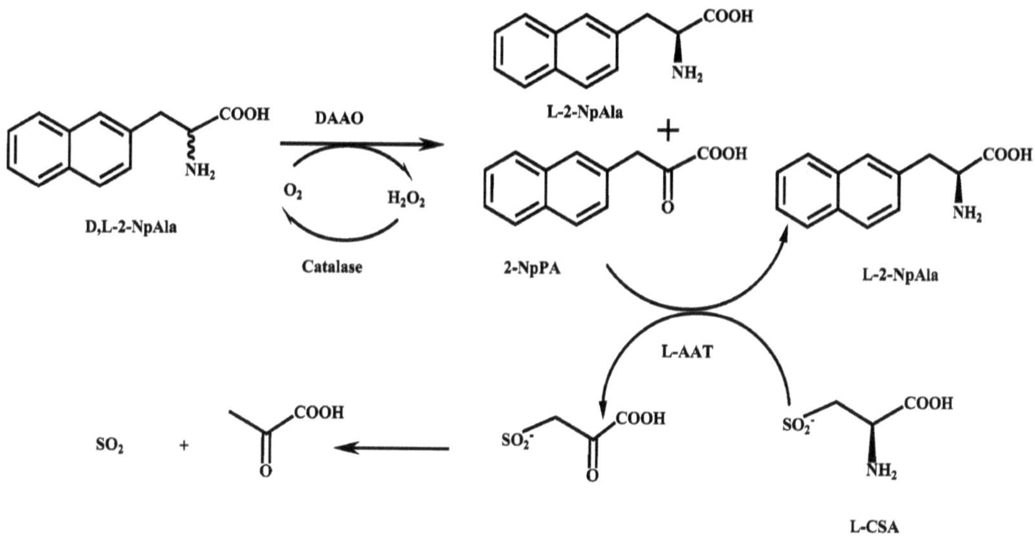

Fig. 1. Multistep enzyme catalysed deracemization of D,L-3-(2-naphthyl)-alanine (D,L-2-NpAla).

the 2-NpPA into L-3-(2-naphthyl)-alanine catalysed by L-AAT using L-cystein sulphinic acid as irreversible amino donor (see Fig. 1) (16).

3.1.1. Enzyme Expression and Purification

1. Recombinant DAAO from *R. gracilis* is expressed and purified from BL21(DE3)pLysS *Escherichia coli* cells as described previously (17).
2. The pure enzyme has a specific activity on D-alanine of 110 U/mg protein.
3. L-Aspartate amino transferase (L-AAT) from *E. coli* is produced and purified following a described procedure from an over-expressing *E. coli* strain TY103 transformed with pUC19-aspC (18–20).

3.1.2. Activity Assay and Kinetic Measurements

1. DAAO activity is assayed with an oxygen electrode at air saturation (0.253 mM O_2) and 25°C, using 28 mM D-alanine as substrate in 75 mM sodium pyrophosphate buffer, pH 8.5. One DAAO unit is defined as the amount of enzyme that converts 1 μmol of D-alanine per minute at 25°C (12, 13, 17).
2. L-AAT is assayed using 2-naphthyl pyruvic acid (2-NpPA) in the following conditions: 1 mL solution of aspartate (40 mM), L-AAT (0.05 U), NADH (0.14 mM), malic dehydrogenase (0.15 U), 2-NpPA (0.2–4 mM) in potassium phosphate buffer 100 mM at pH 8.0; the reaction time course is followed at 340 nm and 25°C.

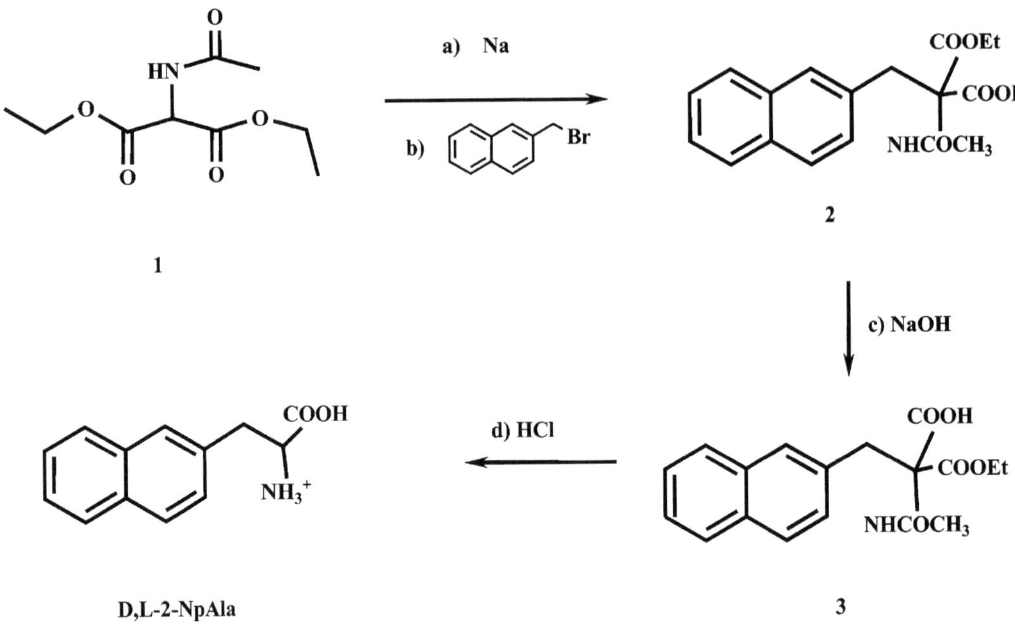

Fig. 2. Synthesis of D,L-2-naphthyl alanine (D,L-2-NpAla).

3.1.3. D,L-2-Naphthyl Alanine Preparation

1. The D,L-2-naphtyl alanine is prepared by a malonic ester synthesis starting from diethyl acetamido malonate **1** alkylated by 2-bromomethylnaphtalene (21).
2. After partial hydrolysis of the diester **2** and subsequent decarboxylation of the monoethylmalonate derivative **3**, the desired product D,L-2-NpAla is obtained (see Fig. 2).

3.1.4. Preparation of Diethyl (2-Naphthylmethyl) Acetamidomalonate 2

1. Prepare a solution of sodium (1.1 g, 46 mmol) in dry ethanol (100 mL).
2. Dissolve diethyl acetamidomalonate **1** (10 g, 46 mmol) in this solution.
3. Add 2-bromomethylnaphtalene (16.3 g, 73.4 mmol) at room temperature.
4. Reflux the reaction mixture for 7 h.
5. Evaporate ethanol and treat the residue with water (200 mL) and extract the product with ethyl acetate (AcOEt).
6. Wash the organic phase twice with water.
7. Dry the organic phase by addition of Na_2SO_4, filter the suspension and evaporate to dryness to obtain diethyl (2-naphtylmethyl) acetamidomalonate **2** (15.5 g, 94.4% yield).
8. The ^{1}H NMR data of compound **2** are: δ 7.83–7.70 (m, 3H), 7.41–7.5 (m, 3H), 7.14–7.1 (m, 1H), 4.296 (q, 7.347Hz, 6H), 3.830 (s, 2H), 2.034 (s, 3H), 1.325 (t, 7.347Hz, 6H).

3.1.5. Preparation of Monoethyl (2-Naphtylmethyl) Acetamidomalonate 4

1. Prepare a solution of diethyl (2-naphtylmethyl)acetamidomalonate **2** (14.15 g, 39.6 mmol) in ethanol (50 mL).
2. Add 200 mL of water and NaOH 10 N (4 mL).
3. Stir the reaction at room temperature and analyse by TLC (eluent: toluene/CH_3CN/CH_3COOH: 6/3/1).
4. After 4 h, the mixture is evaporated.
5. Treat with water (200 mL) and extract with AcOEt (200 mL) at pH 8.
6. Acidify the aqueous phase with 6 N HCl to about pH 2 and extract twice with AcOEt (2 × 150 mL).
7. Collect the organic phases.
8. Dry the organic phases by addition of Na_2SO_4; filter the suspension and evaporate to dryness.
9. The obtained white solid residue (12.5 g) is crystallized from a mixture of ethanol/hexane to give product **3** (6.3 g, 48% yield).
10. The ^{1}H NMR data of compound **3** are: δ 7.9–7.70 (m, 3H), 7.60–7.46 (m, 3H), 7.204 (m, 1H), 4.20 (q, 6.72, 2H), 3.65 (s, 2H), 2.034 (s, 3H), 1.228 (t, 6.72Hz, 3H).

3.1.6. Preparation of D,L-3-(2-Naphthyl)-Alanine

1. Prepare a suspension of monoethyl (2-naphthylmethyl) acetamidomalonate **3** (5 g, 15.2 mmol) (21) in 100 mL water-ethanol (1:1).
2. Add 1 mL of 12 N HCl.
3. Reflux the suspension under magnetic stirring.
4. Monitor the reaction by TLC (AcOEt:*i*-PrOH:AcOH, 2:2:1).
5. After 12 h, add dropwise 10 N NaOH under stirring until a pH of about 6.2 is reached.
6. Collect the white precipitate by filtration.
7. Dry the white solid under vacuum.
8. Recrystallize the solid from AcOEt to obtain D,L-2-NpAla (2.3 g, 70% yield).
9. The ^{1}H NMR data of D,L-2-NpAla are: δ 3.30 (m, 2H), 4.23 (s, 1H), 7.4–7.56 (m, 3H), 7.75–7.92 (m, 4H), 8.37 (br s, 2H) (see Note 1).

3.1.7. Preparation of 2-Naphthyl Pyruvic Acid (2-NpPA) 6

2-NpPA **6** is prepared (see Fig. 3) from the corresponding azalactone **5** (22) which in turn is obtained from the commercial 2-naphtaldehyde **4** as described in the literature (23).

3.1.8. Preparation of Azalactone 5

1. Prepare a solution of 2-naphtaldehyde **4** (10 g, 64 mmol) in acetic anhydride (25 mL).
2. Add *N*-acetylglycine (NAc-gly-OH, 9.75 g, 83 mmol).
3. Add sodium acetate (3.52 g, 70.4 mmol).

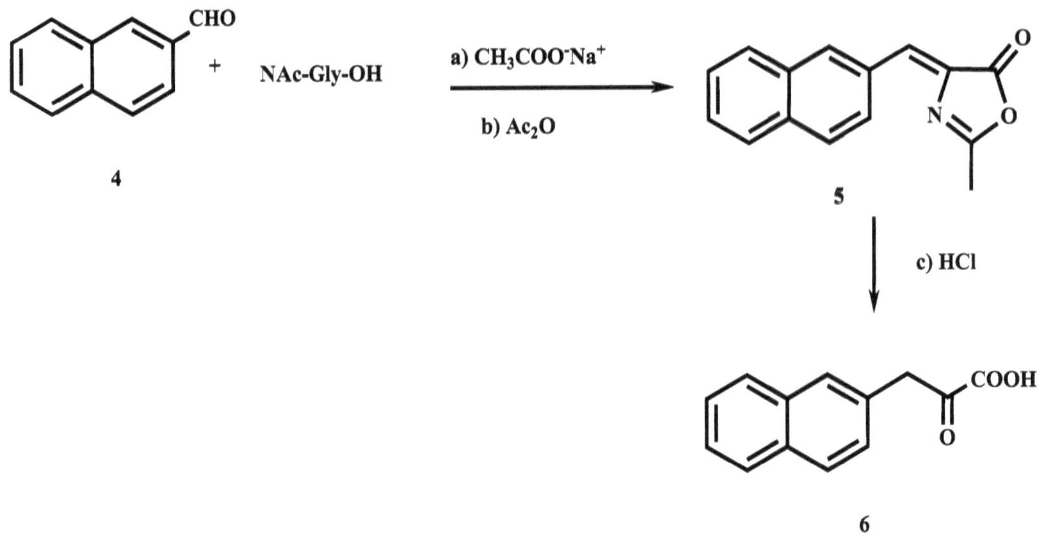

Fig. 3. Synthesis of 2-naphthyl pyruvic acid (2-NpPA) **6**.

4. Heat the mixture to 100°C in an oil bath for 2 h.
5. Stir the reaction mixture at room temperature overnight to allow the corresponding azalactone **5** to precipitate as an yellow solid.
6. Add 25 mL of water.
7. Filter the solid, wash three times with acetone and dry under reduced pression to give product **5** (8 g, 52% yield) of suitable purity.
8. The ^{1}H NMR spectrum data of compound **5** are: δ 12.5 (1H, broad), 9.5 (1H, s), 7.8–8.65 (7H, m); 2.05 (3H, s).

*3.1.9. Preparation of 2-Napthyl Pyruvic Acid **6***

1. Prepare 100 mL of a solution of HCl 3 N: dioxane/1:1.
2. Add azalactone **5** (5 g, 21.1 mmol).
3. Stir the suspension to reflux for 4 h.
4. Cool the solution and isolate the product acid by filtration.
5. Wash the residue with water, dry under reduced pressure and crystallize from ethanol/hexane to obtain product **6** (3.59 g, 80% yield).
6. The ^{1}H NMR spectrum data of compound **6** are: δ 9.2 (1H, broad), 7.5–8.5 (7H, m), 2.5 (2H, s).

3.1.10. Analytical and Preparative Deracemization of D,L-2-NpAla with the DAAO-L-AAT System and L-CSA as Amino Donor

1. Prepare 1 mL of reaction mixture by mixing the substrate D,L-2-NpAla (from 0.2 to 1 mM) and L-cysteine sulfinic acid CSA (40 mM) in 1 mL of water by sonication at 40°C.
2. Add L-AAT (1 U) and catalase (740 U).
3. Perform the enzymatic reaction at pH 8.0 (by addition of 1 M KOH) and 25°C starting by addition of DAAO (0.16 U).

4. The reaction is followed at 254 nm by HPLC analysis.
5. Perform the multienzymatic deracemization of D,L-2-NpAla on a preparative scale employing the three enzymes (DAAO, L-AAT, and catalase), 50 mg of D,L-2-NpAla (0.23 mmol, 1 mM) in 240 mL of water and the other reagents used according to the proportion of the analytical conversion previously described (see Note 2).
6. The reaction is complete in 2 h giving as the only product L-2-NpAla, which is recovered by cationic exchange chromatography in 98% yield and 99.5% ee (see Note 3).
7. Importantly, when the (virtual) concentration is increased tenfold, biotransformation of the insoluble substrate is still possible. In this case, and in order to reach completion of the reaction, add fresh AAT (120 U) and DAAO (20 U) four times at 2 h intervals: the rate limiting step of the reaction appears to be the dissolution process of amino acid.
8. Under optimised conditions, conversion of 1.2 mM of D,L-2-NpAla requires 40 U/L of DAAO per hour and the reaction is complete in 220 min. The enantiomeric excess of the obtained L-2-NpAla is >99.9% and the overall yield >90%.

3.2. Deracemization by Dynamic Kinetic Resolution

Hydrolytic enzymes are the most readily available biocatalysts for synthetic applications, particularly for the kinetic resolution of racemic mixtures (24–26). Like in any resolution method, a maximum of 50% yield of enantiomerically pure product can be obtained, based on racemic starting material. However, when kinetic resolution is coupled with an in situ racemization of the substrate (either chemical or enzymatic), the yield limitation can be overcome leading to a much more efficient process: a deracemization based on a dynamic kinetic resolution (DKR) (10, 27, 28).

Requisites for a successful DKR are an enzyme selective for a single enantiomer, a racemizing system acting on the substrate but not on the product, and a rate of racemization of the substrate higher than the rate of conversion to product. These conditions often require the design of suitable substrates. In thioesters, the acidity of the hydrogen in the α-position is higher in comparison to the corresponding oxo esters, amides or acids. The enzymatic transformation of a thioester into the corresponding carboxylate with a higher pK_a of the α-proton is the basis for a successful DKR, provided that the enzymatic systems resist the basic conditions required for substrate racemization. This concept has been applied in the DKR of α-alkyl thioesters (29).

The L-forms of racemic-*N*-boc-aryl-α-amino acid thioesters were found to be substrates for the subtilisin-catalysed hydrolysis to the corresponding acids. Moreover, the unreacted D-enantiomer was found to be continuously racemized in the presence of an organic base. The combined reactions in a biphasic system allowed the deracemization of the amino acid derivatives based on a dynamic kinetic resolution. Excellent yields and enantioselectivities were achieved (30).

Successively, we made some modifications in the system in order to attain an efficient DKR of α-alkyl-amino acid thioesters too which are not racemized under the previous conditions. In this case, a stronger base is needed and it is necessary to run the reaction in an hydrophilic organic solvent (e.g. *tert*-butanol). The enzyme must be of course compatible with these unusual conditions: we found that subtilisin in CLEA® preparation is particularly effective in carrying out the resolution with proper activity and stability, could be separated at the end of the reaction and potentially reused.

This technology has been applied to a large number of amino acid thioesters, among which the 2-naphthylalanine derivative, shown here as an example, is included.

3.2.1. Saturated Hydrochloric Acid Solution in 1,4-Dioxane

1. Pour 300 mL of 1,4-dioxane in a 500 mL, two necked round-bottomed flask equipped with a water bath and magnetic stirring.
2. The side neck, by a quickfit adapter, is equipped with a sparger immersed in the solvent and connected by PVC tubing to an HCl (g) cylinder (see Note 4), whereas the central neck is connected to an oil bubbler. Set the water bath temperature at 15°C (see Note 5), and switch a vigorous stirring on.
3. Carefully open the cylinder valve in order to assure a small but constant flow of gas into the solution (see Note 6).
4. Once in a while, close the valve, disconnect the cylinder, stopper and weigh the flask. If the weight is not constant, resume the saturation (see Note 7).
5. Once saturated, the dioxane solution is kept in a tightly closed glass bottle stored in a ventilated shelf at room temperature. It is possible to determine the HCl content by dissolving 1 mL of solution in 50 mL of distilled water and titrating with 0.1 N NaOH using phenolphthalein as an indicator (see Note 8).

3.2.2. Enzyme Activation

1. Suspend 1 g of the enzymatic powder in 10 mL of 100 mM sodium borate buffer (pH 9.0) under magnetic stirring (see Note 9).
2. Check the pH using a glass electrode and bring it to 9.0 by a dropwise addition of a 0.1 M sodium hydroxide solution.
3. Then, filter the mixture through a Buchner funnel, wash it twice with 5 mL of ethanol, once with 5 mL of *tert*-butanol and finally vacuum-dry using a water pump (see Note 10).

3.2.3. Synthesis of D,L-Boc-2-NpAla (see Fig. 4)

1. Pour 15 mL of water together with 0.45 g (11.2 mmol) of sodium hydroxide in a 50 mL round-bottomed flask under magnetic stirring.

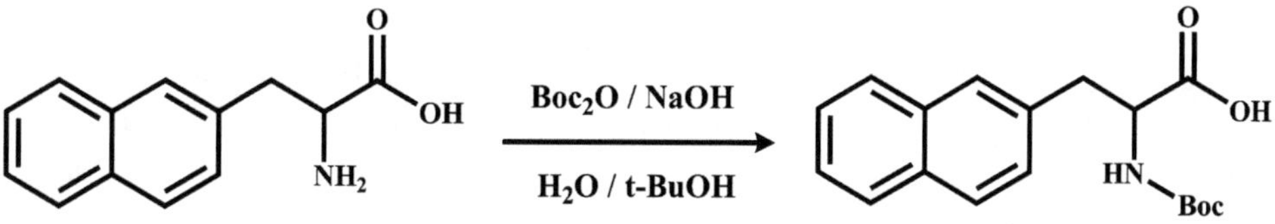

Fig. 4. Synthesis of D,L-boc-2-NpAla.

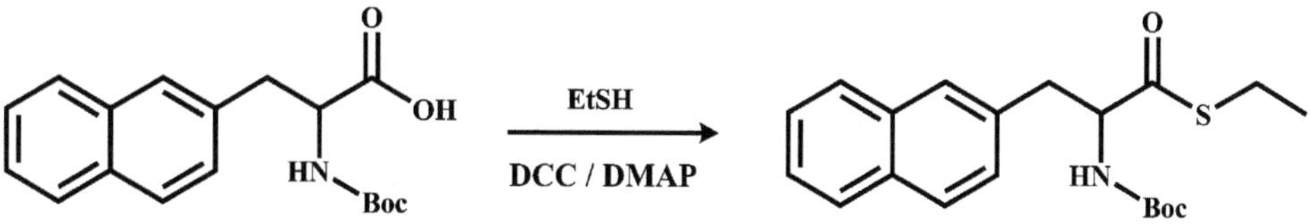

Fig. 5. Synthesis of D,L-boc-2-NpAla-SEt.

2. Then, add 1.5 g (7.5 mmol) of D,L-2-NpAla and 6 mL of *tert*-butanol, obtaining a clear solution.
3. Separately, dissolve 3.32 g (14.9 mmol) of di-*tert*-butyl dicarbonate in 10 mL of *tert*-butanol and slowly drop it into the reaction mixture along 1 h (see Note 11).
4. Let the mixture react overnight.
5. Then (see Note 12), pour the clear solution (see Note 13) in a separatory funnel and extract twice with 50 mL of n-hexane.
6. Extract the organic phases twice with 20 mL of a saturated aqueous sodium bicarbonate solution. Bring together the aqueous layers, under magnetic stirring, to pH 2 by careful addition of 6 N HCl (see Note 14). Extract the resulting milky mixture three times with 100 mL of diethylether.
7. Combine the organic phases together, dry over anhydrous sodium sulfate and filter through a cotton plug.
8. Remove the solvent by means of a rotary evaporator, dissolve the oily residue in 50 mL of n-hexane and leave overnight in a refrigerator.
9. Collect the white, crystalline precipitate by suction-filtration using a Buchner funnel, wash with cold hexane and dry under reduced pressure to give the final product as a white, crystalline powder (2.00 g, 88% yield).

3.2.4. D,L-Boc-2-NpAla-SEt (see Fig. 5)

1. Pour 800 mg (2.5 mmol) of D,L-boc-2-NpAla in a 10-mL flask and dissolve in 4 mL of dichloromethane.
2. Then, add 315 mg (5.1 mmol) of ethanethiol and 31 mg (0.3 mmol) of dimethylaminopyridine (DMAP) (see Note 15).
3. Immerse the flask in an ice bath, and keep under magnetic stirring for 10 min.

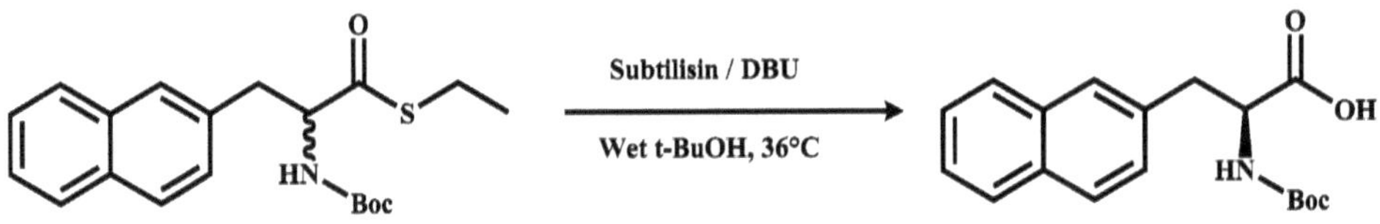

Fig. 6. Dynamic kinetic resolution on D,L-boc-2-NpAla-SEt.

4. At this point, add dropwise a solution of dicyclohexylcarbodiimide (DCC, see Note 16) (1.05 g, 5.1 mmol) in 1 mL of dichloromethane using a Pasteur pipette.
5. Upon completion of the addition, remove the ice bath and leave the reaction mixture overnight at room temperature under stirring.
6. The following day, filter the white mixture through a porous glass in order to eliminate most of the dicyclohexylurea, and evaporate the solvent.
7. Purify the residue by flash chromatography (eluent: hexane/ethyl acetate 95/5), obtaining a white, crystalline solid (802 mg, 88% yield). ^{1}H-NMR: δ=1.169 (t, *J*=7.279 Hz, 3H), 1.321 (s, 9H), 2.826 (q, *J*=7.279, 2H), 3.160 (br s, 1H) 3.259 (br s, 1H), 4.674 (br s, 1H), 4.941 (br s, 1H), 7.225–7.744 (m, 7H).

3.2.5. Dynamic Kinetic Resolution (see Fig. 6)

1. Dissolve the D,L-boc-2-NpAla-SEt (200 mg, 0.56 mmol) in 4 mL of *tert*-butanol, and pour the resulting solution in a 10 mL vial, placed in an orbital shaker thermostated at 36°C.
2. At this point, add 170 μL of DBU (170 mg, 1.12 mmol) and 8 mg of naphthalene. Immediately, withdraw an aliquot from the solution, constituting the t_0 sample for the HPLC analysis.
3. Add the reactivated enzyme Alcalase CLEA® (200 mg) and switch the shaker on.
4. Withdraw aliquots at predetermined intervals in order to monitor the conversion (see Note 17).
5. At the end of the resolution process (t_n = 28 h), centrifuge the reaction mixture (4,000 rpm, 5 min) in order to separate the enzyme from the solution.
6. Carefully, pour the latter in a separate flask, and wash the solid residue (by resuspending and centrifuging again) twice with 2 mL of *tert*-butanol. Keep it in a fridge (4°C) for successive use.
7. Bring the *tert*-butanol solutions together and eliminate most of the solvent using a rotary evaporator.
8. Suspend the residue n 10 mL of water and extract with ethyl ether (5 mL, three times).
9. Separate the aqueous layer and bring it to pH 3 by a dropwise addition of a 6 N HCl solution.

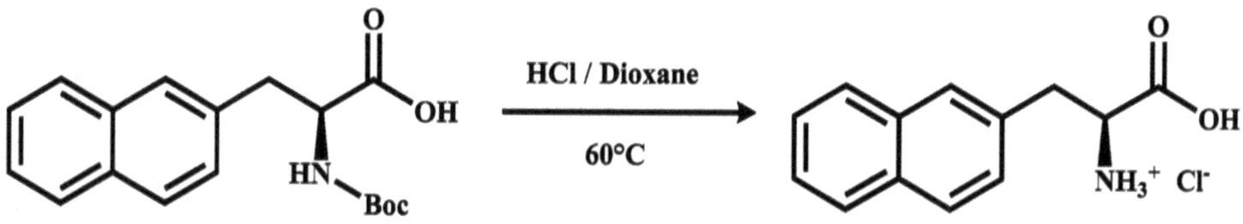

Fig. 7. Deprotection of L-boc-2-NpAla.

10. Extract the resulting dispersion with ethyl acetate (10 mL, three times).
11. Put the organic phases together, dry with sodium sulfate and bring to a rotary evaporator for the complete evaporation of the solvent.
12. A white solid is obtained (165 mg, 93% yield).

3.2.6. Deprotection of L-Boc-2-NpAla (see Fig. 7)

1. Dissolve the obtained L-boc-2-NpAla in 5 mL of a solution of HCl in dioxane (0.8 M).
2. Keep the resulting mixture at 60°C under stirring for 30 min, and then bring to dryness using a rotary evaporator.
3. Resuspend the resulting white solid in 5 mL of diethyl ether and filter through a Buchner funnel, obtaining the final product L-*2-NpAla* whose enantiomeric excess is measured by HPLC.

4. Notes

1. The ^1H NMR spectrum of D,L-2-NpAla has been recorded in DMSO-d_6/TFA-d_4 because of the low solubility of the compound.
2. The large amount of pyruvic acid produced clearly indicates that the use of cysteine sulphinic acid as amino donor was very effective in completely shifting the equilibrium.
3. The main limit of this reaction system is the extremely low solubility of 2-NpAla and thus the low space/time yield of the bioconversion.
4. It is not necessary to obtain a strictly anhydrous solution. For this reason, no on-line devices (like a concentrate sulphuric acid washing bottle) to completely eliminate water are included in this setup.
5. A rigorous temperature control is not required because the saturation process is not strongly exothermic. Nevertheless, the chosen temperature ensures a good saturation level while keeping the solvent, whose melting point is 12°C, in liquid state.

6. During the saturation phase, an excessive flow of HCl can be detected through a steep colour change of an indicator strip placed after the bubbler.
7. While closing the valves and disconnecting the tubings, pay attention to the possible pressure drops which can give rise to leaks or back-suctions.
8. Usually, the final HCl concentration is around 0.8 M.
9. It is advisable to employ a stirring bar with a central ring together with a moderate rotation speed in order to avoid grinding the enzyme.
10. It is not advisable to proceed to complete dehydration of the preparation because some water is necessary to keep active the enzyme. For this reason, water pump is a mild and effective drying device.
11. If the room temperature is too low to keep the *tert*-butanol liquid, it is possible to slightly warm the solution around 30°C or, alternatively, to use acetone or THF in substitution.
12. The solution should be feebly alkaline.
13. Sometimes, little amounts of a white precipitate (a secondary product) are obtained. In this case, it is convenient to filter the mixture before the extraction.
14. A copious development of carbon dioxide occurs. An efficient stirring is crucial in order to prevent the formation of "acidic spots" where the product would decompose.
15. Due to the strongly disagreeable odour of the ethanethiol, the reaction must be run under an efficient hood, and every piece of glassware having been in contact with the thiol should be treated with an oxidizing solution (e.g., a potassium permanganate solution acidified with a few drops of concentrated sulphuric acid).
16. DCC is a potent allergen and sensitizer. Melting the product could be convenient for easy handling without the risk connected to accidental spillages.
17. Sampling must be carried out under an efficient hood, due to the ethanethiol developing from the reaction.

Acknowledgements

This work was supported by Cost Action CM0701 "CASCAT, Cascade Chemoenzymatic Processes. New synergies between chemistry and biochemistry, WG2 Multistep deracemization of multifunctional compounds".

References

1. Senten K, Van der Veken P, De Meester I et al. (2003) Design, synthesis, and SAR of potent and selective dipeptide-derived inhibitors for dipeptidyl peptidases. J Med Chem 46, 5005–5014.
2. Wang L and Schultz P G (2005) Expanding the genetic code. Angew Chem Int Ed Engl 44, 34–66.
3. Sun H, Nikolovska-Coleska Z, Yang C Y et al. (2004) Structure-based design of potent, conformationally constrained Smac mimetics. J Am Chem Soc 126, 16686–16687.
4. Ley S V and Priour A (2002) Total synthesis of the cyclic peptide argyrin B. Eur J Org Chem 23, 3995–4004.
5. Tanaka M (2007) Design and synthesis of chiral alpha,alpha-disubstituted amino acids and conformational study of their oligopeptides. Chem Pharm Bull 55, 349–358.
6. Schneider J P and Kelly J W (1995) Templates that induce alpha-helical, beta-sheet, and loop conformations. Chem Rev 95, 2169–2187.
7. Patel R N (2000) Microbial/enzymatic synthesis of chiral drug intermediates. Adv Appl Microbiol 47, 33–78.
8. Leuchtenberger W, Huthmacher K, and Drauz K (2005) Biotechnological production of amino acids and derivatives: current status and prospects. Appl Microbiol Biot 69, 1–8.
9. Taylor P P, Pantaleone D P, Senkpeil R F et al. (1998) Novel biosynthetic approaches to the production of unnatural amino acids using transaminases. Trends Biotechnol 16, 412–418.
10. Gruber C C, Lavandera I, Faber K et al. (2006) From a racemate to a single enantiomer: Deracemization by stereoinversion. Adv Synth Catal 348, 1789–1805.
11. Curti B, Ronchi S, and Pilone M S (1992) D- and L-amino acid oxidases in: Müller F (ed.) Chemistry and biochemistry of flavoenzymes, CRC Press, Boca Raton, pp. 66–94.
12. Pollegioni L, Sacchi S, Caldinelli L et al. (2007) Engineering the properties of D-amino acid oxidases by a rational and a directed evolution approach. Curr Protein Pept Sci 8, 600–618.
13. Pollegioni L, Piubelli L, Sacchi S et al. (2007) Physiological functions of D-amino acid oxidases: from yeast to humans. Cell Mol Life Sci 64, 1373–1394.
14. Helaine V, Rossi J, and Bolte J (1999) A new access to alkyl-alpha-ketoglutaric acids, precursors of glutamic acid analogues by enzymatic transamination. Application to the synthesis of (2S,4R)-4-propyl-glutamic acid. Tetrahedron Lett 40, 6577–6580.
15. Helaine V, Rossi J, Gefflaut T et al. (2001) Synthesis of 4,4-disubstituted L-glutamic acids by enzymatic transamination. Adv Synth Catal 343, 692–697.
16. Caligiuri A, D'Arrigo P, Gefflaut T et al. (2006) Multistep enzyme catalysed deracemisation of 2-naphthyl alanine. Biocatal Biotransfor 24, 409–413.
17. Fantinato S, Pollegioni L, and Pilone M S (2001) Engineering, expression and purification of a His-tagged chimeric D-amino acid oxidase from *Rhodotorula gracilis*. Enzyme Microb Tech 29, 407–412.
18. Kagamiyama H and Hayashi H (2000) Branched-chain amino-acid aminotransferase of *Escherichia coli*. Method Enzymol 324, 103–113.
19. Kamitori S, Hirotsu K, Higuchi T et al. (1987) Overproduction and preliminary X-ray characterization of aspartate aminotransferase from *Escherichia coli*. J Biochem 101, 813–816.
20. Morino Y, Shimada K, and Kagamiyama H (1990) Mammalian aspartate aminotransferase isozymes. From DNA to protein. Ann N Y Acad Sci 585, 32–47.
21. Berger A, Smolarsky M, Kurn N et al. (1973) A new method for the synthesis of optically active-amino acids and their N derivatives via acylamino malonates. J Org Chem 38, 457–460.
22. Audia J E, Evrard D A, Murdoch G R et al. (1996) Potent, selective tetrahydro-beta-carboline antagonists of the serotonin 2B (5HT2B) contractile receptor in the rat stomach fundus. J Med Chem 39, 2773–2780.
23. Greenstein J P and Winitz M (1961) Chemistry of Amino Acids (Vol. 2). Wiley, New York.
24. Koeller K M and Wong C H (2001) Enzymes for chemical synthesis. Nature 409, 232–240.
25. Kazlauskas R J and Bornscheuer U T (1998) Biotransformations with lipases in: Reem H-J and Reed G (ed.) Biotechnology, Wiley-WCH, Weinheim, 37–191.
26. Williams J M J, Parker R J, and Neri C (2002) Enzymatic kinetic resolution in: Drauz K and Waldmann H (ed.) Enzyme Catalysis in Organic Synthesis, Wiley-VCH, Weinheim 1, 287–310.
27. Pellissier H (2003) Dynamic kinetic resolution. Tetrahedron 59, 8291–8327.
28. Turner N J (2004) Enzyme catalysed deracemisation and dynamic kinetic resolution reactions. Curr. Opin. Chem. Biol. 8, 114–119.
29. Um P J and Drueckhammer D G (1998) Dynamic enzymatic resolution of thioesters. J Am Chem Soc 120, 5605–5610.
30. Arosio D, Caligiuri A, D'Arrigo P et al. (2007) Chemo-enzymatic dynamic kinetic resolution of amino acid thioesters. Adv Synth Catal 349, 1345–1348.

Chapter 3

Enzymatic Production of Enantiopure Amino Acids from Mono-substituted Hydantoin Substrates

Gwynneth F. Matcher, Rosemary A. Dorrington, and Stephanie G. Burton

Abstract

Biocatalytic conversion of 5-substituted hydantoin derivatives is an efficient method for the production of unnatural enantiomerically pure amino acids. The enzymes required to carry out this hydrolysis occur in a wide variety of eubacterial species each of which exhibit variations in substrate selectivity, enantiospecificity, and catalytic efficiency. Screening of the natural environment for bacterial strains capable of utilizing hydantoin as a nutrient source (as opposed to rational protein design of known enzymes) is a cost-effective and valuable approach for isolating microbial species with novel hydantoin-hydrolysing enzyme systems. Once candidate microbial isolates have been identified, characterization and optimization of the activity of target enzyme systems can be achieved by subjecting the hydantoin-hydrolysing system to physicochemical manipulations aimed at the enzymes activity within the natural host cells, expressed in a heterologous host, or as purified enzymes. The latter two options require knowledge of the genes encoding for the hydantoin-hydrolysing enzymes. This chapter describes the methods that can be used in conducting such development of hydantoinase-based biocatalytic routes for production of target amino acids.

Key words: Hydantoinase, *N*-carbamoylase, Optically pure amino acids, Biocatalytic assay, Heterologous genomic library, Insertional mutant library

1. Introduction

Enantioselectivity is an intrinsic characteristic of enzymatic reactions, metabolic processes, and messenger–receptor interactions in nature. Biological metabolic and regulatory pathways are sensitive to the chirality of molecules and in many instances will respond differently depending on which of a pair of enantiomers are present (1–3). Stereochemistry is thus of critical importance when considering the biological activity of xenobiotics such as pharmaceuticals, agrochemicals, food additives, flavours, and fragrances (3). As a

Loredano Pollegioni and Stefano Servi (eds.), *Unnatural Amino Acids: Methods and Protocols*, Methods in Molecular Biology, vol. 794, DOI 10.1007/978-1-61779-331-8_3, © Springer Science+Business Media, LLC 2012

result, interest in the production of chiral precursors for the synthesis of complex enantiomerically pure compounds is increasing. This includes the increased production of enantiomeric pharmaceuticals or pesticides with specific target activity, using unnatural amino acids.

While chemical synthesis of unnatural, chiral amino acids is feasible, the use of enantioselective biocatalysts is an attractive alternative for several reasons. Firstly, enzymes function at a more neutral pH and lower temperatures, produce fewer toxic by-products, exhibit broad substrate-range, are often strictly stereospecific, and potentially result in 100% product yield from a given racemic substrate (4–6). Furthermore, biocatalysts exhibit specific regioselectivity and do not require substrate functional-group protection thus bypassing the additional steps to block and unblock substituents required in organic syntheses (7).

Enzymatic hydrolysis of hydantoin derivatives is of industrial interest for the production of enantiomerically specific natural and unnatural amino acids which is achieved by utilizing hydantoin substrates with suitable substituents on the C-5 position of the hydantoin ring structure. For instance, the use of hydantoin-hydrolysing enzymes has been found to be applicable to the production of highly lipophilic, silicon-containing amino acids resulting in higher lipophilicity, enhanced stability towards proteolytic degradation and a reduction in hydrophobic pocket collapse (8). Further, enantioselective *N*-carbamoylases have been shown to recognise the configuration of the α-carbon as well as the β-carbon of some *N*-carbamylamino acids (9, 10). This is extremely valuable in the synthesis of α,β-diastereomeric amino acids which exist as four stereoisomers, making the stereospecific synthesis of these compounds difficult with conventional chemical or enzymatic methodologies (9–11). To date, D-*p*-hydroxyphenylglycine is the most important compound produced by the hydantoinase process (12) with approximately 2,000 tons produced commercially each year (13). Competing industrial companies produce D-*p*-hydroxyphenylglycine by either using purified enzyme or whole cell biocatalyst (*Bacillus brevis* or *Agrobacterium radiobacter*) (14, 15). An L-stereoselective industrial process for the production of unnatural aromatic L-amino acids from 5-monosubstituted hydantoin derivatives has been developed using whole cells of *Arthrobacter aurescens* DSM 3745 and DSM 3747 (15). Amino acids produced via this method include L-tryptophan, L-phenylalanine, L-*o*-benzylserine, *p*-chloro-phenylalanine, *p*-fluoro-phenylalanine, *p*-nitro-phenylalanine, 1′-naphthylalanine, 2′-naphthylalanine, 3,4-dimethoxyphenylalanine, and 2′-thienylalanine (15).

Hydrolysis of 5-monosubstituted hydantoin derivatives to their corresponding amino acids can be catalysed by a variety of microbial enzyme systems (Fig. 1). Catalysis is initiated by cleavage of the cyclic amide bond at position 2 of the hydantoin ring resulting

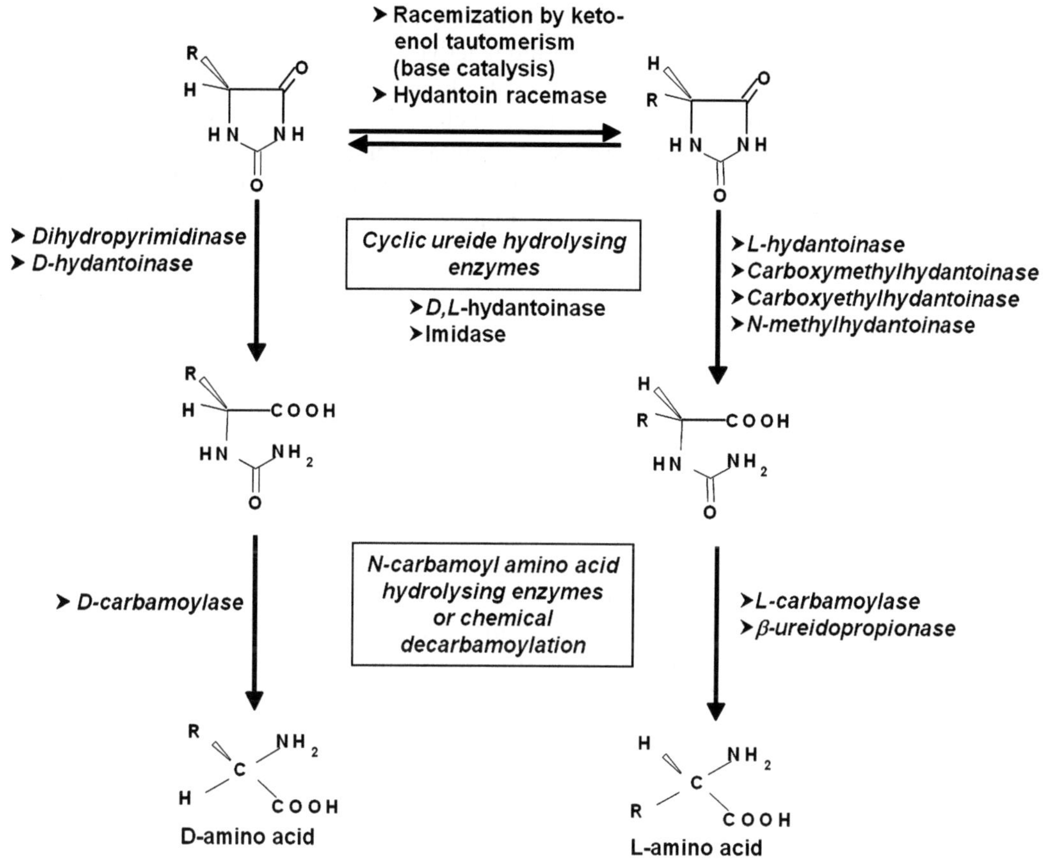

Fig. 1. Enzymatic hydrolysis of hydantoin to form enantiomerically pure amino acids (reproduced with permission G. Matcher) (16).

in the opening of the ring structure of the hydantoin molecule. The resultant *N*-carbamylamino acid is then hydrolysed to the corresponding amino acid either enzymatically or chemically. Depending on the stereospecificities of the enzymes involved in these two reactions, the end product is either a D- or L-enantiomer. Lastly, spontaneous or enzymatic racemisation of the unreacted hydantoin substrate allows for a potential 100% yield of optically pure amino acids (15).

Of the enzymes listed in Fig. 1, the dihydropyrimidinase, hydantoinase (D-, L-, nonstereospecific), *N*-carbamylamino acid amidohydrolase (D- or L-stereospecific), and β-ureidopropionase are of particular interest due to the generally high activity levels and broad substrate-selectivities observed.

Development of new biocatalytic processes with potential industrial applications generally starts with the process of screening for an organism or enzyme that is capable of carrying out the desired reaction. Once a suitable enzyme or organism has been isolated and identified, the optimal reaction conditions for maximum function and reactivity of the biocatalyst must be selected. Characterisation of the genes encoding for the hydantoin-hydrolysing enzymes, as well as the mechanism of regulation of the target gene expression, facilitate this process. This chapter describes the methods which can be used in conducting such development of hydantoinase-based biocatalytic routes for production of target amino acids.

2. Materials

See Note 1.

2.1. Screening of Microorganisms for Hydantoinase Activity

1. 10× M9 Salts: dissolve 60 g Na_2HPO_4, 30 g KH_2PO_4, and 5 g NaCl in 1 L deionised water. Autoclave the solution.
2. Trace elements: dissolve 50 mg boric acid, 40 mg $MnSO_4{\cdot}2H_2O$, 40 mg $ZnSO_4$, 20 mg $(NH_4)_6MO_2O_{24}{\cdot}4H_2O$, 10 mg KI, 4 mg $CuSO_4$ in 980 mL deionised water and autoclave. Dissolve 20 mg $FeCl_3$ in 20 mL deionised water and autoclave. When the solutions are cool, add the 20 mL $FeCl_3$ to the 980 mL containing the rest of the trace elements.
3. 40% Glucose: dissolve 200 g glucose in 500 mL deionised water and autoclave.
4. 1 M $MgSO_4$: dissolve 2.46 g $MgSO_4{\cdot}7H_2O$ in 10 mL deionised water and autoclave.
5. 1 M $CaCl_2$: dissolve 1.11 g $CaCl_2$ in 10 mL deionised water and autoclave.
6. 4% Hydantoin: dissolve 40 g hydantoin in 1 L deionised water and autoclave.
7. 4% *N*-carbamylglycine: dissolve 40 g *N*-carbamylglycine in 1 L deionised water and autoclave.
8. Hydantoin minimal medium (HMM): autoclave 615 mL deionised water in a 1 L Schott bottle. Once all solutions have cooled to touch, in a sterile manner add 250 mL 4% hydantoin, 100 mL 10× M9 salts, 25 mL 40% glucose, 10 mL trace elements, 200 μL 1 M $MgCl_2$, and 200 μL 1 M $CaCl_2$ to the autoclaved water. For agar plates, add 20 g bacteriological agar per litre and autoclave with the water.
9. *N*-carbamylglycine minimal medium (NMM): as for HMM except replace the 250 mL 4% hydantoin with 250 mL 4% *N*-carbamylglycine.

All solutions stored at room temperature.

2.2. Biocatalytic Assays for Hydantoinase and N-carbamoylase Enzyme Activity

1. HMM: see Subheading 2.1.
2. Nutrient broth: dissolve 16 g nutrient broth and 1 g hydantoin in 1 L deionised water and autoclave.
3. Phosphate buffer (0.1 M): prepare 0.1 M KOH and 0.1 M KH_2PO_4 solutions in deionised water. Place ~400 mL KH_2PO_4 into a beaker and pH to 8 using the KOH. Store the solution at 4°C (see Note 2).
4. Substrate: prepare a 100 mM 5-monosubstituted hydantoin and 50 mM *N*-carbamylamino acid solution in 0.1 M phosphate buffer (pH 8). Store solutions for <1 week at 4°C (see Note 3).
5. 12% Trichloroacetic acid (TCA): dissolve 60 g TCA in 500 mL deionised water. Store at 4°C.
6. 4 M Acetate buffer: prepare a 4 M sodium acetate solution and pH to 5.5 with 4 M acetic acid.
7. Ehrlich's reagent: dissolve 50 g *p*-dimethylaminobenzaldehyde in 500 mL 6 N HCl. Store at 4°C.
8. Ninhydrin reagent: dissolve 0.8 g ninhydrin and 0.12 g hydrindantin in 30 mL 2-methoxyethanol. Add 10 mL 4 M acetate buffer (pH 5.5), see Note 4. Store at 4°C until needed.
9. 50% Ethanol.
10. Standard curves: prepare 50 mM *N*-carbamylamino acid and 1 mM amino acid solutions in 0.1 M phosphate buffer. Store at 4°C for <1 week (see Note 3).

All solutions stored at room temperature unless otherwise specified.

2.3. Determining the Enantiospecificity of Hydantoin-Hydrolysing Enzymes

1. Chiralplate: 5 × 20 cm glass plate coated with a reverse phase silica gel impregnated with a chiral selector (a proline derivative DOS 3143726 and EP 0143147) and copper (II) ions.
2. Mobile phase: acetone:methanol:water (10:5:2).
3. TLC ninhydrin reagent: 1% ninhydrin in butan-1-ol.

All solutions stored at room temperature.

2.4. Isolation of Genes Encoding Hydantoin-Hydrolysing Enzymes

2.4.1. Heterologous Genomic Library

1. Preparation of a genomic library (see Notes 5 and 6).
2. Microtire plates: 96-well sterile microtitre plates.
3. Luria broth (LB): dissolve 10 g tryptone powder, 5 g yeast extract, and 5 g NaCl in 1 L deionised water and autoclave.
4. 100 mM Isopropyl β-D-1-thiogalactopyranoside (IPTG): dissolve 0.24 g IPTG in 10 mL deionised water and filter sterilize into a sterile bottle. Aliquot and store −20°C.
5. 40 mg/mL5-Bromo-4-chloro-3-indolyl-β-D-galactopyranoside (X-gal): dissolve 40 mg X-gal in 1 mL dimethylformamide. Store at −20°C.

6. 4% Hydantoin: dissolve 4 g hydantoin in 100 mL deionised water and autoclave.
7. Luria agar plates: add 20 g bacteriological agar to 1 L unsterilized LB. Autoclave and when sufficiently cool, add 10 μL 100 mM IPTG, 1 mL 40 mg/mL X-gal, and the appropriate antibiotic before pouring the agar into petri dishes.
8. Centrifuge with microtitre plate adapters.
9. Phosphate buffer (0.1 M, pH 8) containing 50 mM hydantoin or 30 mM *N*-carbamylglycine (NCG).
10. Assay microtitre plates: 96-well non-sterile microtitre plates.
11. Ehrlich's reagent: see Subheading 2.2.
12. Phosphate buffer: see Subheading 2.2.
13. Sodium acetate buffer: see Subheading 2.2.
14. Ninhydrin reagent: dissolve 0.17 g ninhydrin and 0.17 g hydrindantin in 20 mL 2-methoxyethanol (see Note 4). Store at 4°C until needed.
15. 50% Ethanol.

All Luria broth and Luria agar plates must contain the suitable antibiotic (prepared as per standard laboratory practice) to select for the vector into which the target organisms genomic fragments have been cloned.

All solutions should be stored at room temperature unless otherwise specified.

2.4.2. Insertional Inactivation to Identify Genes Encoding Hydantoin-Hydrolysing Enzymes

1. *Escherichia coli* DH5α (*F*⁻ *endA1 glnV44 thi-1 recA1 relA1 gyrA96 deoR nupG Φ80dlacZΔM15 Δ(lacZYA-argF)U169, hsdR17(r_K^- m_K^+), λ–*) (17) with plasposon.
2. *Escherichia coli* HB101 (*F*⁻ *mcrB mrr hsdS20(r_B^- m_B^-) recA13 leuB6 ara-14 proA2 lacY1 galK2 xyl-5 mtl-1 rpsL20(Sm^R) glnV44 λ*⁻) (17) with pRK2013 (18).
3. Luria broth: see Subheading 2.4.1.
4. Luria agar: add 20 g bacteriological agar to 1 L unsterilized LB. Autoclave and when sufficiently cool, add the appropriate antibiotic before pouring the agar into petri dishes.
5. 10% Glycerol: mix 10 mL glycerol with 90 mL deionised water and filter sterilize the solution into a sterile bottle.
6. Hydantoin minimal medium (HMM): see Subheading 2.1.
7. 10 mM HEPES: dissolve 0.238 g HEPES in 100 mL deionised water and autoclave.
8. Stock solutions for the plates: 4% $(NH_4)_2SO_4$, 4% hydantoin, 10× M9 salts, 40% glucose, trace elements, 1 M $MgCl_2$, 1 M $CaCl_2$. See Subheading 2.1.

9. Control minimal medium plates: autoclave 615 mL deionised water containing 20 g ultrapure agar in a 1 L Schott bottle. In a sterile manner, add 250 mL 4% $(NH_4)_2SO_4$, 100 mL 10× M9 salts, 25 mL 40% glucose, 10 mL trace elements, 200 μL 1 M $MgCl_2$, 200 μL 1 M $CaCl_2$, and suitable antibiotics to select for the transconjugant strains.
10. Selective minimal medium plates: autoclave 615 mL deionised water containing 20 g ultrapure agar in a 1 L Schott bottle. In a sterile manner, add 250 mL 4% hydantoin, 100 mL 10× M9 salts, 25 mL 40% glucose, 10 mL trace elements, 200 μL 1 M $MgCl_2$, 200 μL 1 M $CaCl_2$, and suitable antibiotics to select for the transconjugant strains.
11. Sterile 96-well microtitre plates.
12. 96-Well microtitre plate aseptic replicator.

All solutions should be stored at room temperature unless otherwise specified.

3. Methods

3.1. Screening for Hydantoinase and N-carbamoylase Activity in Microorganisms

The selection of a suitable biocatalyst for the production of enantiospecific amino acids begins with the screening of environmental isolates for the ability to grow utilizing hydantoin or *N*-carbamylamino acids as sole nitrogen sources. This is followed by plate growth assays, scored growth on minimal plates with hydantoin (HMM) or *N*-carbamylamino acids (NMM) as sole nitrogen sources and whole cell biocatalytic assays to determine optimum reaction conditions for the target hydantoin-hydrolysing enzymes.

1. Enrich for hydantoin-hydrolysing bacteria from soil or other biological samples by incubation in HMM, containing hydantoin as the sole nitrogen source, shaking at 25°C.
2. Plate serial dilutions of the enriched culture onto HMM agar plates and incubate at 25°C for 4–6 days.
3. Score the relative hydantoin-hydrolysing activities of isolates from the HMM plates by re-streaking promising isolates on HMM and NMM plates (containing hydantoin or *N*-carbamylglycine as sole sources of nitrogen, respectively).
4. Assay isolates with strong growth on plates with hydantoin or *N*-carbamylamino acids as sole nitrogen sources using the whole cell biocatalytic assay described below.

3.2. Biocatalytic Assay for Hydantoinase and N-carbamoylase Activity

3.2.1. Culture Conditions and Reaction Mixtures

1. Prepare seed cultures by inoculating 50 mL hydantoin minimal medium with a single bacterial colony from an agar plate and incubate at 28°C until sufficient biomass is produced as indicated by OD_{600} of 0.8–1.2 (approximately 4 days).
2. Once sufficient growth has taken place, use the seed culture to inoculate 100 mL nutrient broth containing 0.1% hydantoin as inducer such that a final OD_{600} of 0.02 is achieved.
3. Incubate the cultures, shaking, at 28°C until early stationary phase of growth is reached (see Note 7).
4. Pellet the cells by centrifugation at 7,500×*g* for 10 min in pre-weighed centrifuge tubes.
5. Discard the supernatant and wash the pelleted cells in half the original culture volume with cold potassium phosphate buffer.
6. After re-centrifugation, discard the supernatant and determine the wet cell mass by subtracting the weight of the centrifuge tube (weighed before pelleting the cells) from the final weight of the centrifuge tube containing the cells.
7. Resuspend the pellet such that a final concentration of 40 mg wet cell mass/mL is achieved.
8. Reaction mixtures should then be prepared in McCartney bottles as follows:

Components	Buffer blank (×1)	Substrate blank (×1)	Cell blank (×1)	Samples (×3)
Phosphate buffer (mL)	5	2.5	2.5	–
Substrate dissolved in phosphate buffer (hydantoin or *N*-carbamylamino acid, respectively) (mL)	–	2.5	–	2.5
Cell suspension (mL)	–	–	2.5	2.5

9. Incubate the reaction mixes at 40°C for 3 h, with shaking (see Note 8).
10. Transfer 1.5 mL of each of the reaction mixes to 1.7 mL tubes and centrifuge at maximum speed (typically 15,000 × *g* for standard benchtop microfuges) for 3 min.
11. Test the resultant supernatant for the presence of *N*-carbamylamino acids and amino acids using the Ehrlich's and ninhydrin colorometric assays, respectively.

3.2.2. Ehrlich's Assay for N-carbamylamino Acids Produced by Hydantoinase Activity

1. In a test tube, add 0.5 mL of 12% TCA to 1 mL of the reaction mix supernatant.
2. Vortex the tubes briefly and then add 0.5 mL Ehrlich's reagent and 3 mL triple distilled water.

3. Incubate the tubes at room temperature for 20 min and then measure the absorbance at 420 nm.
4. The concentration of the *N*-carbamylamino acids produced can be calculated using a standard curve of 0–50 mM *N*-carbamylamino acid subjected to the Ehrlich's assay (see Note 9).

3.2.3. Ninhydrin Assay for Amino Acids Produced by N-carbamoylase Activity

1. Aliquot 20 μL of the reaction supernatant into test tubes and add 0.980 mL phosphate buffer followed by 1 mL ninhydrin reagent.
2. Place capped test tubes in a boiling water bath for 15 min after which remove the test tubes from the water bath and cool to room temperature.
3. Once cool, add 3 mL 50% ethanol, vortex briefly and incubate for a further 15 min at room temperature.
4. Read the absorbance at 570 nm and ascertain the amount of amino acid produced from hydantoin or *N*-carbamylglycine substrates using a standard curve of 0–0.5 mM amino acid subjected to the ninhydrin assay (see Note 9).

3.3. Determining the Enantioselectivity of the Hydantoin-Hydrolysing Enzymes

1. The biocatalytic reaction mixtures (Subheading 3.2.1) should be made up and allowed to react. After incubation and centrifugation, the supernatant is analysed by chiral TLC (see Note 10).
2. Spot the supernatant approximately 1.5 cm from the base of the chiral plate with spots approximately 0.6 cm apart.
3. Place the chiral TLC plate into a solvent tank with approximately 0.5 cm depth of mobile phase.
4. Allow the mobile phase to run up the plate until the solvent front reaches approximately 1 cm below the top of the plate (approximately 45 min).
5. Detect the *N*-carbamylamino acids and chiral amino acids by dipping the plate into TLC ninhydrin reagent and drying the plates at 100°C for 15–30 min or 60°C overnight.

3.4. Isolation of Genes Encoding Hydantoin-Hydrolysing Enzymes

Isolation and analysis of the genes encoding for the enzymes responsible for the biocatalytic conversion of hydantoin to enantiopure amino acids is the next logical step in the optimization process. Isolation of encoding genes allows for heterologous expression of enzymes (19–21), alteration of promoter regions, and subsequent optimization of gene expression levels (22–24), as well as co-expression of hydantoinases and *N*-carbamoylases from different wild-type isolates (25). In addition, manipulation of culture media and whole cell reaction conditions allows for a limited potential increase in biocatalytic activity. Beyond this, optimization of the catalytic action of the enzyme itself is required. Directed evolution, via error-prone PCR of the encoding genes of target hydantoinases and *N*-carbamoylases, has been shown to increase thermostability, oxidative stability, activity levels, and even result in enantioselectivity inversion (26–28). Furthermore, availability of encoding genes allows for the construction of fusion proteins to facilitate protein purification or even for the construction of a bifunctional enzyme in which the hydantoinase and *N*-carbamoylase are fused together (29–31).

3.4.1. Screening of a Heterologous Genomic Library

1. Extract genomic DNA from the target bacterial isolate (see Note 5).
2. Generate a genomic library (see Note 6) and screen the resultant recombinants (see Note 11).
3. Prepare a set of master cultures in sterile 96-well microtitre plates by placing appropriate dilutions of *E. coli* cells transformed with the genomic library into the wells containing 150 μL Luria broth, such that each well contains approximately 40 insert-containing recombinant clones.
4. Incubate these microtitre plates overnight at 37°C before placing 60 μL of 50% glycerol into each well and storing these master microtitre plates at −70°C.
5. To screen the library, thaw the master plates and inoculate 20 μL of the culture from each well into the corresponding well of four duplicate plates of which two contain 180 μL Luria broth and two contain 180 μL Luria broth with 1 mM IPTG (see Note 12). Replace the master plates back at −70°C.
6. Incubate these plates at 37°C, with shaking, until late logarithmic to early stationary phase of growth has been reached.
7. Pellet the cells by centrifugation using a microtitre plate adaptor in a benchtop centrifuge.
8. Gently invert and tap the plates in order to remove the supernatant.
9. Resuspend the pellets in 200 μL of 0.1 M phosphate buffer containing 50 mM hydantoin or 25 mM *N*-carbamylglycine and incubate the plates at 40°C for 4–24 h with shaking.
10. Pellet the cells as before and analyse the supernatant using the modified Ehrlich's and ninhydrin assays (see Subheading 3.4.1.1 and 3.4.1.2).
11. Wells in which production of *N*-carbamylglycine or glycine was detected, in duplicated assays, contain approximately 40 insert-containing recombinant clones and individual insert-containing clones need to be isolated and subjected to further screening.
12. To this end, thaw the master plate again and inoculate 5 mL Luria broth with 20 μL from the master plate.

13. Incubate the inoculated culture at 37°C for approximately 5 h before diluting and plating onto Luria agar plates.
14. Select insert-containing colonies (e.g. by blue-white selection for β-galactosidase activity) and grow these pure cultures to stationary phase in 5 mL Luria broth.
15. Place 200 μL of the confluent culture into four duplicate wells of a fresh, non-sterile microtitre plate and pellet the cells as before.
16. After removing the supernatant, resuspend the cells in one of the four duplicate wells with 200 μL phosphate buffer and the remaining three wells with either 50 mM hydantoin or *N*-carbamylglycine.
17. Carry out the biocatalytic reactions as described below.

3.4.1.1. Ehrlich's Assay in Microtitre Plates

1. Place 100 μL of the biocatalytic reaction supernatant into a fresh, non-sterile microtitre plate and add 30 μL Ehrlich's reagent.
2. After adding 100 μL of deionised water, read the absorbance at 420 nm using a microtitre plate reader (see Note 13).

3.4.1.2. Ninhydrin Assay in Microtitre Plates

1. Place 10 μL of the biocatalytic reaction supernatant into a fresh, non-sterile microtitre plate and add 40 μL of 0.1 M phosphate buffer and 50 μL of 4 M sodium acetate buffer.
2. Incubate the microtitre plate at 60°C for 5 min before adding 50 μL ninhydrin solution.
3. Incubate the micotitre plate at 60°C for a further 20 min.
4. Add 100 μL of 50% ethanol and read the absorbance at 570 nm using a microtitre plate reader.

3.4.2. Insertional Inactivation to Identify Genes Encoding Hydantoin-Hydrolysing Enzymes

Utilizing complementation studies in *E. coli* to isolate genes encoding for hydantoin-hydrolysing enzymes has several drawbacks including poor expression of target genes and the formation of inclusion bodies (23, 32–34), reduced growth rates due to the toxic effect of overexpression (22, 23) and instability of the insert DNA resulting in loss of genetic information. As an alternative approach to identifying the genes coding for hydantoin-hydrolysing enzymes, transposon mutagenesis avoids the negative drawbacks associated with screening for heterologous expression in *E. coli*. A tailor-made tool for this methodology is the plasposon construct developed by Dennis and Zylstra (35).

Plasposons consist of a conditional origin of replication (e.g. pMB1 *oriV*), Tn5 transposase gene for random insertion events, Tn5 inverted repeats, origin of transfer (RP4 *oriT*) responsible for the movement of the plasposon into the target organism, and exchangeable antibiotic resistance genes. In order to negate further movement of the transposon once integrated into the host chromosome, the transposase encoding gene has been placed outside of the inverted repeats. As a result, a non-replicating DNA fragment containing the transposase gene is formed after the insertional event and is lost in subsequent cell divisions (35). Transposition thus occurs once only and results in a stable, inheritable mutation which can then be analysed (36). Furthermore, identification of the sequence flanking the transposon site of insertion is possible due to the presence of a conditional origin of replication within the integrated DNA sequence. This is achieved by digestion of the total genomic DNA of a selected insertional mutant strain with a restriction endonuclease nonspecific for the inserted foreign DNA, followed by circularisation of the resultant fragments by self-ligation. The isolation of the genomic fragment containing the plasposon DNA is then accomplished by transformation of the ensuing circular DNA fragments into *E. coli* and selection for the plasposon encoded antibiotic resistance marker. Using primers specific for the terminal ends of the plasposon sequence, the chromosomal DNA flanking the inserted transposon DNA can then be sequenced.

Unlike complementation studies in *E. coli*, insertional mutagenesis requires screening for the desired gene by detection of loss of activity of the gene product expressed under natural conditions rather than relying on expression in a foreign host which may not result in the production of functional proteins or where toxicity may be problematic. In addition to isolating the genes encoding for hydantoin-hydrolysing genes, by screening for a reduction as well as abolition of the target protein activities in mutant strains, this approach also allows for identification of genes and/or metabolic pathways involved in the regulation of the expression or activity of the specific enzymes (37).

3.4.2.1. Introduction of the Plasposon into the Target Bacterial Cells

1. Culture the plasposon (hosted in *E. coli* DH5α), the mobilizing plasmid pRK2013 (hosted in *E. coli* HB101) and the bacterial isolate responsible for hydantoin-hydrolysis in 5 mL Luria broth containing the appropriate antibiotics.
2. Incubate these cultures overnight at optimum growth temperatures.
3. Use these confluent cultures to seed 40 mL Luria broth containing no antibiotic and culture to mid-log phase of growth as determined by their optical densities at OD_{600}.
4. Combine aliquots of each of the cultures, to a final volume of 1 mL, in an Eppendorf tube at ratio's of 1:1:1 and 1:10:10 of target isolate, *E. coli* (plasposon) and *E. coli* (pRK2013), respectively.
5. Pellet the cells by centrifugation (>9,000×*g* on a benchtop microfuge) and resuspend the cells in 150 μL Luria broth.
6. Pipette the cell suspension onto a Luria agar plate (containing no antibiotic) as a pool.

7. Allow the pool to dry before inverting the plates and incubating overnight at the optimal growth temperature of the target bacterial isolate (i.e. the hydantoin-hydrolysing bacterial isolate).
8. Using a sterile loop, scrape the mated cultures off the agar plates and resuspend the cells in 1 mL filter-sterilized 10% glycerol. See Note 14.

3.4.2.2. Screening of the Transposon Mutants

1. To enrich for mutants unable to use hydantoin as a nitrogen source, combine ~40 pools of mutants generated by introduction of plasposon vectors into the target bacterial isolate and place in 50 mL hydantoin minimal medium.
2. Enrich for reduced and/or obviated hydantoin-hydrolysis by incubating the culture at optimal growth temperature for 8 h.
3. Add 1 mg/mL ampicillin and incubate for a further 16 h (see Note 15).
4. Harvest the cells by low speed centrifugation (6,000×*g* for 10 min), wash with 10 mM HEPES and finally combine the pellets in 1 mL 10% glycerol.
5. Plate appropriate dilutions of these cells onto control minimal medium plates containing 0.25% $(NH_4)_2SO_4$ and two antibiotics which in combination should select for transconjugant cells (recipient cells containing the plasposon) and eliminate the donor *E. coli* strains.
6. Aseptically transfer single colonies into individual wells of sterile 96-well microtitre plates containing 150 µL Luria broth and appropriate antibiotics.
7. Incubate the microtitre plates at the optimal growth temperature, shaking gently, for 2 days.
8. Replica plate the resultant cultures, using a 96-well microtitre plate aseptic replicator, onto minimal medium with either 0.25% $(NH_4)_2SO_4$ (positive control) or 1% hydantoin (selective medium) as nitrogen sources (see Note 16).
9. Verify the phenotype of mutant strains which were unable to hydrolyse hydantoin and thus grew poorly or not at all on the selective medium by biocatalytic assays (see Subheading 3.2).

4. Notes

1. In our experience, hydantoin-hydrolysing enzymes are quite sensitive to water quality. If the available water quality is poor, double distilled water, as opposed to deionised water, should be used.

2. Phosphate buffer may be stored at 4°C for up to 2 weeks. The optimal pH at which hydantoin-hydrolysing enzymes function differs depending on the enzyme system and may need optimization (see Note 8). The pH of the phosphate buffer should thus be adjusted accordingly. Typically, the optimal pH falls between 7.0 and 10.0. In our hands, a pH 8.0 was optimal for *Pseudomonas putida* RU-KM3s (38) and the optimal pH for hydantoin-hydrolysing activity in *Agrobacterium tumefaciens* RU-OR varied between 8.0 and 9.0 depending on the substrate used (39). Thus, a pH 8.0 is a suitable pH to use initially.
3. These substrates should not be autoclaved as the high temperatures may result in degradation products which will affect the biocatalytic assay results.
4. Ninhydrin reagent must be freshly prepared for each assay and is light sensitive, and therefore should be stored in a dark bottle. It should be noted that this reagent is carcinogenic; gloves should be worn when working with it and it must be disposed of appropriately.
5. Genomic DNA may be extracted from the bacterial isolate in a number of ways. We have found the genomic extraction protocol of Ausubel *et al.* (40) inexpensive and this method results in adequate quantities and suitable quality DNA for constructing genomic libraries. The use of genomic DNA extraction kits is more expensive but may yield higher quality DNA. If using the Ausubel *et al.* (40) method, note that the CTAB (hexadecyltrimethyl ammonium bromide)/NaCl solution becomes very viscous at room temperature and may need to be warmed in the 65°C waterbath before use.
6. Genomic libraries generated for the isolation of hydantoin-hydrolysing genes have been constructed in plasmid vectors such as pUC18 (21, 22, 41, 42), pBR322 (19), and pBluescript (23), or by construction of cosmid DNA libraries (24) and phage libraries (33, 43, 44). We have had success with the commercial ZAP Express Bam HI/Gigapack III Vector Kit particularly in the case of some difficult to clone genomic DNA such as *Agrobacterium* spp (44).
7. Hydantoin-hydrolysing enzymes are often subject to carbon or nitrogen catabolite repression (37, 44–46) and the highest levels of expression are observed in stationary phase of growth. The time taken to reach this phase varies depending on the isolate but usually falls within the 20–24 h time frame.
8. The optimal conditions for hydantoin-hydrolysing activity differ depending on the enzyme and isolate under investigation and may need to be established. The reaction conditions which impact most significantly on the enzyme activity are temperature, pH, inducer, substrate concentration, and reaction time. As an

example, a substrate concentration of 50 mM was found to be optimal for activity and a biocatalytic reaction time of 3–6 h (*P. putida* vs. *A. tumefaciens*). Highest activity of expressed hydantoinase and NCAAH activity in the biocatalytic reaction was 40°C and pH 8.0–9.0 (38, 39).

9. Provided the same batch of Ehrlich's reagent is used for the standard curve samples and the assay samples, this standard curve need not be repeated. However, the ninhydrin assay standard curves must be redone with each new ninhydrin assay carried out.
10. A 5-monosubstituted hydantoin derivative must be used in this assay as the biocatalytic product of hydantoin is glycine, which is achiral. Suggested substrates include methylhydantoin and hydroxyphenylhydantoin which are hydrolysed to the chiral amino acids alanine and hydroxyphenylglycine, respectively. Of the two, better separation of the hydroxyphenylglycine enantiomers is observed.
11. A critical step in the isolation of genes from genomic libraries is the screening methodology utilized. With respect to hydantoinase and *N*-carbamoylase encoding genes, screening methodologies used include detection of *N*-carbamoylase activity via a change in pH as a result of *N*-carbamylamino acid production using phenol red indicator (19, 44, 47), screening for growth with hydantoins and/or *N*-carbamylamino acids as sole sources of nitrogen for growth (21, 22, 37, 41, 42), or by direct screening for hydantoin-hydrolysing activity using colorimetric assays (23, 24, 48). In our experience, screening for *N*-carbamoylase activity using phenol red plates did not work as some isolates (e.g. *P. putida* and some strains of *E. coli*) resulted in a colour change even in the absence of hydantoin and selection of heterologous expression of hydantoin-hydrolysing enzymes in *E. coli* hosts based on growth with hydantoin or *N*-carbamylglycine as sole nitrogen sources had limited success. Direct screening for hydantoin-hydrolysing activity using colorimetric assays provided the best results in our experience.
12. The presence of the inducer IPTG allows for expression of downstream genes off the strong *lac* promoter. When IPTG is absent, genes present on the cloned chromosomal fragment are dependent on expression driven by their own promoters.
13. Some strains of *E. coli* result in the development of a pink colour on addition of Ehrlich's reagent to the wells. While this colour development does interfere with the OD_{420} readings to a certain extent, a clear positive result can still be readily detected.
14. Instead of conjugal transfer, the plasposon may be transformed directly into the target recipient cells if the isolated hydantoin-hydrolysing bacterium is compatible with transformation or electroporation.

15. The enrichment procedure is based on the inhibitory activity of ampicillin on bacterial cell wall synthesis. Specifically, ampicillin targets the transpeptidase enzyme responsible for cross-linkage of the glycan-linked peptide chains essential for the formation of a strong cell wall in bacteria. In the absence of the transpeptidase activity, the cell wall continues to be formed but gets progressively weaker until cell lysis occurs (49). However, if the cells are dormant, no cell growth occurs and thus, no weak cell wall is synthesized, and no cell lysis occurs. By enriching the plasposon-generated mutant library in minimal medium containing hydantoin as the sole source of nitrogen, mutants in which the hydantoinase and/or *N*-carbamoylase encoding genes have been inactivated will be unable to grow and will consequently survive treatment with ampicillin. The remaining mutants, which are able to hydrolyse and utilize hydantoin as a nutrient source, will actively divide and, due to ampicillin-induced weak cell formation, will lyse. In addition to hydantoinase and *N*-carbamoylase-minus mutants, auxotrophic mutants unable to grow in minimal medium, irrespective of the nitrogen source supplied, also survive the enrichment. In order to eliminate these from the screening process, the mutant strains surviving enrichment should be plated onto minimal medium with ammonium sulphate as a nitrogen source. If the target bacteria under investigation are naturally resistant to ampicillin, an alternative antibiotic with a similar mode of action may be utilised. Whichever antibiotic is selected for the enrichment process, the antibiotic resistance marker present on the plasposon backbone must confer resistance to an antibiotic other than the one selected for the enrichment.
16. As an alternative to minimal medium with hydantoin as a sole source of nitrogen, screening can be done using complete medium containing 5-fluorouracil (5-FU). 5-FU, is toxic to bacterial cells and exhibits structural similarity to dihydropyrimidine which is readily hydrolysed by most hydantoinases (12) to form the non-toxic 5-fluoroureidopropionic acid. In our hands, concentrations of 0.01–0.5% 5-FU in agar plates gave satisfactory results for both *P. putida* and *A. tumefaciens* (37, 44).

References

1. Zaks A and Dodds DR (1997) Application of biocatalysis and biotransformations to the synthesis of pharmaceuticals. *Drug Discovery Today* 2, 513–531.
2. Bommarius AS, Schwarm M and Drauz K (1998) Biocatalysis to amino acid-based chiral pharmaceuticals – examples and perspectives. *J Mol Catal B Enzym* 5, 1–11.
3. Maier NM, Franco P and Lindner W (2001) Separation of enantiomers: needs, challenges, perspectives. *J Chromatogr A* 906, 3–33.
4. Koeller KM and Wong C (2001) Enzymes for chemical synthesis. *Nature* 409, 232–240.
5. Azerad R (1995) Application of biocatalysis in organic synthesis. *Bull Soc Chim Fr* 132, 17–51.

6. Altenbuchner J, Siemann-Herzberg M and Syldatk C. (2001) Hydantoinases and related enzymes as biocatalysts for the synthesis of unnatural chiral amino acids. *Curr Opin Biotechnol* 12, 559–563.
7. Schmid A, Dordick JS, Hauer B, et al. (2001) Industrial biocatalysis today and tomorrow. *Nature* 409, 258–268.
8. Smith R, Pietzsch M, Waniek T, et al. (2001) Enzymatic synthesis of enantiomerically enriched D- and L-3-silated alanines by deracemization of DL-5-silylmethylated hydantoins. *Tetrahedron: Asymmetry* 12, 157–165.
9. Ogawa J, Ryono A, Xie S, et al. (1999) β-carbon stereoselectivity of *N*-carbamoyl-D-α-amino acid amidohydrolase for α,β-diastereomeric amino acids. *Appl Microbiol Biotechnol* 52, 801.
10. Ogawa J, Ryono A, Xie S, et al. (1999) Simultaneous recognition of two chiral centres in α,β-diastereomeric amino acids by an *N*-carbamoyl-L-α-amino acid amidohydrolase from *Alcaligenes xylosoxidans*. *Biotechnol Lett* 21, 711–713.
11. Ogawa J, Ryono A, Xie S, et al. (2001) Separative preparation of the four stereoisomers of β-methylphenylalanine with *N*-carbamoyl amino acid amidohydrolases. *J Mol Catal B Enzym* 12, 71–75.
12. Syldatk C, May O, Altenbuchner J, et al. (1999) Microbial hydantoinases – industrial enzymes from the origin of life? *Appl Microbiol Biotechnol* 51, 293–309.
13. Ogawa J and Shimizu S (2002) Industrial microbial enzymes: their discovery by screening and use in large-scale production of useful chemicals in Japan. *Curr Opin Biotechol* 13, 367–375.
14. Syldatk C, Mackowiak V, Hoke H, et al. (1990) Cell growth and enzyme synthesis of a mutant of *Arthrobacter* sp. (DSM 3747) used for the production of L-amino acids from D,L-5-monosubstituted hydantoins. *J Biotechnol* 14, 345–362.
15. Syldatk C and Pietzsch M (1995) Hydrolysis and formation of hydantoins. in Enzyme catalysis in organic synthesis – A comprehensive handbook. Vol. 1. Eds. Drauz K. & Waldmann H., VCH Publishers Inc. Weinheim, pp. 409–431.
16. Matcher GF (2004) Characterization of the hydantoin-hydrolysing system of *Pseudomonas putida* RU-KM3s. PhD Thesis, Rhodes University, South Africa.
17. Sambrook J, Margolin A, Russell A and Klibanov A (1989) Molecular Cloning – A laboratory manual. Cold Spring Harbor Press, New York.
18. Santos P, Bartolo I, Blatny J, et al. (2001) New broad-host-range promoter probe vectors based on the plasmid RK2 replicon. *FEMS Microbiol Lett* 195, 91–96.
19. Lee DC, Lee SG, Hong SP, et al. (1996) Cloning and overexpression of thermostable D-hydantoinase from thermophile in *E. coli* and its application to the synthesis of optically active D-amino acids. *Ann N Y Acad Sci* 799, 401–405.
20. Chao YP, Fu H, Lo TE, et al. (1999) One-step production of D-p-hydroxyphenylglycine by recombinant *Escherichia coli* strains. *Biotechnol Prog* 15, 1039–1045.
21. Ikenaka Y, Nanba H, Yamada Y, et al. (1998) Screening, characterization, and cloning of the gene for *N*-carbamyl-D-amino acid amidohydrolase from thermotolerant soil bacteria. *Biosci Biotechol Biochem* 62, 882–886.
22. Mukohara Y, Ishikawa T, Watabe K and Nakamura H (1994) A thermostable hydantoinase of *Bacillus stearothermophilus* NS1122A: Cloning, sequencing, and high expression of the enzyme gene, and some properties of the expressed enzyme. *Biosci Biotechol Biochem* 58, 1621–1626.
23. Chien HR, Jih Y, Yang W and Hsu W (1998) Identification of the open reading frame for the *Pseudomonas putida* D-hydantoinase gene and expression of the gene in *Escherichia coli*. *Biochim Biophys Acta* 1395, 68–77.
24. Buson A, Negro A, Grassato L, et al. (1996) Identification, sequencing and mutagenesis of the gene for a D-carbamoylase from *Agrobacterium radiobacter*. *FEMS Microbiol Lett* 145, 55–62.
25. Park JH, Kim GJ and Kim HS (2000) Production of D-amino acid using whole cells of recombinant *Escherichia coli* with separately and coexpressed D-hydantoinase and *N*-carbamoylase. *Biotechnol Prog* 16, 564–570.
26. Oh K, Nam S and Kim HS (2002) Improvement of oxidative and thermostability of *N*-carbamyl-D-amino acid amidohydrolase by directed evolution. *Protein Eng* 15, 689–695.
27. Oh K, Nam S and Kim HS (2002) Directed evolution of *N*-carbamyl-D-amino acid amidohydrolase for simultaneous improvement of oxidative and thermal stability. *Biotechnol Prog* 18, 413–417.
28. May O, Nguyen PT and Arnold FH (2000) Inverting enantioselectivity by directed evolution of hydantoinase for improved production of L-methionine. *Nature Biotechnol* 18, 317–320.
29. Kim GJ, Cheon YH and Kim HS (2000) Directed evolution of a novel *N*-carbamoylase/D-hydantoinase fusion enzyme for functional

expression with enhanced stability. *Biotechnol Bioeng* 68, 211–217.
30. Pietzsch M, Wiese A, Ragnitz K, et al. (2000) Purification of recombinant hydantoinase and L-*N*-carbamoylase from *Arthrobacter aurescens* expressed in *Escherichia coli*: comparison of wild-type and genetically modified proteins. *J Chromatog B* 737, 179–186.
31. Kim GJ, Lee DE and Kim HS (2000) Functional expression and characterization of the two cyclic amidohydrolase enzymes, allontoinase and a novel phenylhydantoinase, from *Escherichia coli. J Bacteriol* 182, 7021–7028.
32. Chao YP, Chiang CJ, Lo TE and Fu H (2000a) Overproduction of D-hydantoinase and carbamoylase in a soluble form in *Escherichia coli. Appl Microbiol Biotechnol* 54, 348–353.
33. Hils M, Munch P, Altenbuchner J, et al. (2001) Cloning and characterization of genes from *Agrobacterium* sp. IP I-671 involved in hydantoin degradation. *Appl Microbiol Biotechnol* 57, 680–688.
34. Wilms B, Wiese A, Syldatk C, et al. (2001) Development of an *Escherichia coli* whole cell biocatalyst for the production of L-amino acids. *J Biotechnol* 86, 19–30.
35. Dennis J and Zylstra G (1998) Plasposons: Modular self-cloning minitransposon derivatives for rapid genetic analysis of gram-negative bacterial genomes. *Appl Environ Microbiol* 64, 2710–2715.
36. DeLorenzo V and Timmis K (1994) Analysis and construction of stable phenotypes in gram-negative bacteria with Tn5- and Tn10-derived minitransposons in *Methods in Enzymology*, Eds. Clark V.L. and Bavoil P.M. Academic Press, San Diego 235, 386–405.
37. Matcher GF, Burton SG and Dorrington RA (2004) Mutational analysis of the hydantoin hydrolysis pathway in *Pseudomonas putida* RU-KM3s. *Appl Microbiol Biotechnol* 65, 391–400.
38. Buchanan K, Burton SG, Dorrington RA, et al. (2001) A novel *Pseudomonas putida* strain with high levels of hydantoin-converting activity, producing L-amino acids. *J Mol Catal B Enzym* 11, 397–406.
39. Hartley CJ, Kirchmann S, Burton SG and Dorrington RA (1998) Production of D-amino acids from D,L-5-substituted hydantoins by an *Agrobacterium tumefaciens* strain and isolation of a mutant with inducer independent expression of hydantoin hydrolysing activity. *Biotechnol Lett* 20, 707–711.
40. Ausubel FM, Brent R, Kingston RE, et al. (1983) Current protocols in molecular biology. Third edition. Wiley-Interscience, New York.
41. Mukohara Y, Ishikawa T, Watabe K and Nakamura H (1993) Molecular cloning and sequencing of the gene for a thermostable *N*-carbamyl-L-amino acid amidohydrolase from *Bacillus stearothermophilus* strain NS1122A. *Biosci Biotechol Biochem* 57, 1935–1937.
42. Nanba H, Ikenaka Y, Yamada Y, et al. (1998) Isolation of *Agrobacterium* sp. strain KNK712 that produces *N*-carbamyl-D-amino amidohydrolase, cloning of the gene for this enzyme, and properties of the enzyme. *Biosci Biotechol Biochem* 62, 875–881.
43. Wiese A, Syldatk C, Mattes R and Altenbuchner J (2001) Organization of genes responsible for the stereospecific conversion of hydantoins to α-amino acids in *Arthrobacter aurescens* DSM 3747. *Arch Microbiol* 176, 187–196.
44. Jiwaji M, Hartley CJ, Clark SA, et al. (2009) Enhanced hydantoin-hydrolyzing enzyme activity in an *Agrobacterium tumefaciens* strain with two distinct *N*-carbamoylases. *Enzyme Microb Techol* 44, 203–209.
45. Hartley CJ, Manford F, Burton SG and Dorrington RA (2001) Over-production of hydantoinase and N-carbamylamino acid amidohydrolase enzymes by regulatory mutants of *Agrobacterium tumefaciens. Appl Microbiol Biotechnol* 57, 43–49.
46. Ishikawa T, Mukohara Y, Watabe K. et al. (1994) Microbial conversion of DL-5-substituted hydantoins to the corresponding L-amino acids by *Bacillus stearothermophilus* NS1122A. *Biosci Biotechol Biochem* 58, 265–270.
47. Kim G.J, Park JH, Lee DC, et al. (1997) Primary structure, sequence analysis, and expression of the thermostable D-hydantoinase from *Bacillus stearothermophilus* SD-1. *Mol Gen Genet* 255, 152–156.
48. LaPointe G, Viau S, Leblanc D, et al. (1994) Cloning, sequencing, and expression in *Escherichia coli* of the D-hydantoinase gene from *Pseudomonas putida* and distribution of homologous genes in other microorganisms. *Appl Environ Microbiol* 60, 888–895.
49. Snyder L and Champness W (2003) Molecular genetics of bacteria. 2nd edition, ASM Press, Washington, 187–207.

Chapter 4

Preparation of Glutamate Analogues by Enzymatic Transamination

Thierry Gefflaut, Zeinab Assaf, and Martine Sancelme

Abstract

Aminotransferases are key enzymes of the metabolism of proteinogenic amino acids. These ubiquitous biocatalysts show high specific activities and relaxed substrate specificities making them valuable tools for the stereoselective synthesis of unnatural amino acids. We describe here the application of aspartate aminotransferase and branched chain aminotransferase from *E. coli* for the synthesis of various glutamate analogues, molecules of particular interest regarding the neuroactive properties of glutamic acid.

Key words: Glutamic acid, Aminotransferase, Transamination, Stereoselective synthesis

1. Introduction

Glutamic acid (Glu) plays a central role in the metabolism of all living organisms. It is also recognized as the major excitatory neurotransmitter within the central nervous system of vertebrates where it interacts with many different receptors and transporters constituting the glutamatergic system. Glu analogues acting as selective ligands are useful tools for neurobiological studies and could also bring some therapeutic effects against a variety of CNS disorders. Glu analogues are also of major interest for the preparation of peptides or peptidomimetics endowed with a large range of biological activities.

Aminotransferases (AT, EC 2.6.1.-) are involved in the metabolism of most proteinogenic amino acids. These pyridoxal phosphate (PLP)-bound enzymes catalyze the stereoselective transfer of the amino group from a donor substrate to an acceptor prochiral carbonyl derivative. ATs usually show high specific activities and

Loredano Pollegioni and Stefano Servi (eds.), *Unnatural Amino Acids: Methods and Protocols*,
Methods in Molecular Biology, vol. 794, DOI 10.1007/978-1-61779-331-8_4, © Springer Science+Business Media, LLC 2012

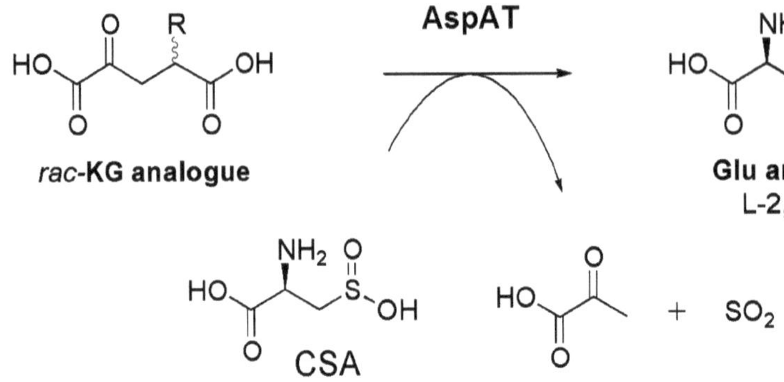

Fig. 1. Synthesis of L-2,4-*syn*-Glu analogues using AspAT and CSA.

relaxed substrate specificity. The development of equilibrium shifted transamination processes allowed the preparation of a variety of biologically active compounds including unnatural L- and D-α-amino acids (1, 2) as well as β-amino acids or even simple amines (3–6).

Interestingly, Glu and its keto acid homologue 2-oxoglutaric acid (KG) are preferred substrates of various ATs which are thus potential tools for the stereoselective preparation of Glu analogues substituted at position 3 or 4. Aspartate aminotransferase (AspAT, EC 2.6.1.1) offers the opportunity to shift the transamination equilibrium through the use of cysteine sulphinic acid (CSA) as the amino donor substrate. CSA, a close analogue of aspartic acid, is converted into the very unstable pyruvyl sulphinic acid which spontaneously decomposes into pyruvic acid and sulphur dioxide, thus providing the equilibrium shift (Fig. 1). AspAT has proven useful for the highly stereoselective preparation of many 4-substituted Glu analogues of L-2,4-*syn* configuration from the corresponding racemic KG analogues (7–9). The enzyme not only gives exclusively the L-amino acid but it also allows the control of the stereochemistry of carbon 4 through a kinetic resolution of the racemic keto acid substrate. In the series of 3-substituted analogues, 3-methyl-KG derivative alone could be efficiently converted to (2*S*,3*R*)-3-methyl-Glu using AspAT. Conversely, branched chain aminotransferase (BCAT, EC 2.6.1.42) are very active with Glu analogues substituted at either position 3 or 4 (10). Contrary to AspAT, BCAT did not show enantiopreference toward any 4-substituted substrates giving a mixture of L-*syn* and L-*anti* diastereomers. The same lack of enantioselection was observed with 3-methyl and ethyl derivatives. However, BCAT allowed the kinetic resolution of KG substituted with bulkier groups at position 3. For example, (2*S*,3*R*)-3-phenyl-Glu was prepared using BCAT as biocatalyst. In that case, as CSA is not a substrate for BCAT, a catalytic amount of Glu was used as the amino donor

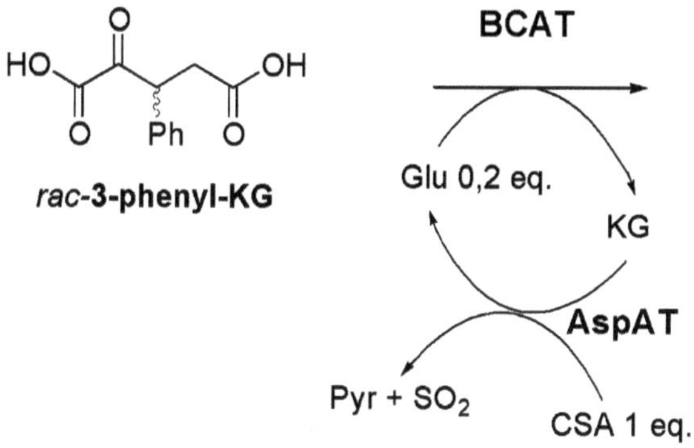

Fig. 2. Synthesis of (2*S*,3*R*)-3-phenyl-Glu using BCAT and AspAT.

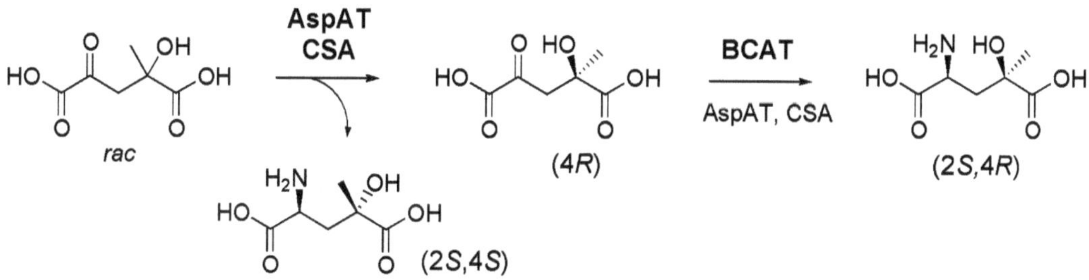

Fig. 3. Synthesis of (2*S*,4*S*)- and (2*S*,4*R*)-4-hydroxy-4-methyl-Glu using AspAT and BCAT.

substrate. Glu was regenerated using AspAT and a stoichiometric amount of CSA thus providing the equilibrium shift (Fig. 2).

AspAT and BCAT appear as complementary tools to access a variety of Glu analogues.

This is exemplified here with the preparation of (2*S*,4*R*)-4-methyl-Glu with AspAT, of (2*S*,3*R*)-3-phenyl-Glu with BCAT and of both stereomers (2*S*,4*S*)- and (2*S*,4*R*)-4-hydroxy-4-methyl-Glu using together AspAT and BCAT in a sequential process (Fig. 3).

2. Materials

2.1. Chemical Synthesis of Keto Acids

Use chemicals and solvents of reagent or analytical grade (>98%):

1. Chemicals: Acetic acid, acetic anhydride, benzaldehyde, celite, 1,4-diazabicyclo(2.2.2)octane (DABCO), lithium hydroxide (LiOH), magnesium sulfate ($MgSO_4$, anhydrous), methyl acrylate, methyl 3-hydroxy-2-methylenebutyrate, propionic acid, pyruvic acid, ruthenium oxide (RuO_2), sodium chloride (NaCl), sodium hydrogenocarbonate ($NaHCO_3$), sodium hydroxide (NaOH), sodium metaperiodate ($NaIO_4$), sodium methoxide (NaOMe), sulphonic resin (Dowex® 50WX8), sulphuric acid (H_2SO_4), triethyl orthoacetate, triethyl orthopropionate.

2. Solvents: Acetonitrile (CH_3CN), carbon tetrachloride (CCl_4), cyclohexane, deuterium oxide (D_2O, for NMR), dichloromethane (CH_2Cl_2), ethyl acetate (EtOAc), methanol (MeOH), toluene (anhydrous).
3. Thin layer chromatography: TLC plates silicagel 60F254 on aluminium sheets. Revelators (a) dissolve $KMnO_4$ (3 g), AcOH (0.25 mL), and K_2CO_3 (20 g) in H_2O (0.3 L); (b) dissolve vanillin (7.5 g) and conc. H_2SO_4 (4 mL) in 95% ethanol (0.25 L).
4. Column chromatography: Silicagel (SiO_2), pore size 60 Å (particle size 40–63 μm).

2.2. Enzymes Production

Prepare all solutions using deionized ultrapure water. Store buffers at 4°C unless otherwise specified.

1. Luria–Bertani (LB)-ampicillin culture medium: 10 g/L Tryptone, 5 g/L yeast extract, 5 g/L NaCl, and 50 mg/L ampicillin. In a 2-L beaker equipped with a stir bar, dissolve 50 g of solid LB mixture (DIFCO) composed of tryptone (20 g), yeast extract (10 g), and NaCl (20 g) in 1.9 L water. Bring to 2 L in a cylinder. Pour 200 mL of solution in each of ten 0.5-L cotton plugged flasks coated with aluminium sheet and sterilize in autoclave for 20 min at 121°C. In each flask, after sterilization add 0.1 mL of 100 mg/mL (0.2 μm filtered solution) of ampicillin (Sigma-Aldrich).
2. Phosphate buffer: 0.1 M potassium phosphate buffer, pH 7. In a 0.5-L beaker, equipped with a stir bar, dissolve 6.8 g KH_2PO_4 in 450 mL water. Adjust pH with 1 M KOH (approximately 25 mL). Bring to 0.5 L in a cylinder.
3. Buffer A: 10 mM phosphate buffer, 10 mM succinate, 5 mM 2-mercaptoethanol, pH 7. In a 0.5-L beaker equipped with a stir bar, dissolve 340 mg KH_2PO_4 and 295 mg succinic acid in 225 mL water. Adjust pH with 1 M KOH (approximately 5 mL). Bring to 250 mL in a cylinder and transfer in a bottle. Add 88 μL of 2-mercaptoethanol (see Note 1).
4. Buffer B: 100 mM phosphate buffer, 40 mM aspartate, pH 7.6. In a 0.25-L beaker equipped with a stir bar, dissolve 1.36 g KH_2PO_4 and 532 mg Asp in 80 mL water. Adjust pH with 1 M KOH (approximately 9 mL). Bring to 0.1 L in a cylinder (see Note 2).
5. Buffer C: 100 mM phosphate buffer, 10 mM succinate, 5 mM 2-mercaptoethanol, 0.05 mM pyridoxal phosphate (PLP), 3.3 M ammonium sulphate (90% saturation), pH 7. In a 2-L beaker equipped with a stir bar, dissolve 27.2 g KH_2PO_4, 2.36 g succinic acid, and 24.8 mg PLP in 1.3 L water. Add by portion 880 g ammonium sulphate until complete dissolution. Adjust pH to 7 with 3 M KOH (approximately 50 mL).

Bring to 2 L in a cylinder and transfer in a bottle. Add 704 μL of 2-mercaptoethanol (see Note 1).

6. Buffer D: 20 mM phosphate buffer, 5 mM 2-oxoglutarate, 5 mM 2-mercaptoethanol, pH 7. In a 0.5-L beaker equipped with a stir bar, dissolve 680 mg KH_2PO_4 and 731 mg 2-oxoglutaric acid in 225 mL water. Adjust pH with 1 M KOH (approximately 5 mL). Bring to 250 mL in a cylinder and transfer in a bottle. Add 88 μL of 2-mercaptoethanol (see Note 1).
7. Buffer E: 100 mM phosphate buffer, 40 mM glutamate, 50 mM ammonium sulfate, 0.1 mM pyridoxal phosphate, pH 7.6. In a 0.25-L beaker equipped with a stir bar, dissolve 1.36 g KH_2PO_4, 588 mg Glu, 660 mg $(NH_4)_2SO_4$, and 2.5 mg PLP in 80 mL water. Adjust pH with 1 M KOH (approximately 10 mL). Bring to 0.1 L in a cylinder (see Note 2).
8. Buffer F: 20 mM phosphate buffer, 5 mM 2-oxoglutarate, 5 mM 2-mercaptoethanol, 0.05 mM pyridoxal phosphate, 3.3 M ammonium sulphate (90% saturation), pH 7. In a 2-L beaker equipped with a stir bar, dissolve 5.44 g KH_2PO_4, 1.46 g 2-oxoglutaric acid, and 24.8 mg PLP in 1.3 L water. Add by portion 880 g ammonium sulphate until complete dissolution. Adjust pH with 1 M KOH (approximately 40 mL). Bring to 2 L in a cylinder and transfer in a bottle. Add 704 μL of 2-mercaptoethanol (see Note 1).
9. Enzymes for titrations and activity measurements: malate dehydrogenase from bovine heart (8,500 IU/mL), lactate dehydrogenase from rabbit muscle (940 IU/mL), glutamate dehydrogenase from bovine liver (11,300 IU/mL). The enzymes are all commercially available as suspensions in 3 M ammonium sulfate (Sigma-Aldrich). Prepare the solution of enzymes just before use: in a 1.5-mL microcentrifuge tube, centrifuge the appropriate volume of suspension at 14,000 × *g* for 5 min. Discard the supernatant and gently dissolve the pellet in the appropriate volume of phosphate buffer, in order to reach the desired final concentration.
10. *Substrates and cofactors.* 0.2 M Potassium 2-oxoglutarate: Dissolve 292 mg 2-oxoglutaric acid in water (5 mL). Adjust pH close to 7.6 with 1 M KOH. Bring to 10 mL in a cylinder. 0.2 M Sodium 4-methyl-2-oxopentanoate in water; 10 mM pyridoxal phosphate in phosphate buffer; 10 mg/mL NADH in phosphate buffer. The solutions can be stored at −18°C for few days.

2.3. Transamination Reactions

1. Sulphonic acid resin column: Pour 30 mL Dowex® 50WX8 (200–400 mesh, H^+ form) in a glass column (2 cm diameter, 20 cm length) equipped with a teflon stopcock and a cotton piece to retain the resin. Wash the resin with water (150 mL).

2. Basic resin column (AcO^- form): Pour 30 mL Dowex® 1×8 (200–400 mesh, Cl^- form) in a glass column (2 cm diameter, 20 cm length) equipped with a teflon stopcock and a cotton piece to retain the resin. Pass through the column 1 M NaOH (300 mL), then H_2O until neutrality (approximately 60 mL), then 1 M AcOH until acidity of the effluent (approximately 60 mL) and finally water until neutrality (approximately 60 mL).
3. Eluents: 1 M NH_3; 0.1, 0.25, and 0.5 M AcOH prepared with ultrapure deionized water.
4. TLC chromatography: Elute with *n*-propanol/water (7:3); reveal by immersion in 2 g/L ninhydrin in EtOH, followed by heating.

3. Methods

3.1. Synthesis of 4-Methyl-2-Oxoglutaric Acid, Dilithium Salt

4-Methyl-2-oxoglutaric acid is prepared from commercially available reagents in three steps including a Claisen–Johnson rearrangement giving a 2-ethyliden-glutarate, an oxidative cleavage affording the 2-oxoglutarate diester and finally a basic hydrolysis of the diester with LiOH (Fig. 4).

1. In a 100-mL round bottom flask equipped with a stir bar, dissolve methyl 3-hydroxy-2-methylenebutyrate (5 g, 38.4 mmol), triethyl orthopropionate (15.4 mL, 76.4 mmol) and propionic acid (0.13 mL, 1.8 mmol) in anhydrous toluene (25 mL). Equip the flask with a distillation apparatus composed of a vigreux column (15–25 cm), a thermometer, and a condenser. Heat carefully the mixture to reflux for 1 h before doing a slow distillation of the azeotrope toluene – EtOH (see Notes 3 and 4).

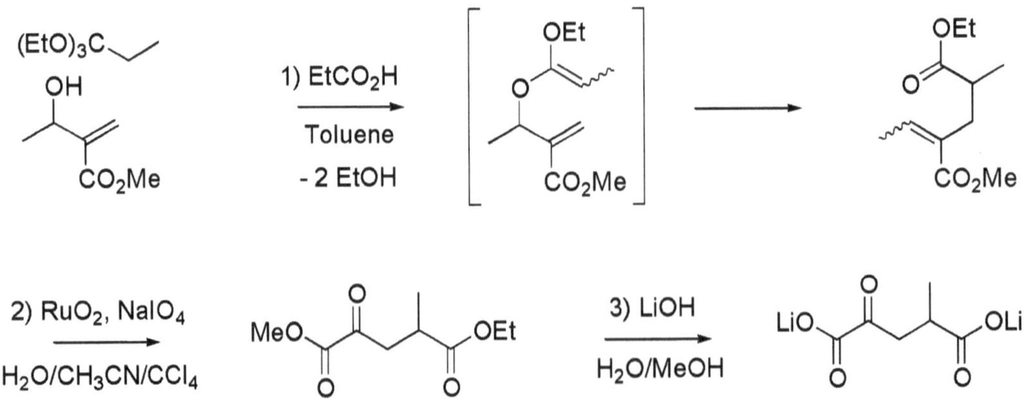

Fig. 4. Synthesis of dilithium 4-methyl-2-oxoglutarate.

Evaporate the residual toluene under reduced pressure using a rotary evaporator. Dissolve the residue in EtOAc (100 mL), transfer the solution in a separating funnel, wash the solution with 10% $NaHCO_3$ (2× 25 mL) and with water (25 mL). Dry the solution by addition of $MgSO_4$ (5 g), filter and concentrate under reduced pressure. The product is isolated as a pale yellow oil.

2. In a 0.5-L round bottom flask equipped with a stir bar dissolve 5 g of the product obtained in step 1 (23.4 mmol) in a mixture of CCl_4 (100 mL), CH_3CN (100 mL), and water (140 mL). Add $NaIO_4$ (20 g, 93 mmol) and RuO_2 (306 mg, 2.3 mmol). Stir the mixture for 20 h at room temperature until complete oxidation of the starting compound (TLC control: cyclohexane/EtOAc, 8/2, revelation with $KMnO_4$) (see Note 5). Filter the reaction mixture on celite (2.5 cm pad in a 100-mL sintered glass filter) and wash the celite pad with CH_2Cl_2 (25 mL). Wash the organic layer with water (100 mL), dry over $MgSO_4$ (5 g) and concentrate under reduced pressure. Purify the product by chromatography on a silica gel column (eluent: cyclohexane/EtOAc, 8/2) (see Note 6). The product is isolated as a colorless oil. An overall yield of 85% for steps 1 and 2 should be obtained.
3. In a 0.25-L round bottom flask equipped with a stir bar, dissolve the diester (3.5 g, 17.3 mmol) in MeOH (89 mL). Add dropwise freshly prepared 0.4 M LiOH (89 mL, 35.6 mmol) over a period of 30 min. Stir at room temperature for 15 h. Evaporate MeOH under reduced pressure. Adjust pH to 7.6 by addition of Dowex® 50WX8 (H^+ form). Remove the resin by filtration and concentrate the solution under reduced pressure (see Note 7). The product is isolated as a white solid in quantitative yield (see Note 8).

3.2. Synthesis of 3-Phenyl-2-Oxoglutaric Acid Dilithium Salt

3-Phenyl-2-oxoglutaric acid is prepared from commercially available reagents in five steps (7) including a Baylis–Hillman condensation between methyl acrylate and benzaldehyde, a thermodynamic favorable rearrangement of the adduct giving methyl 2-hydroxymethylcinnamate and the three-step sequence already described for 4-methyl-2-oxoglutarate (Fig. 5).

1. In a 50-mL round bottom flask equipped with a stir bar, add methyl acrylate (5 mL, 55.5 mmol), benzaldehyde (8.5 mL, 83.6 mmol), and DABCO (314 mg, 2.8 mmol). Stir the mixture at room temperature for 1 week. Dilute with EtOAc (25 mL), transfer in a separating funnel, wash with brine (2× 20 mL) and concentrate under reduced pressure. Purify the adduct by column chromatography (eluent: cyclohexane/EtOAc, 1/1) (see Note 6). The product is isolated as a colorless oil. A yield of 60–75% should be obtained.

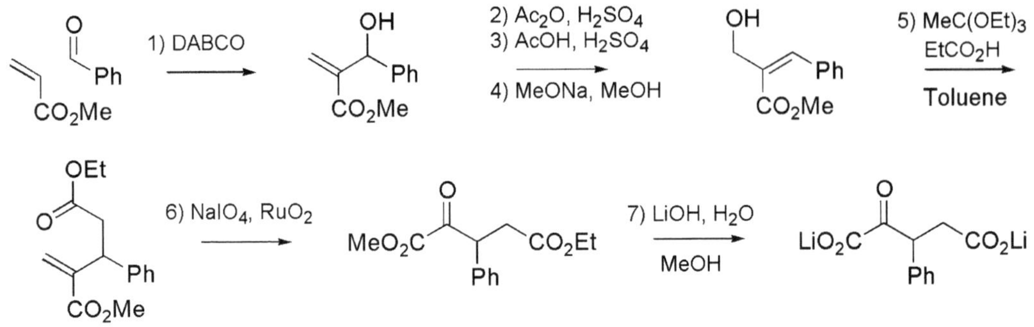

Fig. 5. Synthesis of dilithium 3-phenyl-2-oxoglutarate.

2. In a 100-mL round bottom flask equipped with a stir bar, add 5 g of the product obtained in step 1 (26 mmol), acetic anhydride (30 mL) and conc. H_2SO_4 (0.1 mL). Stir the mixture at room temperature for 30 min. Pour in ice-cold 1 M NaOH (150 mL) and stir for 1 h. Extract the mixture with CH_2Cl_2 (3× 50 mL). Wash the combined extracts with 10% $NaHCO_3$ (4× 50 mL). Dry over $MgSO_4$ (5 g), filter and concentrate under reduced pressure (see Note 9). The product is isolated as a colorless oil.
3. Dissolve the residue in AcOH (7.5 mL, 0.13 mol), add a drop of conc. H_2SO_4 and stir at 60°C for 4 h. Dilute with H_2O (20 mL) and extract with CH_2Cl_2 (2× 20 mL). Wash the extracts with brine (20 mL), dry over $MgSO_4$ and concentrate under reduced pressure. Purify the product by column chromatography (eluant: cyclohexane/EtOAc, 7/3) (see Note 6). The product is isolated as a colorless oil. An overall yield over 90% should be obtained for these steps.
4. In a 250-mL round bottom flask equipped with a stir bar, dissolve the purified compound in MeOH (50 mL), add MeONa (140 mg, 2.6 mmol), and stir at room temperature for 1 h (see Note 10). Concentrate to dryness under reduced pressure. Dissolve the residue in EtOAc (50 mL), transfer in a separating funnel and wash with H_2O (2× 20 mL). Dry over $MgSO_4$ and concentrate under reduced pressure. Methyl 2-hydroxymethyl-3-phenylacrylate is isolated as a viscous colorless oil. A yield of 85% should be obtained.
5. In a 100-mL round bottom flask equipped with a stir bar, dissolve methyl 2-hydroxymethyl-3-phenylacrylate (1.0 g, 5.2 mmol), triethyl orthoacetate (1.3 g, 7.8 mmol), and propionic acid (0.04 mL, 0.5 mmol) in anhydrous toluene (25 mL). Follow the procedure already described in Subheading 3.1 (step 1). The product should be isolated as a pale yellow oil.

6. In a 0.25-L round bottom flask equipped with a stir bar, dissolve the product obtained in step 3 in a mixture of CCl_4 (25 mL), CH_3CN (25 mL), and water (30 mL). Add $NaIO_4$ (4.49 g, 21 mmol) and RuO_2 (69 mg, 0.5 mmol). Follow the procedure already described in Subheading 3.1 (step 2). After the chromatographic step, the product is obtained as a colorless liquid. An overall yield of 70% for steps 5 and 6 should be obtained.
7. In a 100-mL round bottom flask equipped with a stir bar, dissolve the diester obtained in step 6 (0.75 g, 2.8 mmol) in MeOH (14 mL). Add dropwise freshly prepared 0.4 M LiOH (14.5 mL, 5.8 mmol) over a period of 10 min. Stir at room temperature for 15 h. Evaporate MeOH under reduced pressure. Adjust pH to 7.6 by addition of Dowex® 50WX8 (H^+ form). Remove the resin by filtration and concentrate the solution under reduced pressure (see Note 7). The product is isolated as a white solid in a quantitative yield (see Note 11).

3.3. Synthesis of 4-Hydroxy-4-Methyl-2-Oxoglutaric Acid, Dilithium Salt

4-Hydroxy-4-methyl-2-oxoglutaric acid, dilithium salt is simply prepared by autocondensation of pyruvic acid in the presence of LiOH.

1. In a 100-mL round bottom flask equipped with a stir bar, dissolve pyruvic acid (2 g, 22.7 mmol) in H_2O (25 mL). Adjust pH to 12 by addition of solid LiOH monohydrate (approximately 1 g). Stir at room temperature for 30 min. Adjust pH to 7 by addition of Dowex® 50WX8 (H^+ form). Filter and concentrate to dryness under reduced pressure (see Note 7). The product is isolated as a white solid and contains approximately 60% weight of 4-hydroxy-4-methyl-2-oxoglutaric acid, dilithium salt (see Note 12).

3.4. Production of AspAT from E. coli

AspAT from *E. coli* is produced from AspC strain: a recombinant *E. coli* strain JM103 transformed with the plasmid pUC119-*aspC* prepared in Kagamiyama's group (Osaka Medical College) (11). The plasmid contains the *AspC* gene coding for AspAT as well as a gene of ampicillin resistance for selection of transformed cells. Stocks of transformed cells are conserved at −80°C in LB medium (3 mL) supplemented with glycerol (10%). We describe here the cell culture conditions and the preparation of a crude extract of AspAT (see Note 13).

1. Preculture: In a sterilized 0.5-L flask, add a 3-mL stock sample of AspC cells to 100 mL of LB-ampicillin medium. Stir at 200 rpm for 24 h at 30°C (see Note 14).
2. In each of ten sterilized 0.5-L flasks containing 200 mL of LB-ampicillin, add 4 mL of the preculture and shake at 200 rpm for 16 h at 30°C (see Note 14).

3. Centrifuge the culture broth at 11,000 × *g* for 15 min; resuspend and combine the cell pellets in 0.1 M phosphate buffer, pH 7 (200 mL) and centrifuge at 11,000 × *g* for 15 min. Repeat the washing step with phosphate buffer. It is usual to obtain 8–10 g of wet cells from 2 L of culture medium.
4. Resuspend the cell pellet in 100 mL of buffer A and disrupt by sonication at 0°C (see Note 15). Centrifuge at 25,000 × *g* for 20 min at 4°C to remove cell debris.
5. Add 0.5 mL of 10 mM PLP to the supernatant and measure AspAT activity. In a 1.5 mL disposable cuvette, introduce a 10 mg/mL solution of NADH in phosphate buffer (20 μL), a 200 IU/mL solution of malic dehydrogenase (10 μL, 2 IU), an aliquot of the supernatant diluted 25× with phosphate buffer (10 μL) and buffer B (940 μL). Check the stability of the initial absorbance at 340 nm. Add 0.2 M 2-oxoglutarate (20 μL) and measure the absorbance linear decay. The activity of AspAT in the supernatant is calculated from the slope as follows (see Note 16):

$$\text{Activity (IU/mL} = \mu\,\text{mol/min.mL)} = \left[\frac{(\Delta\text{OD}/\Delta t)}{6{,}220}\right] 10^5 \times 25$$

Usually a total activity of 5,000 IU is obtained in the crude extract starting from 2 L culture.
6. Introduce the supernatant in dialysis tubings and immerge in 2 L buffer C. Stir slowly at 4°C for 24 h. Remove the protein suspension from the dialysis tubings and store at 4°C. Usually 25 mL of protein suspension is obtained with an activity close to 200 IU/mL. The suspension can be stored for several months without loss of activity (see Note 17).

3.5. Production of BCAT from E. coli

BCAT from *E. coli* is produced from IlvE strain: A recombinant *E. coli* strain JM103 transformed with the plasmid pUC119-*ilvE* prepared in Kagamiyama's group (Osaka Medical College) (12). The plasmid contains the *IlvE* gene coding for BCAT as well as a gene of ampicillin resistance for selection of transformed cells. Stocks of transformed cells are conserved at −80°C in LB medium (3 mL) supplemented with glycerol (10%). We describe here the cell culture conditions and the preparation of a crude extract of BCAT (see Note13).

1. Perform the culture of IlvE strain as described for AspC strain in Subheading 3.4 (steps 1–3).
2. Resuspend the pellet in 100 mL of buffer D and disrupt by sonication at 0°C (see Note 15). Centrifuge at 25,000 × *g* for 20 min at 4°C to remove cell debris.

3. Add 0.5 mL of 10 mM PLP to the supernatant and measure BCAT activity. In a 1.5-mL disposable cuvette, introduce a 10 mg/mL solution of NADH (20 μL), a 200 IU/mL solution of glutamic dehydrogenase (10 μL, 2 IU), an aliquot of the supernatant diluted five times (10 μL), and buffer E (940 μL). Check the stability of the initial absorbance at 340 nm. Add 0.2 M sodium 4-methyl-2-oxopentanoate (20 μL) and measure the absorbance linear decay. The activity of BCAT in the supernatant is calculated from the slope as follows (see Note 16):

$$\text{Activity (IU/mL} = \mu\,\text{mol/min.mL)} = \left[\frac{(\Delta\text{OD} / \Delta t)}{6{,}220}\right] 10^5 \times 5$$

Usually a total activity of approximately 1,000 IU is obtained in the crude extract starting from 2 L culture.

4. Introduce the supernatant in dialysis tubings and immerge in 2 L of buffer F. Stir slowly at 4°C for 24 h. Remove the protein suspension from the dialysis tubings and store at 4°C. Usually 25 mL of protein suspension is obtained with an activity close to 40 IU/mL. The suspension can be stored for several years without loss of activity (see Note 17).

3.6. Preparation of (2S,4R)-4-Methyl-Glutamic Acid with AspAT

(2*S*,4*R*)-4-Methyl-glutamic acid is obtained through the kinetic resolution of the racemic keto acid substrate (see Fig. 1). The reaction is controlled by titration of pyruvate formed from CSA and stopped near 40% conversion of the racemic substrate.

1. In a 250-mL flask equipped with a stir bar, dissolve dilithium 4-methyl-2-oxoglutarate (0.5 g, 2.9 mmol) and cysteine sulfinic acid (0.45 g, 2.9 mmol) in H_2O (140 mL). Adjust pH to 7.6 by addition of 1 M NaOH (approximately 3 mL).
2. In a 1.5-mL microcentrifuge tube, centrifuge 0.125 mL of the 200 IU/mL suspension of AAT (25 IU) (see Note 18). Discard the supernatant and dissolve gently the pellet in 1 mL of the reaction mixture prepared as described in Subheading 1. Add this solution to the whole reaction mixture and stir very slowly (approximately 60 rpm) at room temperature.
3. Titrate pyruvate formed from CSA every 30 min. In a 1.5-mL disposable cuvette, introduce a 10 mg/mL solution of NADH (20 μL), a 200 IU/mL solution of lactate dehydrogenase (10 μL, 2 IU), and 0.1 M phosphate buffer, pH 7.6 (965 μL). Measure the initial absorbance at 340 nm (OD_i). Add in the cuvette an aliquot of the reaction mixture (10 μL) and measure

the final stable absorbance (OD_f). The concentration of pyruvate in the reaction mixture is calculated as follows:

$$[\text{Pyruvate}] = \left[\frac{(OD_i - OD_f)}{6,220} \right] 100$$

Usually, a concentration of 8 mM (40% conversion) is reached in 2–3 h (see Note 19).

4. Pass the reaction mixture through the column of Dowex50® acidic resin. Collect the effluent in 30-mL glass tubes (20 mL fractions). Wash the column with water and check the fractions content by TLC (eluent: *n*-propanol/H_2O, 70/30; revelation: ninhydrin). Pursue water elution until all the pyruvic acid, residual keto acid substrate and CSA are washed out. Elute with 1 M aqueous ammonia and collect 5 mL fractions. Combine the ninhydrin positive fractions containing the Glu analogue and concentrate to dryness under reduced pressure (see Note 5).
5. Dissolve the residue in water (5 mL) and apply the solution on the top of the Dowex1® basic resin (AcO^- form). Wash the column with water (50 mL) and then elute with 0.1 M AcOH (50 mL), 0.25 M AcOH (100 mL) and 0.5 M AcOH (100 mL). Collect 5 mL fractions and analyse by TLC as described above. Combine the ninhydrin positive fractions containing the Glu analogue and concentrate to dryness under reduced pressure. Add water (5 mL) to the white solid obtained and concentrate to dryness under reduced pressure. Repeat this operation four times to completely remove AcOH (see Note 20). (2*S*,4*R*)-4-Methyl-Glu is obtained as a white solid with a yield close to 40% and a very high chemical and stereochemical purity (>99%) (see Note 21).

3.7. Preparation of (2S,3R)-3-Phenyl-Glutamic Acid with BCAT

(2*S*,3*R*)-3-Phenyl-glutamic acid is obtained through the kinetic resolution of the racemic keto acid substrate (see Fig. 2). The reaction is controlled by titration of pyruvate formed from CSA and stopped near 40% conversion of the racemic substrate.

1. In a 100-mL flask, dissolve dilithium 3-phenyl-2-oxoglutarate (234 mg, 1 mmol), CSA (153 mg, 1 mmol), L-Glu (29 mg, 0.2 mmol), and 10 mM PLP (0.5 mL, 5 μmol) in H_2O (40 mL). Adjust pH to 7.6 with 0.5 M NaOH (approximately 2 mL). Bring to 50 mL with water in a cylinder before transferring the solution in a 100-mL round bottom flask equipped with a stir bar.
2. In a 1.5-mL microcentrifuge tube, centrifuge 0.05 mL of the 200 IU/mL suspension of AAT (10 IU) (see Note 18) and 0.75 mL of the 40 IU/mL suspension of BCAT (30 IU).

Discard the supernatant and dissolve gently the pellet in 1 mL of the reaction mixture prepared as described in Subheading 1. Add this solution to the whole reaction mixture and stir very slowly (approximately 60 rpm) at room temperature.

3. Titrate pyruvate is formed from CSA every 30 min as described previously in Subheading 3.6 (step 3). Usually, a conversion rate of 40% is reached in 2–3 h (see Notes 19 and 22).
4. Purify the product on Dowex50® acidic resin as described in Subheading 3.6 (step 4). At this stage, a mixture of 3-phenyl-Glu and Glu is isolated.
5. Purify the product on Dowex1® basic resin as described in Subheading 3.6 (step 5). Glu is eluted first and completely separated from (2*S*,3*R*)-3-phenyl-Glu obtained as a white solid with a yield close to 40% and a very high chemical and stereochemical purity (>99%) (see Note 23).

3.8. Preparation of (2S,4S)- and (2S,4R)-4-Hydroxy-4-Methyl-Glutamic Acid with AspAT and BCAT

(2*S*,4*S*)-4-Hydroxy-4-methyl-glutamic acid is obtained through the AspAT catalysed kinetic resolution of the racemic keto acid substrate (see Fig. 3). The reaction is controlled by titration of pyruvate formed from CSA and stopped at near 45% conversion of the racemic substrate. The residual keto acid with the major *R* configuration is then converted to (2*S*,4*R*)-4-hydroxy-4-methyl-glutamic acid using BCAT.

1. In a 250-mL flask, dissolve 0.782 g (2.5 mmol) dilithium 4-hydroxy-4-methyl-2-oxoglutarate (60% purity estimated by ^{1}H NMR) and 0.38 g (2.5 mmol) CSA in H_2O (100 mL). Adjust pH to 7.6 with 1 M NaOH (approximately 3 mL). Bring to 125 mL with water in a cylinder before transferring the solution in a 250-mL round bottom flask equipped with a stir bar.
2. In a 1.5-mL microcentrifuge tube, centrifuge 0.25 mL of the 200 IU/mL suspension of AAT (50 IU) (see Note 18). Discard the supernatant and dissolve gently the pellet in 1 mL of the reaction mixture prepared as described in Subheading 1. Add this solution to the whole reaction mixture and stir very slowly (approximately 60 rpm) at room temperature.
3. Titrate pyruvate is formed from CSA every 30 min as described previously in Subheading 3.6 (step 3). Usually, a conversion rate of 40% is reached in 2–3 h (see Note 19).
4. Pass the reaction mixture through the column of Dowex50® acidic resin. Collect the effluent in 30-mL glass tubes (20 mL fractions). Wash the column with water and check the fractions content by TLC (eluent: *n*-propanol/H_2O, 70/30; revelation: ninhydrin). Pursue water elution until all the pyruvic acid, residual keto acid substrate, and CSA are eluted. Combine the

acidic fractions containing the residual keto acid. Adjust pH to 7.6 with 1 M NaOH and concentrate under reduced pressure, at room temperature, to a volume close to 50 mL. Elute the column with 1 M aqueous ammonia and collect 5 mL fractions. Combine the ninhydrin positive fractions containing (2*S*,4*S*)-4-hydroxy-4-methyl-Glu and concentrate to dryness under reduced pressure at room temperature.

5. Purify (2*S*,4*S*)-4-hydroxy-4-methyl-Glu on Dowex1® basic resin as described in Subheading 3.6 (step 5). The product is isolated as a white solid with a yield close to 40% and a very high chemical and stereochemical purity (>98%) (see Note 24).
6. To the solution of the residual keto acid (approximately 1.25 mmol with major *R* configuration), add 190 mg CSA (1.25 mmol), 37 mg Glu (0.25 mmol), and 0.5 mL of 10 mM PLP (5 μmol). Adjust pH to 7.6 with 0.5 M NaOH (approximately 2.5 mL). Bring to 60 mL with water in a cylinder before transferring the solution in a 100-mL round bottom flask equipped with a stir bar.
7. In a 1.5-mL microcentrifuge tube, centrifuge 0.05 mL of the 200 IU/mL suspension of AAT (10 IU) (see Note 18) and 0.5 mL of the 40 IU/mL suspension of BCAT (20 IU). Discard the supernatant and dissolve gently the pellet in 1 mL of the reaction mixture prepared as described in Subheading 1. Add this solution to the whole reaction mixture and stir very slowly (approximately 60 rpm) at room temperature for 15 h.
8. Purify the product on Dowex50® acidic resin as described in Subheading 3.6 (step 4).
9. Purify the product on Dowex1® basic resin as described in Subheading 3.6 (step 5). The product is isolated as a white solid with a yield close to 30% as an approximately 90/10 mixture of (2*S*,4*R*) and (2*S*,4*S*) stereoisomers (see Notes 25 and 26).

4. Notes

1. 2-Mercaptoethanol is added last in the stoppered bottle because of its smell and toxicity.
2. Asp and Glu containing buffers are stored as frozen solutions at −18°C to limit contamination.
3. The mixture can be refluxed for 1 h directly in the distillation apparatus avoiding that the vapors reach the condenser. To start the distillation, increase slowly the heating of the flask. As the toluene-EtOH azeotrope is a 32/68 mixture, the volume of the azeotrope distilled at around 77°C might be approximately 6.5 mL.

If the temperature of the vapors reaches the toluene boiling point (110°C) before this volume has been collected, stop the distillation and leave the reaction mixture under reflux for 30 min longer before starting again the distillation. Repeat this operation if necessary. Stop the distillation when toluene alone is distilling.

4. The reaction completion can be checked by thin layer chromatography: eluant cyclohexane/AcOEt, 8/2, revelation with $KMnO_4$ (the TLC plate is immerged in the $KMnO_4$ solution and then heated with a heat gun or on a heating plate). If the reaction is not complete, add propionic acid (0.13 mL) and repeat heating and distillation.
5. If the reaction is not complete, the mixture can be heated to reflux (80°C) until completion.
6. Column chromatography: pour 250 mL of silica in a 0.5-L beaker. Add eluent while stirring with a glass rod until having an homogeneous and fluid gel. Pour the gel rapidly in a glass column (diameter 4 cm, length 40 cm) equipped with a Teflon stopcock and a piece of cotton to retain silica. Let the eluent flow through the column until the silica bed volume is stabilized and the silica well packed. Stop the flow just before drying the top of the column. Dilute the product to be purified in the eluent (10 mL) and apply carefully the solution on the top of the column. Let it penetrate in the silica gel. Add 5 mL of eluent and let it penetrate in the silica gel. Repeat this operation several times before adding more eluent. Collect the effluent in glass tubes (25 mL fractions) and analyse by TLC. Combine the fractions containing the pure product and concentrate to dryness under reduced pressure.
7. To facilitate water evaporation under reduced pressure, the solution can be heated in a water bath to 40°C without degradation of the product. Alternatively, the product can be isolated by freeze-drying.
8. The structure and purity of the dilithium salt of 4-methyl-2-oxoglutaric acid can be analysed by ^1H and ^{13}C NMR (20 mg of salt in 0.5 mL D_2O): ^1H NMR (400 MHz) δ 2.99 (1H, dd, J=8.0 and 17.5 Hz), 2.70 (1H, dd, J=6.0 and 17.5 Hz), 2.62 (1H, *m*), 1.05 (3H, d, J=7.0 Hz); ^{13}C NMR (100 MHz) δ 205.1, 184.8, 170.2, 43.8, 37.5, 17.4.
9. The reaction can be controlled by TLC: Eluent, cyclohexane/EtOAc, 7/3; revelation with vanillin (the TLC plate is immerged in the vanillin solution and then heated with a heat gun or on a heating plate).
10. The reaction can be controlled by TLC: Eluent, cyclohexane/AcOEt, 1/1; revelation with $KMnO_4$. The product is isolated as a mixture of *Z* and *E* isomers (R_f 0.5 and 0.6).

11. The structure and purity of the dilithium salt of 3-phenyl-2-oxoglutaric acid can be analysed by ^{1}H and ^{13}C NMR (20 mg of salt in 0.5 mL D_2O): ^{1}H NMR (400 MHz) δ 7.39–7.25 (5H, *m*), 4.66 (1H, *t*, $J=7.0$ Hz), 2.89 (1H, dd, $J=7.0$ and 15 Hz), 2.56 (1H, dd, $J=7.0$ and 15 Hz); ^{13}C NMR (100 MHz, D_2O) δ 204.4, 180.0, 169.9, 136.5, 129.1, 128.7, 127.7, 51.8, 39.1.
12. The structure and purity of the dilithium salt of 4-hydroxy-4-methyl-2-oxoglutaric acid can be analysed by ^{1}H and ^{13}C NMR (20 mg of salt in 0.5 mL D_2O): ^{1}H NMR (400 MHz) δ 3.18 (1H, d, $J=18$ Hz), 3.05 (1H, d, $J=18$ Hz), 1.25 (s, 3 H); ^{13}C NMR (100 MHz, D_2O) δ 202.9, 182.2, 169.4, 73.6, 48.6, 26.1.
13. The detailed purification processes of AspAT and BCAT are described in (11, 12). Although a pure enzyme is preferable for kinetic measurements, crude extracts are suitable for synthetic purpose.
14. The cell growth can be checked by OD measurement at 600 nm. If an OD of 4 (measured after 4 times dilution of the culture broth) is not reached, the growth can be prolonged.
15. We used a Bandelin sonopuls model, alternating 30 s on and 30 s off, for 15 min at 50% amplitude. The cell disruption is performed twice, using 50 mL of cell suspension cooled by immersion in a water-ice bath. If a low activity is measured in the crude extract, the disruption can be repeated.
16. The activity of the aminotransferase (AspAT or BCAT) is expressed in international units (IU): 1 IU corresponds to the quantity of enzyme converting 1 μmol of keto acid (KG or 4-methyl-2-oxopentanoic acid) in 1 min in the particular experimental conditions used here. All kinetic measurements are performed at 25°C. Activity is calculated using ε_{NADH} (340 nm) = 6,220/M/cm.
17. The activity can be measured after dialysis: In a 1.5-mL microcentrifuge tube, centrifuge 20 μL of protein suspension at 14,000 × *g* for 5 min. Discard the supernatant and suspend gently the pellet in 1 mL of phosphate buffer (0.2 mL in the case of BCAT). Take 10 μL of this solution for measuring the activity, as described above. Usually, no activity is lost during dialysis.
18. Alternatively, the commercially available pig heart AspAT can be used. However, this latter enzyme is more prone to inhibition by sulphur dioxide formed from CSA and thus more enzyme should be used (five times more units). The addition of acetaldehyde to the reaction mixture (64 mg, 1.45 mmol) partially protects pig heart AspAT from inhibition by SO_2.

19. If for any reason, the time course of the reaction appears slower than indicated, it can be prolonged and/or more enzyme can be added. However, due to the limited stability of pyruvate in solution, its titration after 4–5 h of reaction becomes less reliable for the estimation of the conversion rate.
20. To ensure a high purity, the solution should not be heated above 25°C during concentration. The final compound can be dried under high vacuum. Alternatively, the final residue can be dissolved in minimum volume of water and freeze-dried.
21. The structure and purity of (2*S*,4*R*)-4-methyl-glutamic acid can be determined by the following analyses: mp 178°C; $[\alpha]_D^{25}$ = +24.0° (*c* 1.3, 6 N HCl); ^{1}H and ^{13}C NMR (20 mg in 0.5 mL D_2O): ^{1}H NMR (400 MHz) δ 3.83 (1H, dd, *J* = 4.5 and 8.5 Hz), 2.58 (1H, *m*, *J* = 5.0, 7.0, 8.5), 2.23 (1H, ddd, *J* = 5.0, 8.5 and 14.0 Hz), 1.96 (1H, ddd, *J* = 5.0, 8.5 and 13.5 Hz), 1.25 (3H, d, *J* = 7.0 Hz); ^{13}C NMR (100 MHz) δ 185.3, 178.5, 53.9, 39.5, 36.9, 17.8. The product should correspond to the expected elemental analyses.
22. The various components of the reaction mixture absorb at 340 nm. In order to correct the absorbance changes measured during the reaction, make a blank before the addition of the enzyme.
23. The structure and purity of (2*S*,3*R*)-3-phenyl-glutamic acid can be determined by the following analyses: mp 160°C; $[\alpha]_D^{25}$ = +16.9° (*c* 0.5, 6 N HCl); ^{1}H and ^{13}C NMR (5 mg in 0.5 mL D_2O): ^{1}H NMR (400 MHz) δ 7.32–7.20 (5H, *m*), 3.67 (1H, d, *J* = 6.5 Hz), 3.46 (1H, *m*), 2.62 (2H, *m*); ^{13}C NMR (100 MHz, D_2O) δ 179.6, 174.6, 138.0, 128.9, 128.3, 127.8, 59.9, 44.1, 40.3. The product should correspond to the expected elemental analyses.
24. The structure and purity of (2*S*,4*S*)-4-hydroxy-4-methyl-glutamic acid can be determined by the following analyses: $[\alpha]_D^{20}$ = −3.7° (*c* 0.5, 0.5 N NaOH); ^{1}H NMR (5 mg in 0.5 mL of LiOH 0.1 M in D_2O, 400 MHz) δ 3.31 (1H, dd, *J* = 3.0 and 10.5 Hz), 2.22 (1H, dd, *J* = 3.0 and 14.5 Hz), 1.65 (1H, dd, *J* = 10.5 and 14.5 Hz), 1.38 (3H, *s*). The product should correspond to the expected elemental analyses.
25. Since the (2*S*,4*R*) isomer is eluted first from the column, collection of only the first third of the ninhidrin positive fractions allows the recovery of highly pure product (>98%). In order to increase the quantity of pure (2*S*,4*R*) isomer, the chromatography can then be repeated using the remaining fractions. Alternatively, a longer column can be used but in this case the flow rate is very low.

26. The structure and purity of (2*S*,4*R*)-4-hydroxy-4-methyl-glutamic acid can be determined by the following analyses: $[\alpha]_D^{20} = -7.8°$ (*c* 0.6, 0.5 N NaOH); ^{1}H NMR (5 mg in 0.5 mL of LiOH 0.1 M in D_2O, 400 MHz) *δ* 3.43 (1H, dd, *J* = 5.0 and 8.0 Hz), 2.06 (1H, dd, *J* = 5.0 and 14.5 Hz), 1.9 (1H, dd, *J* = 8.0 and 14.5 Hz), 1.38 (3H, s). The product should correspond to the expected elemental analyses.

Acknowledgment

This work was supported by the French National Center for Scientific Research (CNRS). We are grateful to Kagamiyama's group (Osaka Medical College) for providing us with the AspAT and BCAT overexpressing *E. coli* strains.

References

1. Hwang, B Y et al. (2005) Revisit of aminotransferase in the genomic era and its application to biocatalysis. *J Mol Cat B, Enzymatic* 37, 47–55.
2. Ager, D J et al. (2001) Novel biosynthetic routes to non-proteinogenic amino acids as chiral pharmaceutical intermediates. *J Mol Cat B, Enzymatic* 11, 199–205.
3. Yun, H et al. (2004) Kinetic resolution of (R, S)-sec-butylamine using omega-transaminase from *Vibrio fluvialis* JS17 under reduced pressure. *Biotechnol Bioeng* 87, 772–778.
4. Shin, J S, Kim B G (2001) Comparison of the omega-transaminases from different microorganisms and application to production of chiral amines. *Biosci Biotechnol Biochem* 65, 1782–1788.
5. Iwasaki, A et al. (2003) Microbial synthesis of (R)- and (S)-3,4-dimethoxyamphetamines through stereoselective transamination. *Biotechnol Lett* 25, 1843–1846.
6. Yun, H et al. (2004) ω-Amino acid: pyruvate transaminase from *Alcaligenes denitrificans* Y2k-2: a new catalyst for kinetic resolution of β-amino acids and amines. *Appl Environ Microbiol* 70, 2529–2534.
7. Alaux, S et al. (2005) Chemoenzymatic synthesis of a series of 4-substituted glutamate analogues and pharmacological characterization at human glutamate transporters subtypes 1–3. *J Med Chem* 48, 7980–7992.
8. Sagot, E et al. (2008) Chemoenzymatic synthesis of (2S,4R)-2-amino-4-(3-(2,2-diphenylethylamino)-3-oxopropyl)pentanedioic acid: a novel selective inhibitor of human excitatory amino acid transporter subtype 2. *J Med Chem* 51, 4085–4092.
9. Sagot, E et al. (2008) Chemo-enzymatic synthesis of a series of 2,4-syn-functionalized (S)-glutamate analogues: new insight into the structure-activity relation of ionotropic glutamate receptor subtypes 5, 6, and 7. *J Med Chem* 51, 4093–4103.
10. Xian, M et al. (2007) Chemoenzymatic synthesis of glutamic acid analogues: substrate specificity and synthetic applications of branched chain aminotransferase from *Escherichia coli*. *J Org Chem* 72, 7560–7566.
11. Kamitori, S. et al. (1987) Overproduction and preliminary X-Ray characterization of Asparatate aminotransferase from *Escherichia coli*. *J Biochem* 101, 813–816.
12. Kagamiyama, H, Hayashi, H (2000) Branched-chain amino-acid aminotransferase of *Escherichia coli*. *Meth Enzymol* 324, 103–113.

Chapter 5

Carbon–Carbon Bond-Forming Enzymes for the Synthesis of Non-natural Amino Acids

Pere Clapés, Jesús Joglar, and Mariana Gutiérrez

Abstract

An enzymatic methodology for the preparation of β-hydroxy-α-amino acid derivatives is described. The method consists of the stereoselective aldol addition reaction of glycine to *N*-Cbz-amino aldehydes furnishing 3-hydroxy-2,4-diaminobutyric derivatives.

Key words: Lyases, Biotransformation, Aldol reaction, Low specific L-*allo*-threonine aldolases, Serine hydroxymethyl transferase

1. Introduction

Pyridoxal-5′-phosphate (PLP)-dependent lyases involved in β-hydroxy-α-amino acid metabolism have received much attention due to their ability to catalyze the reversible asymmetric aldol addition of glycine to a variety of acceptor aldehydes leading to rare or non-natural amino acids (1–6). β-Hydroxy-α-amino acids constitute a class of non-canonical amino acids with interesting biological activity on their own. Moreover, some of them can be identified as components of many naturally occurring complex compounds and analogues such as antibiotics, immunosuppressants among others (1, 3, 6–12). Furthermore, these polyfunctional compounds may be useful building blocks for peptidomimetics and other non-proteinogenic peptide-like structures of biological interest.

Threonine aldolases (ThrA, EC 4.1.2.5) and serine hydroxymethyl transferases (SHMT, EC 2.1.2.1) catalyze reversibly the aldol addition of glycine to aldehydes with a complete control of the α-carbon stereochemistry but, in many instances, with moderate to low stereoselectivity at the β-carbon (13, 14). Interestingly, they

Loredano Pollegioni and Stefano Servi (eds.), *Unnatural Amino Acids: Methods and Protocols*,
Methods in Molecular Biology, vol. 794, DOI 10.1007/978-1-61779-331-8_5, © Springer Science+Business Media, LLC 2012

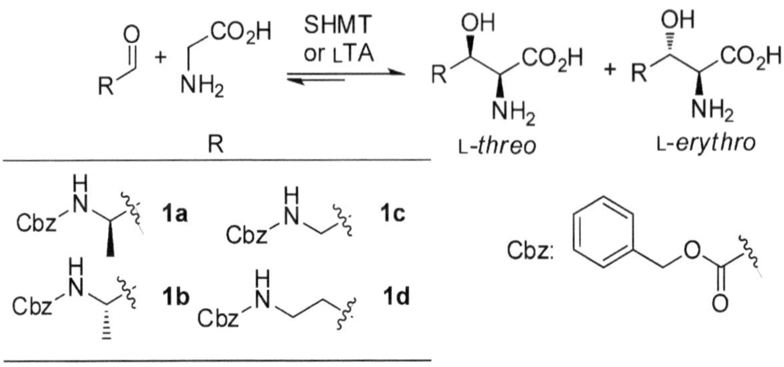

Fig. 1. Serine hydroxymethyltransferase (SHMT) from *Streptococcus thermophilus* and L-threonine aldolase from *Escerichia coli* (LTA) catalyzed synthesis of 3-hydroxy-2,4-diaminobutyric acid derivatives.

are highly stereoselective for the retroaldol reactions being potent tools for the resolution of β-hydroxy-α-amino acid racemates. In the synthetic direction, since two new stereogenic centers are formed, four different products with complementary stereochemistry can be obtained formally from a single aldehyde by using L-threonine aldolase, D-threonine aldolase, L-*allo*-threonine aldolase, and D-*allo*-threonine aldolase as well as by manipulating the reaction conditions towards the kinetic or thermodynamic product (14).

Here, we present a methodology to obtain derivatives of 3-hydroxy-2,4-diaminobutyric acid by means of a stereoselective chemo-enzymatic approach (Fig. 1). The key step of this approach consisted of the aldol addition of glycine to *N*-protected amino aldehydes catalyzed by threonine aldolases. We have recently cloned a novel serine hydroxymethyltransferase (EC 2.1.2.1) from *Streptococcus thermophilus* (15) (SHMT) and L-threonine aldolase (EC 4.1.2.5) from *Escerichia coli* (LTA) (1), both overexpressed in *E. coli* as a recombinant His6-tagged proteins. Adehydes such as *N*-(*R*)-Cbz-alaninal (**1a**), *N*-(*S*)-Cbz-alaninal (**1b**), *N*-Cbz-3-aminopropanal (**1c**), and *N*-Cbz-2-aminoethanal (**1d**) were the acceptor substrates. The aldol addition of glycine to *N*-Cbz-3-aminopropanal will furnish a stereoisomeric precursor of β-hydroxyornithine, a relevant building block for the β-lactamase inhibitor clavulanic acid and the antibiotic and anticancer acivicin (16).

2. Materials

1. Prepare all solutions using ultrapure water (i.e. obtained by purifying deionized water to attain a sensitivity of 18 MΩ cm at 25°C) and analytical grade reagents.

2. Prepare and store all reagents at room temperature (unless indicated otherwise).
3. Diligently follow all waste disposal regulations when disposing waste materials.

2.1. Reagents

1. L-Threonine, L-*allo*-threonine, yeast alcohol dehydrogenase (ADH), pyridoxal-5′-phophate, glycine, and benzyloxyacetaldehyde were from Sigma-Aldrich.
2. *N*-Benzyloxycarbonyl-3-aminopropanal (*N*-Cbz-3-aminopropanal), *N*-benzyloxycarbonyl-2-aminoethanal (*N*-Cbz-glycinal), and (*R*)-, and (*S*)-*N*-benzyloxycarbonyl alaninal were synthesized by previously described procedures (see below) (17, 18).
3. Serine hydroxymethyl transferase from *S. thermophilus* (SHMT) and L-threonine aldolase from *E. coli* (LTA) were produced and purified in our lab and in a fermentation plant facility following previously described methodologies (see below) (15).
4. All chemicals and reagents used in this work are of analytical-grade.

2.2. Solutions and Buffers

1. PCR-amplification solution: 50 μL of commercially supplied Tris buffer (0.12 mM Tris–HCl, pH 8.0, 6 mM $(NH_4)_2SO_4$, 10 mM KCl, 0.01% Triton X-100, 0.001% bovine serum albumin, and 1 mM $MgCl_2$), 0.2 mmol of each nucleotide triphosphate, 0.4 μmol of each primer, 1 μg of genomic DNA, and 1.25 U of *Thermococcus kodakaraensis* polymerase (KOD).
2. Enzyme production LB medium: 1% (w/v) tryptone, 0.5% (w/v) yeast extract, 1% (w/v) NaCl, pH 7.2, 0.1 mg/mL ampicillin and 0.025 mg/mL kanamycin. Dissolve 10 g of bactotryptone, 5 g of yeast extract, and 10 g of sodium chloride in 1 L of water. Adjust the pH to 7.0 with 1 M NaOH and sterilize for 20 min in autoclave.
3. Enzyme purification starting buffer: 20 mM sodium phosphate, 20 mM imidazole, 0.5 M NaCl, at pH 7.4. Dissolve 2.76 g of $NaH_2PO_4{\cdot}H_2O$, 1.37 g of imidazole, and 29.22 g of NaCl in 1 L of water. Adjust the pH to 7.4.
4. Enzyme purification elution solution: 20 mM sodium phosphate, 300 mM imidazole, 0.5 M NaCl at pH 7.4. Dissolve 2.76 g of $NaH_2PO_4{\cdot}H_2O$, 20.64 g of imidazole, and 29.22 g of NaCl in 1 L of water. Adjust the pH to 7.4.
5. Enzyme dialysis solution: PBS buffer, pH 8.0, containing 0.01 mM PLP. Take 52.5 g of K_2HPO_4 and 39.8 mg of PLP and dissolve them in 1.5 L of water. Adjust the pH to 8. Dilute 1:10 before use.

6. Potassium phosphate buffer, 200 mM, pH 6.5 (for SHMT activity): mix 9.1 mL of 1 M K_2HPO_4 (prepared from 17.42 g of K_2HPO_4 dissolved in 100 mL of water) with 10.9 mL of 1 M KH_2PO_4 (prepared from 13.61 g of KH_2PO_4 dissolved in 100 mL of water) and adjust it to 100 mL with water; the pH of the solution must be 6.5.
7. L-Threonine, 500 mM (for SHMT and LTA activity): dissolve 297.7 mg of L-threonine in 5 mL of water.
8. Pyridoxal-5′-phosphate, 2 mM (for SHMT and LTA activity): dissolve 1 mg in 2 mL of water (see Note 1).
9. NADH, 0.15 M (for SHMT and LTA activity): dissolve 2.23 mg of NADH in 1 mL of degassed water (see Note 2).
10. Yeast alcohol dehydrogenase suspension (ADH, for SHMT and LTA activity): take 26 mg of ADH (481 U/mg) and dissolve in 5 mL of water (see Note 3).
11. Tris–HCl, 200 mM pH 8 (for LTA activity): dissolve 485 mg of Tris base in 18 mL of degassed water, adjust pH to 8 with 6 N HCl.
12. Preparation of *N*-benzyloxycarbonyl amino alcohols. The protection reaction of the starting aminoalcohols was carried out in a 4:1 (v/v) mixture of dioxane:water. The required solutions are: 5% (w/v) aqueous solution of $NaHCO_3$ (5 g of sodium bicarbonate dissolved in 995 mL of water); 5% (w/v) aqueous solution of citric acid (5 g of citric acid dissolved in 995 mL of water); brine (i.e., a saturated solution of NaCl in water).
13. Preparation of *N*-benzyloxycarbonyl amino aldehydes. The oxidation reaction of protected aminoalcohols to the corresponding aminoaldehydes was carried out in ethyl acetate (EtOAc). The required solutions are: 5% (w/v) aqueous solution of $NaHCO_3$ (5 g of sodium bicarbonate dissolved in 995 mL of water), and brine (i.e., a saturated solution of NaCl in water).
14. For the enzymatic aldol additions in highly concentrated water-in-oil emulsions, the solutions required are: glycine, 13.1 mg of glycine in 2 mL of 10 mM phosphate buffer pH 8 for LTA (or pH 6.5 for SHMT); 10 mM phosphate buffer: prepared by dissolving 138 mg of NaH_2PO_4 in 100 mL of water and adjusted to pH 8 or 6.5 with 1 M NaOH; pyrydoxal-5′-phosphate, 2.2 mg of PLP in 1.5 mL of 10 mM phosphate buffer pH 8 for LTA (or pH 6.5 for SHMT).
15. For the enzymatic aldol additions in dimethylformamide (DMF)/H_2O 1:4 (v/v) (for the amounts see Table 1) the solutions required are: glycine solution in 10 mM phosphate buffer pH 6.5 for SHMT or pH 8 for LTA (see above); pyrydoxal-5′-phosphate, 2.2 mg of PLP in 1.5 mL of 10 mM phosphate buffer pH 8 for LTA (or pH 6.5 for SHMT).

Table 1
SHMT and LTA-catalyzed aldol addition reaction of glycine to aldehydes 1a–d to furnish β-hydroxy-α-amino acid derivatives 3a–h

Entry	Acceptor aldehyde	Aldolase (U/mL)[a]	Product	Conversion % (time, h)	Temp. (°C)	Gly/aldehyde mmol (mL)[d]	Isolated[e] yield %	L-*erythro*:L-*threo* dr
1	**2a**	SHMT (8)	**3a**	60 (30)[b]	4	2.4/2.0 (34)	30	100:0
2	**2a**	LTA (12)	**3a/3b**	40 (20)[b]	4	2.6/2.2 (37)	20	16:84
3	**2b**	SHMT (8)	**3c**	48 (20)[a]	4	15.0/3.5 (58)	30	100:0
4	**2b**	LTA (12)	**3c/3d**	54 (20)[c]	25	5.9/2.5 (42)	27	18:82
5	**2c**	SHMT (8)	**3e/3f**	35 (20)[b]	4	2.8/2.5 (39)	13	86:14
6	**2c**	LTA (12)	**3e/3f**	60 (20)[c]	25	7.7/3.4 (54)	18	30:70
7	**2d**	SHMT (8)	**3g/3h**	34 (25)[b]	25	2.5/2.1 (36)	10	50:50
8	**2d**	LTA (12)	**3g/3h**	49 (25)[c]	25	2.9/2.5 (42)	11	50:50

[a]Unit definition: 1 U of TA activity is defined as the amount of enzyme that catalyzed the formation of 1 μmol of acetaldehyde (i.e. 1 μmol of NADH oxidized) per minute at 37°C
[b]DMF/H_2O 1:4 (v/v)
[c]High water content emulsions, which consist of H_2O/hexadecane/$C_{14}E_4$ 90/6/4 w%
[d]Reaction volume
[e]Purification procedures were not optimized

3. Methods

3.1. PCR-Amplification of the Threonine Aldolase Genes (glyA and ltaE) and Construction of the Plasmids pQEglyA1 and pQEltaE for Protein Expression

1. Isolation of the complete coding region of the *glyA* gene from *S. thermophilus* YKA-184 and *ltaE* gene from *E. coli* K-12 was obtained by PCR-amplification of purified genomic DNA from these strains.
2. The reaction was carried out in a mixture containing 50 μL of 0.12 mM Tris–HCl, pH 8.0, 6 mM $(NH_4)_2SO_4$, 10 mM KCl, 0.01% Triton X-100, 0.001% bovine serum albumin, 1 mM $MgCl_2$, 0.2 mmol of each nucleotide triphosphate, 0.4 μmol of each primer, 1 μg of genomic DNA, and 1.25 U of KOD polymerase.
3. PCR-amplification was conducted at 95°C for 15 s, 46°C for 30 s and 72°C for 1 min for a total of 30 cycles.
4. For the SHMT gene, the 5′ primer (5glyA1) containing the ATG initiation codon (lowercase letters) and the 3′ primer (3glyA1) with the TAA termination codon (lowercase letters) had the following sequences, respectively: 5′-ATCGA*GGATCC*atgATTTTTGATAAAGAAG-3′ and 5′-AATGT*GTCGACT*taaTAGAGTGGGAAAGCA-3′; for the LTA where, respectively: 5′-GTAC*GGATCC*atgATTGATTTACGCAGTGA-3′ and 5′-ATAC*AAGCTTT*taaCGCGCCAGGAATGCAC-3′.
5. To facilitate the cloning, an additional restriction site (sequence in italics) was incorporated in each primer.
6. The amplified PCR product was separated by agarose gel electrophoresis, digested with *Bam*HI and *Sal*I for the SHMT and *Bam*HI and *Hin*dIII for the LTA, purified and inserted downstream of the T5 promoter of pQE40.
7. This ligation mixture was used to transform M15(pREP4) competent cells.
8. The resulting plasmids were named pQEglyA1 and pQEltaE, for SHMT and LTA, respectively.
9. To confirm that no errors had been introduced during PCR amplification, the nucleotide sequences of the *glyA* and *ltaE* genes from pQEGlyA1 and pQEltaE were determined.

3.2. Production of Serine Hydroxymethyl Transferase from S. thermophilus YKA-184 (SHMT) and L-Threonine Aldolase from E. coli (LTA)

1. Recombinant *E. coli* M15 strain cells transformed with the pQEglyA1 plasmid containing the gene encoding the SHMT from *S. thermophilus* or with pQEltaE for L-threonine aldolase from *E. coli* were cultivated at 37°C in LB medium: 1% (w/v) tryptone, 0.5% (w/v) yeast extract, 1% (w/v) NaCl, pH 7.2, 0.1 mg/mL ampicillin, and 0.025 mg/mL kanamycin.
2. Cultivation conditions for expression were as follows: preinoculum cultures of *E. coli* M15 (pQEglyA and pQEltaE) were

grown from a frozen stock overnight in LB with 0.1 mg/mL ampicillin and 0.025 mg/mL kanamycin at 37°C on a rotary shaker at 250 minute. A final optical density at 600 nm (OD_{600}) of 3–4 was typically achieved.

3. An aliquot of 25 mL of pre-inoculum was transferred to a 2 L shake-flask containing 800 mL of LB medium with the corresponding antibiotics (*vide supra*) and incubated at 37°C with shaking at 250 minute.
4. For induction, 0.05 mM IPTG was added during the middle exponential phase growth (OD_{600} = 0.6) and the cultivation temperature decreased to 30°C to minimise inclusion bodies formation.

3.3. Purification of Serine Hydroxymethyl Transferase from S. thermophilus YKA-184 (SHMT) and L-Threonine Aldolase from E. coli (LTA)

1. Cells from four flasks containing 800 mL of induced-culture broths were withdrawn and centrifuged at 12,000 × *g* for 10 min at 4°C.
2. The pellet was suspended with starting buffer: 20 mM sodium phosphate, 0.5 M NaCl, 10 mM imidazole, at pH 7.4 to a final cell density (OD_{600}) of 50.
3. The cell suspension was lysed using a TS 0.75 kW 40 K Cell Disrupter (Constant Systems).
4. Cellular debris was removed by centrifugation at 12,000 × *g* for 30 min at 4°C.
5. The clear supernatant was applied to a cooled HR 16/40 column (GE Healthcare) containing affinity beads (50 mL) (see Note 4) and it was washed with starting buffer (150 mL).
6. The column was washed with 5× column vol. of starting buffer and the recombinant protein was eluted with 300 mM imidazole in 20 mM sodium phosphate, pH 7.4, containing 0.5 M NaCl, at 3 mL/min flow rate (see Note 5).
7. Fractions with threonine aldolase activity were pooled and checked by SDS–polyacrylamide gel electrophoresis (SDS–PAGE), dialyzed against 250 vol. of PBS buffer, pH 8.0, containing 0.01 mM PLP at 4°C for 24 h.
8. The dialyzed solution was frozen at −80°C (see Note 6) and lyophilized using a benchtop 5-L lyophilizer.
9. The powder obtained was stored at −20°C (see Note 7).

3.4. Determination of Threonine Aldolase Activity

1. The catalytic activity for the cleavage of L-threonine was determined with a coupled continuous assay. In the first step, L-threonine is cleaved to acetaldehyde and glycine. In a second step, acetaldehyde was reduced using yeast alcohol dehydrogenase (ADH) and the consumption of NADH was monitored spectrophotometrically at 340 nm.

2. In the assay, potassium phosphate buffer pH 6.5 solution (500 μL of a 200 mM solution) for SHMT or Tris–HCl pH 8 solution for LTA (500 μL of a 200 mM solution), L-threonine (100 μL of a 500 mM solution), pyridoxal-5′-phosphate (25 μL of a 2 mM solution), ADH (30 U, 12 μL of a 5.2 mg/mL solution), NADH (50 μL of a 0.15 M NADH), and plain water (303 μL), was incubated for 10 min at 37°C (final volume 1 mL).
3. The reaction was started adding SHMT (10 μL of a 3.4 mg protein/mL solution) or LTA (10 μL of a 3.4 mg protein/mL solution).
4. One unit of threonine aldolase (TA) was defined as the amount of enzyme that catalyzed the formation of 1 μmol/min of acetaldehyde at 37°C (see Note 8).

3.5. Preparation of N-Benzyloxycarbonylaminoaldehyde Derivatives

3.5.1. Preparation of N-Benzyloxy-carbonylaminoalcohols (2a–d, Fig. 2)

1. To 150 mL of a solution of aminoalcohols (55.5 mmol) in dioxane/H_2O 4:1 (v/v), *N*-(benzyloxycarbonyloxy)succinimide (12.3 g, 49.4 mmol) dissolved in 50 mL of dioxane/H_2O 4:1 (v/v) was added dropwise at 25°C (see Note 9).
2. The reaction mixture was stirred overnight at room temperature, then evaporated to dryness under reduced pressure.
3. The residue was dissolved in EtOAc and washed with an 5% (w/v) aqueous solution of $NaHCO_3$ (2 × 50 mL), an 5% (w/v) aqueous solution of citric acid (2 × 50 mL) and brine (2 × 50 mL).
4. The organic layer was dried over anhydrous Na_2SO_4, filtered, and evaporated under vacuum to dryness furnishing the corresponding *N*-Cbz-amino alcohols (**2a–d**).
5. Specific rotation and ^1H and ^{13}C NMR spectra of these products were consistent with those reported in the literature (19–21).
6. Example: *N*-benzyloxycarbonyl-3-amino-1-propanol (2d) (yield 90%). ^1H NMR (500 MHz; $CDCl_3$) δ (ppm) = 7.35

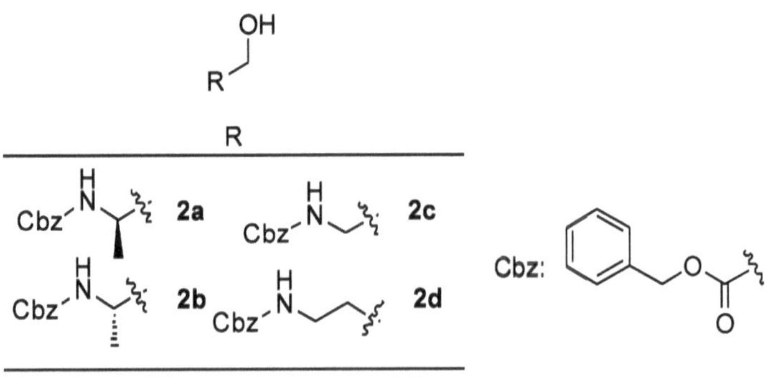

Fig. 2. Amino alcohol precursors synthesized.

(m, Ar, 5H), 5.10 (s, $PhCH_2$, 2H), 5.07 (s, NH, 1H), 3.67 (t, *J*= 5.5, $-CH_2OH$, 2H), 3.36 (dd, *J*= 12.3, 6.1, $NHC\underline{H}_2$, 2H), 1.70 (td, *J*= 11.9, 5.8 and 5.8, $CH_2C\underline{H}_2CH_2$, 2H).

3.5.2. Preparation of N-Benzyloxy-carbonylami-noaldehydes (1a–d, Fig. 1)

1. 2-Iodoxybenzoic acid (IBX) (9.7 g, 34.5 mmol) was added to a solution of *N*-benzyloxycarbonylaminoalcohol (22.3 mmol) in EtOAc (150 mL) and the mixture was heated at reflux (~77°C) for 6 h (see Note 10), then cooled to room temperature and filtered.
2. The filtrate was washed with a 5% (w/v) aqueous solution of $NaHCO_3$ (2 × 90 mL) and brine (2 × 90 mL) and the organic phase was dried with anhydrous sodium sulphate, filtered, and the solvent evaporated under vacuum to dryness affording the corresponding *N*-Cbz-amino aldehydes (**1a–d**, Fig. 1), see Note 11.
3. The specific rotation in chiral compounds and ^1H and ^{13}C NMR spectra were consistent with those reported in the literature (17, 20, 22), see Note 12.
4. Example: *N*-benzyloxycarbonyl-3-amino-1-propanal (3.77 g, isolated yield 82%). ^1H NMR (500 MHz; $CDCl_3$) δ(ppm) = 9.80 (s, CHO, 1H), 7.34 (m, Ar, 5H,), 5.17 (s, $PhCH_2$, 2H), 5.08 (s, NH, 1H), 3.49 (dd, *J*= 11.9, 6.0, $NHC\underline{H}_2$, 2H), 2.75 (t, *J*= 5.6, $-C\underline{H}_2CHO$, 2H).

3.6. Preparation of β-Hydroxyamino Acid Derivatives (Fig. 3)

3.6.1. Enzymatic Aldol Additions in Highly Concentrated Water-in-Oil Emulsions (see Table 1 Entries 4, 6 and 8), see Note 13

1. The *N*-Cbz-amino aldehyde (see Table 1 entries 4, 6 and 8), hexadecane (2.5, 3.3 and 2.5 mL entries 4, 6 and 8, respectively, Table 1), and technical grade poly(oxyethylene) tetradecyl ether surfactant, with an average of 4 mol of ethylene oxide per surfactant molecule (1.7, 2.2 and 1.7 mL entries 4, 6 and 8, respectively, Table 1) were mixed by a vortex for 15 s (2,400 minute).
2. Then, a glycine solution (37.8, 48.5 and 37.8 mL containing 5.9, 7.7, and 2.9 mmol glycine, respectively, entries 4, 6, and 8, respectively; see Table 1) in 10 mM phosphate buffer pH 8 for LTA (or pH 6.5 for SHMT, not used in the examples showed in Table 1), containing pyridoxal-5′-phosphate (3.5, 3.6 and 3.5 mg entries 4, 6 and 8, respectively, Table 1) was added dropwise while stirring at 4°C with a vortex (see Note 14).
3. Finally, the reaction was started by adding the enzyme (see Table 1 entries 4, 6 and 8 for the units of enzyme used).
4. The final reaction volume is indicated in Table 1 for each reaction.
5. The reaction mixtures were placed in a reciprocal shaker (50 minute) at constant temperature.

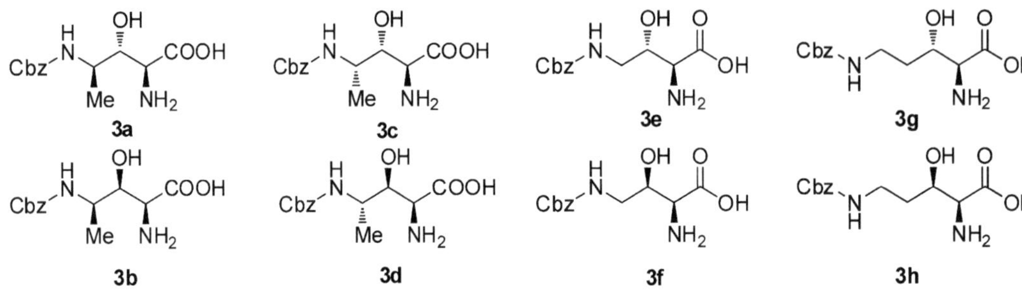

Fig. 3. Structure of the β-hydroxy-α-amino acid derivatives.

6. The reactions were followed by HPLC until the peak of the product reached a maximum.

3.6.2. Enzymatic Aldol Additions in DMF/H_2O 1:4

1. The aldehyde (for the mmol see entries 1–3, 5 and 7 Table 1) was dissolved in DMF/H_2O 1:4 (v/v) (respect to the total reaction volume, see Table 1).
2. Then, a glycine solution in 10 mM phosphate buffer pH 6.5 for SHMT or pH 8 for LTA (for the mmol see entries 1–3, 5 and 7 Table 1), containing pyridoxal-5′-phosphate (0.09 mg/mL reaction volume), was added while stirring at 4°C with a vortex (see Note 15).
3. Finally, the reaction was started by adding the enzyme (for the units of TA activity employed in each case, see entries 1–3, 5 and 7 in Table 1).
4. The final reaction volume is reported in Table 1 for each reaction.
5. The reaction mixtures were placed in a reciprocal shaker (50 minute) at constant temperature.
6. The reactions were followed by HPLC until the peak of the product reached a maximum.

3.6.3. Purification

1. The enzymatic reactions were stopped by the addition of MeOH/HOAc 39:1 (v/v) and the crude was filtrated through Celite/activated charcoal 95:5 (w/w), see Note 16.
2. The filtration cake was cleaned with MeOH to recover the product that might be adsorbed.
3. The filtrate was evaporated under vacuum to eliminate MeOH.
4. The residue thus obtained was dissolved in H_2O and the non-reacted aldehyde and other hydrophobic impurities were extracted with ethyl acetate (3 × 150 mL).
5. The aqueous layer that contained the product was freeze-dried and then purified by semipreparative RP-HPLC on a XTerra

Prep MSC18 OBDTM 250×19 mm column, filled with C18, 10 μm type stationary phase, by using the following procedure.

6. First, the column was equilibrated with 0.1% (v/v) TFA in H_2O.
7. Then, the sample was dissolved in water and adjusted to pH 2–3 with TFA and loaded onto the column.
8. The salts were eliminated by washing with of 0.1% (v/v) TFA in H_2O (100 mL).
9. Then, the product was eluted with a gradient of 0.1% (v/v) TFA in CH_3CN (8–48% CH_3CN in 30 min).
10. The flow rate was 10 mL/min, and the products were detected at 215 nm.
11. The pure fractions were pooled and lyophilized for further characterization by NMR spectroscopy.

4. Notes

1. Protect from light.
2. It is very convenient to degas the water with a stream of helium for 10–15 min.
3. ADH must be freshly prepared.
4. Ni Sepharose™ 6 Fast Flow from GE Healthcare was used.
5. It is recommendable to connect the detector to an analogical signal recorder or digital computer. Thus, the column wash and protein elution can be easily monitored. For instance, the protein can be eluted when the absorbance reaches its initial value (the one that corresponds to the starting buffer).
6. The sample must be frozen with a mixture of dry ice with ethanol or better with liquid nitrogen.
7. The enzymes can also be precipitated with ammonium sulphate and stored as suspensions. To do this, a fraction containing the recombinant protein was brought to 40% saturation with ammonium sulfate at 4°C for 24 h. The suspension was centrifuged at 13,000×*g* for 30 min. The supernatant was discarded (i.e., eliminating the imidazole containing buffer) and the precipitated material was collected, suspended in ice-cold ammonium sulfate (40%), and stored at 4°C in a dark container.
8. For precipitated enzyme samples (i.e., ammonium sulphate suspension), an aliquot of 0.2 mL of the suspension was centrifuged at 13,000×*g* for 5 min, the supernatant discarded, and

the precipitate dissolved in 0.5 mL of 200 mM potassium phosphate, pH 8.0. For lyophilized samples, approximately 1 mg of enzyme powder was dissolved in 1 mL of cold distilled water. In order to measure the enzyme activity, appropriate dilutions with deionised cold water were done.

9. The reaction can be monitored by TLC in mixtures hexane:ethyl acetate in a proportion that depends on the aminoalcohol. If the progress of the reaction is slow, it can be heated smoothly to 30–40°C.
10. The reaction reached completion after 6 h; however, depending on the IBX quality it may take longer time. In some circumstances, specially when the IBX is old (i.e. after months at 4°C), it is necessary to add an extra amount (i.e. typically 2 g, 7 mmol) of IBX to force the reaction to completion.
11. Aldehydes are usually reactive molecules (i.e. strong electrophiles) and must be handled with special care. It is very convenient to store them at −20°C under nitrogen or argon to preserve from oxidation with O_2.
12. For new synthesis, these data must be consistent with the published data.
13. In the literature, highly concentrated emulsions as reaction media are also referred to as high internal-phase-ratio emulsions (HIPRE) or gel emulsions. This emulsion is characterized by volume fractions of dispersed phase (i.e. water in the present case) higher than 0.73, the critical value of close-packed monodispersed spheres. For more information, see ref. 23.
14. Gel emulsions were prepared by slow addition of the internal or dispersed phase (water or an aqueous solution) to a homogeneous mixture of the external or continuous phase (surfactant, oil, and oil-soluble reactants) while stirring continuously with a vortex mixer. All emulsions were prepared by using the same rates of addition and stirring to obtain reproducible emulsions.
15. When the aqueous mixture of glycine is mixed with the DMF solution of the aldehyde, the solution is heated up. It is important to allow the reaction mixture to reach the reaction temperature before adding the enzyme solution.
16. This step is useful to remove the enzyme, which usually precipitates on the HPLC column during purification. It is very important to keep the percentage of activated charcoal at 5% (w/w). Higher amounts of activated charcoal may cause a decrease of product recovery, because the aromatic moiety is tightly adsorbed in the charcoal and it can hardly be recovered even with methanol.

References

1. Kimura T, Vassilev VP, Shen GJ, et al. (1997) Enzymatic synthesis of β-hydroxy-α-amino acids based on recombinant D- and L-threonine aldolases. J Am Chem Soc 119, 11734–11742.
2. Liu JQ, Odani M, Yasuoka T, et al. (2000) Gene cloning and overproduction of low-specificity D-threonine aldolase from *Alcaligenes xylosoxidans* and its application for production of a key intermediate for parkinsonism drug. Appl Microbiol Biotechnol 54, 44–51.
3. Liu JQ, Odani M, Dairi T, et al. (1999) A new route to L-threo-3-[4-(methylthio)phenylserine], a key intermediate for the synthesis of antibiotics: recombinant low-specificity D-threonine aldolase-catalyzed stereospecific resolution. Appl Microbiol Biotechnol 51, 586–591.
4. Herbert RB, Wilkinson B, Ellames GJ, et al. (1993) Stereospecific lysis of a range of β-hydroxy-α-amino acids catalyzed by a novel aldolase from *Streptomyces amakusaensis*. Chem Commun 205–206.
5. Machajewski TD, and Wong C-H (2000) The catalytic asymmetric aldol reaction. Angew Chem Int Ed 39, 1353–1374.
6. Liu JQ, Dairi T, Itoh N, et al. (2000) Diversity of microbial threonine aldolases and their application. J Mol Catal B: Enzym 10, 107–115.
7. Saeed A, and Young DW (1992) Synthesis of L-β-hydroxy amino acids using serine hydroxymethyltransferase. Tetrahedron 48, 2507–2514.
8. Shibata K, Shingu K, Vassilev VP, et al. (1996) Kinetic and thermodynamic control of L-threonine aldolase catalyzed reaction and its application to the synthesis of mycestericin D. Tetrahedron Lett 37, 2791–2794.
9. Miura T, and Kajimoto T (2001) Application of L-threonine aldolase-catalyzed reaction to the preparation of protected 3*R*,5*R*-dihydroxy-L-homoproline as a mimetic of idulonic acid. Chirality 13, 577–580.
10. Hiranuma S, Vassilev VP, Kajimoto T, et al. (1995) Design of oligosaccharide mimetics with higher stability and simpler structures than parent oligosaccharides. RIKEN Review 8, 19–20.
11. Fujii M, Miura T, Kajimoto T, et al. (2000) Facile synthesis of 3,4-dihydroxyprolines as an application of the L-threonine aldolase-catalyzed aldol reaction. Synlett 1046–1048.
12. Vassilev VP, Uchiyama T, Kajimoto T, et al. (1995) An efficient chemo-enzymatic synthesis of α-amino-β-hydroxy-γ-butyrolactone. Tetrahedron Lett 36, 5063–5064.
13. Clapés P, Fessner W-D, Sprenger GA, et al. (2010) Recent progress in stereoselective synthesis with aldolases. Curr Opin Chem Biol 14, 154–167.
14. Steinreiber J, Fesko K, Reisinger C, et al. (2007) Threonine aldolases – an emerging tool for organic synthesis. Tetrahedron 63, 918–926.
15. Vidal L, Calveras J, Clapés P, et al. (2005) Recombinant production of serine hydroxymethyl transferase from *Streptococcus thermophilus* and its preliminary evaluation as a biocatalyst. Appl Microbiol Biotechnol 68, 489–497.
16. DeMong DE, and Williams RM (2001) An efficient asymmetric synthesis of (2*S*,3*S*)- and (2*R*,3*R*)-β-hydroxyornithine. Tetrahedron Lett 42, 183–185
17. Espelt L, Parella T, Bujons J, et al. (2003) Stereoselective aldol additions catalyzed by dihydroxyacetone phosphate dependent aldolases in emulsion systems: preparation and structural characterization of linear and cyclic aminopolyols from aminoaldehydes. Chem Eur J 9, 4887–4899.
18. Castillo JA, Calveras J, Casas J, et al. (2006) Fructose-6-phosphate aldolase in organic synthesis: preparation of D-fagomine, *N*-alkylated derivatives, and preliminary biological assays. Org Lett 8, 6067–6070.
19. Harada H, Asano O, Kawata T, et al. (2001) 2-Alkynyl-8-aryladenines possessing an amide moiety: their synthesis and structure–activity relationships of effects on hepatic glucose production induced via agonism of the A2B adenosine receptor. Bioorg Med Chem 9, 2709–2726.
20. Geall AJ, and Blagbrough IS (2000) Homologation of polyamines in the rapid synthesis of lipospermine conjugates and related lipoplexes. Tetrahedron 56, 2449–2460.
21. Fernandez S, Brieva R, Rebolledo F, et al. (1992) Lipase-catalyzed enantioselective acylation of *N*-protected or unprotected 2-aminoalkan-1-ols. J Chem Soc Perkin Trans 1, 2885–2889.
22. Bischofberger N, Waldmann H, Saito T, et al. (1988) Synthesis of analogs of 1,3-dihydroxyacetone phosphate and glyceraldehyde 3-phosphate for use in studies of fructose-1,6-diphosphate aldolase. J Org Chem 53, 3457–3465.
23. Espelt L, Clapés P, Esquena J, et al. (2003) Enzymatic carbon–carbon bond formation in water-in-oil highly concentrated (gel emulsions). Langmuir 19, 1337–1346.

Chapter 6

Engineering Cyclic Amidases for Non-natural Amino Acid Synthesis

Francisco Javier Las Heras-Vázquez, Josefa María Clemente-Jiménez, Sergio Martínez-Rodríguez, and Felipe Rodríguez-Vico

Abstract

Hydantoinases/dihydropyrimidinases are important biotechnological enzymes involved in the production of α- and β-amino acids. Their isolation from new sources with different substrate specificities, improved activity, enantioselectivity, or higher stability continues to be of great industrial interest. Here, we provide a detailed description of how to produce high quantities of the recombinant hydantoinase/dihydropyrimidinase enzyme from *Sinorhizobium meliloti* CECT4114 (SmeDhp). Several techniques are combined to obtain this goal, from cloning to activity measurement by HPLC.

Key words: α- and β-Amino acids, Hydantoinases/dihydropyrimidinases, Cloning, Overexpression, Metal-affinity, HPLC

1. Introduction

Cyclic amidases, also named cyclic amidohydrolases (EC 3.5.2. –), have been grouped in a superfamily based on the functional and structural similarity of the related enzymes and there is evidence that they have evolved from a common ancestor (1, 2). This superfamily includes allantoinases (EC 3.5.2.5), dihydroorotases (EC 3.5.2.3) and dihydropyrimidinases (EC 3.5.2.2). The three enzymes have been described in the hydrolysis of cyclic amide bonds of five- or six-membered rings in nucleotide metabolism (3). Dihydropyrimidinases are involved in the reductive pathway of pyrimidine degradation, catalyzing the hydrolysis of 5,6-dihydrouracil and 5,6-dihydrothymine to the corresponding *N*-carbamoyl-β-amino acids in eukaryotic species (4). Besides this natural function, dihydropyrimidinases also catalyze the hydrolysis of a variety of

Loredano Pollegioni and Stefano Servi (eds.), *Unnatural Amino Acids: Methods and Protocols*,
Methods in Molecular Biology, vol. 794, DOI 10.1007/978-1-61779-331-8_6, © Springer Science+Business Media, LLC 2012

5-monosubstituted hydantoins (5), which led to the hypothesis that dihydropyrimidinase is identical to hydantoinase, and hence to the synonymous use of both names in the EC nomenclature. The exact metabolic function and physiological substrate of hydantoinases have not yet been elucidated. However, they have been used in the production of non-natural amino acids. The so-called "hydantoinase process" allows the production of optically pure D-amino acids starting from racemic mixtures of 5-monosubstituted hydantoins, using a hydantoinase together with an enantiospecific carbamoylase and a hydantoin racemase (6, 7). The isolation of new hydantoinases/dihydropyrimidinases with different substrate specificities, improved activity, enantioselectivity or higher stability continues to be of great industrial interest.

Genome databases provide information on a variety of genomes, complete chromosomes, sequence maps with contigs, and integrated genetic and physical maps. Additionally, the availability of numerous completely sequenced genomes in these databases has led to the development of computationally based strategies for cloning new genes from the putative sequences.

Thus, our group has cloned and overexpressed genes, and purified the recombinant enzymes involved in the "hydantoinase process", for biochemical and biophysical characterization (8, 9). Here, we present a description of the complete process, from cloning to characterization of the hydantoinase/dihydropyrimidinase enzyme from *Sinorhizobium meliloti* CECT4114 (SmeDhp).

2. Materials

2.1. Cloning and Expression of a Putative New Gene

1. Vertical laminar flow bench (model V-100) (Telstar, Barcelona, Spain).
2. Autoclave Stericlav 75 (Raypa, Barcelona, Spain).
3. Refrigerated Incubator (model CIR350) (Ingeniera Climas, Barcelona, Spain).
4. Orbital shaker with adhesive matting (model Labotron) (Infors-HT, Bottmingen, Switzerland).
5. Growth medium for *Escherichia coli* BL21 and *Sinorhizobium meliloti* CECT4114 (Luria–Bertani, LB): place 700 mL of deionized water in a graduated cylinder. Weigh 10 g of bactotryptone, 5 g of yeast extract, and 5 g of NaCl. Mix and adjust pH with NaOH and HCl (see Note 1). Make up to 1 L with water. Transfer the medium to an autoclavable bottle and autoclave it in a liquid cycle. After cooling, store at 4°C.
6. LB solid medium: follow the instructions as in the previous step and add 15 g of agar to 1 L of LB medium; autoclave it on

a liquid cycle. Pour this LB medium onto Petri dishes (20 mL per plate) inside a flow bench. Wait until the medium solidifies and store the Petri plates at 4°C.

7. LB solid medium with antibiotic: add 2 mL ampicillin (Fluka, Madrid, Spain) at stock concentration of 50 mg/mL when the medium is warm and mix it before pouring onto Petri dishes (see Note 2).
8. PCR primers for *Smedhp* gene amplification: the PCR primers designed for *Smedhp* gene amplification are based on GenBank sequence accession No. NC003047 (see Note 3). The primers are Smedhp5 (5′-CG*TCTAGA*GTGAC**AGGAAA**AACGCC **ATG**AGCACTGTCATCAAGGG-3′) and Smedhp3 (5′-CG *AAGCTT***TTAATGATGATGATGATGATG**GACGC CGCTTGCGGGAATG-C3′) (see Note 4).
9. PCR Primers Stock Solution (PPSS): dissolve the primers in double-distilled water to a final concentration of 1 μg/μL (see Note 5). PCR Primers Working Solution (PPWS): dilute the PPSS in double-distilled water until a final concentration of 200 ng/μL (see Note 6). Store both solutions at –20°C.
10. KAPA HiFi DNA Polymerase (1 U/μL), 5× KAPA HiFi Fidelity Buffer, $MgCl_2$ (25 mM) and KAPA dNTP Mix (10 mM each dNTP), all supplied as KAPA HiFi PCR Kit (Kapabiosystems, Boston, MA, USA).
11. Personal-sized 96-well thermal cycler 2720 (Applied Biosystems, Carlsbad, CA, USA).
12. Tris-Acetate-EDTA (TAE 1×) buffer: 40 mM Tris acetate, 1 mM EDTA, pH 8.0 (see Note 7).
13. SYBR Green solution: SYBR Green I nucleic acid gel stain (Invitrogen) at 1,000× (see Note 8).
14. Wide Mini-Sub cell GT DNA electrophoresis cell and PowerPac 300 power supply (Bio-Rad).
15. Gel casting: close the gel tray with adhesive tape and place the comb (see Note 9).
16. Loading buffer (5×): 0.10% Bromophenol Blue, 49.9% glycerol and 50% TAE 1× buffer.
17. Non-UV Transilluminator Dark Reader (DR) 46B, with amber viewing screen and a pair of DR glasses (Clare Chemical Research, Denver, CO, USA).
18. pBluescript II SK(+) phagemid vector (pBSK): 20 μg pBSK (1 μg/μL) (Agilent Technologies, Inc., Stratagene Products Division, CA, USA).
19. T4 DNA ligase, *Xba*I and *Hind*III enzymes (Roche Diagnostics, Mannheim, Germany).
20. Gel Cutting Tips: GelX6.5 tips for 1 mL pipette (Cleaver Scientific Ltd., Warwickshire, UK).

21. Thermomixer compact and MiniSpin Plus microcentrifuge (Eppendorf).
22. E.Z.N.A. Gel Extraction Kit (Omega Bio-Tek, Inc., Norcross, GA, USA).
23. Revco Elite Plus –86°C Upright Freezer (model ULT1786-6-V) (Thermo Sci., Aseville, NC, USA).
24. DH5α competent cells (DH5αcc): Subcloning Efficiency DH5α Chemically Competent *E. coli* cells (Invitrogen) (see Note 10).
25. Isopropyl β-D-1-thiogalactopyranoside (IPTG) 0.8 M stock solution: dissolve 0.19 g of IPTG in 1 mL of double-distilled water and filter through a 0.22 μm filter with a syringe (see Note 11). Store at –20°C.
26. 5-Bromo-4-chloro-3-indolyl-beta-D-galacto-pyranoside (X-gal) 20 mg/mL stock solution: dissolve 20 mg of X-gal in 1 mL dimethylformamide (DMF). Store at –20°C (see Note 12).
27. E.Z.N.A. Plasmid mini Kit (Omega Bio-Tek, Inc.).
28. BL21-Gold Competent Cells (BL21cc) (Agilent Technologies, Inc., Stratagene Products Division) (see Note 13).
29. High performance centrifuge JA2-21 with JA-14 and JA-20 rotors (Beckman Coulter Inc).

2.2. Substrate Synthesis

1. Racemic mixtures of α- and β-amino acids (α-methionine, α-valine, β-leucine, β-homoleucine, potassium cyanate) (Fluka, Madrid, Spain).
2. Short neck round bottom flask (100 mL) (POBEL, Madrid, Spain).
3. Magnetic stirrer with heating (IKAMAG RCT basic, IKA, Staufen, Germany), electronic contact thermometer (ETS-D 4 fuzzy IKATRON, IKA, Staufen, Germany) and air condenser (POBEL, Madrid, Spain).
4. Buchner funnel and water jet filter pump (Brand, Wertheim, Germany).
5. Crystallizing dish (900 mL) (POBEL, Madrid, Spain).
6. Desiccator (POBEL, Madrid, Spain).
7. UV spectrophotometer (Spectronic Helios β, Thermo, Bonsai Technologies S.A., Madrid, Spain).

2.3. Recombinant Enzyme Purification and Activity Assay

1. Wash buffer: 300 mM NaCl, 50 mM sodium phosphate, pH 7.0. Weigh 17.54 g of NaCl and add 500 mL of distilled water until it is dissolved. Add 250 mL of 200 mM of sodium phosphate pH 7.0 and measure the pH, adjusting it to pH 7.0 (see Note 14). Make up to 1 L with distilled water.

2. Elution buffer: 100 mM NaCl, 300 mM imidazole and 2 mM Tris(hydroxymethyl)aminomethane (Tris), pH 8.0. Weigh 5.85 g of NaCl and 20.40 g of imidazole and add 500 mL of distilled water until it is dissolved. Add 10 mL of 200 mM Tris buffer (see Note 15), measure and adjust the pH if necessary and make it up to 1 L with distilled water.
3. TALON™ resin: metal-affinity resin (Clontech Laboratories, Inc., Nucliber, Madrid, Spain).
4. Glass column (Sigma-Aldrich, Madrid, Spain).
5. Sonicator: UP 200 S Ultrasonic Processor (Dr. Hielschder GmbH, Germany).
6. Dialysis membrane: dialysis tubing Medicell International Ltd London, England. The membrane has a pore size of 12,000–14,000 Da and two diameters (6.3 or 21.5 mm) depending on the volume to dialyze.
7. For concentration use Centripep^R YM-10 (Amicom, Millipore Iberica, Madrid, Spain) for proteins of molecular mass up to 10,000 Da and maximum volume of 15 mL.
8. Sodium phosphate buffer pH 8.0: 200 mM sodium dihydrogen phosphate anhydrous (NaH_2PO_4) and 200 mM disodium hydrogen phosphate anhydrous (Na_2HPO_4). Weigh 12 g NaH_2PO_4 and transfer to a flask, add distilled water to a final volume of 100 mL. Weigh 14.2 g Na_2HPO_4 and make up to 100 mL with distilled water. Mix 6.8 mL of NaH_2PO_4 and 93.2 mL of Na_2HPO_4 and dilute with distilled water up to 1 L (see Note 16).
9. Synthesized hydantoins and dihydrouracils: dissolve each substrate at a final concentration of 100 mM in 10 mL of 200 mM sodium phosphate buffer pH 8.0 (see Note 17) (Fig. 1).
10. Phosphoric acid 1%: mix 1 mL of H_3PO_4 and 99 mL of distilled water.
11. Mobile Phase Filtration System (MPFS): filterSys consisting of a 300 mL funnel with a diameter of 47 mm and 1 L vacuum flask (Phenomenex, Torrance, CA, USA).
12. Mobile phase: different proportions of 20 mM H_3PO_4 pH 3.2 and methanol. For 20 mM H_3PO_4, add about 900 mL of MilliQ water to a 1-L glass and 1,348 μL of H_3PO_4 (85%) (Merck). Mix and adjust pH to 3.2 with HCl and make up to 1 L with water. Filter the 1 L solution in the MPFS through a nitrocellulose filter of 0.2 μm with a diameter of 47 mm. Filter 1 L methanol in the MPFS through a teflon (PTFE) filter of 0.2 μm with a diameter of 47 mm (see Note 18).
13. The HPLC system (Waters 1515 isocratic HPLC pump, Waters 2487 UV detector and Waters 717 plus autosampler). LUNA C18-2 column (4.6–250 mm, 5 μm particle size) (Phenomenex). MiliQ water, both for liquid chromatography.

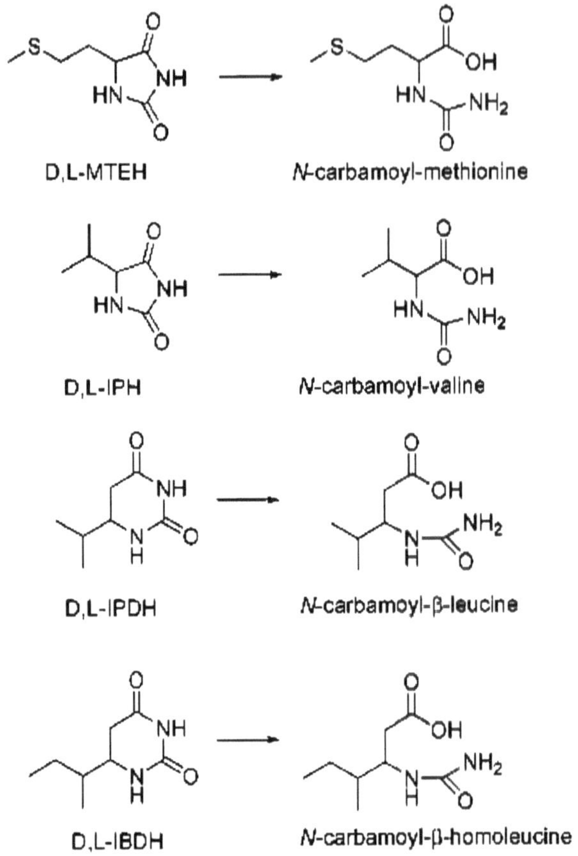

Fig. 1. Dihydrouracils and hydantoins used as SmeDhp substrates and their respective *N*-carbamoyl-α- or β-amino acids.

3. Methods

3.1. Cloning and Expression of a New Smedhp

1. Dissolve the lyophilized *S. meliloti* CECT4114 in 200 μL LB medium (see Note 19). Mix and add all the volume to a Petri plate with solid LB medium. Spread the liquid and incubate at 30°C for 24 h.
2. Add a single colony of *S. meliloti* CECT4114 to 50 μL double-distilled water. Boil the cells at 100°C for 10 min and cool for 5 min (see Note 20). Take 5 μL of the supernatant after centrifugation at 12,500 × *g* for 10 min in a bench-top microfuge.
3. Mix 29.5 μL of double-distilled water, 10 μL of 5× KAPA HiFi Fidelity buffer (containing Cl_2Mg), 5 μL of template DNA from the previous step, 1.5 μL of each primer, 1.5 μL of KAPA dNTP mix, and 1 μL of KAPA HiFi DNA polymerase in a 0.2 mL Eppendorf tube. Add 50 μL of mineral oil (see Note 21).

4. Place the PCR tube in the thermal-cycler and start the reaction with an initial cycle of 4 min at 95°C for DNA denaturation. Continue the reaction with 30 cycles of DNA denaturation (98°C for 30 s), primer annealing (55°C for 15 s), and extension (72°C for 90 s). After these 30 cycles, make a final extension cycle at 72°C for 4 min. Store at 4°C (see Note 22).
5. Add 0.5 g of agarose to 50 mL TAE buffer (1×) in a 200 mL conical flask. Heat the solution to boiling point in the microwave to dissolve the agarose (see Note 23). Add 50 μL of SYBR Green solution to the dissolved agarose and mix (see Note 24).
6. Pour the melted agarose solution onto the gel tray. Let the gel cool to room temperature for 15 min. Put the gel tray in the electrophoresis chamber after removing out the sticky tape (see Note 25). Carefully remove the comb. Mix 25 μL PCR product and 5 μL loading buffer in an Eppendorf tube (see Note 26). Load the mixture into the wells using a micropipette. Load in the first lane 7 μL molecular mass marker. Run the separation of the samples at 100 V for 60 min or until the dye has migrated for at least 6.0 cm. Remove the gel tray with the agarose gel and place on the DR transilluminator. Visualize the fluorescently stained PCR product (see Note 27).
7. Mix 6 μL of double-distilled water, 10 μL of PCR product, 2 μL of restriction buffer B (10×), 1 μL of *Xba*I, and 1 μL of *Hind*III in an Eppendorf tube. In a separate tube, mix 10 μL of double-distilled water, 4 μL of pBSK plasmid, 2 μL of restriction buffer B, 1 μL of *Xba*I and 1 μL of *Hind*III. Incubate at 37°C for 3 h (see Note 28).
8. Prepare a new gel for DNA electrophoresis as in steps 5 and 6. Mix 20 μL of double-digested PCR product (insert) and 4 μL of loading buffer in an Eppendorf tube. In a second Eppendorf tube, mix 20 μL of double-digested pBSK and 4 μL of loading buffer. Load each mixture in a different lane. Run the separation of the samples as in step 7.
9. Following the electrophoresis, place the agarose gel on the transilluminator. Visualize the bands and cut the insert and pBSK plasmid agarose bands using GelX cutting tips (see Note 29). Place each slice in a different Eppendorf tube.
10. Use the E.Z.N.A. Gel Extraction Kit to purify the DNA from the gel slices. Add 300 μL of binding buffer XP2 to each gel slice tube (digested insert and plasmid). Incubate the mixture in the thermomixer with shaking, at 60°C for 10 min. Place a HiBind DNA mini column in a 2-mL tube, add the DNA/agarose solution and centrifuge for 1 min at room temperature. Discard the flow-through liquid and place the column into the same tube. Add 300 μL of binding buffer XP2, centrifuge as before and discard the flow-through. Add 700 μL of

SPW Wash Buffer, centrifuge and discard twice. Centrifuge the empty column for 2 min to dry the column matrix. Place the column in a new tube before adding 30 μL double-distilled water directly onto the column. Wait for 2 min and centrifuge as before to elute DNA (see Note 30).

11. Mix 13 μL of insert, 4 μL of plasmid, 2 μL of ligation buffer (10×), and 1 μL of T4 DNA ligase in an Eppendorf tube (ligation tube). In a second tube, mix the same components adding 13 μL of double-distilled water instead of the insert (ligation control tube). Incubate both tubes at 4°C for 16 h (see Note 31).

12. Add 5 μL of the reaction mixture from the ligation tube or ligation control tube and 5 μL of double-distilled water (transformation control), respectively, to three tubes containing DH5αcc cells. Incubate on ice for 30 min. Place the tubes in a thermomixer for 2 min at 42°C and leave them on ice for 2 min. Add 980 μL of pre-warmed LB medium to each tube and incubate at 37°C for 1 h at 200 rpm. During incubation, add X-gal to LB plates with ampicillin to a final concentration of 50 μg/mL and IPTG to a final concentration of 1 mM. Following incubation, spread 100 and 900 μL from each transformation (ligation and ligation control tubes) on four LB plates with ampicillin. Spread 100 μL from the transformation control tube on a LB plate (positive control) and 100 μL on a plate of LB with ampicillin (negative control) (see Note 32). Incubate the plates overnight at 37°C.

13. After transformation, separate the white colonies on a new Petri plate (maximum 50 colonies). Inoculate a colony in a tube with 2 mL of LB medium supplemented with 4 μL of ampicillin. Carry out this operation with the first six separate colonies (see Note 33). Incubate overnight at 37°C both the plates and the tubes (shaking the latter).

14. Centrifuge the six tubes with 2 mL cultures for 3 min and resuspend the bacterial pellet with 250 μL of Solution I (E.Z.N.A. plasmid kit) by vortexing. Add 250 μL of Solution II and mix gently by inverting. Incubate at room temperature for 3 min. Add 350 μL of Solution III and mix by inverting. Centrifuge for 10 min at room temperature and add the supernatant to a prepared HiBind DNA Mini Column. Put the column in a tube and centrifuge for 1 min. Add 500 μL of HB Buffer and centrifuge 1 min. Add 700 μL of DNA Wash Buffer and centrifuge for 1 min. Place the column in a new tube and add 30 μL double-distilled water directly onto the column. Wait for 2 min and centrifuge as before to elute DNA. Store at −20°C (see Note 34).

15. Check for the presence the *Smedhp* gene in the six isolated plasmids by digestion with the *Xba*I and *Hind*III enzymes as

explained in step 8. Visualize the digested fragments as mentioned in steps 5, 6, and 9. Load 10 μL double-digested plasmid and 2 μL loading buffer. Select the plasmid containing the *Smedhp* gene to continue with the expression trials (see Note 35).

16. Add 2 μL of pSER38 to one thawed BL21cc tube. Continue the transformation as shown in step 12. After transformation, separate the colonies on a new Petri plate with ampicillin (see Note 36).
17. Inoculate a BL21 pSER38 colony in 10 mL of LB culture and 20 μL ampicillin in a 100 mL flask. Incubate overnight at 37°C shaking at 200 rpm. Then transfer 5 mL pre-culture to 500 mL of LB and 1 mL of ampicillin in a 2 L flask. After 2 h of incubation at 37°C with shaking, add 125 μL IPTG. After additional incubation with shaking for 4 h at 37°C, collect the cells by centrifugation at 10,000 ×*g* at 4°C for 20 min. Store the cells at −20°C (see Note 37).

3.2. Hydantoins and Dihydrouracils Racemic Mixtures Synthesis

The synthesis reaction is the same for all hydantoins and dihydrouracils unless otherwise specified. Carry out all procedures inside a laboratory gas cabinet.

1. Mix 20 mmol of the amino acid and 25 mmol of potassium cyanate to a short neck round bottom flask with 15 mL of H_2O (see Note 38).
2. Heat reaction mixture to reflux and stir slightly at 60–80°C for 1 h.
3. Add 5 mL of hydrochloric acid (36.5%) to the stirred solution and continue heating for a further 15 h (see Note 39).
4. Cool the reaction mixture on ice until the product is crystallized (see Note 40).
5. Vacuum filter the product in a Buchner funnel.
6. Add 20 mL of cold 20% ethanol (v/v) solution to the Buchner funnel containing the product to wash it (see Note 41).
7. Recrystallize the 5-monosubstituted hydantoin from hot 20% ethanol (v/v).
8. Dry the powder in a filter paper bag, placed in a desiccator for 1 day.
9. Measure the maximum wavelength in the spectrophotometer of each 1 mM 5-monosubstituted hydantoin, dihydrouracil, and carbamoyl-amino acid (see Note 42).
10. Store the substrates as powder at 4°C.

3.3. Single-Step Purification of SmeDhp and Activity Monitoring

1. Resuspend the pellet of BL21 pSER38 stored at −20°C in 30 mL wash buffer (300 mM NaCl, 50 mM sodium phosphate, pH 7.0) at room temperature (see Note 43).
2. Disrupt the cell by sonication on ice (see Note 44) for six periods of 30 s at pulse mode = 0.5 and sonic power = 60%.
3. Add 3 mL of TALON™ resin to an empty glass column (see Note 45) and wash with five volumes of wash buffer.
4. Sediment the cell debris by centrifugation at 12,000 × *g*, 4°C and 20 min. Filter the supernatant through a paper filter and eliminate cell residues (see Note 46).
5. Apply the cellular extract with the overproduced SmeDhp to the resin, taking care not to disrupt the top of the matrix.
6. When all the cellular extract has passed through the resin, wash with ten volumes of wash buffer.
7. Elute the protein very slowly with elution buffer (see Note 47) and collect in 2 mL aliquots (see Fig. 2). Measure the protein concentration at 280 nm and take aliquots until absorbance is below of 0.5.
8. Take the most concentrated aliquots and place in a dialysis tube. Introduce the tube in a flask and dialyze against 100 volumes of 200 mM sodium phosphate buffer (pH 8.0) by three changes of buffer in constant stirring, at 4°C. Recombinant His-tagged (see Note 48) SmeDhp is stored at 4°C.

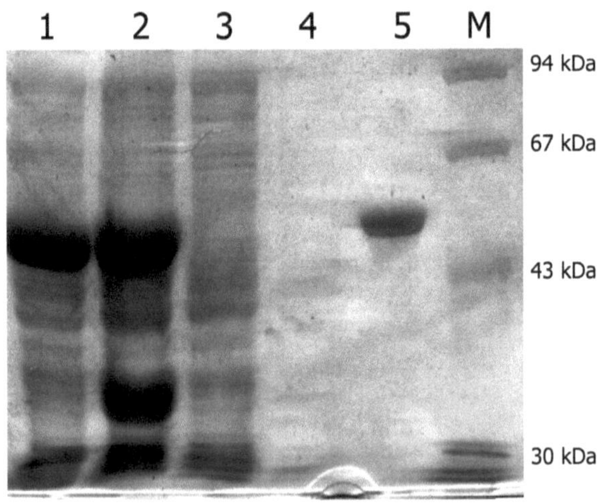

Fig. 2. SDS–PAGE analysis of each purification step of SmeDhp from *E. coli* BL21 harboring the pSER38 plasmid. *Lanes 1* and *2*, supernatant and pellet of the resuspended crude extract after cell sonication; *lane 3*, eluate after adding the sonicated supernatant to the metal-affinity column; *lane 4*, flowthrough after washing the metal-affinity column with buffer; *lane 5*, purified SmelDhp; *lane M*, low molecular mass markers (kDa).

9. Concentrate the protein by a Centripep filter: add the protein solution to the sample container up to the line drawn on the container. Insert the capped filtrate collector and centrifuge at 3,000 × *g* at 4°C, until the desired volume is reached. The concentrated protein is in the outer container. The buffer without protein passes through the polycarbonate membrane to the inner container.
10. Mix purified SmeDhp (final concentration 0.5–2.5 μM) with hydantoin or dihydrouracil (final concentration 20 mM) dissolved in 200 mM sodium phosphate buffer (pH 8.0) in 1,000 μL of reaction volume in a 1.5-mL Eppendorf tube (see Note 49).
11. Incubate the reaction mixture for 2 h at 40°C in a thermomixer and take out aliquots of 25 μL each at different intervals. Stop the reaction of each aliquot by adding it to 475 μL of 1% H_3PO_4 and centrifuge at 10,000 × *g* for 10 min in a microcentrifuge. Transfer the supernatant to a HPLC vial (see Note 50).
12. Place the vials in the autosampler. Inject 20 μL of sample into the HPLC with a C18 column and run it for 30 min under isocratic conditions at a flow rate of 0.5 mL/min, using the right mobile phase (see Note 19) and wavelength of detection (see Note 51). Compare the chromatograms obtained at different reactions time to check the activity (Fig. 3).

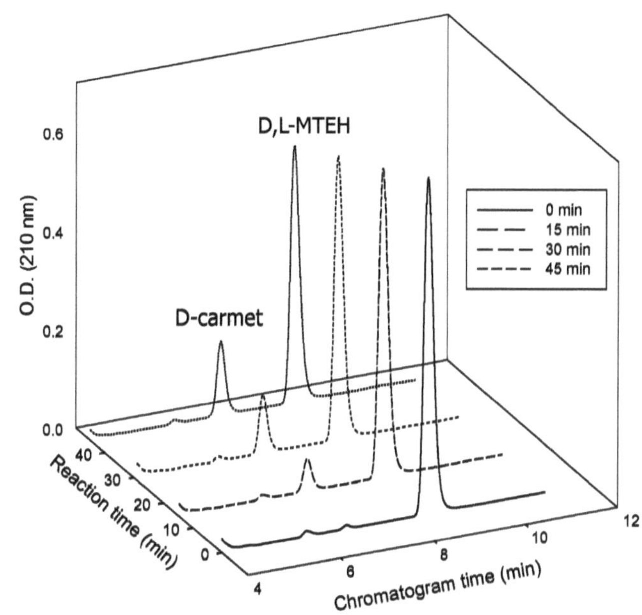

Fig. 3. Enzymatic hydrolysis of D,L-MTEH by SmeDhp for conversion to D-*N*-carbamoyl-methionine (D-carmet). Activity was monitored by HPLC at 0, 15, 30, and 45 min. The D,L-MTEH peak decreased while the D-carmet peak appeared (and it increased with time).

4. Notes

1. Adjust pH using diluted NaOH and HCl (1 M). Dilution and pH measurements should be carried out in a laboratory gas cabinet.
2. To check that the medium is not too hot, touch the bottle. Add the ampicillin when you can touch the bottle without burning yourself. Final concentration of ampicillin should be 100 μM.
3. Nucleotide sequences of dihydropyrimidinase/hydantoinase genes that codified enzymes with detected activity were used as templates to search for the gene in new microorganisms, see refs. 10, 11. A DNA fragment from *S. meliloti* strain 1021 chromosome sequence evidenced a very high sequence homology against these templates. Both PCR primers were designed to amplify this DNA fragment. For an in-depth explanation of sequence similarity searching see Chapter 1 of volume 537 of this collection, see refs. 12, 13.
4. Smedhp5 primer includes the *Xba*I site in italics, and the ribosome-binding site and start codon in bold. Smedhp3 primer includes the *Hind*III site in italics and the polyhistidine tag (His6 tag) before the stop codon, both in bold. Add two nucleotides before restriction enzyme sites (underlined) to guarantee fixing and cleaving of DNA.
5. To prepare the PPSS, dissolve the lyophilized primers in the same volume (in μL) as the microgram of primers in the commercial tube. Thus the final concentration is 1 μg/μL.
6. To prepare PPWS transfer 10 μL of PPSS to an Eppendorf tube with 40 μL of double-distilled water. Do not dilute all the PPSS because freeze/thaw cycles degrade PPWS and hence more must be prepared as required.
7. Prepare 50× native buffer (2 M Tris-acetate, 50 mM EDTA). Weigh 242 g of Tris base, 18.6 g of $EDTANa_2 \cdot 2H_2O$ and add 57.2 mL of glacial acetic acid. Mix and make up to 1 L with water. Dilute 100 mL of 50× TAE buffer up to 5 L with double-distilled water. Store at 25°C.
8. Dilute 100 μL of 10,000× SYBR Green I nucleic acid gel stain to 900 μL TAE buffer. The commercial SYBR Green vial (10,000×) must be warmed to room temperature to ensure that the DMSO is completely thawed and that the solution is homogeneous.
9. Make sure that the comb is as close as possible to the "black" electrode.

10. The kit must be shipped on dry ice. Handle DH5αcc gently as they are highly sensitive to changes in temperature or mechanical lysis caused by pipetting. Place 40 tubes of 1.5 mL on ice, while simultaneously thawing on ice the four tubes of DH5αcc. Make aliquots of 50 μL of cells in each empty tube. Store the aliquots at −80°C. Do not store in liquid nitrogen.
11. IPTG solution must be prepared and handled inside a laminar flow bench to avoid contamination.
12. DMF is toxic, allow X-Gal solution containing dimethylformamide to completely dry before plating bacteria on agar plates.
13. Handle BL21cc gently. Make 20 aliquots of 50 μL/each of cells operating as with DH5αcc. Store at −80°C.
14. The pH values will not vary from 7.0 by more than ±0.12 even at 37°C, but it is recommended to measure and adjust if necessary with HCl or NaOH. The pH 7.0 is necessary in the wash buffer to avoid non-specific binding, which decreases resin capacity and the final purity of the protein as well as resin efficiency. All the buffers must be filtered through a 0.22-μm filter.
15. To prepare 200 mM Tris dissolve 24.2 g of Tris (Roche Diagnostics) in 1 L of distilled water. Mix 50 mL of this solution with 26.8 mL of 200 mM HCl and make it up to 200 mL.
16. Adjust pH with NaH_2PO_4 and Na_2HPO_4. The concentration of monobasic and dibasic sodium phosphate solutions is 1 M, and after dilution the working concentration is 200 mM.
17. The solubility of the substrates D,L-5-methyl-thio-ethylhydantoin (D,L-MTEH) (MW = 174.22), D,L-5-iso-propylhydantoin (D,L-IPH) (MW = 156.19), and D,L-6-iso-propyldihydrouracil (D,L-IPDH) (MW = 156.18) is higher than 100 mM. However, for D,L-6-iso-butyldihydrouracil (D,L-IBDH) (MW = 170.21) the maximum solubility is 15 mM, see ref. 12.
18. Mobile phases should be degassed prior to use. A different mobile phase is necessary for each substrate. If we name 20 mM H_3PO_4 pH 3.2 as compound A and methanol as compound B, the proportions are: D,L-MTEH is 80% A and 20% B; D,L-IPH is 90% A and 10% B; D,L-IPDH and D,L-IBDH are 85% A and 15% B. The respective product (*N*-carbamoyl-α- or β-amino acid) of each substrate is detected in the same mobile phase. For 80% A:20% B, mix 800 mL of 20 mM H_3PO_4 pH 3.2 with 200 mL of methanol.
19. Culture collections deliver lyophilized microorganisms because they are stable for years at room temperature. However, they must be transferred to a growth medium for additional use.

20. Take a single colony from the LB plate using a sterile inoculating loop inside a flow chamber to avoid contamination. After boiling, chill the sample on ice immediately to favour cell debris precipitation. The supernatant containing genomic DNA is used for PCR amplification of the gene encoding *Smedhp*.
21. Add the components to the tube in the above order, to avoid the precipitation of any component. PCR components are thawed on ice and the PCR reaction must be set up on ice. The mineral oil avoids the evaporation during the cycling protocol in the thermal cycler.
22. Initial denaturation at 98°C is not recommended as it may lead to template damage. However, in each cycle, denaturation must be performed at 98°C for a short period (30 s). Store at 4°C for short periods, but for periods of over 24 h, store at −20°C.
23. For 1% agarose (1% w/v), add 0.5 g of agarose to 50 mL of 1× TAE. For other percentages of agarose, adjust the ratio accordingly. To avoid sample evaporation, cover with an inverted flask. To dissolve 50 mL, it should take about 90 s at maximum power. To avoid distorted bands, check that no solid particles are present.
24. Add SYBR Green solution just before pouring the gel. Make sure that the agarose solution is not too hot.
25. Check that there are no air bubbles in the gel. If there are, move them to the lateral walls of the gel tray with a tip. Air bubble in the middle of the gel can distort the DNA bands.
26. Shortly spin the tube in the microfuge to ensure mixing.
27. Check the gel while it is running to make sure it is not getting too hot, as this would distort the bands or melt the agarose. Use the amber viewing screen to see the DNA bands on the transilluminator. Verify the molecular mass of the PCR products by comparison with the molecular mass markers.
28. For DNA digestion, add the components in the above order to avoid precipitation of the component. Shortly spin the tube in the microfuge to ensure mixing.
29. Insert the GelX cutting tip onto a 1 mL pipette and set it to 1,000 μL. Place on the DR glasses and select the insert or plasmid band in the agarose gel. Press the tip into the gel without pressing the pipette plunger. Wiggle the tip to make sure that the slice is completely cut. Remove the band by lifting the tip very slowly. Depress the plunger quickly to eject the fragment to the tube. It is important to minimize the size of the gel slice.
30. The colour of the Gel/Binding Buffer mixture should be light yellow (pH > 8). If the colour of the mixture becomes orange

or red, add 5 μL of 5 M sodium acetate (pH 5.2) to lower the pH. SPW Wash Buffer must be diluted with absolute ethanol before use. Do not skip the centrifugation of the empty column; it is crucial for the removal of ethanol. DNA recovery is approximately 70% of bound DNA. All the centrifugations are at 10,000 × *g*.

31. For DNA ligation, add the components in the tube in this order to avoid the precipitation of components. Shortly spin the tube in the microfuge to ensure mixing. For plasmid re-ligation control, prepare a second tube with same components with 13 μL double-distilled water instead of the insert.
32. Thaw on ice three tubes of DH5αcc. After adding DNA to the tubes, mix gently, but not by pipetting up and down. Avoid shaking during the heat shock at 42°C. Spread IPTG and X-gal immediately after adding to the plate, in order to ensure that both compounds are distributed on the plate. The transformation control allows verifying whether the cells are alive after the transformation process (positive control) and they are unable to grow in the presence of ampicillin without plasmid (negative control).
33. To make the separation and culture preparation for plasmid extraction simultaneously, pick up the colony from the transformation plate with a yellow tip and spot in the new plate. Introduce the same tip in the 2-mL LB broth with ampicillin. Although the bacterial quantity is very little, it is enough for both steps. In this way you can save one working day. Make sure that the plate spread with ligation control has no colonies.
34. All centrifugations are at 10,000 × *g*. Add the vial of RNase A supplied with the kit to the bottle of Solution I and store at 2–8°C. Complete re-suspension of cell pellet and Solution I is vital to obtain good yields. Avoid vigorous mixing that could decrease plasmid purity after adding solutions II and III. To prepare HiBind DNA Mini Column, place it into a 2-mL collection tube, add 100 μL of Equilibration Buffer, and centrifuge for 30–60 s. Discard the flow-through liquid after adding HB and washing buffers. Repeat the washing step twice with Wash Buffer and then centrifuge the empty HiBind Mini Column for 2 min to ensure no ethanol is present in the column.
35. The transformant containing the plasmid with the *Smedhp* gene is named DH5α pSER38. Sequence the cloned fragment at least twice. If none of the six isolated plasmids contains the *Smedhp* gene, grow the next six transformants separated in step 14.
36. For *Smedhp* expression, it is necessary to change the host to BL21. For this transformation, no control is necessary.

37. Ampicillin final concentration in LB medium is 100 μg/mL. After 2 h of incubation at 37°C with vigorous shaking, the OD_{600} of the resulting culture is 0.3–0.5. IPTG starts the *Smedhp* gene expression and the final concentration in the 2 L flask is 0.2 mM. Induced cells are collected by centrifugation in 200 mL bottles in a JA-20 rotor.
38. Some aromatic or long-chain aliphatic amino acids might not directly solubilize using 20 mmol. This is not a big issue since the reaction will also take place; as the soluble fraction of the amino acid is converted in the reaction, the remaining non-solubilized substrate will dissolve.
39. It is not necessary to stop heating. Remove the air condenser and carefully add HCl (drop-by-drop), as vapours will be generated. Once the HCl is added, place the air condenser once again and continue with the protocol. For the synthesis of carbamoyl-amino acids, cool the solution on ice at this step. Once cooled, add HCl drop-by-drop.
40. Some cyclic amides might not crystallize directly. In this case, store the flask at 4°C. If even after some days at this temperature the product is not obtained, the use of a rotavapor might be used to evaporate the solution.
41. Repeat this step twice.
42. The substrates synthesized are: D,L-MTEH, D,L-IPH, D,L-IPDH, D,L-IBDH, and their respective *N*-carbamoyl-α- or β-amino acids. Prepare 1 mL of a 10 mM solution of each 5-monosubstituted hydantoin, dihydrouracil, and *N*-carbamoyl-α- or β-amino acid in a 200-mM sodium phosphate buffer (pH 8.0). Then, dilute ten times with 1% H_3PO_4 up to a final volume of 1 mL.
43. The proportion is 30 mL of wash buffer per 500 mL of induced culture.
44. From this moment, the samples must be on ice. Samples at room temperature result in elevated proteolytic activity and over-heating during sonication.
45. Before starting the procedure, the resin must be transferred to a column. Thoroughly resuspend the TALON™ resin in the bottle provided, and immediately transfer the required amount to an empty glass column with a filter pore size of 10–20 μm, avoiding introducing air bubbles. The binding capacity of TALON™ resin is 5–10 mg protein/mL resin, but it is supplied in ethanol at 20% as a suspension. Taking into account the volume of the ethanol plus the resin, use 2 mL of suspension per 3 mg of polyhistidine tagged protein. Allow the resin to settle, which takes several minutes. Allow the ethanol to flow through and start washing and the equilibration of the resin.
46. To obtain a clarified sample, transfer the supernatant carefully without disturbing the pellet. A small portion of the clarified

sample can be reserved for SDS–PAGE analysis to estimate the protein retained in the TALON™ resin (see Fig. 2).

47. Elution occurs when the bound polyhistidine-tagged protein is competitively eluted simply by adding imidazole to the elution buffer (see Fig. 2). Alternatively, the histidine nitrogen can be protonated generating a positively charged ammonium ion which is repelled by the positively charged metal ion.

48. Although three histidines may bind transition metals, six histidines reliably bind transition metals in the presence of strong denaturants such as guanidinium, and even proteins from denatured inclusion bodies can be purified.

49. D,L-IBDH final concentration in the reaction is only 10 mM, due to its low solubility. Preincubate the substrate and the buffer in the reaction tube for 10 min in a thermomixer at 40°C, before adding the SmeDhp enzyme.

50. Take aliquots every 15 min for 2 h. It is recommendable to prepare the tubes to stop the reaction with 475 μL of 1% H_3PO_4 before starting the reaction. After centrifuging, transfer the supernatant to the vial as soon as possible to avoid resuspension of the pellet. Repeat the reaction at least three times.

51. The maximum wavelength (λ_{max}) of absorption of each substrate and product couple is different (the λ_{max} is instead the same for the substrate and the product). The λ_{max} are: 210 nm for D,L-MTEH, 205 nm for D,L-IPH and 200 nm for D,L-IPDH, and D,L-IBDH. A calibration curve with the known concentrations of substrate and product is necessary to determine their concentration in the reaction.

Acknowledgements

This work was supported by the Spanish Ministry of Education and Science and European Regional Development Fund (ERDF) through the project BIO2007-67009, the Andalusian Regional Council of Innovation, Science and Technology through the projects CV7-02651 and TEP4691 and the European Science Foundation COST Action CM0701.

References

1. Kim G J, Kim H S (1998) Identification of the structural similarity in the functionally related amidohydrolases acting on the cyclic amide ring. Biochem J 330, 295–302.
2. May O, Habenicht A, Mattes R et al. (1998) Molecular evolution of hydantoinases. Biol Chem 379, 743–747.
3. Nam S H, Park H S, Kim H S (2005) Evolutionary relationship and application of a superfamily of cyclic amidohydrolase enzymes. The Chemical Record 5, 298–307.
4. Lohkamp B, Andersen B, Piskur J, Dobritzsch D (2006) The crystal structures of dihydropyrimidinases reaffirm the close relationship

between cyclic amidohydrolases and explain their substrate specificity. J Biol Chem 281, 13762–13776.

5. Syldatk C, May O, Altenbuchner J et al. (1999) Microbial hydantoinases-industrial enzymes from the origin of life? Appl Microbiol Biotechnol 51, 293–309.
6. Martínez-Rodríguez S, Las Heras-Vázquez F J, Clemente-Jiménez J M et al. (2002) Complete conversion of D,L-5-monosubstituted hydantoins with a low velocity of chemical racemisation into D-amino acids using whole cells of recombinant *Escherichia coli*. Biotechnol Prog 18, 1201–1206.
7. Martínez-Gómez A I, Martínez-Rodríguez S, Clemente-Jiménez J M et al. (2007) Recombinant polycistronic structure of hydantoinase process genes in *Escherichia coli* for the production of optically pure D-amino acids. Appl Environ Microbiol 73, 1525–1531.
8. Las Heras-Vázquez F J, Martínez-Rodríguez S, Mingorance-Cazorla L et al. (2003) Overexpression and characterization of hydantoin racemase from *Agrobacterium tumefaciens* C58. Biochem Biophys Res Commun 303, 541–547.
9. Martínez-Rodríguez S, Clemente-Jiménez J M, Rodríguez-Vico F, Las Heras-Vázquez F J (2005) Molecular cloning and biochemical characterization of L-*N*-carbamoylase from *Sinorhizobium meliloti* CECT4114. J Mol Microbiol Biotechnol 9, 16–25.
10. Capela D, Barloy-Hubler F, Gouzy J et al. (2001) From the Cover: Analysis of the chromosome sequence of the legume symbiont *Sinorhizobium meliloti* strain 1021. Proc Natl Acad Sci USA 98, 9877–9882.
11. Galibert F, Finan T M, Long S R et al. (2001) The composite genome of the legume symbiont *Sinorhizobium meliloti*. Science 293, 668–672.
12. Menlove K J, Clement M, Crandall K A (2009) Similarity searching using BLAST. In: Posada, D. Ed. Bioinformatics for DNA Sequence Analysis. Methods in Molecular Biology series Vol. 537. Humana Press, a part of Springer Science, New York.
13. Martínez-Rodríguez S, Martínez-Gómez A I, Clemente-Jiménez J M et al. (2010) Structure of dihydropyrimidinase from *Sinorhizobium meliloti* CECT4114: new features in an amidohydrolase family member. J Struct Biol 169, 200–208.

Part II

Application of Unnatural Amino Acids

Chapter 7

NMR Analysis of Unnatural Amino Acids in Natural Antibiotics

Franca Castiglione

Abstract

A large number of modified amino acids other than the canonical amino acid residues can be found in natural products, especially antibiotics. The structure of these peptide-based compounds is investigated using modern two-dimensional NMR techniques. The automatic assignment of the 2D NMR proton spectra and consequent determination of the primary and 3D structure of peptides or small size proteins containing natural amino acids is nowadays routine. However, a deficiency in the ability to readily sequence peptides containing unnatural amino acids still remains and a great human effort and time is required. The experimental methods and the protocols of manual analysis of the data are described in the following sections.

Key words: Unnatural amino acids, NMR spectroscopy, Structure, Antibiotics

1. Introduction

Analysis of unnatural amino acids has long been a key component in the research of new antibiotics produced by natural microorganisms. Nuclear Magnetic Resonance (NMR) spectroscopy is an established and very powerful biophysical method to study the primary, 3D structure, dynamics, and function of proteins and their complexes approaching a megadalton in size. This technique has been successfully used to study unnatural amino acids incorporated into large proteins providing insights into structure, ligand-binding sites, and conformational changes (1–5). The majority of modern applications to the study of biological macromolecules involve the analysis of two (or higher) dimensional (2D) spectra. The assignment of resonances to individual nuclei is the essential first step in the identification of the unnatural amino acid components

Loredano Pollegioni and Stefano Servi (eds.), *Unnatural Amino Acids: Methods and Protocols*,
Methods in Molecular Biology, vol. 794, DOI 10.1007/978-1-61779-331-8_7, © Springer Science+Business Media, LLC 2012

in antibiotic macromolecules and is performed using principally bi-dimensional ^1H homonuclear correlation experiments. Nevertheless, for large proteins this is still the size-limited step. The structural information is then obtained from measurements of the Nuclear overhauser effect (NOE) in the 2D map which provide the peptide sequential assignment and constrains on the internuclear distances. The primary and 3D structure of a substantial number of natural antibiotics containing modified amino acids have been determined (6–8) by the analysis of several 2D NMR experiments.

Here the methodology is outlined, describing the general procedures to follow with practical instructions on how to choose the right experiment to obtain the desired information, and consequently how to analyse the data. Significant recent applications of this approach are described in details together with the major improvements in the experimental method.

2. Materials

Sample preparation is a key step in the NMR analysis of unnatural amino acids forming a high molecular mass protein. The sample solution is prepared at room temperature using a mixture of H_2O/D_2O 9:1 with analytical grade water as solvent. In order to prepare a "*good NMR sample*" many practical aspects might be taken into account.

2.1. Sample Quantity

1. The fundamental problem in the biological applications of the solution-state NMR technique is its intrinsically poor sensitivity.
2. The intensity of the signal is proportional to the amount of material dissolved in the "sensitive volume" of the spectrometer.
3. The optimum sample volume is around 400 μL.
4. The solution concentration needed for structural studies of a protein is at least 1 mM to acquire a one-dimensional spectrum, while for multi-dimensional experiments 3–5 mM is required.
5. When dealing with samples of limited quantities, volume-limiting NMR tube such as Shigemi tubes can be used in combination with high sensitivity cold probes.

2.2. Sample Solubility and Stability

1. Sample solubility is another important factor when ample amount of protein is available; the signal-to-noise ratio of the spectrum can be increased by increasing the concentration of the material.

2. However, aggregation of the macromolecule must also be minimized mainly acting on pH, ionic strength, buffer and temperature: this can make a real difference to the quality of the spectrum.
3. Several samples can be prepared varying these parameters and the 1D spectrum is acquired.
4. Moreover, the addition of organic solvents to the solution can increase solubility, decrease aggregation and improve the spectral resolution. The concentration of the organic co-solvent should be less than 10% to avoid direct binding with the protein and other techniques should be used to determine whether a change in the protein structure is occurring under these conditions.
5. The commonly used organic co-solvents are alcohols (especially trifluoro-ethanol), acetone, and DMSO (see Note 1).
6. The effect of the different experimental conditions on both the intensity and the resolution of the proton spectrum provides information as to the optimal conditions to be used in running the more time-consuming multi-dimensional spectra.
7. The stability of the protein in solution must be also checked (see Note 2).

2.3. pH Adjustment and Ionic Strength

1. The optimal value of the pH for the stability of a protein solution is around the neutral value, while a better resolution of the amide protons in the 1D spectrum is observed in acid conditions.
2. The pH of the solution is varied adding a few microliters of a diluted HCl solution up to a value of pH = 3–4 avoiding the protein to precipitate and maintaining clear the solution.
3. If the pH is varied for a sample dissolved in D_2O, a few additional precautions need to be taken (see Note 3).
4. The solubility of a protein in water solution increases as the salt concentration is increased, due to the increase of the dielectric constant of the solution. A high dielectric constant assists a charge separation of the individual protein molecule from each other.
5. Sodium chloride is commonly used as a 5-M solution.
6. The pH, the ionic strength, and the temperature are always varied in concert to maximize their effects on the protein solubility and stability in solution.

2.4. Choice of Buffer and Detergents

1. The most suitable buffers for acquiring ^1H NMR spectra are those containing no exchangeable protons (simple inorganic buffers) or more complex molecules (Tris and acetate) with deuterium at non-labile positions.

2. The most commonly used buffers are:
   - acetate, $pK_a = 4.76$, pH range (3.7–5.6)
   - Tris, $pK_a = 8.06$, pH range (7.1–8.9)
   - phosphate, $pK_a = 7.20$, (2.15, 12.33), pH range (5.8–8.0)
   - bicarbonate, $pK_a = 6.35$ (10.33), pH range (6.0–8.0)
   - succinate, $pK_a = 4.21$ (5.64), pH range (3.5–6.0).
3. The buffer concentration should be around 10–50 mM.
4. The detergents (usually SDS micelles) can stabilize the protein native conformation. These solutions are often prepared for 3D structure studies of a protein.

3. Methods

3.1. Useful Parameters from NMR Measurements

Structural information is readily obtained from the parameters measured in the simple 1D experimental spectrum. These include the chemical shift (δppm), spin–spin coupling constant (*J*Hz), coupling pattern (***multiplicity***), *NOE,* and signal intensity. There are many NMR references that discuss these parameters in detail (9–12); only a brief survey will be presented here with emphasis on the analysis of unnatural amino acids.

1. The chemical shift (δ) is a measure of the chemical environment of a given nucleus and is therefore indicative of the types of chemical groups that are present in a molecule. It is measured relative to a reference compound (usually tetramethylsilane TMS, see Note 4).
2. Spin–spin coupling constant (*J*) characterizes the scalar interactions through bond between nuclei linked via a small number of covalent bonds in a chemical structure. The J value depends on several factors including the number of bonds, bond order, bond length, bond angle, types of coupled nuclei, and substituents' electronegativity. For example, *cis* or *trans* olefinic protons can be differentiated according to the observed *J* value (15–18 and 8–12 Hz, respectively). Moreover, *J* couplings for geminal protons ($^2J_{H-H}$) are usually negative and larger than those for vicinal protons ($^3J_{H-H}$) that are positive.
3. Spin multiplicity is another important parameter for structure determination. The intensities of the components in the first-order coupling patterns for spin ½ nuclei (^1H, ^{13}C, ^{15}N, ^{31}P) follow the rule of the Pascal triangle. Then doublets, triplets. and quartets with intensity ratios of 1:1, 1:2:1, and 1:3:3:1 represent coupling from one, two, or three chemically equivalent nuclei, respectively. The spectrum of 2-fluoroPhe, reported in Fig. 1c, shows a triplet for the protons 4 and 5. Coupling

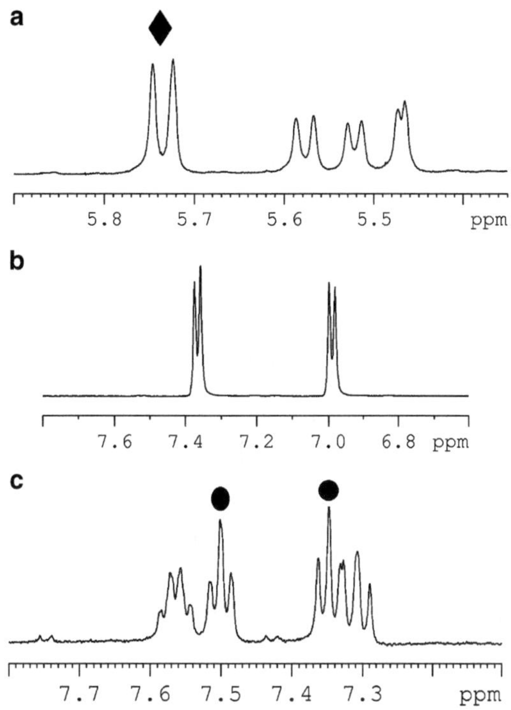

Fig. 1. 1D ^{1}H Spectra of a mixture compound (**a**), 4-hydroxyphenylglycine (**b**), 2-fluorophenylglycine (**c**). The line (*filled diamond*) in (a) has double intensity; two triplets (*filled circle*) in (**c**) are indicated.

between chemically non-equivalent geminal nuclei gives doublets of doublets splitting.

4. Nuclear Overhauser effects are due to dipolar interactions (through space) between different nuclei and lead to a change in signal intensity in a double irradiation experiment. Because the NOE is correlated with the inverse sixth power of the internuclear distance, it is only sensitive to neighbour nuclei.
5. The intensity of the NMR peaks, in a normal 1D spectrum, reflects the number of nuclei manifested by these lines. For example, in the spectrum in Fig. 1b, 4-hydroxyPhe gives rise to two proton lines (2, 3, 5, 6) of double intensity.

3.2. 1D NMR Methods

3.2.1. ^{1}H NMR

1. 1D Method allows the determination of important parameters such as chemical shift and J coupling constants directly from the spectrum obtained using a simple one-pulse experiment with solvent presaturation.
2. An example ^{1}H NMR spectrum of enduracidin (13) is shown in Fig. 2. Various signals resonating between 0.9 and 9.2 ppm were observed, many of which show first order or pseudo-first-order multiplicity. However, some spectral regions (e.g., 2.9–4.0 and

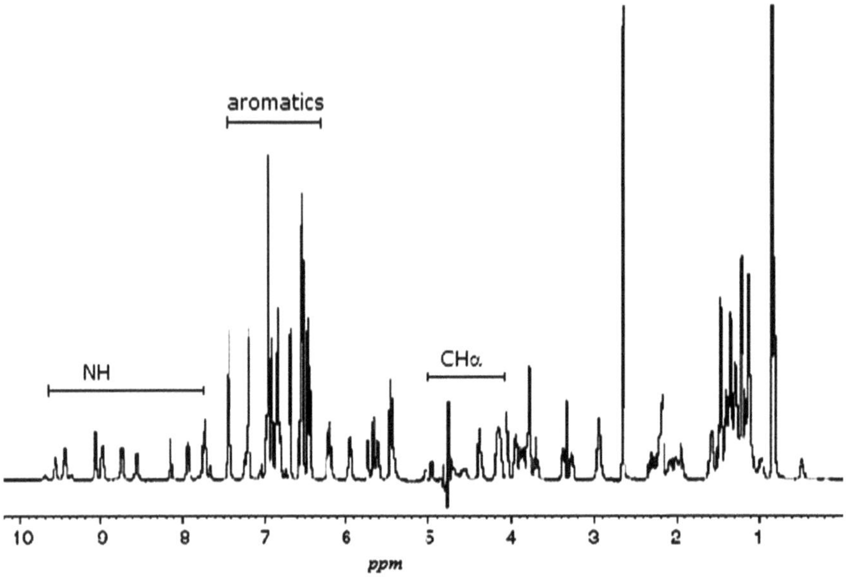

Fig. 2. 1D ^{1}H Spectrum of enduracidin in H_2O–DMSO-*d6* (4:1) acquired at 303°K.

7–8.4 ppm) usually are too crowded in large proteins and the peaks cannot be assigned (see Fig. 2). Table 1 lists the ^{1}H and ^{13}C chemical shifts of the commonly occurring unnatural amino acids referred to the TMS and that can be used to easily recognize some modified amino acid residues in the ^{1}H spectrum.

3. It is important to observe the change in chemical shifts as the pH of the solution changes from basic to acid values.
4. Moreover, the change in the chemical shifts of the amide protons as the temperature is slowly varied gives us important information. A low dependency of the amide chemical shift on temperature in an aqueous environment is usually indicative of the presence of a stable hydrogen bond, whereas solvent-accessible amide protons, indicative of a protein region exposed to the solvent, are more sensitive to the temperature changes. Assessing the temperature dependency of the amide proton chemical shifts represents a simple means of gaining information about the secondary structure of the investigated protein.

3.2.2. ^{13}C NMR

1. ^{13}C NMR is the most used heteronuclear technique for biochemical studies despite its low sensitivity and low natural abundance (1.1%). This is mainly because of the wide spread of its chemical shifts (0–220 ppm), which allows a large number of signals to be resolved and assigned (14).
2. The ^{1}H decoupled ^{13}C experiment is the simplest method employed to measure the carbon chemical shifts with improved

Table 1
^{1}H and ^{13}C chemical shifts (ppm) of unnatural amino acids

	α-position		β-position		Others	
Residue	**^{1}H**	**^{13}C**	**^{1}H**	**^{13}C**	**^{1}H**	**^{13}C**
Carnitine	CH_2 3.45	43.90	CH 4.58	71.4	$CH_2γ$ 3.43 N–CH_3 3.23	65.1 54.90 COO^- 175
Citrulline	CH 3.76		CH_2 1.88		$CH_2γ$ 1.58 $CH_2δ$ 3.15	
Creatine	–		–		N–CH_3 3.04 N–CH_2 3.94	37.25 54.18 COO^- 174.93 N=C 157.47
Creatinine	–		CH 3.32		N–CH_3 3.05 N–CH_2 4.10	30.44 56.6 COO^- 189.09 N=C 169.59
Ornithine	CH 3.8		CH_2 1.94		CHγ 1.75–1.81 $CH_2δ$ 3.05	
Sarcosine	CH_2 3.6		–		N-CH_3 2.73	
α-Aminobutyric acid	CH 4.92	58.6	CH 3.47	50.23	CH_3 1.2	22.3
2,3-Didehydroalanine		135	CH 5.26–5.75	109.5		COO^- 167.6
2,3-Didehydrobutyrine		129.1	CH 6.5	133.8	CH_3 1.7	12.07 COO^- 167.4
2-Aminovinyl-cysteine	CH 6.93	130.5	CH 5.63	105.3		

(continued)

Table 1 (continued)

Residue	α-position		β-position		Others	
	^{1}H	^{13}C	^{1}H	^{13}C	^{1}H	^{13}C
Enduracidine			CH 2.21		CHγ 4.06 CHδ 3.26/3.69	
4-Hydroxy-phenylglycine	CH 6.94		–		b/f 7.18 c/e 6.67	
3,5-Dichloro 4-hydroxy-phenylglycine	CH 5.72		–		b/f 6.95	
4-Hydroxy-proline	CH 4.54		CH_2 2.37/1.93		CH_2δ 3.6/3.86	

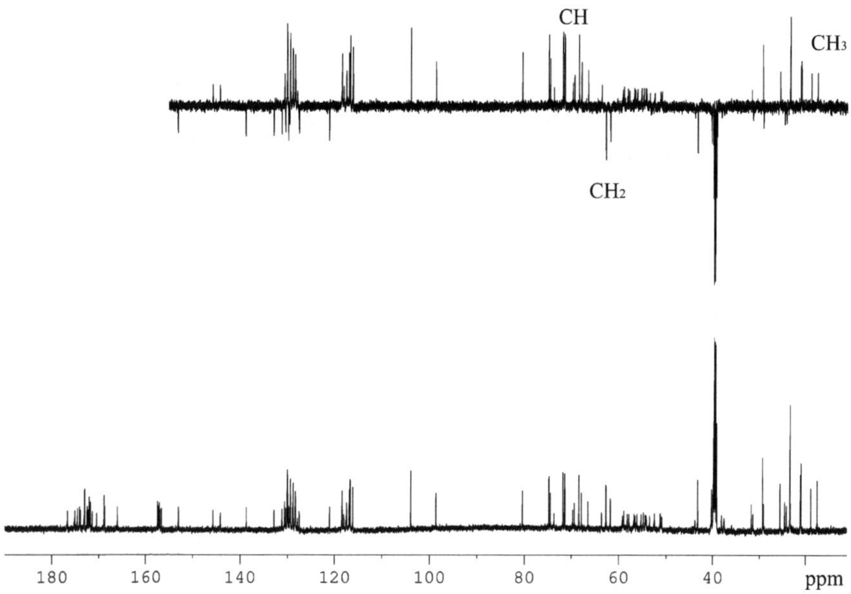

Fig. 3. ^{13}C Spectrum (proton decoupled) (*bottom*) and ^{13}C DEPT spectrum (*top*) of enduracidin in H_2O–DMSO-*d6* (4:1) acquired at 303°K.

signal intensity due to the coalescence of the multiplets (^1H couplings) and NOE effects (see Note 5).

3. Nevertheless, the simple ^1H decoupled ^{13}C spectrum does not allow to determine the number of protons directly bonded to the carbon, then ^{13}C spectral editing methods such as INEPT (Insensitive Nuclei Enhanced by Polarization Transfer) or DEPT are used (15).
4. These types of experiments, which utilize polarization transfer from protons to carbon, are applied for distinguishing CH_3, CH_2, and CH groups. In the DEPT spectrum the signals of CH_2 show negative intensity, while those of CH_3 and CH groups remain positive.
5. The peak intensity sign (positive or negative) indicates the number of protons directly bond to the carbon, while signals from quaternary carbons are always absent.
6. In Fig. 3 is reported the ^{13}C spectrum (proton decoupled) and ^{13}C DEPT spectrum of the antibiotic enduracidin. Also the ^{13}C chemical shifts depend on the solvent or solvent mixture used and in aqueous solution they are sensitive to the pH and ionic strength of the solution.

3.3. 2D NMR Methods

Usually, the signal overlap in the 1D ^1H spectrum is severe for large proteins and prevents the determination without ambiguities of the chemical shifts and spin coupling parameters from the simple one-pulse experiment. Two- and three-dimensional methods (16) have been developed to overcome these natural limitations of the 1D experiment. These methodologies are also valuable tools for structure determination of low molecular mass compounds and for conformational analysis. However, 2D measurements typically require much longer experimental time than 1D spectra.

3.3.1. COSY

1. The first 2D ^{1}H technique developed was the homonuclear shift correlation spectroscopy (*COSY*) performed in magnitude mode (17). Figure 4 shows an example of the ^{1}H COSY spectrum along with the high resolution 1D spectrum acquired for a 24-amino acid residues lantibiotic.
2. This 2D map, in addition to the chemical shifts, contains the two and three bond scalar coupling connectivities (correlations) as off-diagonal peaks which are symmetric with respect to the diagonal.
3. The complete 1D spectrum can be recognized on the diagonal from the upper right to the lower left.
4. Coupled diagonal peaks can be identified by lines through the cross peaks and are parallel to the two frequencies axes.

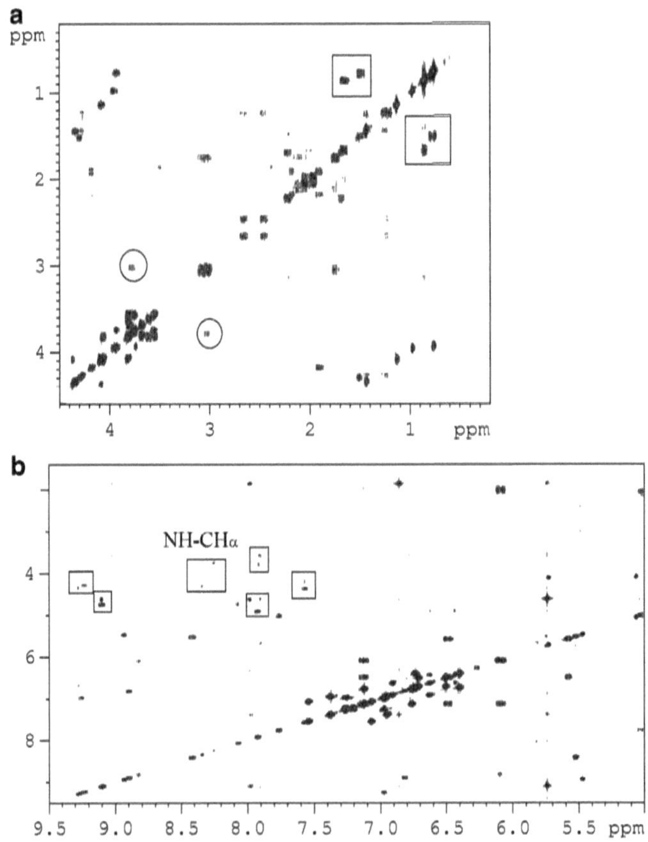

Fig. 4. ^{1}H 2D COSY Spectrum of a 24-amino acid residues lantibiotic dissolved in H_2O/D_2O (9:1) acquired at 303°K. (**a**) Aliphatic region and (**b**) amide region containing NH–CHα connectivities.

5. In Fig. 4 this is indicated above the diagonal with solid lines for the connectivities of the highest field Leu methyl resonance with its CHγ and the lowest field NH with its CHα.
6. Also the coupling among the β, β′, and α protons of lanthionine as noted by the circled cross peaks are readily discernible.
7. In general, the connectivities between the following protons *a–e* can be found in a single COSY spectrum of a protein containing unnatural amino acids:
   - CHα–CHβ of citrulline (Cit), ornithine (Orn), α-aminobutyric acid (Abu), 3,3-diphenyl-alanine (Dpa), Phe(4-NO_2), Phe(3-Cl), Phe(4-NH_2), Phe(4-Me)
   - CHδ–NH of Cit, CHδ–NH_2 of Orn, CHβ-NH_2 of α,β-diaminopropionic acid (Dpr)
   - CHα–CHβ, CHα–CHβ′ of Phe(4-NO_2), Orn, α,γ-diaminopropionic acid (Dbu), Abu, Tyr(Me)
   - aromatic ring protons of phenylglycine (Phg), Phe(4-NO_2), Phe(3-Cl), Phe(4-NH_2), Phe(4-Me)
   - backbone CHα–NH
8. The backbone CHα-NH region in H_2O spectra is often of special interest, since it presents a fingerprint of the amino acid sequence. Each amino acid residue gives a single CHα-NH cross peak (as represented in Fig. 3), while no cross peaks come from the N-terminal residues.

3.3.2. TOCSY

1. Another important ^{1}H 2D method is the homonuclear total correlation spectroscopy (TOCSY) (18). In these experiments, correlations are created between all protons within a given spin system, not just between geminal or vicinal protons as in COSY spectra (see Fig. 5 in which TOCSY and COSY correlations are depicted).

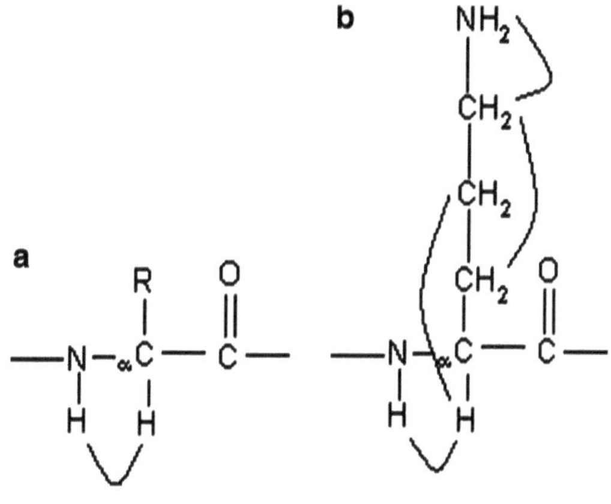

Fig. 5. Description of the ^{1}H–^{1}H vicinal correlation COSY (**a**) and the total correlation TOCSY (**b**).

2. Correlations are seen between distant protons up to five or six bonds as long as successive protons are coupled.
3. The presence of hetero-atoms, such as oxygen, usually disrupts TOCSY transfer.
4. Also the TOCSY spectrum is 2D map with the 1D spectrum on the diagonal and cross peaks that connect the diagonal correlated resonances.
5. An example of the ^{1}H TOCSY spectrum is shown in Fig. 6. It is clear that the spin connectivities from the α through β, and CH_3γ protons of α-aminobutyric acid (Abu) moiety are fully observed in this spectrum, thereby providing unequivocal identification of the Abu moiety in this protein.
6. Also hydroxyproline shows connectivities from the α through β, β′, and γ to δ, δ′ protons; 2,3-didehydrobutyrine (Dhb) shows cross peaks between β and CH_3γ.
7. Instead no correlations are observed between β and the aromatic protons of diphenylalanine.

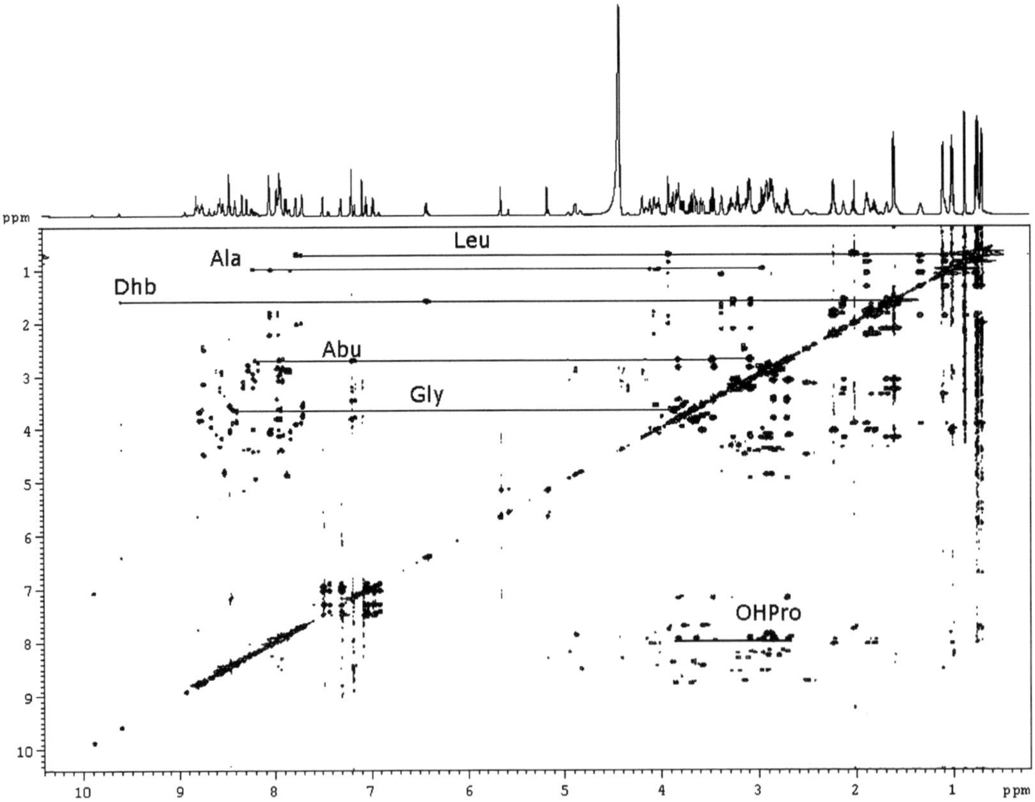

Fig. 6. ^{1}H 2D TOCSY Spectrum of a 24-amino acid residues lantibiotic dissolved in H_2O/D_2O (9:1) acquired at 303°K. The region above the diagonal *solid lines* define distinct amino acids.

8. It should be noted that COSY and TOCSY experiments provide a particularly powerful combination sufficient to identify and assign most of the natural and unnatural amino acids present in a large protein providing the primary structure of the macromolecule.

3.3.3. ¹⁵N NMR

1. The very low intrinsic sensitivity of ^{15}N NMR together with its low natural abundance (0.37%) has precluded the acquisition of a 1D ^{15}N NMR spectrum of a protein. Nevertheless, ^{15}N NMR has advantages similar to those of ^{13}C NMR in terms of wide chemical shift range and selective enrichment with ^{15}N labelled precursors.
2. The enrichment approach improves the detection limit of at least two orders of magnitude.
3. Moreover, recently developed 2D methodologies allow the observation of [^{15}N–^{1}H] heteronuclear correlation (HSQC) (19) in inverse detection using the ^{1}H sensitivity together with cold-probes (see Note 6). This 2D method detects ^{15}N via its bonded protons.
4. Example HSQC spectrum using ^{15}N natural abundance of a 24 amino acids lantibiotic is shown in Fig. 7. In this experiment, since the ^{15}N amide resonance is correlated with that of its directly bonded proton, a wider range of chemical shifts (for ^{15}N) allow to resolve the peak overlap occurring in the amide region of the normal ^{1}H spectrum.

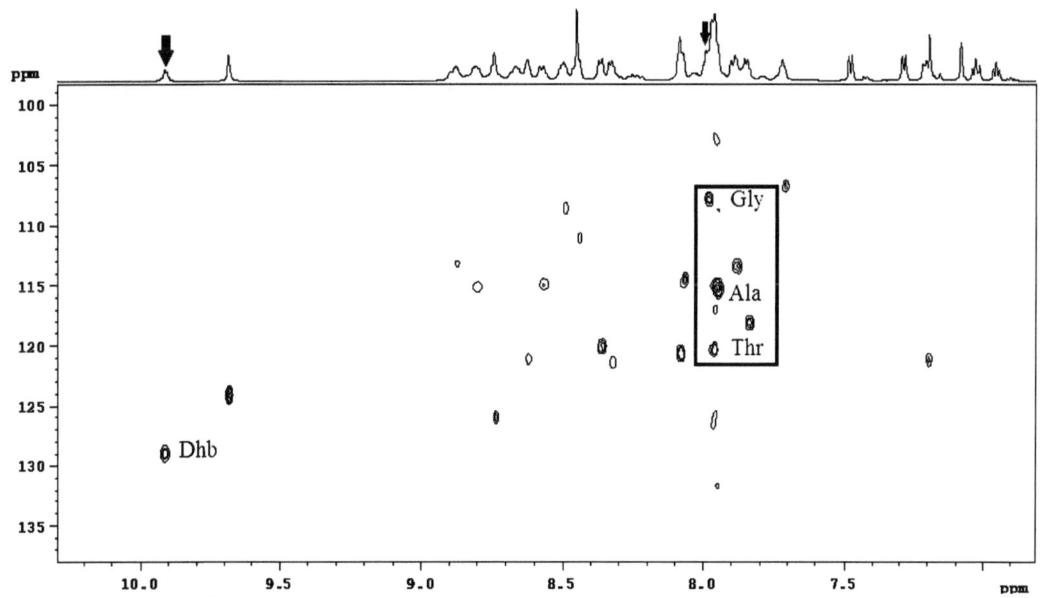

Fig. 7. [^{15}N–^{1}H] HSQC Spectrum of a 24 amino acid residues lantibiotic dissolved in H_2O/D_2O (9:1) acquired at 303°K. The solid line shows the separation in the ^{15}N dimension of overlapped protons.

5. If the protein is folded, the peaks are usually well dispersed, and most of the individual amide signals can be distinguished.
6. Indeed, as described in Fig. 6, it is possible to separate in the ^{15}N dimension the amide resonance of alanine from the NH of threonine and glycine.

3.3.4. ^{1}H NOESY

1. Sequence-specific assignment of the amino acid spin system to a specific residue in the protein sequence relies on short-range dipolar interaction of spins (the NOE) observed in the NOESY spectra (20).
2. The intensity of the NOE is in first approximation proportional to $1/r^6$, with r being the distance between the two protons, and falls off very rapidly increasing their distance. Consequently, NOEs are observed only for proton pairs separated by 5 Å.
3. The neighbouring residues are inherently close in space, so the assignments can be made by searching the cross peaks between the CHα, CHβ, or amide proton of (i) residue and the amide proton of the adjacent (i+1) residue in the peptide sequence.
4. This step of the analysis involves the identification of sequential NOE effects beginning at the N-terminal amino acid and continuing to the C-terminal residue.
5. Nowadays, this sequential assignment is automated for small size proteins with natural amino acid residues (21).
6. In practice, breaks in the assignment are frequently observed due to several reasons.
7. First, peaks overlap prevents unambiguous residue assignment; second, each hydroxyproline and each thiazolidine residue gives a break; thirdly, multiple unnatural amino acids are not recognized. Consequently, the sequential assignment process of a protein containing unnatural amino acids, is performed in shorter peptide segments following the traditional procedure.
8. In Fig. 8 is reported the NOESY spectrum of enduracidin together with the sequential assignment of the 17 amino acids. Moreover, NOE effects are not restricted only to adjacent amino acids in the sequence, but these cross peaks can be used to identify regions of a defined secondary structure (see Note 7).

3.4. Experimental Protocols

3.4.1. Preparation of the NMR Sample

1. Prepare a stock solution of H_2O/D_2O (9:1) and an acid solution HCl (0.1 N).
2. Dissolve the protein in 400 μL of the H_2O/D_2O solution (conc. 1 mM), spinning the sample for a few seconds in the micro-centrifuge.
3. Measure the pH of the sample. If the solution is clear, run the 1D proton spectrum.

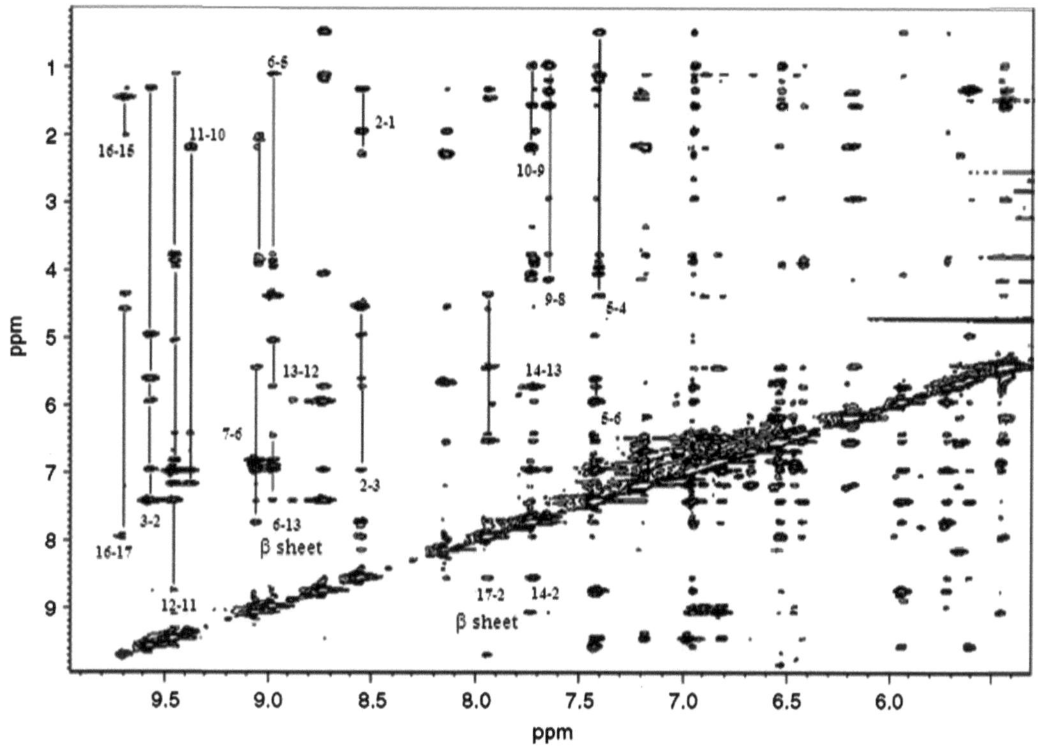

Fig. 8. ^{1}H 2D NOESY Spectrum (amide expanded region) of enduracidin in H_2O-DMSO-*d6* (4:1) acquired at 303°K. Sequential inter-residue NOE contacts are also shown.

4. Adjust the pH to the value of 3–4 adding slowly a few microliters of the acid solution.
5. Spin the sample in the micro-centrifuge and run the 1D proton spectrum.
6. If the solution is not clear and a precipitate appears, add a co-solvent (trifluoroethanol) at 10% concentration.
7. Run the ^{1}H spectrum and compare with previous one, look for a better resolution of the signals as evidence for decreased aggregation.
8. Repeat step 7 increasing the temperature in step of 5°C up to 40°C.
9. If the solution is still not clear, increase the ionic strength by adding a small amount of highly concentrated buffered salt solution (5 M NaCl); repeat steps 7 and 8 until there is no further improvement in the spectrum.
10. Make up a fresh protein solution and dissolve in SDS micelles according to (22).

3.4.2. 2D NMR Analysis of a Protein Sample

1. Acquire the ^{1}H spectrum using a water suppression sequence (23) with 32–64 scans.
2. For highly concentrated samples, acquire the ^{13}C spectrum overnight.
3. Analyse the proton and carbon spectra searching for the chemical shifts characteristic of peculiar amino acids (see Table 1).
4. Acquire the ^{1}H 2D COSY spectrum with 48 scans, sweep width of 10 ppm, 1,024 increments in the t_1 dimension and 4,096 complex data point in t_2.
5. Acquire the ^{1}H 2D TOCSY spectrum using the same conditions described at step 4 and a mixing time of 70 ms.
6. Analyse the 2D data observing the direct and long-range correlations between protons. Many of the unnatural amino acids have unique spin system topologies and are easily distinguishable in the TOCSY spectrum. The protein amino acid composition is determined in this step.
7. If the signal overlap in the proton spectrum is severe, repeat at different temperature or pH.

3.4.3. 2D NMR Sequential Assignment

1. Acquire the ^{1}H 2D NOESY spectra with 48 scans, sweep width of 10 ppm, 1,024 increments in the t_1 dimension and 4,096 complex data point in t_2 and a mixing time of 120, 180, and 300 ms.
2. Expand the amide region (6.5–10 ppm) of the spectrum in the F_2 dimension while retaining the entire ^{1}H spectrum in the F_1 dimension.
3. Identify intra-residue NOE effects.
4. Starting from the most down field NH signal, locate the NOE cross peak with its CHα and its CHβ in the spectrum. This is the residue *i*.
5. Search along the NH(*i*) shift for an inter residue cross peak with the CHα ($i-1$), than the residue $i-1$ can be assigned.
6. Look along the CHα(*i*) shift for strong NOE with the NH belonging to the residue $i+1$. When a cross peak is found, the residue $i+1$ in the sequence can be assigned.
7. Find along the NH($i+1$) shift of this peak for an intra residue NH–CHα NOE.
8. Confirm the ($i-1$) and ($i+1$) residues just found searching for CHβ(*i*)–NH($i-1$) and CHβ($i+1$)–NH(*i*) NOEs.
9. Repeat steps 6 and 8 for the next ($i+1$) residue.
10. In the case of large proteins, spectral peaks are usually overlapped also in the amide region. This seriously hinders amino acid sequencing, then several experiments must be performed (see Note 8).

4. Notes

1. The addition of organic solvents allows to perform the NMR experiments at temperatures below 0°C. Methanol and DMSO are used as co-solvents to extend the fluid range to −50°C. However, at these low temperatures the viscosity of the solution seriously affects the spectral resolution.
2. The protein solution must be stable during the course of all the NMR experiments. To check the sample stability a quick 1D spectrum is acquired after a long experimental session and compared with the starting data.
3. In order to prevent proton contamination of the D_2O solution, after referencing the pH electrode in the H_2O buffer, rinse it with D_2O and then soak it in a vial containing D_2O. ^2HCl solutions in ^2H_2O arc commercially available at several concentrations.
4. The chemical shift reference can be either external or internal. The external reference is contained in a capillary within the NMR tube and does not contact the protein sample directly. The internal reference is dissolved directly in the sample buffer and should be biochemically inert; also the residual water peak can be used and the chemical shift is assigned at 4.75 ppm at room temperature. This value is very sensitive to temperature changes, varying of 0.005 ppm/°C.
5. When observing ^{13}C or ^{15}N nuclei, the broadband proton decoupling sequence must be carefully chosen. WALTZ-16 gives the better quality decoupling for biological applications without causing excessive sample heating.
6. [^{13}C–^1H] Heteronuclear correlation (HSQC) experiment can also be performed for the ^{13}C-labeled proteins. The resulting spectrum is 2D map with one axis for the ^1H and the other for the ^{13}C nuclei. This spectrum contains a cross-peak for each unique proton attached to the carbon considered. Thus, if the chemical shift of a specific proton is known, the chemical shift of the coupled carbon can be determined, and vice versa.
7. [^1H–^1H] NOESY cross peaks are used to identify the protein secondary structure. Indeed, the α helix is characterized by close distance between residues i and $(i+3)$ and between residues i and $(i+4)$. In the 3_{10} helix, short distances are observed between residues i and $(i+2)$, i and $(i+3)$. Also the β sheets are characterized by short interstrand ^1H–^1H distances (see Fig. 8).
8. When an overlap occurs, a list of adjacent NOE is difficult to compile, then performing 2D experiments at different temperatures (10°C above or below the original temperature)

causes a change in the NH chemical shift and leads to more resolved spectra. This new set of COSY and NOESY spectra can be analysed following protocol 3.4.3. The residue assignment obtained at different temperatures can be compared in order to resolve ambiguities and to confirm the obtained peptide structure.

References

1. Jones, D. H., et al. (2010) Site-specific labeling of proteins with NMR-active unnatural amino acids. *J Biomol NMR* 46, 89–100.
2. England, P. M. (2004) Unnatural amino acid mutagenesis: a precise tool for probing protein structure and function. *Biochemistry* 43, 11623–11629.
3. Cellitti, S. E., et al. (2008) *In vivo* incorporation of unnatural amino acids to probe structure, dynamics, and ligand binding in a large protein by NMR. *J Am Chem Soc* 130, 9268–9281.
4. Lo, M. C., et al. (2001) A new structure for the substrate-binding antibiotic ramoplanin. *J Am Chem Soc* 123, 8640–8641.
5. Hsu, S., et al. (2004) The nisin-lipid II complex reveals a pyrophosphate cage that provides a blueprint for novel antibiotics. *Nature Struct Mol. Biology* 11, 963–967.
6. Kettenring, J. K., et al. (1990) Sequence determination of actagardine, a novel lantibiotic, by homonuclear 2D NMR spectroscopy. *Antibiotics* 9, 1082–1088.
7. Kurz, M., and Guba, W. (1996) 3D structure of ramoplanin: a potent inhibitor of bacterial cell wall synthesis. *Biochemistry* 35, 12570–12575.
8. Pearce, C. M., and Williams, D. H. (1995) Complete assignment of the ^{13}C NMR spectrum of vancomycin. *J Chem Soc Perkin Trans 2*, 153–157.
9. Fan, T. W.-M. (1996) Metabolite profiling by one- and two-dimensional NMR analysis of complex mixtures. *Prog NMR Spectrosc* 28, 161–219.
10. Roberts G. C. K. (1995) NMR of macromolecules, a practical approach. Oxford University Press.
11. Wuthrich, K. (1986) NMR of proteins and nucleic acids. John Wiley & Sons, New York.
12. Levy, G. C., and Lichter, R. L. (1979) Nitrogen-15 nuclear magnetic resonance spectroscopy. John Wiley & Sons, New York.
13. Castiglione, F., et al. (2005) Structure elucidation and 3D solution conformation of the antibiotic enduracidin determined by NMR spectroscopy and molecular dynamics. *Magn Reson Chemistry* 43, 603–610.
14. Kalinowski, H.-O., Berger, S., and Braun, S. (1988) Carbon-13 NMR spectroscopy. Wiley Chichester.
15. Schenker, K. V., and Philipsborn, W. (1986) ^{13}C editing distortionless enhancement by polarization transfer techniques. *J Magn Reson* 66, 219–229.
16. Clore, G. M., and Gronenborg, A. M. (1991) Application of three and four dimensional heteronuclear NMR spectroscopy to protein structure determination. *Prog NMR Spectrosc* 23, 43–92.
17. States, D. J., Haberkorn, R. A., and Ruben, D. J. (1982) A two-dimensional nuclear Overhauser experiment with pure absorption phase in four quadrants. *J Magn Reson* 48, 286–292.
18. Bax, A., and Davis, D. G. (1985) MLEV-17-based two-dimensional homonuclear magnetization transfer spectroscopy. *J Magn Reson* 65, 355–360.
19. A Otting, G., Messerle, B. A., and Soler, L. P., (1996) ^{1}H detected gradient enhanced ^{15}N and ^{13}C NMR experiments for the measurements of small heteronuclear coupling constants and isotopic shifts. *J Am Chem Soc* 118, 5096–5102.
20. Neuhaus, D., and Williamson, M. (1989) The Nuclear Overhauser effect in structural and conformational analysis, VCH Weinheim, 253–305.
21. Oezguen, N., et al. (2002) Automated assignment and 3D structure calculation using combination of 2D homonuclear and 3D heteronuclear NOESY spectra. *J Biomol NMR* 22, 249–263.
22. Lee, K. H., Fitton, J. E., and Wuthrich, K. (1987) Nuclear magnetic resonance investigation of the conformation of d-haemolysin bound to dodecylphosphocholine micelles. *Biochim Biophys Acta* 911, 144–150.
23. Hwang, T.-L., and Shaka, A. J. (1995) Water suppression that works, excitation, sculpting using arbitrary wave-forms and pulsed field gradients. *J Magn Reson Ser A* 112, 275–279.

Chapter 8

Site-Specific Incorporation of Unnatural Amino Acids as Probes for Protein Conformational Changes

Jennifer C. Peeler and Ryan A. Mehl

Abstract

Site-specific *in vivo* incorporation of unnatural amino acids provides powerful tools for the study of protein interaction and dynamics. Here, we provide a protocol for the incorporation of six such UAA probes into a GFP reporter system, expressed in *Escherichia coli* from both arabinose and lactose-inducible expression plasmids using an autoinduction media.

Key words: Unnatural amino acid probe, Autoinduction, GFP reporter, Amber codon suppression, pDule2 plasmid

1. Introduction

The site-specific incorporation of unnatural amino acid (UAA) probes into full-length proteins provides a powerful tool for a variety of protein studies, though improvements in the ease and yield of incorporation must be made for widespread utility. Site-specific incorporation of UAAs can be achieved *in vivo* through amber codon suppression with the expression of an orthogonal aminoacyl-tRNA synthetase (RS)/tRNA pair (1, 2). A number of RSs have been evolved to incorporate UAAs that act as probes of protein interaction and dynamics, but many investigators have experienced limitations in terms of protein production. The UAAs para-trifluoromethylphenylalanine (ptfmF) and para-trifluoromethyl-meta-fluorophenylalanine (ptfmmfF) have both been utilized as ^{19}F-NMR probes for monitoring conformational changes and can be incorporated using the ptfmF A65V/S158A RS (Fig. 1a) (3–6). The UAAs para-cyanophenylalanine (pCNF), para-ethynylphenylalanine (pENF), and para-azidophenylalanine (pN_3F)

Loredano Pollegioni and Stefano Servi (eds.), *Unnatural Amino Acids: Methods and Protocols*,
Methods in Molecular Biology, vol. 794, DOI 10.1007/978-1-61779-331-8_8, © Springer Science+Business Media, LLC 2012

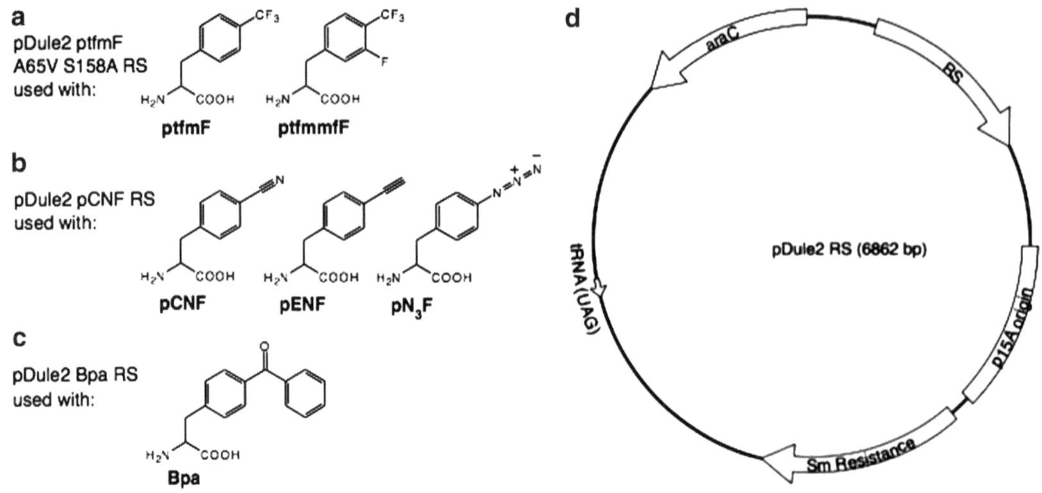

Fig. 1. UAA probes and the pDule2 plasmid used for incorporation. (**a**) UAAs para-trifluoromethylphenylalanine (ptfmF) and para-trifluoromethyl-meta-fluorophenylalanine (ptfmmfF) should be incorporated using the pDule2 ptfmF A65V S158A RS plasmid. (**b**) UAAs para-cyanophenylalanine (pCNF), para-ethynylphenylalanine (pENF), and para-azidophenylalanine (pN_3F) should be incorporated using the pDule2 pCNF RS plasmid. (**c**) UAA benzoylphenylalanine (Bpa) should be incorporated using the pDule2 Bpa RS plasmid. (**d**) The general pDule2 plasmid contains the RS open reading frame (bases 194–1114) and $tRNA_{UAG}$ (bases 4855–4947) under constitutively active expression (lpp promoter) as well as the p15A origin of replication (bases 1116–2436), araC open reading frame (bases 5680–6858), and *aadA* gene (bases 2664–3455) for spectinomycin/streptomycin resistance.

have been used as infrared spectroscopy probes; pCNF and pENF have been shown to work as FRET pairs with naturally occurring tryptophan, and pN_3F provides copper(I) catalized Huisgen cycloaddition chemistry capability (7–9). All three of these UAAs can be incorporated by the pCNF RS with high efficiency and fidelity (Fig. 1b). Additionally, the UAA benzoylphenylalanine (Bpa) has been widely used as a photocrosslinking probe and can be incorporated into proteins using the Bpa RS (Fig. 1c) (10, 11).

This protocol provides a number of improvements to previous approaches to UAA incorporation through amber codon suppression. First, in order to improve protein yield, this protocol calls for expression in *Escherichia coli* BL21-ai cells (Invitrogen). These cells are optimized for protein production by the removal of the *lon* and OmpT proteases. However, their tetracycline resistance prevents their use with the pDule plasmid (also tetracycline resistant) that has previously been used to express the RS gene and $tRNA_{UAG}$ (3, 6). This work reports the use of a pDule2 plasmid, which contains the RS open reading frame and $tRNA_{UAG}$ under constitutively active expression as well as the p15A origin of replication but with the *aadA* gene conferring spectinomycin/streptomycin resistance instead of the tetracycline resistance marker (Fig. 1d). Additionally, controlled autoinduction medium with an N-Z-Amine (bovine casein enzymatic hydrolysate) amino acid source has been developed to prevent the need for manual induction. This medium induces expression with both lactose and arabinose and has been shown to work with both pBad and pET overexpression plasmids (6, 12).

This work presents the incorporation of six different UAA probes in the GFP reporter gene at residue Asn150, expressed in the pBad plasmid. This reporter system allows for qualitative and quantitative evaluation of expression of full-length protein without purification or SDS-PAGE analysis. While this machinery can be used to incorporate UAAs in nearly any protein and at nearly any residue, we recommend the use of the GFP reporter as a control expression since it allows for facile visual confirmation of expression times for the autoinducition media and RS/tRNA machinery effectiveness (3, 10).

2. Materials

All water purity should be 18 MΩ grade. All media stock components should be autoclaved and stored at room temperature unless otherwise noted.

2.1. Media Component Stocks (12)

1. 50× 5052 solution: 125 g glycerol, 12.5 g glucose, 50 g α-lactose, 365 mL water (see Note 1).
2. $MgSO_4$,1 M: 60.18 g $MgSO_4$, 500 mL water.
3. Glucose, 40% (w/w): 20 g D-(+)-glucose, 30 mL water.
4. Arabinose, 20% (w/w): 2 g L-arabinose, 8 mL water. This solution should be sterile filtered and stored at −20°C.
5. 25× M: 44.36 g Na_2HPO_4, 42.55 g KH_2PO_4, 33.45 g NH_4Cl, 8.9 g Na_2SO_4, 500 mL water (see Note 2).
6. ZY media: 5 g N-Z-amine AS (bovine casein enzymatic hydrolysate), 2.5 g yeast extract, 500 mL water.
7. 5,000× trace metals: Individual stock solutions of metals should be made in 30 mL aliquots as directed in Table 1. Then, combined for a 50-mL final solution as shown in Table 1.
8. NaOH, 8 M: Dissolve 320 g NaOH in 1 L of water.

2.2. Cell Lines, Plasmids, and UAAs

1. *E. coli* BL21-ai cell line (Invitrogen) (see Note 3).
2. The UAAs verified for this protocol include the following UAAs from the following vendors: ptfmF (Peptech), ptfmmfF (synthesized), pCNF (Bachem), pENF (synthesized as HCl salt), pN_3F (Bachem), Bpa (Peptech). Structures of these UAAs are shown in Fig. 1a–c.
3. pDule2 RS plasmid: Contains the RS open reading frame for the UAA of interest as well as the UAG tRNA, p15A origin of

Table 1
Preparation of individual trace metal stock solutions and 5,000× stock solution

Amount for 30-mL stock solution	Amount of 30-mL stock for 50-mL (5,000×) solution	1× Media concentration (μM)
$CaCl_2 \cdot 2H_2O$ (8.82 g)	500 μL	4
$MnCl_2 \cdot 4H_2O$ (5.93 g)	500 μL	2
$ZnSO_4 \cdot 7H_2O$ (8.62 g)	500 μL	2
$CoCl_2 \cdot 6H_2O$ (1.32 g)	500 μL	0.4
$CuCl_2$ (807 mg)	500 μL	0.4
$NiCl_2$ (777 mg)	500 μL	0.4
$Na_2MoO_4 \cdot 2H_2O$ (1.45 g)	500 μL	0.4
Na_2SeO_3 (1.03 g)	500 μL	0.4
H_3BO_3 (371 mg)	500 μL	0.4
$FeCl_3$ (486 mg)	25 mL	10

Dilute to 50 mL with sterile water

replication, *aadA* gene for spectinomycin/streptomycin resistance, and araC gene. The RS required for each UAA is listed in Fig. 1a–c and the plasmid map for pDule2RS is shown in Fig. 1d (see Notes 4 and 5).

4. Overexpression plasmid containing GFP gene. Either pBad (Invitrogen) or pET28a (Novagen) is recommended (see Note 6).
5. QuikChange Site-Directed Mutagenesis Kit (Stratagene) (see Note 7).
6. Electroporator.
7. SOC rescue media: 9.75 mL 2× YT, 200 μL 18% glucose, 50 μL 2 M $MgCl_2$.

2.3. Expression

1. Selective media: LB agar, antibiotics (ampicillin, kanamycin, spectinomycin) (see Note 8).
2. 15-mL culture tubes.
3. Plastic baffled flasks with membrane lid for aeration (see Note 9).

2.4. Evaluation

1. BD Talon Metal Affinity Resin (BD Sciences); lysis buffer: 50 mM sodium phosphate, 300 mM sodium chloride,

0.25 mg/mL lysozyme (pH 8); elution buffer: 50 mM sodium phosphate, 300 mM sodium chloride, 150 mM imidazole (pH 8); disposable gravity columns (see Note 10).

2. PAGEr Gold Precast Gels: 10–20%, 9 cm × 10 cm, 1 mm thick, Tris–glycine (Lonza) (see Note 11).
3. PAGE-gel rig apparatus and power supply.
4. InstantBlue Stain (Accurate Chemical & Scientific Corp.) (see Note 12).
5. PBS: 10 mM sodium phosphate, 140 mM sodium chloride (pH 7.2).
6. HORIBA Jobin Yvon FluoroMax-4.

3. Methods

3.1. Plasmid Preparation

1. Acquire a pBad vector containing the ORF for GFP or your protein of interest (see Note 13).
2. Obtain pBad-GFP-150TAG. To your gene of interest in pBad or pET vector add an amber codon (TAG) in the location that an UAA is desired (see Notes 7 and 14).
3. Generate *E. coli* cell lines combining the pBad-GFP plasmids with the pDule2 plasmid containing the gene for the RS capable of incorporating the UAA of interest (Fig. 1a–c). Two cell lines of *E. coli* BL21-ai should be produced: one containing pBad-GFP-wt (without a TAG site) and the pDule2-RS plasmid, and another containing pBad-GFP-150TAG and the pDule2-RS plasmid. A cotransformation via electroporation (2.5 kV) followed by rescue in 1 mL SOC medium for 1 h and then selection on LB agar with 100 μg/mL Amp and 100 μg/mL Spec is sufficient (see Note 8).

3.2. Expressions

1. Prepare ZY noninducing media using the medium stocks as shown in Table 2 (12). For each transformed strain, a single colony should be grown to saturation (37°C, shaking at 300 rpm) in 5 mL of noninducing media with appropriate antibiotics in 15-mL culture tubes (see Notes 15 and 16).
2. Prepare 150-mL ZY autoinduction media, sufficient for three expression trials of 50 mL each, using the media stocks as shown in Table 2, with appropriate antibiotics. Aliquot into three 250-mL baffled flasks (see Notes 9 and 17).
3. Inoculate a 50-mL aliquot of medium (in a 250-mL baffled flask) with 500 μL of saturated noninducing medium containing pBad-GFP wt and pDule2-RS for the expression of wild-type GFP.

Table 2
Preparation of 150 mL of ZY noninducing and autoinducing media

Stock solution	Noninducing media	Autoinducing media
1 M $MgSO_4$	300 μL	300 μL
25× M	6 mL	6 mL
5,000× Trace metals	30 μL	30 μL
50× 5052	–	3 mL
20% Arabinose	–	375 μL
40% Glucose	1.875 mL	–

Dilute to 150 mL with ZY stock solution

4. Resuspend enough UAA for a final expression concentration of 1 mM in 0.5 mL of sterile water. Add 8 M NaOH dropwise until the UAA has dissolved with vortexing. For the UAAs verified for this protocol, the following masses are appropriate for 50 mL expressions: ptfmF: 11.65 mg, ptfmmfF: 14.3 mg, pCNF: 9.5 mg, pENF: 11.25 mg, pN_3F: 10.3 mg, Bpa: 13.5 mg (see Notes 18 and 19).
5. Add dissolved UAA to one 50-mL aliquot of autoinduction media and inoculate with 500 μL of saturated noninducing media containing pBad-GFP-150TAG and pDule2-RS for the expression of GFP-UAA (see Note 20).
6. Inoculate the final flask of autoinduction media (without UAA) with 500 μL of saturated noninducing media containing pBad-GFP-150TAG and pDule2-RS as a negative control.
7. Allow all cultures to grow at 37°C, shaking at 300 rpm for 30 h before harvesting via centrifugation at 5,525 × *g* for 10 min (see Notes 21 and 22).

3.3. Expression Evaluation

1. Measure the fluorescence of culture conducted with the GFP control. Dilute culture 100× in PBS. Excite the culture at 488 nm and monitor emission from 500 to 520 nm.

   Fluorescence can be compared between different expression trials by graphing the sum of fluorescence counts between 500 and 520 nm.
2. Purify protein as necessary for further experimentation, including the negative control without UAA added (see Note 10).
3. SDS-PAGE analysis should be used to confirm sufficient expression of GFP-UAA as well as a lack of expression of GFP-TAG without UAA added, as shown in Fig. 2 (see Note 23).

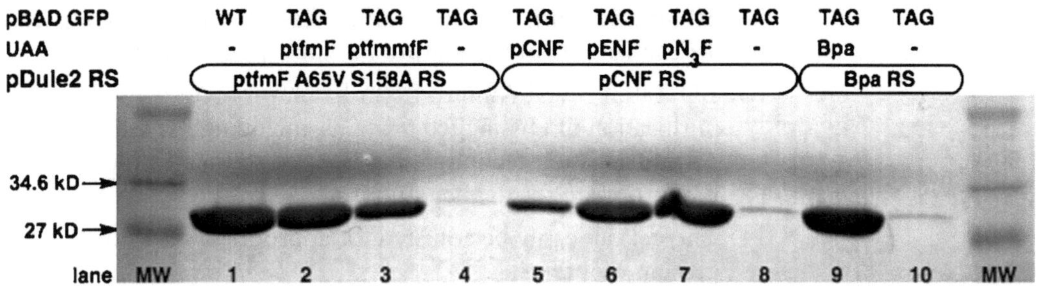

Fig. 2. SDS-PAGE results of purified GFP-wt, GFP-UAA, and GFP-TAG without UAA. All proteins were expressed using the pBAD vector and purified using the BD Talon batch method, binding with 500-μL bed volume of resin and eluting with 2 mL of elution buffer. All samples were then diluted 20-fold before SDS-PAGE using a 10–20% Tris–glycine gel. The gel was stained using InstantBlue Stain. *Lane* 1 shows GFP-wt (27 kDa) expression in the presence of pDule2 tfmF A65V S158A RS. *Lanes* 2–3, 5–7, and 9 show GFP containing each of the 6 UAA probes, expressed with their appropriate pDule2 RS. *Lanes* 4, 8, and 10 show negative controls for each pDule2 RS, expressed in the absence of UAA.

4. Notes

1. Heating at 55°C is required for complete dissolution, but the solution will then remain dissolved at room temperature for years.
2. Previous literature recommends using a 50× M stock, but the 25× solution is much less likely to recrystalize after storage at room temperature.
3. Previous reports used the DH10B cell line. While both can be used, BL21-ai is better tailored for protein production and needs to be used with pET Vectors.
4. Previous reports used the pDule plasmid with Tet resistance. However, Tet resistance made pDule incompatible with the Tet-resistant BL21-ai cell line and therefore Sm-resistant pDule2 was developed.
5. Available from Ryan A. Mehl (ryan.mehl@fandm.edu).
6. Both pBad and pET28a GFP constructs are available from Ryan A. Mehl (ryan.mehl@fandm.edu).
7. Any method of site directed mutagenesis is acceptable, but the Stratagene QuikChange kit has previously been successfully used.
8. When using the pDule2 plasmid the appropriate medium concentration of spectinomycin is 100 μg/mL. For pBad vectors, an ampicillin concentration of 100 μg/mL should be used; for pET28a vectors, a kanamycin concentration of 25 μg/mL should be used.
9. The appropriate size of the baffled flask is dependent upon the scale of expression. Generally, growths are performed between

50 and 500 mL and use baffled flasks with volume at least 4× the volume of the culture for appropriate aeration.

10. BD Talon Metal Affinity Resin batch protocol is used for our purification of His-tagged proteins, including the GFP control. Any purification technique can be used so that purified protein is appropriate for further study.
11. The gel can easily be adjusted for appropriate resolution of the protein of interest.
12. Any gel staining method is acceptable, but gels presented in this work were stained with InstantBlue.
13. Previous reports suggest that only pBad vectors were usable. However, the use of the BL21-ai cell line and the ZY media allow the T7-controlled pET vectors to work as well. Expressions from the pET vector exhibit higher yields than those from pBad for wt proteins, but lower yields for GFP-UAA expressions (Fig. 3). Additionally, the extent of difference in expression levels for pBad versus pET vectors seems to depend upon the UAA probe.

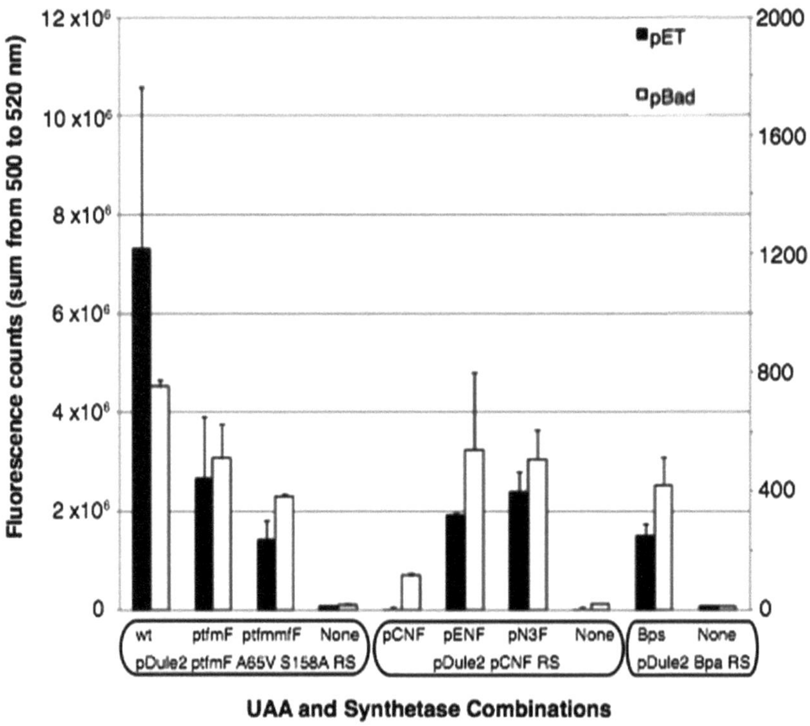

Fig. 3. Comparison of GFP expression for pBad versus pET28a vectors for UAA probes. Expression cultures of 5 mL were performed for GFP-wt and GFP-UAA for each of the six probes, as well as negative controls for each of the pDule2 RSs. After 33 h of expression, crude cell expressions were diluted 100-fold in PBS before fluorescence analysis. Samples were excited at 488 nm, and fluorescence was recorded from 500 to 520 nm using an HORIBA Jobin Yvon FluoroMax-4. All expression trials were performed in duplicate, and error bars represent standard deviation.

14. In general, the sites that are most permissive to tyrosine-based UAA probes are aromatic residues. However, predicting permissivity of a residue is quite difficult, and it is recommended that a variety of residues be tested in parallel.
15. 12 h is generally sufficient to reach saturation ($OD_{600} > 1$).
16. The noninducing medium contains sufficient levels of glucose and no lactose or arabinose in order to prevent protein induction.
17. Autoinduction medium provides an appropriate mixture of sugars such that glucose will prevent premature induction but will be depleted at mid-log growth at which point lactose or arabinose will induce overexpression without manual addition.
18. Excessive addition of 8 M NaOH can cause reduced cell growth, therefore dropwise addition of base to the UAA in water with mixing is necessary to minimize the amount needed.
19. In order to enhance stability, we recommend storing all UAA as solids and only dissolving immediately before addition to media.
20. If poor cell growth in the presence of UAA is seen and UAA toxicity is suspected, inoculate autoinduction medium first, and then add UAA after 30–60 min of growth.
21. We observe increase in expression levels up to 30 h and then a plateau from 30 to 40 h for GFP. Any growth (expression) length between 30 and 40 h will be acceptable, though ideal expression times may vary by protein.
22. If using the GFP control plasmids, the fluorescent intensity of GFP expression can be used visually to evaluate successful induction qualitatively.
23. Background incorporation will be dictated by the RS fidelity and the medium and should be consistent from one expression protein to another.

References

1. Liu C C, Schultz P G (2010) Adding new chemistries to the genetic code. Annu Rev Biochem 79, 413–444.
2. Xie J, Schultz P G (2005) An expanding genetic code. Methods 36, 227–238.
3. Miyake-Stoner S J et al. (2010) Generating permissive site-specific unnatural aminoacyl-tRNA synthetases. Biochemistry 49, 1667–1677.
4. Li C G et al. (2010) Protein F-19 NMR in *Escherichia coli*. J Am Chem Soc 132, 321–327.
5. Jackson J C, Hammill J T, Mehl R A (2007) Site-specific incorporation of a F-19-amino acid into proteins as an NMR probe for characterizing protein structure and reactivity. J Am Chem Soc129, 1160–1166.
6. Hammill J T et al. (2007) Preparation of site-specifically labeled fluorinated proteins for F-19-NMR structural characterization. Nat Protoc 2, 2601–2607.
7. Taskent-Sezgin H et al. (2009) Interpretation of p-cyanophenylalanine fluorescence in proteins in terms of solvent exposure and contribution of side-chain quenchers: a combined fluorescence, IR and molecular dynamics study. Biochemistry 48, 9040–9046.

8. Miyake-Stoner S J et al. (2009) Probing protein folding using site-specifically encoded unnatural amino acids as FRET Donors with tryptophan. Biochemistry 48, 5953–5962.
9. Chin J W et al. (2002) Addition of p-Azido-L-phenylalanine to the genetic code of *Escherichia coli*. J Am Chem Soc 124, 9026–9027.
10. Stokes A L et al. (2009) Enhancing the utility of unnatural amino acid synthetases by manipulating broad substrate specificity. Mol Biosyst 5, 1032–1038.
11. Chin J W et al. (2002) Addition of a photocrosslinking amino acid to the genetic code of *Escherichia coli*. Proc Natl Acad Sci USA 99, 11020–11024.
12. Studier F W (2005) Protein production by auto-induction in high density shaking cultures. Protein Express Purif 41, 207–234.

Chapter 9

Application of Unnatural Amino Acids to the *De Novo* Design of Selective Antibiotic Peptides

Rickey P. Hicks and Amanda L. Russell

Abstract

Because of their unique mechanism of cytotoxicity against bacteria and other microorganisms, antimicrobial peptides have received a great deal of attention as possible therapeutic agents. Incorporation of unnatural amino acids into the peptide sequences has the potential to improve the organism selectivity and potency of these peptides as well as increase their metabolic stability. This protocol outlines the logic used to selectively incorporate unnatural amino acid into a peptide sequence in an attempt to obtain peptides with increased therapeutic potential as antibiotic agents.

Key words: Antimicrobial peptides, Unnatural amino acids, Organism selectivity, *De novo*

1. Introduction

Most chemists and biologists are very familiar with the use and function of the 20 naturally occurring RNA-encoded amino acids. However, the incorporation of unnatural amino acids into peptides and proteins is making a major impact on several areas of biochemical research and are becoming more common place (1–4). Research involving the incorporation of unnatural amino acids has focused on a wide array of applications from drug discovery (1) to site(s)-specific incorporation of unnatural amino into proteins (2–4). This protocol focuses on the rationale and logic used to incorporate unnatural amino acids into the design of novel antimicrobial peptides.

As a defense mechanism against invading microorganisms, almost every class of living organism has developed its own family of antimicrobial peptides (AMPs) (5, 6). These evolutionary AMPs have been isolated and characterized from almost all known classes

Loredano Pollegioni and Stefano Servi (eds.), *Unnatural Amino Acids: Methods and Protocols*,
Methods in Molecular Biology, vol. 794, DOI 10.1007/978-1-61779-331-8_9, © Springer Science+Business Media, LLC 2012

of organisms including humans (7), amphibians (8), insects, mammals, birds, fish, and plants (6). In an effort to improve the therapeutic usefulness of these compounds, a large number of analogs of naturally occurring peptides, as well as novel peptides, have been synthesized (9–14). As of January 2009, more than 1,330 (15) natural and synthetic antimicrobial peptides have been characterized and have exhibited a wide range of antimicrobial activity (16). AMPs are generally small (10–50 amino acid residues) and highly positively charged (+3 to +9) (17) amphipathic molecules with well-defined regions of hydrophobicity and hydrophilicity (18, 19). Currently, AMPs are divided into two mechanistic classes, membrane-disruptors and non-membrane-disruptors (20, 21). Membrane-disrupting peptides can be divided even further into several different structural classes, but for the purpose of this discussion we will focus on linear amphipathic peptides. These peptides normally exhibit characteristics of a random coil conformation in aqueous solutions or organic solvents; however, upon binding to micelles or membranes they adopt an ordered amphipathic secondary structure (22).

Membrane-disruptors can be divided into two sub-classifications based on biological activity: cell selective and non-selective (23). Cell-selective AMPs, as the name implies, exhibit potent activity against bacteria cells while remaining relatively non-toxic against mammalian cells. Non-selective AMPs exhibit cytotoxic activity against bacteria as well as mammalian cells. A large number of investigations have focused on the development of cell selective AMPs (23–26) with varying degrees of success. In general, hemolytic activity (the ability to kill or lysis red blood cells) correlates with high hydrophobicity, amphipathicity, and helicity (18, 27). Selective incorporation of D-amino acids, proline, peptoids, and amino acids with differing hydrophobicity can reduce the resulting hemolytic activity by inducing changes in the secondary structure that will affect the amphipathicity and α-helicity of the peptide (23, 26, 28).

The selectivity of AMPs for prokaryotic versus eukaryotic cells is believed to be derived from the differences in the chemical compositions of their respective cell membranes (6, 18). Bacteria cells contain a high percentage of negatively charged phospholipids while mammalian cells contain a much higher concentration of zwitterionic phospholipids (29). Other differences between the two types of cells include the incorporation of sterols, lipopolysaccharides, peptidoglycan, etc., into the membrane composition of bacteria cell which induces changes in membrane structure (5), transmembrane potential, and polarizability (5, 6, 30). The lipid bilayer of Gram-positive bacteria is covered by a porous layer of peptidoglycan, while the structure of Gram-negative bacteria is more complex with two lipid membranes containing in addition to normal lipids, lipopolysaccharides, and porins (6, 30). The outer membrane of mycobacteria is the most complex of the three types consisting of a very thick mycolate-rich outer coat (31) which is very difficult to penetrate. There is an increasing amount of evidence in the literature supporting the concept that the selectivity and potency of a specific AMP is determined, in a large measure, by the chemical composition of the target cell's membrane (20, 32). Thus it is reasonable to postulate that the interactions that occur between a membrane's physicochemical surface properties and the physicochemical surface properties of the AMP will define organism selectivity (6, 20, 30, 33). Antimicrobial peptides have been extensively investigated as therapeutic agents to possibly replace current antibiotics because of their novel mechanism of killing bacteria (13, 14, 18, 34). An international health care crisis has arisen due to the dramatic and continued evolution of drug-resistant strains of bacteria (35, 36). This ever-evolving increase in resistance has been the driving force behind the intensive research effort to develop new classes of compounds that exhibit novel mechanisms of antibacterial activity (14, 35, 37, 38). Both natural and synthetic antimicrobial peptides exhibit a very high potential as new therapeutic agents because of their novel mechanisms of antibiotic activity coupled with the difficulty of bacteria to develop resistance to them (14, 18, 39, 40). The most promising applications of AMPs thus far are focused on their clinical development as topical antibiotics (40).

The use of unnatural amino acids offers several major advantages over the 20 naturally occurring amino acids, including limited conformational flexibility (1, 41–45), improved enzymatic stability (46), pharmacodynamics (1), as well as control over charge distribution (47–49) and electron density (3) in the development of novel AMPs. Having such control over these properties of unnatural amino acids will allow for the design and synthesis of AMPs that will interact in specific ways with bacteria membranes. In the case of our research, AMPs have been designed that interact with high potency and selectivity toward targeted bacteria strains (47, 50).

2. Materials

2.1. POPC and POPC/POPG Liposomes

1. 1-Palmitoyl-2-oleoyl-*sn*-glycero-3-phosphocholine (POPC) and 1-palmitoyl-2-oleoyl-*sn*-glycero-3-(phospho-*rac*-(1-glycerol)) (sodium salt) (POPG) are available from Avanti Polar Lipid Inc.
2. Mini-Extruder and the 100-nm pore size polycarbonate membrane filters are available from Avanti Polar Lipids.
3. 40 mM Sodium phosphate buffer, pH 6.8, is available from Sigma-Aldrich.

2.2. Calcein Leakage Assays

1. 70 mM High purity calcein (Invitrogen Incorporated) in 10 mM Bis-Tris buffer, 150 mM NaCl, 1 mM EDTA, pH 7.2. The pH is corrected using 1 mM NaOH. All chemicals except calcein are available from Sigma-Aldrich.
2. Fluorescence spectrometer and 1 cm quartz cuvette.
3. G50-Superfine Sephadex gel filtration column; eluent 10 mM Bis-Tris buffer, 150 mM NaCl, 1 mM EDTA, pH 7.2.
4. 1.0 mM Stock solution of peptide in buffer (10 mM Bis-Tris, 150 mM NaCl, 1 mM EDTA, pH 7.2). Small aliquots of 10–50 μL peptide solution are added to the cuvette containing calcein-encapsulated liposomes (36.6 μM lipid concentration) to give an overall peptide concentration of 4–20 μM.

2.3. Circular Dichroism

1. CD Spectrometer and 1 cm cuvette and 0.1 mm cylindrical quartz cell (Starna Cells, Atascadero, CA)
2. Solution of 1 mg/mL peptide in 40 mM sodium phosphate buffer.
3. Solution of 1 mg/mL peptide in 40 mM sodium phosphate buffer with 1.75 mM POPC liposomes.
4. Solution of 1 mg/mL peptide in 40 mM sodium phosphate buffer with 1.75 mM 4:1 POPC/POPG (mol-to-mol) liposomes.

2.4. Micelles

1. Sodium dodecyl sulfate (SDS) is available from Sigma-Aldrich Chemical Company and dodecylphosphocholine (DPC) is available from Avanti Polar Lipids.

3. Methods

3.1. Design Process

3.1.1. Selection and Investigation of Model Natural Antimicrobial Peptide and/or Target

The first step in the *de novo* design of selective antimicrobial peptides incorporating unnatural amino acids involves selecting a naturally occurring antimicrobial peptide as a model. In this approach, the primary amino acid sequence is not a major focus. However, knowledge of the secondary structure adopted by the peptide on interacting with a zwitterionic and an anionic micelle or a liposome and the resulting physiochemical surface properties generated as a result of that conformation is very critical for a successful outcome in the design process. It is a very difficult process to sit down with a blank sheet of paper and design a series of potent selective antimicrobial peptides without an appropriate three-dimensional physicochemical model as a guide. Thus, the starting point for such a design process is either to determine the SDS and DPC micelle-bound conformations of an antimicrobial peptide in your own laboratory by employing 2D-NMR and molecular modeling techniques (50) or use similar structural data reported in the literature.

1. In our laboratory we selected an analog of magainin-2 amide which incorporated three Ala replacements (Ala^{8,13,18})magainin-2 amide, as our model because this peptide is active against Gram-positive and Gram-negative bacteria, fungi, and protozoa while exhibiting very little mammalian cell toxicity (19). The magainin family of AMPs has been extensively investigated and has been characterized as α-helical amphipathic cell-selective membrane-disruptors (51–53). The investigation of AMPs with model membrane systems has provided a great deal of insight into the role played by physicochemical properties in the mechanisms of the antibacterial activity of these compounds (21). For our investigation of (Ala^{8,13,18})magainin-2 amide, DPC micelles (54) were selected as a simple model for zwitterionic lipids and SDS micelles (55) were selected as a simple model for anionic lipids. Due to our previous experience in investigating peptide–micelle interactions (56–59), we selected NMR as the method of choice for this study. Two-dimensional NMR and molecular modeling (22) investigations indicated that (Ala^{8,13,18})magainin-2 amide bound to DPC micelles adopts an α-helical structure involving residues 2–16 with the four C-terminal residues converging to a loose β-turn like structure, while (Ala^{8,13,18})magainin-2 amide bound to SDS micelles adopts an α-helical structure involving residues 7–18 with the C- and N-terminal residues exhibiting a great deal of conformational flexibility. The most plausible explanation for (Ala^{8,13,18})magainin-2 amide adopting different conformations on binding with SDS and DPC micelles is that different non-covalent (electrostatic and hydrophobic) interactions are occurring between the surface of the peptide and the two micelle surfaces (22). The observation of different binding conformations on interaction with anionic and zwitterionic micelles and or liposomes is critical for the development of a cell-selective AMP (22, 60). The ultimate goal of the design process is to develop AMPs that will form membrane-disrupting pores in anionic liposomes while leaving zwitterionic liposomes intact (50).
2. Using the NMR-determined SDS and DPC micelle-bound conformations of (Ala^{8,13,18})magainin-2 amide, the electrostatic surface potential maps (ESP) for these conformers were calculated indicating that the surface electron density of these peptides are very different and conformationally dependent. Computer-generated ESP maps provide a visualization of the physicochemical properties of the molecule to be replicated.
3. In order to obtain the desired potency and selectivity for prokaryotic and eukaryotic cells, it is critical that the peptide interact with zwitterionic and anionic membrane models via different mechanisms, i.e. they must adopt different conformations on binding to zwitterionic and anionic membrane models (13, 47,

60, 61). The physicochemical properties that lead to different mechanisms of membrane binding are not clearly understood (61). However, factors such as hydrophobicity, structure, electronic charge, hydrophobic moment, etc., appear to play major roles in defining selectivity (30, 47, 61, 62). Therefore in our design methodology, it is more important to focus on the overall shape of the molecule and the spatial distribution of physicochemical properties, such as charge density and hydrophobicity, rather than the amino acid composition or sequence (47, 50).

4. In our laboratory we have focused on developing AMPs that contain unnatural amino acids to control the conformation and physicochemical properties of the resulting peptide (50) while also taking advantage of the inherent metabolic stability of unnatural amino acids (18). The guiding hypothesis for our work has evolved from the assertion that the 3D-physicochemical surface properties of a cell's membrane (bacterial or mammalian) interact with the 3D-physicochemical surface properties of the approaching AMP in a very specific way, thus defining the resulting organism selectivity and potency (47, 50). For the AMPs developed in our laboratory, this hypothesis was supported by the development of 3D-QSARs which defined the physicochemical properties required for activity against *Staphylococcus aureus* (*SA*) and *Mycobacterium ranae* (*MR*) bacteria. These peptides exhibited different *in vitro* activity against *SA* and *MR* bacteria, and we hypothesized that the differences in the observed biological activity was a direct manifestation of different physicochemical interactions that occur between these peptides and the two different types of cell membranes. For a validation of this hypothesis, different physicochemical descriptors must correlate with the antibacterial activity of these compounds against *SA* and *MR* bacteria. There are six physicochemical properties specific to the *SA* 3D-QSAR model, while there are five different physicochemical properties specific to the *MR* 3D-QSAR model. These results support the hypothesis that for any particular AMP, organism selectivity and potency are controlled by the chemical composition of the target cell's membrane (47). Therefore, by varying the physicochemical and conformational properties of a peptide, it should be possible to design potency and selectivity for a particular membrane's chemical composition.

3.1.2. Making a Cartoon Model

The next step in the design process is to identify which of the following functionalities are critical for the desired biological potency and selectivity.

- Secondary structure.
- Molecular flexibility between regions of limited conformational flexibility.

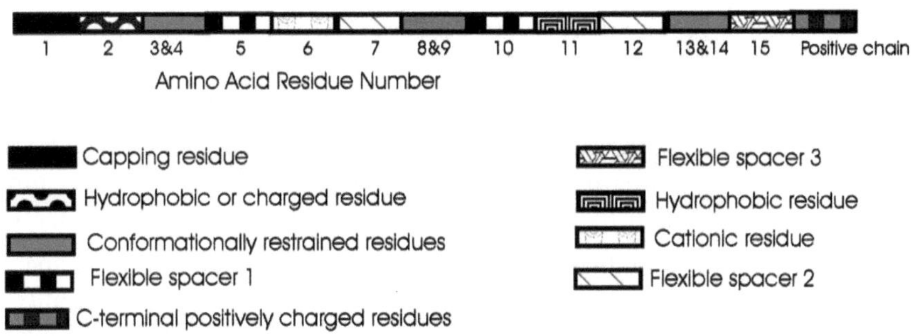

Fig. 1. A cartoon representation of the important structural and physicochemical features to be incorporated into the new antimicrobial peptide is shown to aid in the design process.

- Molecular charge.
- Charge density.
- Flexibility between the regions of limited conformational flexibility and charge clusters.

In Fig. 1 an example of this cartoon representation is given.

1. The cartoon starts at the far left with an N-capping residue. This residue is normally a natural amino acid of neutral charge and may exist as the free amine or be acetylated.
2. The second residue in the sequence is a natural amino acid that is either positively charged (Arg, Lys, His) or hydrophobic in nature (Phe, Trp, Tyr, Leu). This particular residue defines the physicochemical character of the N-terminus.
3. The next two amino acids, residues 3 and 4, are unnatural amino acids that form a conformationally restrained dipeptide unit which is used to control local secondary structure and molecular flexibility. See Note 1.
4. If included, the fifth amino acid residue also referred to as a spacer, can be a Gly residue or an unnatural amino acid such as β-Ala, GABA or longer amino acid. This spacer, if included in the sequence, provides molecular flexibility between the preceding conformationally restrained dipeptide and the following positively charged amino acid. See Note 2.
5. The next amino acid, residue 6, is a positively charged residue which may be either a natural or unnatural amino acid. The selection of this residue is critically important and is discussed in greater detail in Note 3.
6. The seventh amino acid residue is another spacer and, if included, is the second spacer in the sequence. As with the previous spacer, it provides molecular flexibility between the preceding cationic residue and residues 8 and 9, the second conformationally restrained dipeptide unit.

7. The next amino acid residue, 10, is also an optional spacer that provides molecular flexibility between the preceding dipeptide and the following hydrophobic amino acid, residue 11.
8. Residue 11, the hydrophobic residue which may be natural or unnatural, is critical, and size, aromatic character, and electron density are factors that must be taken into consideration.
9. The hydrophobic residue is followed by another spacer, which can be incorporated to provide molecular flexibility between the preceding hydrophobic residue and the following dipeptide unit, which is the third conformationally restrained unit of the sequence and incorporates residues 13 and 14. The amino acid sequence from the first dipeptide unit to the third dipeptide unit is considered a repeat unit.
10. This repeat unit may recur 1–2 more times and is followed by the fifth spacer, residue 15.
11. Residue 15 is the last spacer of the sequence and if included provides molecular flexibility between the last conformationally restrained dipeptide and a cluster of positively charged amino acids at the C-terminus. The last 3–6 amino acids of the sequence make up a highly positively charged cluster at the C-terminus.
12. The position of the positively charged and hydrophobic residues may be switched depending on the charge and hydrophobic character of the physicochemical surface properties of the AMP being modeled. See Note 4.

3.1.3. Converting the Cartoon Model into Amino Acid Sequence

Once a cartoon representation of the desired structure and physicochemical properties is created, it must be converted to an amino acid sequence. The basic skeleton of the AMPs developed in our laboratory is given in Fig. 2 and how the amino acid sequence was determined will be described in the following sections.

3.1.4. Selection of Conformationally Restrained Dipeptide Units

1. Select a conformationally restrained dipeptide unit to induce an ordered structure onto the peptide. Our original skeletal design of an unnatural AMP incorporated placement of multiple L-Tic–L-Oic dipeptide units into the polypeptide backbone (50). Kyle and co-workers reported, using NMR and molecular modeling methods, that the dipeptide consisting of the un-natural amino acids (D)-tetrahydroisoquinolinecarboxylic acid (Tic), and (L)-octahydroindolecarboxylic acid (Oic) when placed in positions $i+1$ and $i+2$ of a four amino acid sequence induced a β-turn (44). The structure of the (L)-Tic–(L)-Oic dipeptide conformationally restrained unit used in the AMPs designed in our laboratory is given in Fig. 3.

   The Tic–Oic dipeptide is by no means the only choice for a conformationally restrained dipeptide unit. The Cα-tetrasubstituted α-amino acids are another possible choice. Of the available Cα-tetrasubstituted α-amino acids, α-amnoisobutyric

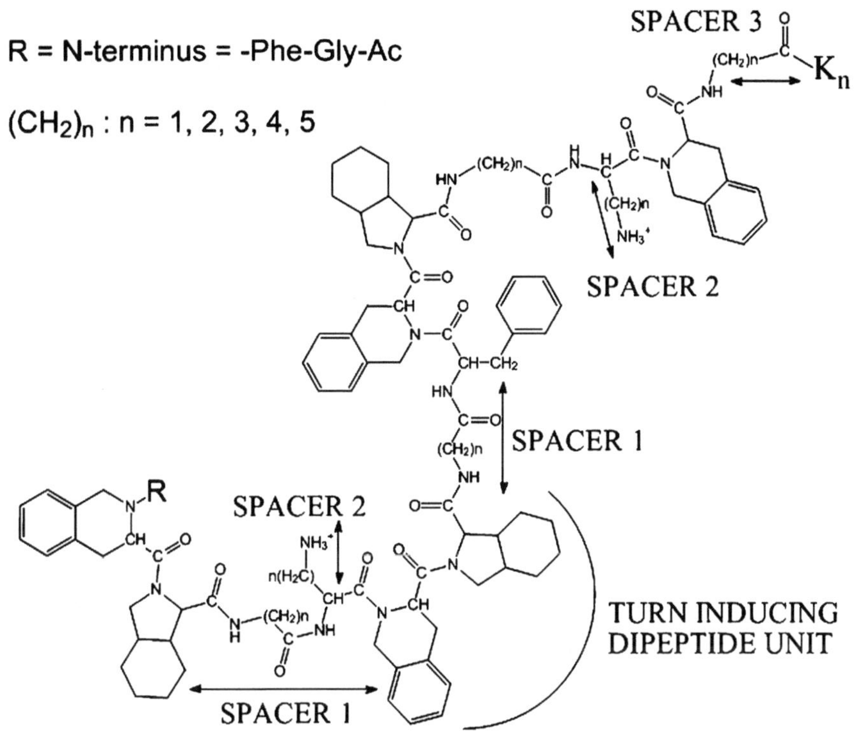

Fig. 2. The basic skeleton of the AMPs developed in our laboratory.

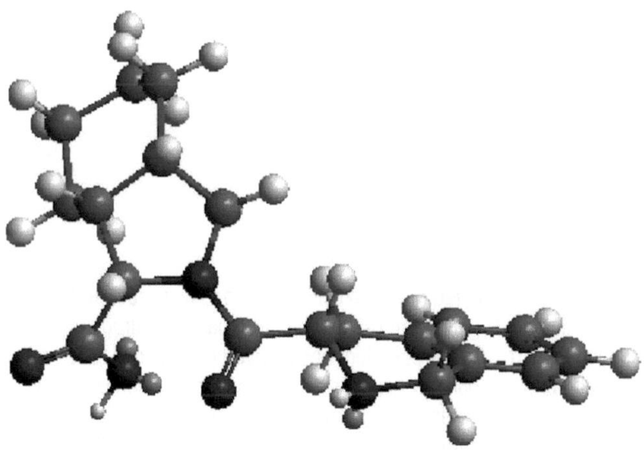

Fig. 3. The structure of the Tic–Oic dipeptide conformationally restrained unit used in the AMPs designed in our laboratory.

acid (Aib) is the most commonly employed and studied residue and its allowed conformations are limited to constrained right- and left-handed helical structures (63). However, additional Cα-tetrasubstituted α-amino acids are being developed and investigated which may offer the investigation of additional conformational space. Grauer and Konig (43) have reported

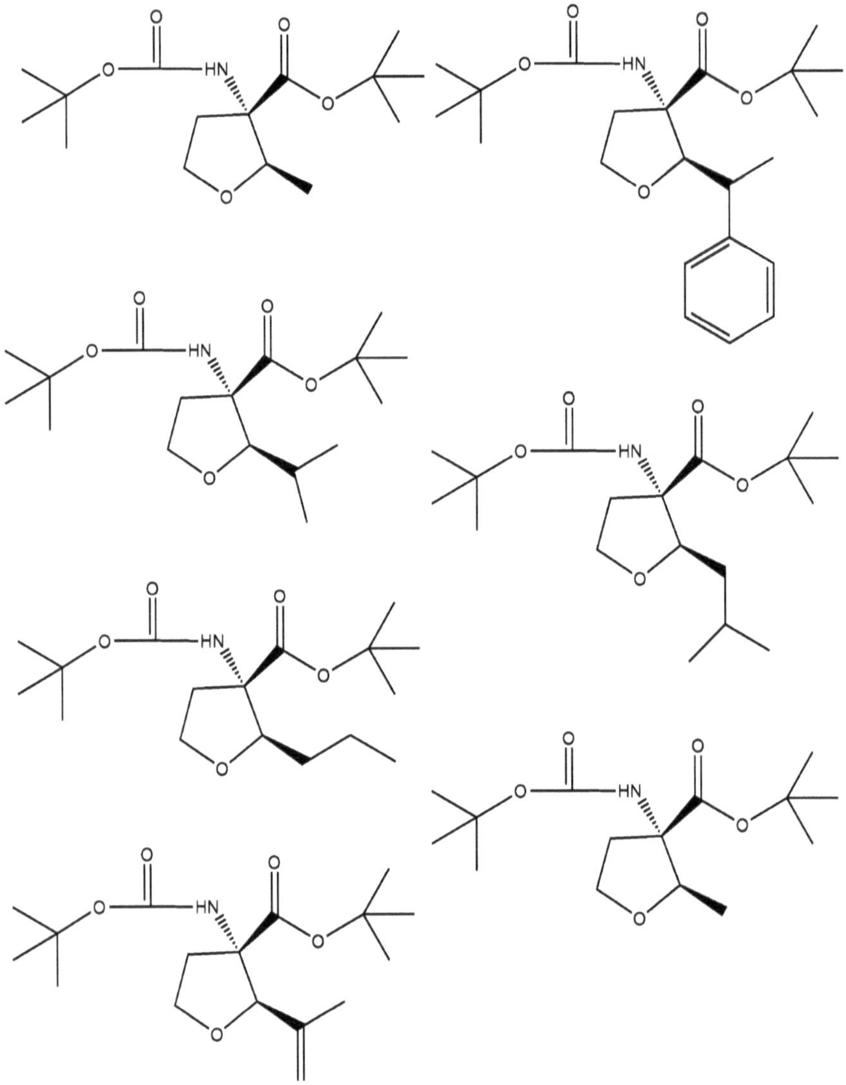

Fig. 4. Representative novel Cα-tetrasubstituted α-amino acids developed by Grauer and Konig (43). Cα-tetrasubstituted α-amino acids are unnatural amino acids where the Cα-carbon atom has one nitrogen atom, one carbonyl carbon, and two alkyl carbon atoms bonded to it. The Cα-carbon atom has no hydrogen bonded to it, as would be the case with the naturally occurring amino acids normally found in peptides or proteins. The nomenclature used for these specific compounds is *tert*-butyl 3-(*tert*-butoxycarbonylamino)-2-X-tetrahydrofuran-3-carboxylate, where X = the substituent on the tetrahydrofuran ring. For example, the compound at the *top left corner* of the figure is named *tert*-butyl 3-(*tert*-butoxycarbonylamino)-2-methyltetrahydrofuran-3-carboxylate.

the syntheses of a series of novel Cα-tetrasubstituted α-amino acids to reduce the conformational flexibility of a peptide, several of which are given in Fig. 4. Incorporation of specific substituted Cα-tetrasubstituted α-amino acids derivatives into a peptide sequence can induce stable β-turns (67) or 3_{10} or α-helices (43, 63, 68).

There are several Cα-tetrasubstituted α-amino acids that are commercially available as either the Fmoc of tBoc protected

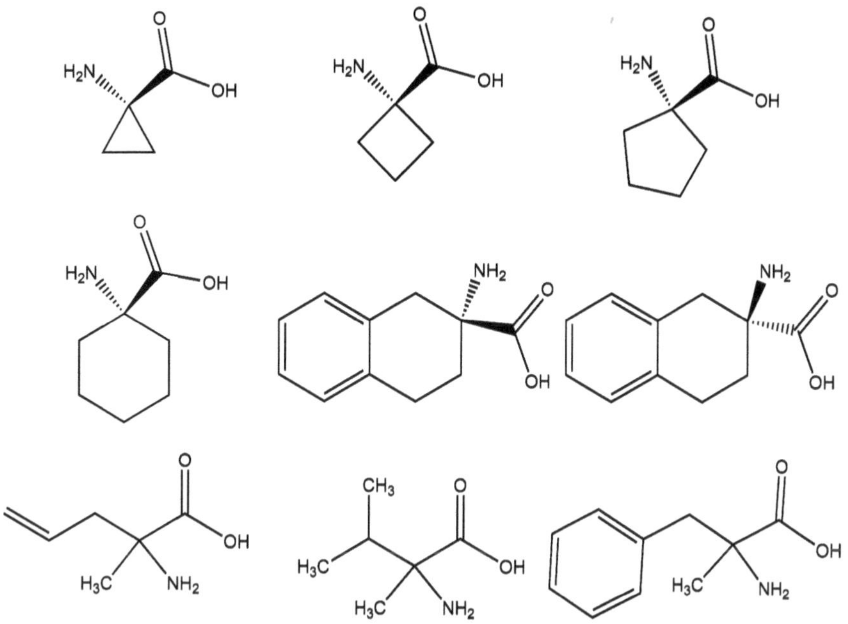

Fig. 5. Alicyclic α-amino acids commercially available as either the Fmoc of tBoc protected analogs or the free amino acids.

analogs or the free amino acids and these are shown in Fig. 5. (R)-2-Amino-1,2,3,4,-tetrahydronaphthalene-2-caroxylic acid and (S)-2-amino-1,2,3,4,-tetrahydronaphthalene-2-caroxylic acid shown in Fig. 5 are very interesting alicyclic α-amino acids and have the potential to act as conformationally restrained dipeptide units. See Notes 5 and 6.

2. Determine the stereochemistry and placement of the dipeptide unit needed to induce the desired conformation. The purpose of the conformationally restrained dipeptide unit is to induce an ordered structure onto the peptide and the stereochemistry of the amino acids used (D- or L-isomers), as well as the placement of the amino acid residues in the sequence may be of great importance in determining the overall conformation of the dipeptide unit. With this in mind, the first logical step in the selection of a conformationally restrained dipeptide unit would be constructing a simple tetra peptide to determine whether or not the dipeptide in fact provides the desired conformational rigidity via circular dichroism. The importance of stereochemistry on the resulting conformation of a tetrapeptide is illustrated by the CD spectra of the three tetra peptides Ac-Ala–Tic–Oic–Ala-NH_2, Ac-Ala–D-Tic–Oic–Ala-NH_2, and Ac-Ala–Oic–D-Tic–Ala-NH_2 are shown in Fig. 6. These CD spectra clearly demonstrate that the stereochemistry of the alpha carbon can affect the solution conformation of the conformationally restrained dipeptide units and must be taken into consideration in the design of these peptides.

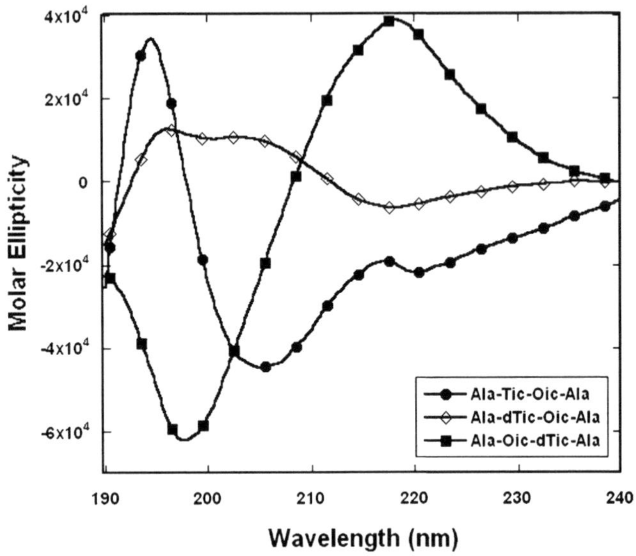

Fig. 6. The Far-UV circular dichroism spectra of the three tetrapeptides (1 mg/mL) dissolved in 40 mM phosphate buffer (pH 6.83). All CD spectra were obtained by acquiring eight scans on a Jasco J-815 CD spectrometer using a 0.1-mm cylindrical quartz cell (Starna Cells, Atascadero, CA) from 260 to 178 nm at 20 nm/min, 1 nm bandwidth, data pitch 0.2 nm, response time 0.25 s and 5 mdeg sensitivity at room temperature (~25°C). The CD spectrum (shown in *circles*) of the tetrapeptide Ac-Ala–Tic–Oic–Ala-NH_2 exhibits an ordered turn containing secondary structure as indicated by the maximum at 195 nm and the minimum at 205 nm in the CD spectrum which is consistent with the presence of β-turn and helical conformers (64–66). The CD spectrum (shown in *diamonds*) of the tetrapeptide Ac-Ala–D-Tic–Oic–Ala-NH_2 exhibits an ordered turn containing secondary structure as indicated by the double maximum in the CD spectrum at about 195 and 205 nm, and a minimum at 185 nm, which is consistent with the presence of possible β-turn conformers (64–66). The CD spectrum (shown in *squares*) of the tetrapeptide Ac-Ala–Oic–D-Tic–Ala-NH_2 exhibits a maximum in the CD spectrum at about 220 nm, and a minimum at about 198 nm, which is consistent with a random coil conformers (64, 65).

3. Determine how many dipeptide units to incorporate. The cartoon representation discussed a repeating unit that started and stopped with the dipeptide unit. However, in the case of our peptides, the end of the last repeat unit only includes the first residue of the dipeptide unit, Tic. Therefore, our peptide contains one repeat unit, which concludes with half a dipeptide unit.
4. The AMPs developed in our laboratory have several Tic–Oic dipeptide units connected via one or two amino acid spacers with defined properties of charge and hydrophobicity. Knowing such properties about the amino acids allows for the design of peptides with well-defined physiochemical properties that are able to maintain sufficient conformational flexibility, allowing the peptide to adopt different conformations when interacting with membranes of different chemical composition (47, 49, 50).

3.1.5. Selection of Number, Position and Type of Flexible Spacers

1. Spacer #1 in Fig. 2 defines the distance between successive Tic–Oic dipeptide units, and refers to amino acids 5 and 10 of the cartoon representation. It is involved in defining the flexibility of any induced turn or helical structure.
2. The amino acid residues employed as the flexible spacers in our previous work were the natural amino acid glycine, which contains one carbon atom between the carbonyl carbon and the amide nitrogen, the β-peptide β-Ala with two carbons, the unnatural amino acids Gaba (gamma-aminobutyric acid) and Ahx (6-aminohexanoic acid) with three and five carbons atoms, respectively.
3. The average distance between the carbonyl carbon atom and the amide nitrogen atom is 2.46, 2.94, 4.68, and 6.91 Å for Gly, β-Ala, Gaba and Ahx, respectively.
4. Longer alkyl spacers include the commercially available ones; 9-aminooctanoic acid (9-Aoa), 10-aminodecanoic acid (10Ada), 12-aminododecanoic acid (12-Adda), and 16-aminopalmitic acid (16-Apa) could also be used. See Notes 7 and 8.
5. A number of different β-amino acids have been used to control the conformation of antimicrobial peptides (61, 69, 70). In our approach, we limit their use to the role of a flexible spacer. In addition to β-Ala, analogs of β-Leu and β-Phe (see Note 9) are shown in Fig. 7, and could also be considered as flexible spacers. These analogs, which are commercially available, have the additional advantage of introducing hydrophobicity into the spacer, along with flexibility.
6. These spacers can also be used to induce a rigid region onto the peptide backbone using derivatives of amino benzoic acid shown in Fig. 8. These analogs provide both rigidity and aromatic character, or hydrophobicity, into the backbone. Incorporation of these residues as spacers, however, should be done with care.

3.1.6. Selection of Number and Position of Hydrophobic Properties

The next step in the design process is the selection of hydrophobic residues to incorporate into the peptide. There are two basic types of hydrophobic residues: aromatic and non-aromatic. See Note 10.

1. Hydrophobic residues are known to play a role in anchoring an antimicrobial peptide to the membrane by inserting itself into the hydrophobic core of the membrane (48).
2. Of all the naturally occurring amino acids. Wimley and White determined tryptophan to be the most hydrophobic (48, 71, 72). It is believed that the tryptophan residue remains located near the membrane interface and does not partition deep into the hydrophobic core of the lipid (48).

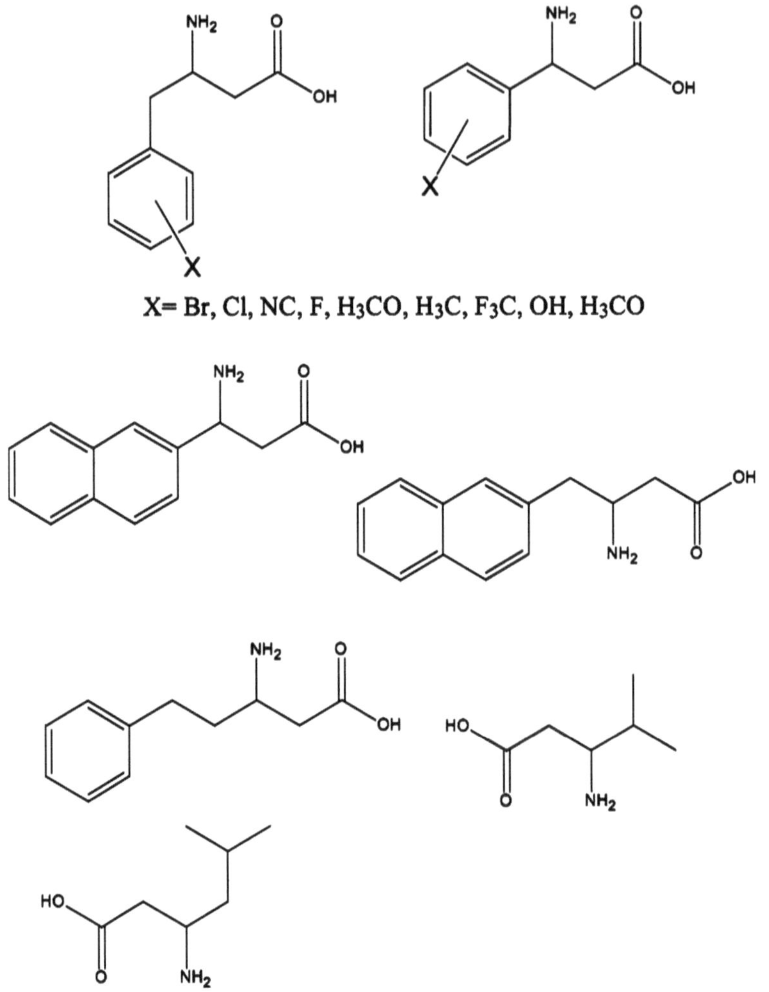

Fig. 7. Commercially available analogs of β-Leu and β-Phe.

3. Haug and co-workers developed a series of short lactoferricin-based antimicrobial peptides incorporating different analogs of tryptophan (48). The net result of this investigation was the development of a novel class of shorter peptides with high antibacterial activity against several resistant strains of bacteria (48).
4. Some of the hydrophobic and tryptophan analogs used by Haug and co-workers are given in Fig. 9 (48).
5. Figure 10 gives a partial list of commercially available substituted unnatural amino acids which may be used to modify the electronic character of the aromatic ring of the amino acid Phe.

3.1.7. Selection of Number and Position of Electrostatic Properties

1. In our original work, the positively charged amino acid residue was referred to as Spacer #2 and is residue 6 of the cartoon representation. Spacer #2 defines the distance between the polypeptide backbone residues and the positively charged side

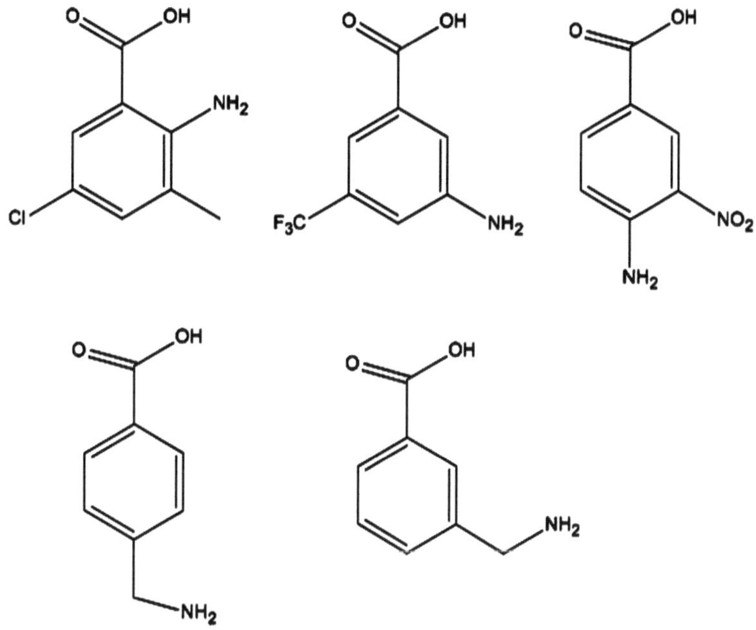

Fig. 8. A partial listing of the commercially available rigid spacers derived from benzoic acid.

chain amine. It participates in determining the overall surface charge density as well as the distance between the membrane surface and the polypeptide backbone. Amino acids such as Lys, Orn, Dpr, or Dab, each decreasing the distance from the side chain nitrogen to the α-carbon atom of the peptide backbone, were used as Spacer #2. Table 1 gives the average distances for each amino acid residues (calculated using Chem-Draw 3D, Cambridge Software).

2. Oh and Lee reported (46) the synthesis and biological evaluation of several antimicrobial peptides containing unnatural analogs of the amino acid Lys. These unnatural analogs of Lys are shown in Fig. 11 and were designed to incorporate more positive charge (increased number of amino groups) and increase the side chain bulk (46). The increase in positive charge should result in an increase in the electrostatic attraction of the peptide to the surface of a bacterial membrane and thus greater antibiotic activity (73, 74). See Note 11.

3.1.8. Positive Charge Density at the C-terminus

1. We have shown that by increasing the number of basic amino acid (47) residues at the C-terminus, the net positive charge density at the C-terminus also increases.
2. It has been previously observed that increased positive charge density, to a point, at the C-terminus can increase antibacterial activity while decreasing hemolytic activity (47, 50).
3. Selectivity for different bacteria strains has also been observed (49, 50).

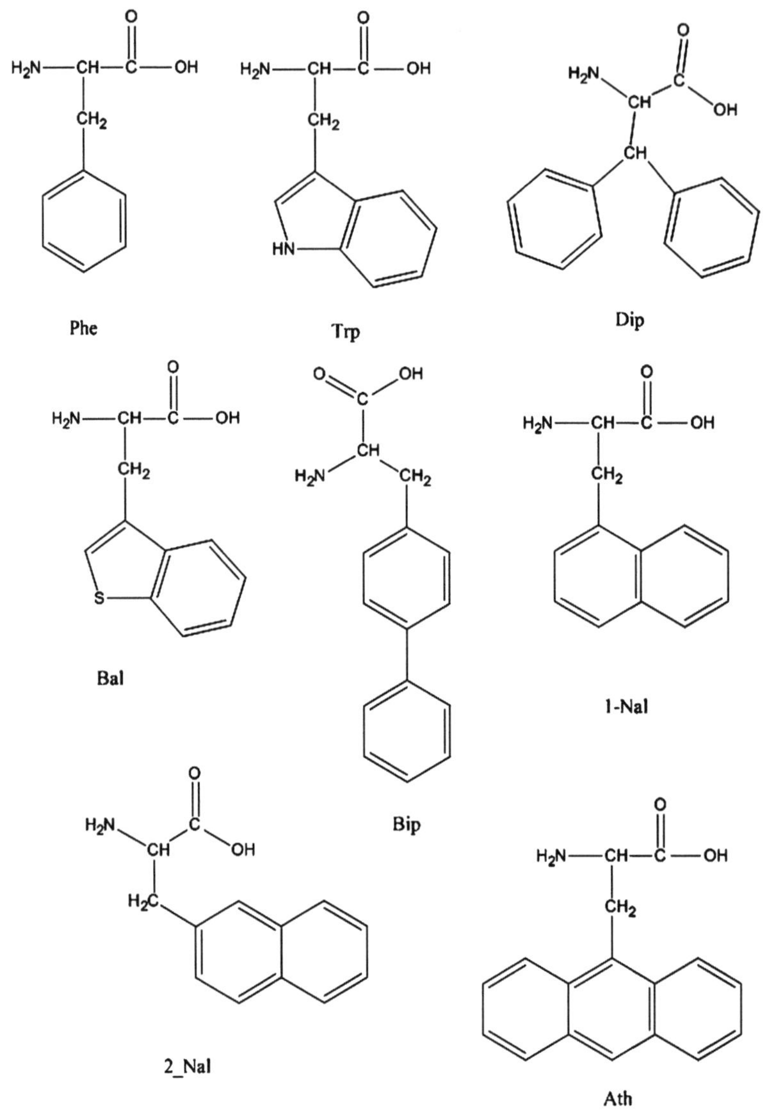

Fig. 9. Some of the hydrophobic and tryptophan analogs used by Haug and co-workers (48) to develop a series of short lactoferricin-based antimicrobial with high antibacterials activity against several resistant strains of bacteria. Lactoferricin is a cyclic peptide containing 49 amino acid residues that exhibited potent antimicrobial activity. *Bal* β-(benzothien-3-yl) alanine, 1-*Nal* β-(naphtha-1-yl)alanine, 2-*Nal* β-(naphtha-2-yl)alanine, *Dip* diphenylalanine, *Bip* biphenylalamine, *Ath* anthrenealanine.

4. Spacer #3 follows the last repeat unit and defines the distance between the last Tic residue and the positive charge density at the C-terminus, providing additional conformational flexibility on surface binding (50).

3.1.9. Assemble the Amino Acids to Obtain the Desired Structural and Physicochemical Properties

In Fig. 12, the amino acid sequence of a peptide, compound 46 (amino acid sequence; Ac-βAla–Fpa–Tic–Oic–βAla–Dpr–Tic–Oic–βAla–Fpa–Tic–Oic–βAla–Dpr–Tic–Dpr–Dpr–Dpr–Dpr-$CONH_2$) that was designed and synthesized in our lab based on components discussed herein, is given. Although this particular peptide is

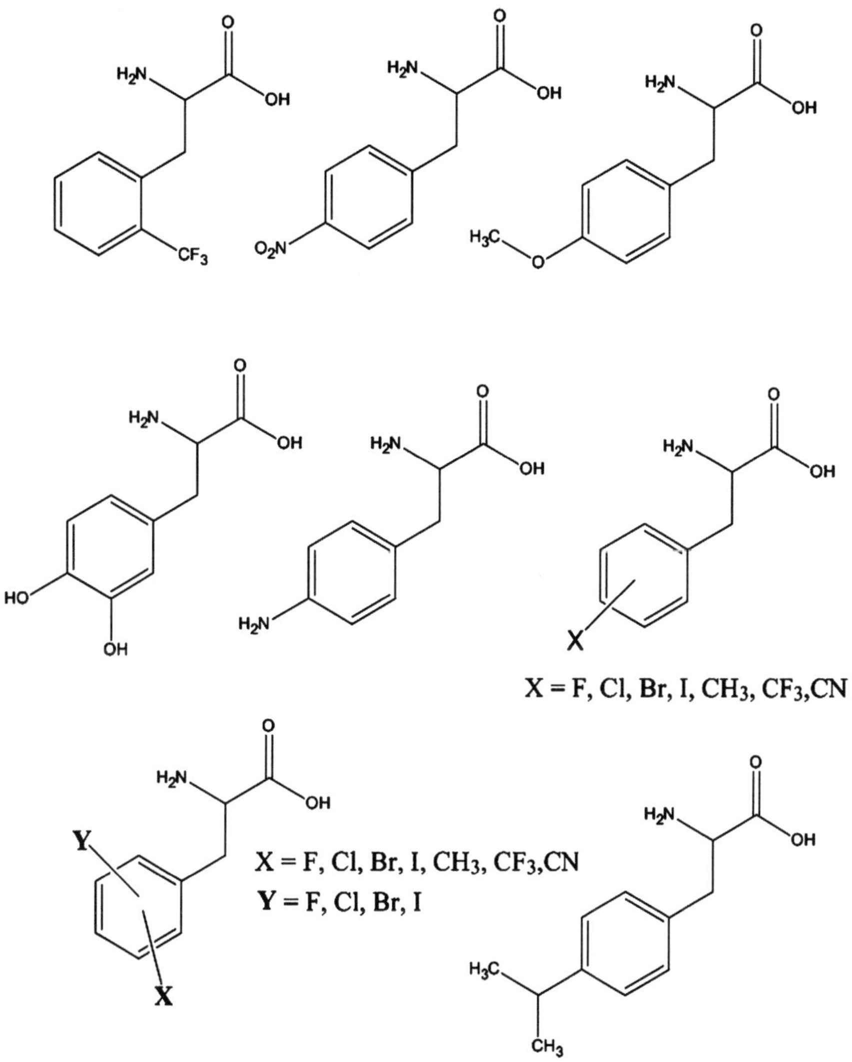

Fig. 10. Partial list of commercially available hydrophobic amino acids which may be used to modify the electronic character of the aromatic ring of the amino acid phenylalanine. Nomenclature for these compounds would be that used for mono and di-substituted phenylalanine. For example, the compound in the *top left corner* of the figure is named 2-trifluoromethylphenylanine.

Table 1
Distance from side chain nitrogen to α-carbon atom of the peptide backbone

Amino acid	Lys	Orn	Dab	Dpr
Average distance, Å	4.76	3.55	2.99	2.56
Number of carbons in side chain	4	3	2	1

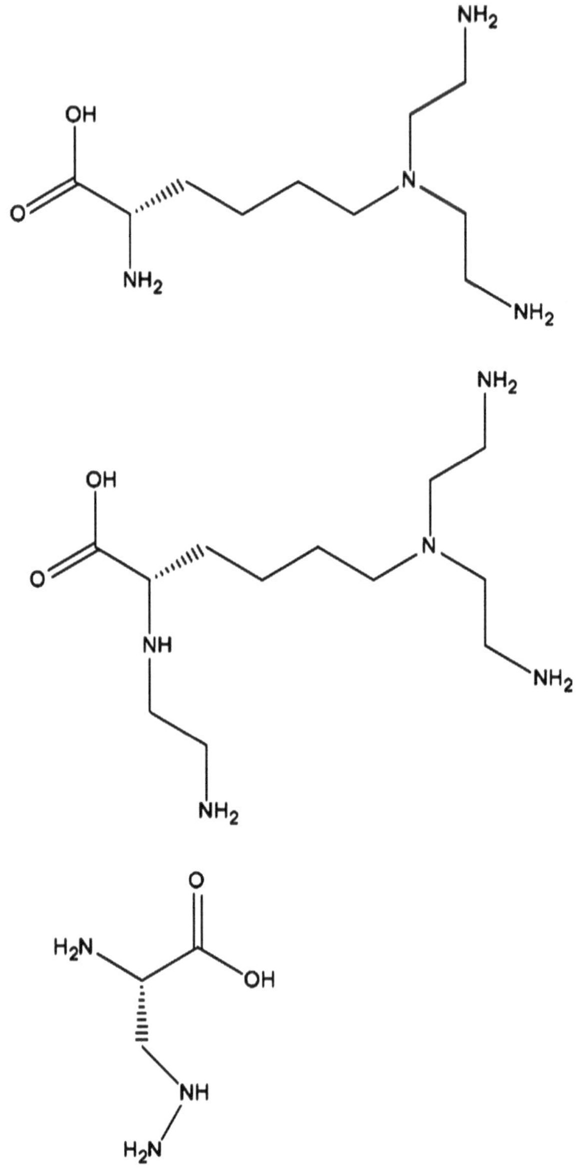

Fig. 11. Analogs of the amino acid lysine with increased positive charge and side chain bulk developed by Oh and Lee (46). The authors did not provide systematic nomenclature for these unnatural amino acids.

biologically active (activity given in Table 2), this protocol does not guarantee that all peptides designed in this manner will be biologically active, and this peptide is used to illustrate the final step before peptide synthesis and purification. In the proceeding sections, the process for selecting each amino acid residue with the appropriate properties has been discussed.

1. The first residue, as previously mentioned, is normally a natural amino acid with neutral charge and it may be the free amine or

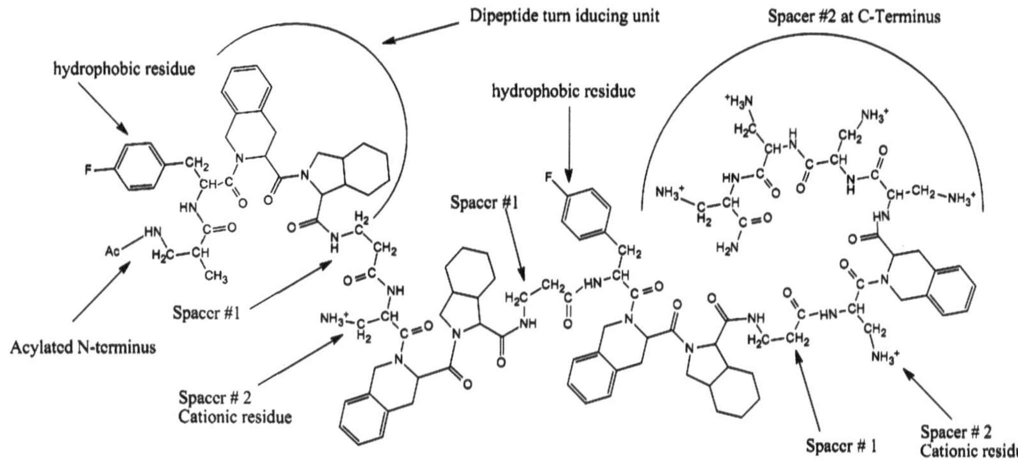

Fig. 12. Designed amino acid sequence of compound 46 incorporating the desired structural and physicochemical properties.

acetylated. In the case of our example, the N-capping residue, and residue 1, is β-Ala and is acetylated.

2. The second residue defines the physicochemical character of the N-terminus and is a natural amino acid residue that is either positively charged (Arg, Lys, His, or an unnatural basic residue) or is hydrophobic in nature (Phe, Trp, Tyr, Leu). See Note 12.
3. Residues 3 and 4 are unnatural amino acids that form a conformationally restrained dipeptide unit. For compound 46, we selected Tic and Oic, as the dipeptide unit and together they control the local secondary structure and molecular flexibility of the peptide. See Note 13.
4. The fifth residue is an optional spacer, as are all the spacers. In our series of peptides, the spacer immediately following the conformationally restrained dipeptide is referred to as Spacer #1 and can be a Gly residue or an unnatural amino acid such as β-Ala, GABA or longer amino acid, as previously mentioned. See Note 14.
5. Residue 6 is either a positively charged natural or unnatural amino acid, and is also referred to as Spacer #2. See Note 15.
6. Another optional spacer follows the positively charged residue. The seventh residue, although excluded in compound 46, would precede the dipeptide unit and provide molecular flexibility between the positively charged residue and the next dipeptide unit residues 8 and 9.
7. As before, the dipeptide is followed by a spacer. This spacer, residue 10, is the second Spacer #1 of the sequence and is also β-Ala. It provides molecular flexibility between the preceding dipeptide and the following hydrophobic amino acid, residue 11. See Note 16.

Table 2
***In vitro* activity (μg/mL) of compound 46 against specific bacterial strains (49, 50)**

Acinetobacter baumanii ATCC 19606	*Acinetobacter baumanii* drug resistant	*Staphylococcus aureus* ATCC 33591 – MRSA	*Staphylococcus aureus* ME/GM/TC resistant	*Salmonella typhimurium*	*Bacillus subtillis*	*Bacillus anthracis* AMES	*Mycobacterium ranae*
7.8	7.8	250	66	6.6	0.66	15.6	228

8. The hydrophobic residue is followed by another spacer, residue 12, which can be incorporated to provide additional molecular flexibility between the hydrophobic residue and the third Tic–Oic dipeptide unit, residues 13 and 14. Residue 12, like residue 7, was excluded from compound 46.
9. As mentioned previously, the section of the amino acid residues from the first dipeptide unit to the third dipeptide unit is considered one repeat unit and a peptide can consist of up to three repeat units, although our peptides only incorporates one repeat unit.
10. Following the last repeat unit is the final spacer, residue 15, and it is referred to as Spacer #3 in our series of peptides. Spacer #3 is not always included, as in compound 46, but when it is included it usually consists of the same amino acid residue that is used for spacer #1. This particular spacer provides molecular flexibility between the last conformationally restrained dipeptide and the cluster of positively charged amino acid residues located at the C-terminus. See Note 17.

3.2. Preliminary Evaluation

1. Once the peptide(s) have been synthesized and purified their potential as antimicrobial peptides must be evaluated.
2. *In vitro* screening against several strains of bacteria is critical in determining the activity of the new compounds.
3. However, in most chemical research laboratories, these assays are not readily available. Therefore samples must be submitted to commercial research organizations for testing.
4. This process can take 4–6 weeks to obtain results and depending on the number of peptides and assays to be screened, this can prove to be a relatively expensive operation.
5. We recommend the use of CD spectroscopy and calcein leakage studies as a first evaluation of new peptides.
6. If the CD spectra of the peptides do not exhibit different conformations in membrane model systems than they do in buffer, it is most likely that they are not interacting with the membrane models, and therefore, it is unlikely they will interact with the membrane of a bacteria cell (60).
7. The results from the calcein leakage assays are critical in determining if the peptide has the potential to disrupt a cell membrane (49, 60, 75). Failure to induce calcein leakage, however, does not mean that the peptide will exhibit no antibiotic activity, but rather that any observed *in vitro* activity is derived interacting with an intracellular target.

3.2.1. Preparation of POPC for CD Experiments

1. Weigh out a defined amount of dried POPC to give a total lipid concentration of 10 mM.

2. Suspend in 40 mM sodium phosphate buffer, pH 6.8, and vortex for 2–3 min or stir for approximately 5 min, until solution appears milky.
3. LUVs were prepared by extrusion using a Mini-Extruder (Avanti Polar Lipid Inc) (76, 77). The solution was passed through a 100-nm pore size polycarbonate membrane 21 times.
4. Allow LUV solution to sit for a minimum of two hours in the refrigerator before using.
5. The final lipid concentration was calculated based on the weight of the dried lipid (78–82).

3.2.2. Preparation of POPC/POPG Liposomes for CD Experiments

1. Weigh out a defined amount of dried POPC to give a total lipid concentration of 10 mM. Suspend in chloroform.
2. Weigh out a defined amount of dried POPG to give a total lipid concentration of 10 mM. Suspend in chloroform.
3. Mix the POPC and POPG solutions in a 4:1 ratio in a glass test tube. Vortex.
4. Put test tube under a direct, steady flow of nitrogen for approximately 3 h to remove all chloroform. Time under nitrogen flow varies depending on volume of solution.
5. Suspend dried sample in 40 mM sodium phosphate buffer, pH 6.8, to give desired concentration. Vortex or stir until solution appears milky.
6. LUVs were prepared by extrusion using a Mini-Extruder (Avanti Polar Lipid Inc) (76, 77). The solution was passed through a 100-nm pore size polycarbonate membrane 21 times.
7. Allow LUV solution to sit for a minimum of two hours in the refrigerator before using.

3.2.3. Preparation of POPC Liposomes for Dye Release Experiments

1. Weigh out a defined amount of dried POPC to give a total lipid concentration of 35 mM.
2. Suspend in calcein-containing buffer (70 mM calcein, 10 mM Bis-Tris, 150 mM NaCl, 1 mM EDTA, pH 7.2, the pH was corrected using 1 mM NaOH) and vortex for 2–3 min or stir for approximately 5 min, until the solution is homogeneous.
3. LUVs were prepared by extrusion using a Mini-Extruder (Avanti Polar Lipid Inc) (76, 77). The solution was passed through a 100-nm pore size polycarbonate membrane 21 times.
4. Allow LUV solution to sit for a minimum of two hours in the refrigerator before running gel filtration column.
5. Unencapsulated calcein is removed by gel filtrations on a Sephadex G50 column (eluent: buffer containing 10 mM Bis-Tris, 150 mM NaCl, 1 mM EDTA, pH 7.2).

6. Collect the entire first band to come off the column (orange) and note the volume. This fraction contains the LUVs with encapsulated calcein.
7. The final concentration of liposomes is calculated based on the beginning concentrations of lipids and the volume before and after the column.

3.2.4. Preparation of 4:1 POPC/POPG Liposomes for Dye Release Experiments

1. Weigh out a defined amount of dried POPC to give a total lipid concentration of 35 mM. Suspend in chloroform.
2. Weigh out a defined amount of dried POPG to give a total lipid concentration of 35 mM. Suspend in chloroform.
3. Mix the POPC and POPG solutions in a 4:1 ratio in a glass test tube. Vortex.
4. Put test tube under a direct, steady flow of nitrogen for approximately 3 h to remove all chloroform. Time under nitrogen flow varies depending on volume of solution.
5. Suspend in calcein-containing buffer (70 mM calcein, 10 mM Bis-Tris, 150 mM NaCl, 1 mM EDTA, pH 7.2, the pH was corrected using 1 mM NaOH) and vortex for 2–3 min or stir for approximately 5 min, until solution is homogeneous.
6. LUVs were prepared by extrusion using a Mini-Extruder (Avanti Polar Lipid Inc) (76, 77). The solution was passed through a 100-nm pore size polycarbonate membrane 21 times. Note the volume following extrusion.
7. Allow LUV solution to sit for a minimum of two hours in the refrigerator before running gel filtration column.
8. Unencapsulated calcein is removed by gel filtrations on a Sephadex G50 column (eluent: buffer containing 10 mM Bis-Tris, 150 mM NaCl, 1 mM EDTA, pH 7.2).
9. Collect the entire first band to come off the column (orange) and note the volume. This fraction contains the LUVs with encapsulated calcein.
10. The final concentration of liposomes is calculated based on the beginning concentrations of lipids and the volume before and after the column.

3.2.5. Circular Dichroism

1. All CD spectra were obtained by acquiring eight scans on a Jasco J-815 CD Spectrometer using a 0.1-mm cylindrical quartz cell (Starna Cells, Atascadero, CA) from 260 to 178 nm at 20 nm/min, 1 nm bandwidth, data pitch 0.2 nm, response time 0.25 s and 5 mdeg sensitivity at room temperature (~25°C).
2. Dissolve peptides (1 mg/mL) in either 40 mM phosphate buffer (pH 6.83), 80 mM SDS or DPC micelles in buffer, or 10 mM POPC or POPC/POPG (4:1) LUVs in buffer.

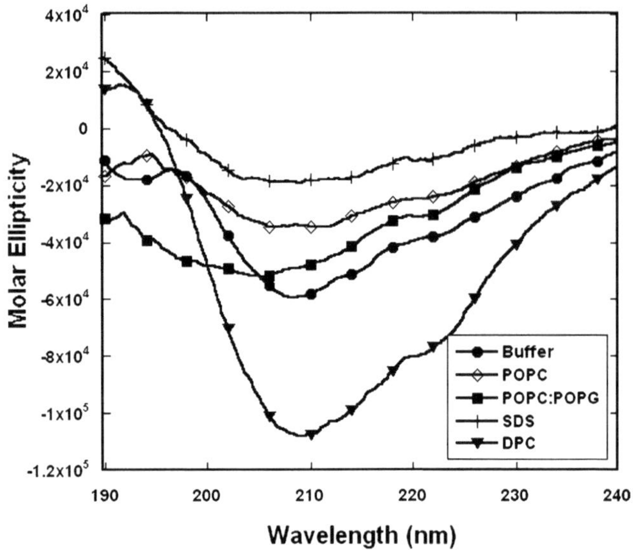

Fig. 13. The Far-UV circular dichroism spectra of compound 46 (1 mg/mL) dissolved in either 40 mM phosphate buffer (pH 6.83), 80 mM SDS or DPC micelles in buffer, or 10 mM POPC or POPC/POPG (4:1) LUVs in buffer. All CD spectra were obtained by acquiring eight scans on a Jasco J-815 CD spectrometer using a 0.1-mm cylindrical quartz cell (Starna Cells, Atascadero, CA) from 260 to 178 nm at 20 nm/min, 1 nm bandwidth, data pitch 0.2 nm, response time 0.25 s and 5 mdeg sensitivity at room temperature (~25°C).

3. Subtract the lipid spectra of the corresponding peptide-free suspensions to eliminate contributions due to micelles and LUVs.
4. Convert CD spectra to molar ellipiticty using the JASCO Spectra Analysis program (76, 83).
5. Smooth spectra, as needed, using means movement with convolution width of 5–25.
6. Analyze data. For an example of the observed differences in the CD spectra in these five different environments for compound 46, please see Fig. 13.

3.2.6. Calcein Leakage Assays

Peptide-induced calcein leakage was investigated using an ISS PC1 photon counting spectrofluormeter (ILC Technology) at an excitation wavelength of 494 nm and an emission wavelength of 518 nm.

1. Fill 1 cm quartz cuvette with buffer (10 mM Bis-Tris, 150 mM NaCl, 1 mM EDTA, pH 7.2) and a stir bar.
2. Add the appropriate amount of calcein-encapsulated liposomes to give an overall cell concentration of 36.6 μM lipid. Allow to mix thoroughly, then measure the fluorescence of the liposomes without peptide, F_0.

3. Add 10–50 μL of 10% (v/v) Triton X. Allow to mix for approximately 1–3 min and measure the fluorescence, F_T.
4. Calculate the self-quenching of the liposomes using the following equation:

$$[1-(F_0/F_T)]\times 100\%$$

Do not use solutions that are less than 80% self-quenching.

5. Once self-quenching has been determined, prepare a fresh liposome sample in a clean cuvette and measure the background fluorescence, F_0.
6. Add a 10–50 μL aliquot of 1.0 mM stock solution of peptide in buffer (10 mM Bis-Tris, 150 mM NaCl, 1 mM EDTA, pH 7.2) to the cell containing liposomes to give a final peptide concentration of 4–20 μM.
7. Take a measurement every minute for the first 20 min of the experiment and every 10 min after until the emission intensity showed no further significant increase (approximately 90 min). See Note 18 (76, 78, 84–86).
8. Calculate the apparent percent leakage using the following equation: refs. 76, 78, 84–86.

$$\%\text{leakage} = [1-(F_I - F_0)/F_T)]\times 100\%$$

where F_I is the intensity measured and F_0 and F_T are the initial fluorescence before introduction of peptide and after the addition of Triton-X100, respectively.

9. For an example of the observed differences in calcein leakage in POPC and 4:1 POPC/POPG liposomes for compound 46 please see Fig. 14. See Note 19.

3.3. Peptide Synthesis

Peptide synthesis was performed manually using tBOC chemistry (49, 87–91) or FMOC chemistry (49, 50, 89–93). Selection of synthetic method to synthesize the peptides was based on the techniques used by the specific facility.

4. Notes

Making a cartoon model

1. The use of conformationally restrained amino acids, such as Tic and Oic residues in our case, reduces the flexibility of the local peptide backbone and thus reduces the total conformational space that may be sampled by the peptide during lipid binding.

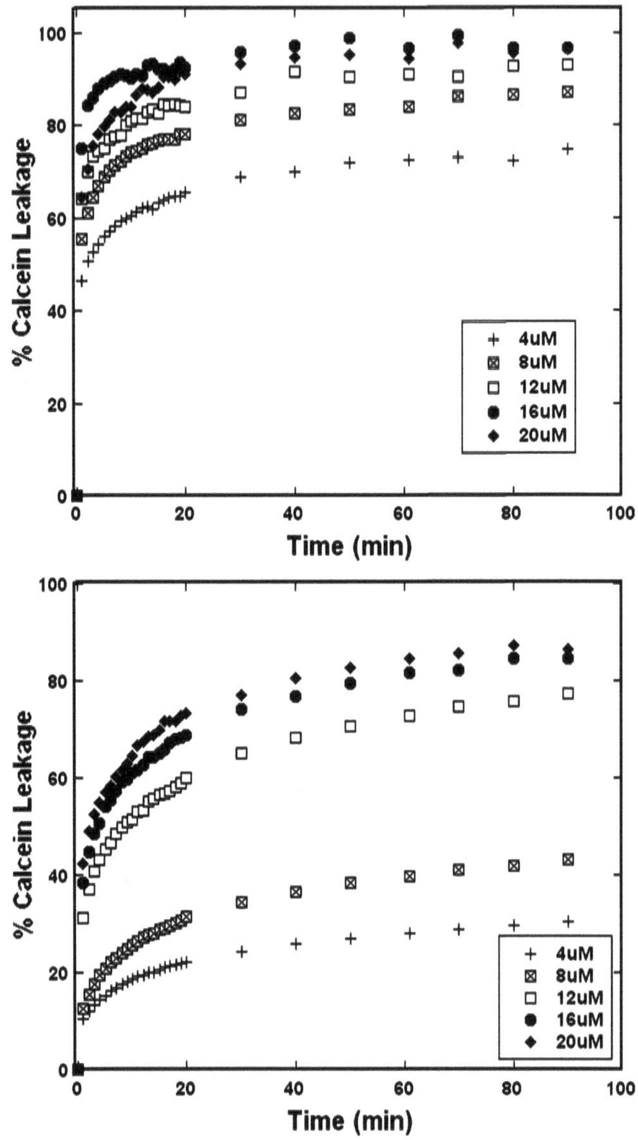

Fig. 14. Calcein leakage data for compound 46 from POPC (*top*) and from 4:1 POPC/POPG liposomes (*bottom*).

2. Increasing the number of $-CH_2-$ groups in the spacer, in our case by including unnatural amino acids such as β-Ala, Gaba or Ahx, the length of the peptide backbone increases, thus increasing the molecular flexibility of the local peptide chain. This results in an increase in the total conformational space that may be sampled by the peptide during lipid binding. By combining regions containing conformationally restrained amino acids and regions of conformational flexibility into the

peptide backbone, regions of high and low molecular flexibility are created providing a fine-tuning of the conformational space that may be sampled by the peptide during lipid binding.

3. By varying the number of $-CH_2-$ groups in the positively charged residue from 1 to 4 in the side chain of the basic residues (Lys, Orn, Dab, Dpr residues), the distance between the positive charge and the peptide backbone will decrease. This results in a less flexible side chain during binding, which is more important in the binding with zwitterionic lipids than with anionic lipids, and the positive charge density will reside closer to the peptide backbone. The reason that side chain flexibility is more important in the binding of zwitterionic lipids than with anionic lipids is the following. Electrostatic interactions are responsible for peptide-membrane binding while hydrophobic interactions are responsible for inducing a stable secondary structure onto the peptide (94, 95). The electrostatic interactions that occur between cationic peptides and zwitterionic lipids are more "complex" than those that occur with anionic lipids (95). The counter ions in solution are covalently bonded to zwitterionic lipids, but are free flowing in solution (i.e. are not covalently bonded to the lipid) for an anionic lipid, thus their positive and negative charges have limited freedom of motion. As the positive charges of the incoming peptide approaches the surface of the zwitterionic lipid, the positive counter ions of the lipid cannot be displaced from the surface and diffuse into solution away from the lipid, as is the case with the free flowing ions in anionic lipids. Therefore, both attractive electrostatic interactions between the positive charges of the peptide and the negative charges on the lipid, as well as the repulsive electrostatic interactions between the positive charges on the peptide and the positive charges of the lipid will be inversely synergistic in nature (95). Consequently, the binding process will require the conformation of the incoming peptide to adapt in response to these interactions so as to maximize the attractive interactions while concurrently minimizing these repulsive interactions.
4. The net result of these modifications is to induce semi-rigid conformations onto the peptide backbone, as well as to induce specific regions of hydrophobicity and charge in hopes of dramatically changing the physicochemical surface properties presented to the lipid on binding.

Selection of conformationally restrained dipeptide units

5. The advantage the 2-amino-1,2,3,4,-tetrahydronaphthalene-2-caroxylic acid has over the Tic residue is that it contains an amide proton when incorporated into the peptide which makes NMR three-dimensional structural studies much easier.

6. For an excellent review on the structure of extended and expanded amino acids including Cα-tetrasubstituted α-amino acids, see ref. 63.

Selection of number, position, and type of flexible spacers

7. The longer the amino acid backbone, the more flexible the amino acid residue will be which allows it to occupy a larger amount of conformational space and participate in different intra-molecular hydrogen-bonding sequences (63).
8. An excellent review of the structure of extended and expanded amino acids has been provided by Vasudev et al. (63).
9. The substituted analogs of β-Phe are particularly interesting. The availability of both electron withdrawing and donating groups on the phenyl ring allows for the control of the electron density of the aromatic ring, thus providing the β-Phe analogs the ability to play a dual role as both a flexible spacer and as well as a hydrophobic residue with varying electronic properties.

Selection of number and position of hydrophobic properties

10. Non-aromatic residues include the natural amino acids Leu, and Ile. Also derivatives of β-Leu shown in Figs. 6 and 7 could be used as hydrophobic residues. However, it has been our experience that aromatic residues are preferable due to their enhanced ability to insert into the hydrophobic core of the lipid (47, 50).

Selection of number and position of electrostatic properties

11. One must remember, however, that not all cationic charges are created equal. Arg residues contain a guanidinium group at the end of their side chain. In this planar group, the positive charge is delocalized over two different nitrogen atoms. Therefore, the positive charge is more dispersed as compared to the localized positive charge on the one nitrogen of the Lys side chain amino group (48, 96). Also, the work of Haug and co-workers has shown that incorporation of quaternary ammonium groups leads to analogs with similar or lower antimicrobial activity compared to analogs containing Lys residues (48); one possible explanation for this behavior is that by having both positive charge and hydrophobic character at the same site, an environment is created that interferes with the efficiency of the electrostatic interaction between the peptide and lipid head groups (48).

Assemble the amino acids to obtain the desired structural and physicochemical properties

12. Residue 2 can also be a derivative of any of the substituted Phe analogs, such as Fpa (4-fluorophenylalanine) as in compound 46.
13. The use of conformationally restrained amino acids, such as Tic and Oic, reduces the flexibility of the peptide backbone, consequently decreasing the total conformational space that

may be sampled by the peptide. It should be noted that the two amino acids used as the conformationally restrained unit may be the same or different depending on the situation.

14. There are two Spacer #1 amino acids in the cartoon representation, residues 5 and 10, and when included in the sequence they provide molecular flexibility between the preceding conformationally restrained dipeptide and the following. By increasing the number of $-CH_2-$ groups in the spacer, in our case by including unnatural amino acids such as, β-Ala, Gaba or Ahx, the length of the peptide backbone increases, thus increasing the molecular flexibility of the local peptide chain. Longer alkyl spacers that may also be employed include the commercially available, 9-aminooctanoic acid (9-Aoa), 10-aminodecanoic acid (10Ada), 12-aminododecanoic acid (12-Adda), or 16-aminopalmitic acid (16-Apa). The longer the amino acid backbone is, the more flexible the peptide will be, allowing it to occupy a larger conformational space and participate in different intra-molecular hydrogen bonding sequences (63). For compound 46, the two Spacer #1 amino acids (residues 5 and 10, royal blue) are β-Ala which introduces two $-CH_2-$ groups into the peptide backbone.
15. By varying the number of $-CH_2-$ groups in the positively charged residue from 1 to 4 in the side chain of the basic residues (Lys, Orn, Dab, Dpr residues), the distance between the positive charge and the peptide backbone will decrease. Dpr, or 2,3-diaminopropionic acid, has only one methylene group in its side chain. Therefore for compound 46, the flexibility of Spacer #2 is greatly decreased compared to flexibility of the side chain of the naturally occurring amino acid Lys. As the length of the side chain decreases, the positive charge density will reside closer to the peptide backbone resulting in a less flexible side chain during membrane binding. As previously discussed, this flexibility is more important in the binding with zwitterionic lipids than with anionic lipids.
16. When selecting a hydrophobic residue, which may be natural or unnatural, factors including size, aromatic character, and electron density must be considered. For compound 46, Fpa (4-flurophenylalanine) was chosen, and it must be noted that Fpa was also the hydrophobic amino acid chosen for residue 2. These two hydrophobic residues need not be the same.
17. The last 3–6 amino acids of the sequence make up a highly positively charged cluster. These residues may be selected from the common basic residues (Lys, Orn, Dab, Dpr residues) or other unnatural basic amino acid residues and usually, but not always, the same as the amino acid used for Spacer #2, as is the case for compound 46.

Calcein leakage assays

18. One hundred percent leakage was determined with the addition of 50 μL of 10% Triton X. The apparent percent leakage was calculated using the following equation:

$$\%\text{leakage} = \left[1 - (F_0/F_T)\right] \times 100\%$$

where F_0 and F_T are the initial fluorescence before introduction of peptide and after the addition of Triton-X100, respectively (76, 78, 84–86).

19. In this protocol, the procedures and logic used in our laboratory to design antimicrobial peptides containing unnatural amino acids has been outlined. This protocol does not guarantee to lead to potential selective antimicrobial agents on the first attempt. However, careful selection and placement of amino acids with specific physicochemical properties in an iterative process of refinement based on information obtained from the CD and calcein leakage studies should ultimately lead to analogs exhibiting antimicrobial activity.

Acknowledgements

The authors would like to acknowledge funding from the Bacterial Therapeutics Program 2.1 of the Defense Threat Reduction Agency, contract # W81XWH-08-2-0095. The authors would also like to acknowledge funding from the North Carolina Biotechnology Center grant number 2006-FRG-1015 and from East Carolina University.

References

1. Ma J S (2003) Unnatural amino acids in drug discovery. CHIMICA OGGI Chemistry Today, June, 65–68.
2. Hendrickson T L, de Crecy-Lagard V, Schimmel P (2004) Incorporation of nonnatural amino acids into proteins. Annu Rev Biochem 73, 147–176.
3. Padgett C L. et al. (2007) Unnatural amino acid mutagenesis of GABAa receptor binding site residues reveals a novel cation-pie interaction between GABA and b2 Tyr97. J Neurosci 27, 886–892.
4. Rodriquez E A, Lester H A, Dougherty D A (2006) In vivo incorporation of multiple unnatual amino acids through nonsense and frame shift suppression. PNAS 103, 8650–8655.
5. Yeaman M R, Yount N Y (2003) Mechanisms of antimicrobial peptide action and resistance. Pharmacol Rev 55, 27–55.
6. Dennison S R et al. (2005) Amphiphilic α-helical antimicrobial peptides and their structure/function relationships. Prot Pept Lett 12, 31–39.
7. Ganz T (2003) Defensins: antimicrobial peptides of innate immunity. Nature Reviews, Immun 3, 710–720.
8. Simmaco M, Mignogna G, Barra D (1999) Antimicrobial peptides from amphibian skin: what do they tell us? Biopolymers 47, 435–450.
9. Lee D G, et al. (2004) Structure-antiviral activity relationships of cecropin A-magainin 2 hybrid peptide and its analogues. J Pept Sci 10, 298–303.

10. Patch J A, Barron A E (2003) Helical peptoid mimics of magainin-2 amide. J Am Chem Soc 125, 12092–12093.
11. Jasir A et al. (2003) New antimicrobial cystatin C-based peptide active against gram-positive bacterial pathogens, including methicillin-resistant *Staphylococcus aureus* and multiresistant coagulase-negative staphylococci. APMIS 111, 1004–1010.
12. Deslouches B et al. (2005) Activity of the De Novo engineered antimicrobial peptide WLBU2 against pseudomonas aeruginosa in human serum and whole blood for systemic applications. Antimicro Agents Chemother 49, 3208–3216.
13. Conlon J M, Abraham B, Leprince J (2007) Strategies for transformation of naturally-occurring amphibian antimicrobial peptides into therapeutically valuable anti-infective agents. Methods 42, 349–357.
14. Shlaes D M, Projan S J, Edwards J E (2004) Antibiotic discovery: state of the state. ASM News 70, 275–281.
15. Wang G, Li X, Wang Z (2009) APD2: the updated antimicrobial peptide database and its application in peptide design. Nucleic Acids Res. 37, 933–937.
16. Khandelia H I, Mouritsen O G (2008) The impact of peptides on lipid membranes. Biochim Biophys Acta 1778, 1528–1536.
17. Hancock R E W, Lehrer R (1998) Cationic peptides: a new source of antibiotics. Trends Biotechnol 16, 82–88.
18. Toke O (2005) Antimicrobial peptides; new candidates in the fight against bacterial infections. Biopolymers 80, 717–735.
19. Zasloff M (2002) Antimicrobial peptides of multicellular organisms. Nature 415, 389–395.
20. Powers J-P S, Hancock R E W (2003) The relationship between peptide structure and antibacterial activity. Peptides 24, 1681–1691.
21. Brogden K A (2005) Antimicrobial peptides: pore formers or metabolic inhibitors in bacteria? Nature Reviews Microbiol 3, 238–250.
22. Hicks R P et al. (2003) Comparison of the conformation and electrostatic surface properties of magainin peptides bound to SDS and DPC micelles: Insight into possible modes on antimicrobial activity. Biopolymers 68, 459–470.
23. Song Y M et al. (2005) Cell selectivity and mechanism of action of antimicrobial model peptides containing peptoid residues. Biochemistry 44, 12094–12106.
24. Papo N, Shai Y (2004) Effects of drastic sequence alteration and D-amino acid incorporation on the membrane binding behavior of lytic peptides. Biochemistry 43, 6393–6403.
25. Dathe M et al. (2002) General aspects of peptide selectivity towards lipid bilayers and cell membrane studied by variation of structural parameters of amphipathic α-helical model peptides. Biochim Biophys Acta 1558, 171–186.
26. Chen Y et al. (2005) Rational design of α-helical antimicrobial peptides with enhanced activities and specificity/therapeutic index. J Biol Chem 280, 12316–12329.
27. Oren Z, Hong J, Shai Y (1997) A repertoire of novel antibacterial diastereomeric peptides with selective cytolytic activity. J Biol Chem 272, 14643–14649.
28. Dathe M, Wiencek J M (1999) Structural features of helical antimicrobial peptides: their potential to modulate activity on model membranes and biological cells. Biochim Biophys Acta 1462, 71–87.
29. Papo N, Shai Y (2003) New lytic peptides based on the D,L amphipathic helix motif preferentially kill tumor cells compared to normal cells. Biochemistry 42, 9346–9354.
30. Giangaspero A, Sandri L, Tossi A (2001) Amphipathic α-helical antimicrobial peptides. Eur J Biochem 268, 5589–5600.
31. Azuma I et al. (1973) Cell wall of *Mycobacterium lepraemurium* strain Hawaii. J Bacteriology 113, 515–518.
32. Yeaman N R, Yount N Y (2003) Mechanisms of antimicrobial peptide action and resistance. Pharmacol Rev 55, 27–55.
33. Glukhov E et al. (2005) Basis for selectivity of cationic antimicrobial peptides for bacterial versus mammalian membranes. J Biol Chem 280, 33960–33967.
34. Zhang L, Harris S C, Falla T J (2005) Therapeutic application of innate immunity peptides. Horizon Bioscience, San Diego.
35. Bush K (2004) Why it is important to continue antibacterial drug discovery. ASM News 70, 282–287.
36. Klevens R M et al. (2007) Estimating health care-associated infections and deaths in U.S. hospitals.2002. Public Health Reports 122, 160–166.
37. Wang G L X, Wang Z (2009) APD2: the updated antimicrobial peptide database and its application in peptide design. Nucleic Acids Res 37, D933–937.
38. Hancock R E, Patrzykat A (2002) Clinical development of cationic antimicrobial peptides: from natural to novel antibiotics. Curr Drug Targets Infect Disord 2, 79–83.
39. Kamysz W (2005) Are antimicrobial peptides an alternative for conventional antibiotics. Nuclear Med Reviews 8, 78–86.

40. Zhang L, Falla T J (2009) Host defense peptides for use as potential therapeutics. Curr Opin Investig Drugs 10, 164–171.
41. Nowick J S et al. (2000) An unnatural amino acid that mimics a tripeptide β-strand and forms β-sheet like hydrogen-bonded dimers. J Am Chem Soc 122, 7654–7661.
42. Nowick J S (2008) Exploring β-sheet structure and interactions with chemical model systems. Acc Chem Res 41, 1319–1330.
43. Grauer A A, Konig B (2009) Synthesis of new C^{α} tetrasubstituted α-amino acids. Beilstein J Org Chem 5, 5.
44. Kyle D J et al. (1993) NMR and computational evidence that high-affinity bradykinin receptor antagonists adopt C-terminal beta-turns. J Med Chem 36, 1450–1460.
45. Kyle D J et al. (1992) A novel beta-turn mimic useful for mapping the unknown topology of peptide receptors. Pept Res 5, 206–209.
46. Oh J E, Lee K H (1999) Synthesis of novel unnatural amino acid as a building block and its incorporation into an antimicrobial peptide. Bioorg & Med Chem 7, 2985–2990.
47. Bhonsle J B et al. (2007) Application of 3D-QSAR for identification of descriptors defining bioactivity of antimicrobial peptides. J Med Chem 50, 6545--6553.
48. Haug B E, Strom M B, Svendsen J S M (2007) The medicinal chemistry of short lactoferricin-based antibacterial peptides. Curr Med Chem 14, 1–18.
49. Venugopal D et al. (2010) Novel antimicrobial peptides that exhibit activity against select agents and other drug resistant bacteria. Bioorg & Med Chem 18, 5137–5147.
50. Hicks R P et al. (2007) De Novo design of selective antibiotic peptides by incorporation of unnatural amino acids. J Med Chem 50, 3026–3036.
51. Matsuzaki K. (1998) Magainins as paradigm for the mode of action of pore forming polypeptides. Biochim Biophys Acta 1376, 391–400.
52. Wenk M R, Seelig J (1998) Magainin 2 amide interaction with lipid membranes: calorimetric detection of peptide binding and pore formation. Biochemistry 37, 3909–3916.
53. Yang L et al. (2000) Crystallization of antimicrobial pores in membranes: magainin and protegrin. Biochemical J 79, 2002–2009.
54. Jing W et al. (2003) The structure of the antimicrobial peptide Ac-RRWWRF-NH_2 bound to micelles and its interactions with phospholipid bilayers. J Pept Res 61, 219–229.
55. Watson R M et al. (2001) Conformational changes in pediocin AcH upon vesicle binding and approximation of the membrane-bound structure in detergent micelles. Biochemistry 40, 14037–14046.
56. Whitehead T L et al. (1998) Membrane-induced secondary structures of neuropeptides: a comparison of the solution conformations adopted by agonists and antagonists of the mammalian tachykinin NKl receptor. J Med Chem 41, 1497–1506.
57. Whitehead T L, Hicks R P (2001) Rationale for using simple and complex micelles in polypeptide conformational analysis. Recent Res Devel Med Chem 1, 213–233.
58. Perrine S A et al. (2000) Solution structures in SDS micelles and functional activity at the bullfrog substance P receptor of ranatachykinin peptides. J Med Chem 43, 1741–1753.
59. Young J K, Hicks R P (1994) NMR and molecular modeling investigations of the neuropeptide bradykinin in three different solvent systems: DMSO, 9:1 dioxane/water, and in the presence of 7.4 mM lysophosphatidylcholine micelles. Biopolymers 34, 611–623.
60. Russell A L et al. (2010) Spectroscopic and thermodynamic evidence for antimicrobial peptide membrane selectivity. Chem Phys Lipids 163, 488–497.
61. Porter E A, Weisblum B, Gellman S H (2002) Mimicry of host defense peptides by unnatural oligomers: antimicrobial β-peptides. J Am Chem Soc 124, 7324–7330.
62. Tossi A, Sandri L, Glangaspero A. (2000) Amphipathic α-helical antimicrobial peptides. Biopolymers 55, 4–30.
63. Vasudev P G et al. (2010) Structural chemistry of peptides containing backbone expanded amino acid residues: Conformational features of β, ε and hybrid peptides. Chem Reviews 111, 657–687.
64. Turner J et al. (1998) Activites of LL-37, a cathelin-associated antimicrobial peptide of human neutrophilis. Antimicr Agents Chemother 42, 2206–2214.
65. Fuchs P F J et al. (2006) Kinetics and thermodynamics of type VIII β-turn formation: A CD, NMR and microsecond explicit molecular dynamics study of the GDNP tetrapeptide. Biophysical J 90, 2745–2759.
66. Perczel A, Fasman G D (1992) Quantitative analysis of cyclic β-turn models, Prot Sci 1, 378–395.
67. Crisma M et al. (2005) Turn stabilization in short peptides by C^{α}-methylated α-amino acids. Biopolymers 80, 279–293.
68. Toniolo C et al. (2002) Control of peptide conformation by the Thorpe-Ingold effect (C^{α}-tetrasubstitution). Pept Sci 60, 396–419.
69. Epand R F et al. (2004) Antimicrobial 14-helical β-peptides: Potent bilayer disrupting agents. Biochemistry 43, 9527–9535.

70. Seebach D, Matthews J L (1997) β-Peptides: a surprise at every turn. Chem Commun 2015–2022.
71. White S H, Wimley W C (1998) Hydrophobic interactions of peptides with membrane interfaces. Biochim Biophys Acta 1376, 339–352.
72. Wimley W C, White S H (1993) Membrane partitioning: distinguishing bilayer effects from hydrophobic effect. Biochemistry 32, 6307–6312.
73. Dathe M et al. (1997) Hydrophobicity, hydrophobic moment and angle subtended by charged residues modulate antibacterial and haemolytic activity of amphipathic helical peptides. FEBS Letter 403, 208–213.
74. Kiyota T, Lee S, Suqihara G (1996) Design and synthesis of amphiphilic alpha-helical model peptides with systematically varied hydrophobic–hydrophilic balance and their interaction with lipid- and bio-membranes. Biochemistry 35, 13196–13201.
75. Medina M L et al. (2002) Transient vesicle leakage initiated by synthetic apoptotic peptides derived from the death domain of neurotrophin receptor p75NTR. J Pept Sci 59, 149–158.
76. Wei S-T (2006) Solution structure of a novel tryptophan-rich peptide with bidirectional antimicrobial activity. J Bacteriology 188, 328–334.
77. Hunter H N et al. (2005) The interactions of antimicrobial peptides derived from lysozyme with model membrane systems. Biochim Biophys Acta, 1668, 175–189.
78. Wieprecht T et al. (2000) Membrane binding and pore formation of the antibacterial peptide PGLa: thermodynamic and mechanistic aspects. Biochemistry 39, 442–452.
79. Wieprecht T, Apostolov O, Seelig J (2000) Binding of the antibacterial peptide magainin 2 amide to small and large unilamellar vesicles. Biophys Chem 85, 187–198.
80. Wieprecht T, Beyermann M, Seelig J (2002) Thermodynamics of the coil α-helix transition of amphipathic peptides in a membrane environment: role of vesicle curvature. Biophys Chem 96, 191–201.
81. Wieprecht T, Seelig J (2002) Isothermal titration calorimetry for studying interactions between peptides and lipid membranes. In: Simon S A, McIntosh T J (eds), Peptide–Lipid Interactions, Academic Press, San Diego, pp 32–58.
82. Wen S (2007) Dicynthaurin (ala) monomer interaction with phospholipid bilayers studied by fluorescence leakages and isothermal titration calorimetry. J Phys Chem 111, 6280–6287.
83. Bringezu F et al. (2007) The insertion of the antimicrobial peptide dicynthaurin monomer in model membranes: thermodynamic and structural characterization. Biochemistry 46, 5678–5686.
84. Wieprecht T et al. (1996) Conformation and functional study of magainin 2 in model membrane environment using the new approach of systematic double D-amino acid replacement. Biochemistry 35, 10844–10853.
85. Dathe M (1996) Peptide helicity and membrane surface charge modulate the balance of electrostatic and hydrophobic interactions with lipid bilayers and biological membranes. Biochemistry 35, 12612–12622.
86. Tamba Y, Yamazaki M (2005) Single giant unilamellar vesicle method reveals effect of antimicrobial peptide Magainin 2 on membrane permeability. Biochemistry 44, 15823–15833.
87. Mahe E et al. (1998) Solid-phase synthesis, conformational analysis, and biological activity of AVR9 elicitor peptides of the fungal tomato pathogen *Cladosporium fulvum*. J Pept Res 52, 482–494.
88. Muri E M et al. (2004) N-t-Boc-amino acid esters of isomannide potential inhibitors of serine proteases. Amino Acids 27, 153–159.
89. Grant G A (2002) Synthetic Peptides, A user's guide, 2nd ed., Oxford University Press, New York, NY.
90. Birr C, Lochinger W, Wieland T (1969) Peptide synthesis. XLII. Comparison of several coupling methods. Justus Liebigs Annalen der Chemie 729, 213–216.
91. Bodanszky M (1985) In search of new methods in peptide synthesis. A review of the last three decades. Internat J Pep Prot Res 25, 449–474.
92. Benoiton N L (2006) Chemistry of Peptide Synthesis, Taylor and Francis (CRC Press), Boca-Raton, FL.
93. Gobbo M et al. (2002) Antimicrobial peptides: synthesis and antibacterial activity of linear and cyclic drosocin and apidaecin 1b analogues. J Med Chem 45, 4494–4505.
94. Whitehead T L, Jones L M, Hicks R P (2001) Effects of the incorporation of CHAPS into SDS micelles on neuropeptide-micelle binding: separation of the role of electrostatic interactions from hydrophobic interactions. Biopolymers 58, 593–605.
95. Whitehead T L, Jones L M, Hicks R P (2004) PFG-NMR investigations of the binding of cationic neuropeptides to anionic and zwitterionic micelles. J Biomol Struct Dyn 21, 567–576.
96. Vogel H J et al. (2002) Towards a structure–function analysis of bovine lactoferricin and related tryptophan- and arginine-containing peptides. Biochem Cell Bio 80, 49–63.

Chapter 10

Use of Unnatural Amino Acids to Probe Structure–Activity Relationships and Mode-of-Action of Antimicrobial Peptides

Alessandro Tossi, Marco Scocchi, Sotir Zahariev, and Renato Gennaro

Abstract

Endogenous antimicrobial peptides (AMPs) can have multimodal mechanisms of bacterial inactivation, such as membrane lysis, interference with cell wall biosynthesis or membrane-based protein machineries, or translocation through the membrane to intracellular targets. The controlled variation of side-chain characteristics in their amino acid residues can provide much useful information on structure–activity relationships and mode-of-action, and also lead to improved activities. The small size and relatively low complexity of AMPs make them amenable to solid-phase peptide synthesis, facilitating the use of nonproteinogenic amino acids and vastly increasing the accessible molecular diversity of side chains. Here, we describe how such residues can be used to modulate such key parameters as cationicity, hydrophobicity, steric factors conformational stability, and H-bonding.

Key words: Unnatural amino acids, Nonproteinogenic amino acids, Antimicrobial peptide, Host defence peptide, SAR studies, Solid-phase peptide synthesis

Abbreviations

AA	Amino acid
DBU	1,8-Diazabicyclo[5.4.0]undec-7-ene
DCM	Dichloromethane
DIPCDI	Diisopropyl carbodiimide
DKP	Diketopiperazine
DMF	N,N-dimethylformamide
DODT	3,6-Dioxa-1,8-octanedithiol
EDT	Ethaneditiol
Fmoc	Fluorenylmethyloxycarbonyl
HFIP	Hexafluoroisopropanol
HOBt	Hydroxybenzotriazole
ivDde	1-(4,4-Dimethyl-2,6-dioxo-cyclohexylidene)-3-methyl-butyl

Loredano Pollegioni and Stefano Servi (eds.), *Unnatural Amino Acids: Methods and Protocols*,
Methods in Molecular Biology, vol. 794, DOI 10.1007/978-1-61779-331-8_10, © Springer Science+Business Media, LLC 2012

Mmt	Monomethoxytrityl
NMP	*N*-methylpyrrolidone
OSu	Succinimidyl carbonate
PEG-PS	Polyethyleneglycol-polystyrene
PG	Side-chain protecting group
PIP	Piperidine
PyBOP	(Benzotriazol-1-yl-oxy)tripyrrolidinophosphonium hexafluoro phosphate
s.r.v.	Swelled resin volume
SPPS	Solid-phase peptide synthesis
TIPS	Triisopropylsilane

1. Introduction

Antimicrobial peptides (AMPs), also known as host defence peptides (HDPs), are held to have a considerable potential as leads for the development of novel therapeutic agents, considering also the increasingly serious problem of antibiotic resistance (1). They have co-evolved with the host producer's microbial biota and the pathogenic microorganisms it encounters, and can use multiple mechanisms (2) for microbial inactivation that make it difficult for bacteria to develop efficient evasion strategies (3). Interaction of AMPs with the microbial membrane is an important part of their mechanism of action, leading to membrane breaching as a key inactivation mechanism, and/or to interference with vital, membrane located protein machinery, to self-promoted or assisted transport into the cytoplasm where they can inhibit vital bacterial processes (2, 4, 5). Some AMPs can switch from one to another of these strategies, depending on the microbial target and/or the concentration reached.

AMPs are often cationic and quite selective for microbial with respect to host cells, attributed to the presence in the former of anionic phospholipids and lack of cholesterol (6). Membrane-interaction, however, also requires a high proportion of hydrophobic residues, arranged so as to form amphipathic conformations with respect to polar ones. Residue side-chain characteristics can thus both determine molecular interactions with specific molecular targets and more generalised interactions with microbial membranes.

The rational variation of residue side chains allows for the systematic and sometimes independent variation of key global features such as cationicity, hydrophobicity, amphipathicity, or conformational stability, as well as more specific features such as H-bonding capacity, and is a prerequisite for SAR studies. Noninvasive incorporation of spectroscopic features can also help elucidate the mode-of-action of AMPs.

In this chapter, we briefly describe different examples of side-chain manipulation used for both types of studies. The methods of incorporation of unnatural amino acids are analogous to those used for normal ones, although reduced reactivity, increased steric hindrance, and increased susceptibility to racemisation must be considered. The presence of unnatural residues can also affect the methods used for quantifying physico-chemical properties and analytical techniques used to study the peptides.

2. Materials

Many different unnatural (nonproteinogenic) amino acids (AAs) can be obtained commercially in α-amino and side-chain protected forms, suitable for solid-phase peptide synthesis (SPPS). Apart from side-chain variation, the peptides' main chain can also be altered, using protected β-homo-AA or statins, for example. Otherwise, and for introduction of side chains on the amide nitrogen (peptoids), appropriate protection procedures have to be implemented. All reagents and solvents used must be peptide synthesis grade. Only the more facile and generally used Fmoc chemistry is described.

2.1. Peptide Synthesis

SPPS is normally carried out on automated synthesisers, with dedicated software that controls all aspects of the synthesis, including weights and volumes of reagents and solvents to be used. It can, however, be carried out manually using a simple setup consisting of a syringe with a Teflon frit, placed on a solvent-resistant vacuum station (see Fig. 1; Note 1).

2.1.1. Manual SPPS

Requires the following steps: STEP (1) resin washing/swelling; STEP (2) Fmoc deprotection; STEP (3) solvent wash; STEP (4) AA coupling. Back to STEP (1). At synthesis end, STEPs (2) and (3) are followed by STEP (5) deprotection/cleavage from resin. Reagents/procedures for each step are described.

2.1.2. Resins

1. Polyethylene glycol-polystyrene resins (PEG-PS) are widely used for AMP synthesis, and can be acquired commercially already functionalised with the C-terminal AA.
2. To avoid diketopiperazine (DKP) formation (especially if Pro or Gly are the C-terminal or previous AA) or racemisation (especially with C-terminal Cys), Trityl resins are preferred (see Note 2).
3. Synthesis of peptide amides on amide-handle PEG-PS avoids DKP formation and does not require resin functionalisation with C-terminal AA.
4. STEP (1): For 0.1-mmol synthesis scale with 0.2 mmol/g resin functionalisation (see Note 3), place 0.5 g resin in fritted syringe. Use 4–5 swelled resin volumes (s.r.v.) for wash/deprotection/coupling, sufficient to completely cover the resin.

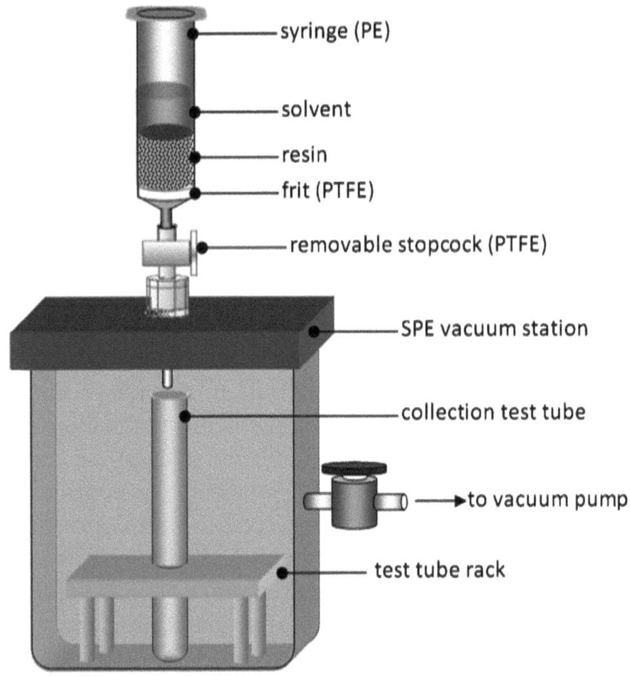

Fig. 1. Setup for manual SPPS. Polyethylene (PE) syringes and polytetrafluoroethylene (PTFE; Teflon) frits and stopcocks are resistant to synthesis and cleavage conditions. Solid-phase extraction (SPE) vacuum stations can be used if components are also resistant (i.e., glass, Teflon, etc.). These normally have multiple syringe ports. Removable stopcocks allow syringe disconnection for agitation/heating during the coupling reaction.

2.1.3. Solvents

1. The default solvent in Fmoc SPPS is *N*,*N*-dimethylformamide (DMF, anhydrous, amine-free).
2. Mixtures of DMF with *N*-methylpyrrolidone (NMP) and/or dichloromethane (DCM) (e.g., 1:1:1) may also be used to improve swelling/solvation.
3. STEP (1): Swell resin for 5–10 min in 5 s.r.v. of DCM; drain to vacuum station.

2.1.4. Fmoc Deprotection

1. Use 20% piperidine (PIP) (synthesis grade) (see Note 1) in DMF or NMP. 1,8-Diazabicyclo[5.4.0]undec-7-ene (DBU) can also be used at only 2% (alone or in combination with PIP) to overcome steric hindrance problems and to avoid PIP adducts (see Note 4).
2. STEP (2): Prepare 100 mL of 20% PIP in DMF (sufficient for multiple deprotections). Treat resin twice for 5 min with 2× s.r.v. of solution; drain to vacuum station (see Note 5).
3. STEP (3): Wash at least six times with DMF (2× s.r.v).

2.1.5. Protected AA

1. A wide selection of unnatural Fmoc-protected AAs with side-chain protecting groups (PG) is available from Anaspec,

Bachem, Iris, Novobiochem, Peptides International, Polypeptide Laboratories, Senn, Sigma-Aldrich, and several others.

2. Each AA is added with at least a 4-M excess on synthesis scale. For hindered side chains, a higher excess and/or double coupling is recommended.
3. STEP (4): Calculate protected AA weight [i.e., Fmoc-AA(PG)-OH] using MW provided by manufacturer and considering 4–8× excess on synthesis scale. Dissolve in DMF (or DMF/NMP/DCM mixture) to 0.25–0.3 M. Add activator and base and use immediately (see below).

2.1.6. Activators

1. (Benzotriazol-1-yl-oxy)tris-pyrrolidinophosphonium hexafluoro phosphate (PyBOP) is commonly used; it has a reasonable cost and allows extended coupling times.
2. It is added at a 1:1 molar ratio with AA, with the base *N,N*-diisopropylethylamine (DIPEA) added at a 1.7–2 M ratio.
3. To increase coupling efficiency, double couplings can be effected using PyBOP or alternative activators (e.g., phosphonium or uronium coupling agents, carbodiimides, protected AA chlorides or fluorides, symmetrical anhydrides) (7).
4. STEP (4): for a 4× AA excess on a 0.1-mmol synthesis scale, add 0.21 g PyBOP (MW = 520), and 0.12–0.14 mL DIPEA (MW = 129, density = 0.74 g/mL) to the AA solution. Add solution to resin, leave at least 1 h then drain to vacuum station. Mild agitation and/or heating (~50°C) can aid coupling (requires removable syringe). For double coupling, repeat STEP (4), without washing, using fresh coupling solution. Appropriately modify mixture components if required.
5. For more expensive Fmoc-AAs, 10-min preactivation with the carbodiimide DIPCDI and hydroxybenzotriazole (HOBt) (1:1.2:1.2–1.5 AA:DIPCDI:HOBt) allows coupling with only 1.5× excess of AAs, as the coupling time can be significantly extended with acceptably low levels of racemisation (see Note 1).
6. Modify weights in STEP (4) to give 0.15 mmol AA, 0.18 mmol DIPCDI, 0.18–0.22 mmol HOBt in the minimum volume of DMF required to cover the resin. After 10-min preactivation, add to syringe and agitate gently for 2–12 h; drain to vacuum station.

2.2. Protecting Groups

If the required unnatural AA is not available in Fmoc- and side-chain protected form, these protecting groups can be introduced.

2.2.1. Fmoc Protection

Fmoc-succinimidyl carbonate (Fmoc-OSu) allows facile α-amino protection, while extensive literature is available for the many orthogonal side-chain protection strategies, see ref. 7 for a comprehensive review.

2.2.2. In Situ Side-Chain Modifications

1. Some natural AA side chains can be modified *in situ* during SPPS. Ser, Thr, Tyr, Gln, Asn, introduced without side-chain protection using DIPCDI/HOBt allow phosphorylation/glycosylation, etc.
2. Basic or acidic residues are introduced with orthogonal protection (i.e., not removed by base or strong acid in the case of the Fmoc chemistry).
3. Deprotection then leaves the side-chain amine or acid moiety free for modifications on resin, such as cyclisation, acylation/amidation, dye introduction, biotinylation, etc.
4. Appropriate protection, if necessary, of the modified side chain needs to be considered. After modification, Fmoc deprotection allows the synthesis to continue (7).

2.3. Cleavage and Side-Chain Deprotection

Cleavage/deprotection of peptides from the resin requires trifluoroacetic acid (TFA, normally at >90%) with suitable scavengers to prevent side reactions with reactive deprotection intermediates (mainly carbocations). Side-chain characteristics (presence of reactive moieties, double bonds, etc.) determine which scavengers are used, see ref. 7.

2.3.1. TFA Cleavage

1. Commonly used scavengers are phenol, ethanedithiol (EDT), or 3,6-dioxa-1,8-octanedithiol (DODT, reduced odour), thioanisole, triisopropylsilane (TIPS), H_2O, etc. each usually added at between 1 and 5%.
2. Cleavage can be carried out in the same syringe used for manual SPPS, after washing, STEP (3) (see Note 1).
3. STEP (5): Prepare cleavage cocktail with volume 30× resin weight. Add volume 20× resin weight to syringe and leave for 1–2 h under mild agitation. Increase cleavage time if Arg residues are present (sulphonyl-based protecting groups are removed slowly). Drain to tube in vacuum station, wash with leftovert volume for a few minutes, and drain again. Precipitate peptide with at least ten volumes of *t*-butyl ether kept at –20°C.

2.3.2. Mild Acid Cleavage

If required, fully protected peptide acids can be cleaved from Trityl resins using 30% hexafluoroisopropanol (HFIP) in DCM for 10 min, treating the resin with 2 s.r.v mixture 3–5 times, and then evaporating the collected solution (see Note 1).

3. Methods

Substitutions of natural with unnatural AA in AMPs are decided according to the supposed mechanism of action of a lead peptide, to alter a desired feature. Prediction of how physical properties vary obviously requires estimating these properties also for unnatural residue/s. SPPS protocols are altered to take into account steric factors, reactive side-chain moieties, or the need for in situ modifications.

3.1. Charge, Hydrophobicity, and Surface properties

3.1.1. Charge

In linear helical AMPs, it can be gradually changed (e.g., from +1 to +9 in a 20 residue peptide) by interchanging Orn with Gln and Glu residues on the polar face (8–10). These side chains have similar size and hydrophobicity, although charge and H-bonding capacity vary.

3.1.2. Hydrophobicity

This can be gradually varied by choosing amongst linear aliphatic residues of increasing chain length, from Ala to isobutyric acid (Abu), to norvaline (Nva), and to norleucine (Nle) (Fig. 2). It could be further increased by using α-branched aliphatic residues [e.g., diethylglycine (Deg) or dipropylglycine (Dpg)] (see Notes 6–8) or aromatic ones (see Note 9).

3.1.3. Side-Chain Size

This can also be varied (see Note 10) so as to modulate the surface properties of AMPs. For helical ones (Fig. 2), variations in the hydrophobic sector are effected as in Subheading 3.1.2. In the polar sector, switching from Lys→ornithine (Orn)→diaminobutyric acid (Dab)→diaminopropionic acid (Dap) (see Note 11) decreases cationic residue size. Switching from Gln→homoserine (Hse)→Ser decreases neutral residue size (Fig. 2). This affects how deeply AMPs sink into the membrane while lying orthogonal to its surface ("snorkel effect") (8–10). Similar variations have probed how surface characteristics of defensins (β-sheet AMPs) affect membrane interactions (11).

3.2. Quantifying Hydrophobicity

Hydrophobicity is a fundamental property of AMPs and is quantified as the mean per residue value, calculated from individual side-chain hydrophobicity index (H_i) values, most often using the Eisenberg or Kyte and Doolittle scales (12, 13). Unnatural AAs are not included in these or most other scales (see Note 12), but H_i can be estimated if the scale used correlates well with side-chain log $P_{o/w}$ (octanol/water partition coefficient).

1. First correlate the H_i values for the 20 natural AAs, using the scale of choice (e.g., our CCS scale, see Note 12) against their published log $P_{o/w}$ values (14) (entry FAUJ830101 at http://www.genome.jp/aaindex/; note: they are in the *N*-acetyl-amino-acid-amide or Ac-AA-NH_2 form).
2. Calculate the log $P_{o/w}$ value (ClogP) for the unnatural AA of interest (in the Ac-AA-NH_2 form) using one or more of the many available online facilities. One can build the AA structure

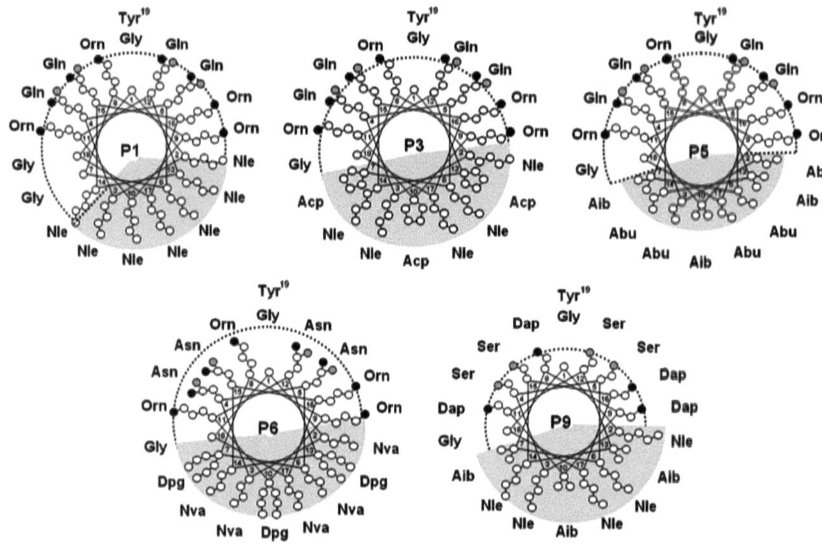

P1	Gly-Nle-Nle-Gln-Gln-Nle-Gly-Orn-Orn-Nle-Orn-Gln-Nle-Nle-Gln-Orn-Nle-Gly-Tyr
P3	Gly-Nle-Nle-Gln-Gln-Nle-Acp-Orn-Orn-Acp-Orn-Gln-Acp-Nle-Gln-Orn-Nle-Gly-Tyr
P5	Gly-Abu-Abu-Gln-Gln-Abu-Aib-Orn-Orn-Aib-Orn-Gln-Aib-Abu-Gln-Orn-Abu-Gly-Tyr
P6	Gly-Nva-Nva-Asn-Asn-Nva-Dpg-Orn-Orn-Dpg-Orn-Asn-Dpg-Nva-Asn-Orn-Nva-Gly-Tyr
P9	Gly-Nle-Nle-Ser-Ser-Nle-Aib-Dap-Dap-Aib-Dap-Ser-Aib-Nle-Ser-Dap-Nle-Gly-Tyr

Fig. 2. Use of unnatural amino acids to modulate structures of model helical AMPs. Sequences are displayed on helical wheel projections with side chains shown schematically. The lead peptide P1 is composed mainly of Orn, Gln, and Nle; structural stability was increased in P3 by Nle → Acp substitution, the hydrophobic sector was made shallower (less hydrophobic) in P5 by Nle → Abu substitution; hydrophobicity was increased by Nle → Dpg substitution in P6; the polar sector was made shallower by Orn → Dap and Gln → Ser substitution in P9.

directly on a graphical interface (e.g., using the OSIRIS property explorer) or enter it as a SMILES code.

3. Extrapolate an estimated H_i value from the correlation plot. One should also calculate the ClogP of all the normal AAs to check the correlation (see Note 13).

3.3. Conformational Stability

AMPs often have well-defined active structures, which can be modified in a controlled manner using unnatural AAs. We have found these particularly useful for modulating the structuring propensity in helical AMPs (Fig. 2), which normally undergo the random coil to helix transition only at the membrane surface. The structuring propensity affects both antimicrobial activity and selectivity (cytotoxicity to host cells) (8–10).

3.3.1. α-Branched Amino Acids

Aib is well known to stabilise helical conformations (10). It is abundant in nonribosomally synthesised peptaibol AMPs such as alamethicin. The cyclic branched residue 1-amino-1-cylopentanecarboxylic acid (Acp or cycloleucine) is even better, and the presence of just three residues in a 19-mer AMP promotes structuring and oligomerisation also in an aqueous environment (8). Other α-branched residues are less effective (see Note 7). If achiral, these residues affect the peptides' circular dichroism spectra, often used to monitor helix formation (9) (see Note 8).

3.3.2. D-Amino Acids

These are well known to destabilise helical AMPs without affecting other properties (charge, hydrophobicity, etc.). Double D-AA insertions (two successive residues) are required to significantly affect structuring, especially when induced by membrane interactions (15, 16). Diastereomeric peptides (mixed L- and D-AA) can show reduced cytotoxicity and increased stability to degradation while maintaining antimicrobial activity (17). Complete enantiomerisation (all-*D* AMPs) results in mirror image structures, also increases stability, while not reducing antimicrobial or some host cell modulating activities of helical AMPs (15, 17). Interactions with medium and cell wall components can however be affected and cytotoxicity can increase for both bacterial and host cells (18, 19). Enantiomerisation instead markedly reduces the antimicrobial activity of nonlytic, proline-rich AMPs which have an active transmembrane transport and internal targets (20).

3.4. Arg/Lys Analogues to Vary H-Bonding Capacity

Arg and Lys provide the positive charge essential for the activity/selectivity of many AMPs at the bacterial membrane surface. Their H-bonding interactions can also be relevant. Both these features can be modulated using numerous available analogues (Fig. 3).

3.4.1. Lysine

Participates as donor in one or two H-bonding interactions. These are eliminated in N^{ε}-dimethyl- or trimethyl lysine without altering charge (11, 21). Basicity, hydrophobicity, and steric bulk all increase however. H-Bonding may also be affected by shortening the side chain (see Subheading 3.1.3) or changing side-chain orientation using the D-enantiomer.

3.4.2. Arginine

Can undergo up to five H-bonding interactions (Fig. 3) as donor, as well as cation–π interactions with aromatic side chains, both of which are important also for membrane interactions (22–24). Charged, symmetrically or asymmetrically mono-, di-, and tri-methylated Arg residues have been be used to test the relevance of H-bonding in AMPs (see Fig. 3), including defensins (11), protegrins (22, 24), and proline-rich ones (Tossi et al., unpublished). Chain length can be increased (homoarginine or Hag) (11) or decreased (α-amino-3-guanidino-propionic acid or Agp) (25), and this may affect H-bonding/antimicrobial potency, depending on the mechanism. This also renders AMPs less prone to enzymatic degradation (25, 26). In citrulline (Cit) and nitroarginine (Noa,) both charge and H-bonding are affected. All these available AAs are Fmoc-protected for SPPS.

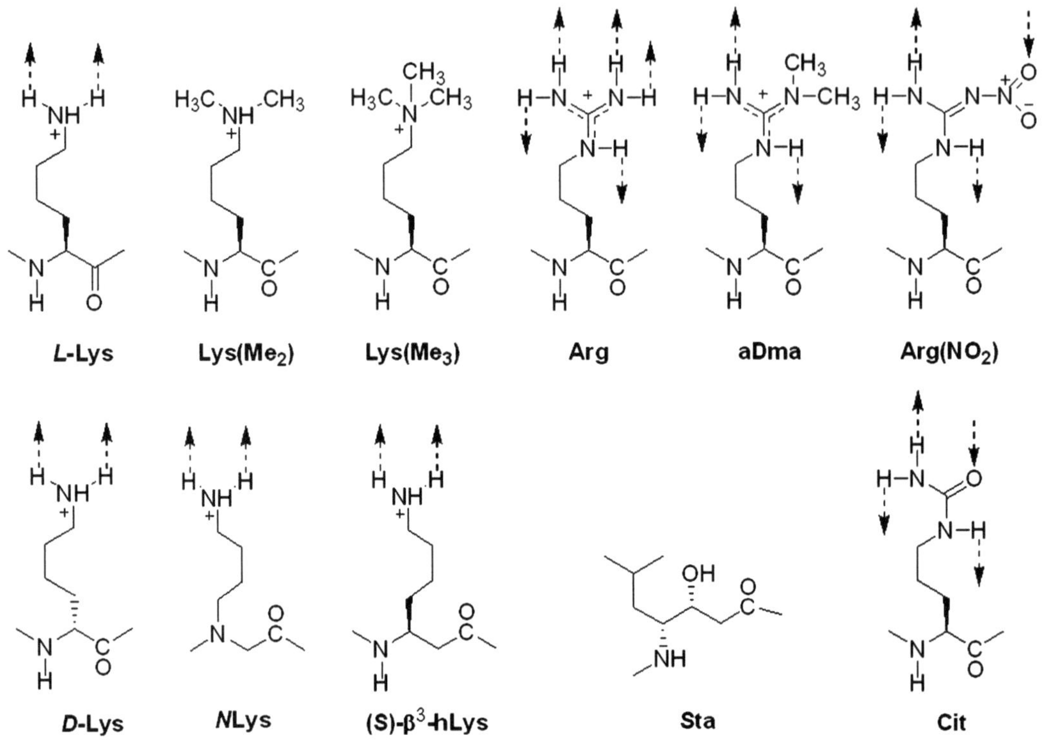

Fig. 3. Some examples of unnatural amino acids, based mainly on Lys and Arg. Several variants with other types of side chains are also available. L*-Lys* L-lysine (proteinogenic), D*-Lys* D-lysine, *Lys(Me$_2$)* (6-N,6-N)dimethyllysine or 2-amino-6-(dimethylamino)hexanoic acid, *Lys(Me$_3$)* (6-N,6-N,6-)trimethyllysine or 2-amino-6-(trimethylamino)hexanoic acid, *Arg* L-arginine (proteinogenic), *aDma* N^{ω},N^{ω}-dimethyl-L-arginine (asymmetrical dimethylarginine); Arg(NO$_2$) or Noa: N^{ω}-nitroarginine; *Cit* citrulline or 2-amino-5-(carbamoylamino)pentanoic acid, *NLys* *N*-(4-aminobutyl)glycine, *(S)-β3-hLys* L-β-homolysine or (S)-3,7-diaminoheptanoic acid, *Sta* statine or (3*S*,4*S*)-4-amino-3-hydroxy-6-methylheptanoic acid.

3.5. Main Chain Modification

The peptide main chain can be modified by using Fmoc-protected β-amino acids or statins. These both extend the chain and make it more flexible, as well as resistant to proteolysis. The side chain can also be shifted from C_{α} to N_{α} in peptoid–peptide chimaeras.

3.5.1. β- and γ-Amino Acids

β-Alanine (β-homo-glycine) and statins have both been used to introduce flexibility into helical AMPs (9). Other Fmoc-protected β-homo-AA are available with a wide range of both natural and unnatural side chains on the α-carbon (Fig. 3). Their effect on AMPs have not been tested. Statins are γ-amino acids with a hydroxyl group on the β-carbon (Fig. 3). They are present in natural peptides with acid protease inhibiting activity and might introduce this type of activity also into AMPs.

3.5.2. Peptoids

These moieties can be introduced into AMPs by using the submonomer method (27). First, DIPCDI-activated α-bromoacetic acid is used to acetylate the free amine of a growing peptide. The bromide is then displaced with a primary amine bearing the side chain of choice, leading to an *N*-substituted glycine residue. For example, mono Boc-protected 1-4-diaminobutane leads directly to the protected peptoid residue NLys (Fig. 3), while 1,3-diaminopropane followed by guanidinylation, or direct use of protected guanidino propylamine, leads to *N*Arg (27). "Ampetoids" (antimicrobial peptoid oligomers) can form biostable, amphipathic helical structures (28), where polar and hydrophobic residues are aligned in well-defined ridges, with a mechanism of action analogous to that of canonical helical AMPs. Pro-rich ampetoids show increased resistance to proteolytic degradation, and the peptoid moiety modulates internalisation (27).

3.6. Branched Peptides

Some natural AMPs are homo or heterodimers, linked by intermolecular disulphide bridges. This can be mimicked in so-called "branched," "multivalent," or "dendrimeric" AMPs, on which a large body of literature exists (see refs. 29, 30 for reviews).

3.6.1. Asymmetric Homodimers

Peptide branches are synthesised from both the α- and ε-amino groups of a C-terminal lysine/ornithine. Shorter side-chain length diamino acids (Subheading 3.1.3) could also be used as long as they are Fmoc-protected on both amine groups, but are more sterically hindered.

3.6.2. Symmetric Homodimers

Peptides are synthesised starting from the symmetrically branched *N*-substituted *N*,*N*-bis (*N*′-Fmoc-3-aminopropyl)-glycine.

3.6.3. Heterodimers

1. Peptides are synthesised starting from C-terminal Fmoc-Lys(X)-OH, where X indicates 1-(4,4-dimethyl-2,6-dioxocyclohexylidene)-3-methyl-butyl (ivDde) or monomethoxytrityl (Mmt) protecting groups. The first peptide is synthesised after removing Fmoc. When it is completed, it is N-terminally Boc-protected and the second peptide is initiated after orthogonal deprotection of the ε-amine from ivDde or Mmt (see Note 14).
2. The Universal NovaTag™ resin from NovaBiochem® is preloaded with an orthogonally protected Fmoc/Mmt diamine core for the synthesis of asymmetric branched peptides as well as C-terminal dye or biotin-labelled ones.

3.7. Spectroscopic Probes

It is common to label AMPs with fluorophores to follow membrane interactions or internalisation into bacteria or eukaryotic cells, particularly if they have cell penetrating properties (5, 31). This can however significantly alter the hydrophobicity and also the cationicity of the AMP (e.g., if fluorescein is used), so altering activity. Alternatively, tryptophan is introduced as the only natural fluorescent AA, which can however also affect the mode-of-action (see Note 9). Some unnatural amino acids can be used as minimally invasive spectroscopic probes.

3.7.1. Fluorescence Probes

Phe or Tyr residues can be replaced with the sterically similar *p*-cyanophenylalanine Phe(CN), whose fluorescence is five times stronger than that of Phe (λ_{ex} = 240 nm, λ_{em} = 300 nm). It is significantly decreased in a hydrophobic environment compared to water, and is quenched by chloride ions, making it a sensitive probe for the local environment (32).

3.7.2. IR Probes

The cyano group vibrational absorption band is outside peptide absorption regions and red-shifts significantly on going from an organic (~2,228 cm^{-1}) to an aqueous (~2,238 cm^{-1}) environment. This makes it a valuable IR probe for studying AMP/membrane interactions using supported membranes in ATR-FTIR (33). Fmoc-Ala(CN) is also available and can be used in this manner (32), substituting for a similar size proteinogenic residue such as Ala or Ser.

3.7.3. Reference Peptides

To better understand the environmental effects on Phe(CN) fluorescence, it is advisable to use a short, soluble peptide such as H–Lys–Phe(CN)–Gly–NH_2, which cannot aggregate or interact with membranes, as reference. Phe(CN) undergoes FRET with vicinal Trp and Tyr residues, and this can be accounted for by including either one of these residues in the reference peptide.

4. Notes

1. All reagents used for SPPS should be considered as toxic. All operations should be carried out in a fume hood, and gloves and protective eyewear worn. In particular, piperidine and TFA are highly corrosive. Carbodiimides are both toxic and skin irritants. Benzotriazoles such as HOBt can be allergenic, and can result in explosions. HFIP and hydrazine are quite toxic. Cleavage scavengers can have very unpleasant odours.
2. If Trityl resins are used in MW-assisted SPPS with DCM in the solvent mixture, heat at most to 45°C to avoid premature peptide detachment from the resin.
3. Resin substitution varies batch-to-batch. Resins are also available with higher substitution, but those >0.3 mmol/g should be avoided for longer peptides (over 20 residues) as this could lead to peptide aggregation and steric hindrance problems.
4. If deprotection is carried out at higher temperature (e.g., MW-assisted SPPS), hydroxybenzotriazole (HOBt) should be added to avoid racemisation and other side reactions. DBU can exacerbate side reactions if Asp or Asn are present. The presence of HOBt can help diminish this.
5. Collect the solution from Fmoc deprotection in a clean test tube. The efficacy of the previous AA coupling reaction can be

estimated using the extinction coefficient of the released Fmoc piperidine adduct (ε=7,800/M/cm at 301 nm), after appropriate dilution. Synthesis scale (mmol) = (Absorption × Volume × Dilution)/(ε × Path length). If this is close to the initial synthesis scale (Resin weight × Resin substitution), then the synthesis is likely to be proceeding well.

6. α-Branched amino acids are sterically hindered. Extended coupling, more powerful activation, or double coupling, are recommended for insertion both of this and the following residue.
7. The presence of α-branched side chains affects conformational stability in different ways. Aib and Acp increase helix stability, Dpg and Deg do not (7, 10).
8. If α-branched side chains are symmetrical, then the residue is achiral, and this must be taken into account when estimating helix content from the molar per residue ellipticity obtained using CD spectroscopy, see ref. 9.
9. Bulky, polarisable aromatic side chains tend to prefer membrane interfaces (especially Trp), which may affect AMP/ membrane interactions additionally to their increased hydrophobicity.
10. AA nomenclature is rather varied. The prefix "homo-" indicates one more methylene in the side chain of a natural AA while "nor-" one less. "β-homo-" is used to indicate the extra main-chain methylene in β-AAs.
11. Altering the side chain in Lys analogues causes a variation in the pK_a, so peptide cationicity may become more pH sensitive.
12. We have developed a consensus scale which lists the H_i values for a few of the unnatural AAs mentioned in this chapter, and tools to calculate the hydrophobicity and hydrophobic moment for linear peptides (see HydroMCalc at http://www.bbcm.units.it/~antimic/tools.html).
13. It is best not to use ClogP calculation tools as black boxes. Check the calculated values against published ones for all the proteinogenic AA as well. This can help point out problems.
14. ivDde can be removed with 2% hydrazine in DMF (see Note 1). Mmt can be removed with soft acid (e.g., 1% TFA in DCM) or using 0.6 M HOBt in DCM/TFE (1:1).

Acknowledgment

This work was supported by the Friuli Venezia Giulia LR 26 regional grant R_3A_2.

References

1. Hancock R E, Sahl H G (2006) Antimicrobial and host-defense peptides as new anti-infective therapeutic strategies. Nat Biotechnol 24, 1551–1557.
2. Brogden K A (2005) Antimicrobial peptides: pore formers or metabolic inhibitors in bacteria? Nat Rev Microbiol 3, 238–250.
3. Peschel A, Sahl H G (2006) The co-evolution of host cationic antimicrobial peptides and microbial resistance. Nat Rev Microbiol 4, 529–536.
4. Lai Y, Gallo R L (2009) AMPed up immunity: how antimicrobial peptides have multiple roles in immune defense. Trends Immunol 30, 131–141.
5. Mattiuzzo M et al. (2007) Role of the *Escherichia coli* SbmA in the antimicrobial activity of proline-rich peptides. Mol Microbiol 66, 151–163.
6. Matsuzaki K (2009) Control of cell selectivity of antimicrobial peptides. Biochim Biophys Acta 1788, 1687–1692.
7. Isidro-Llobet A et al. (2009) Amino acid-protecting groups. Chem Rev 109, 2455–2504.
8. Zelezetsky I et al. (2005) Controlled alteration of the shape and conformational stability of alpha-helical cell-lytic peptides: effect on mode of action and cell specificity. Biochem J 390, 177–188.
9. Zelezetsky I et al. (2005) Tuning the biological properties of amphipathic alpha-helical antimicrobial peptides: rational use of minimal amino acid substitutions. Peptides 26, 2368–2376.
10. Zelezetsky I, Tossi A (2006) Alpha-helical antimicrobial peptides - using a sequence template to guide structure-activity relationship studies. Biochim Biophys Acta 1758, 1436–1449.
11. Zou G et al. (2007) Toward understanding the cationicity of defensins. Arg and Lys versus their noncoded analogs. J Biol Chem 282, 19653–19665.
12. Eisenberg D, McLachlan A D (1986) Solvation energy in protein folding and binding. Nature 319, 199–203.
13. Kyte J, Doolittle R F (1982) A simple method for displaying the hydropathic character of a protein. J Mol Biol 157, 105–132.
14. Fauchere J L, Pliska V (1983) Hydrophobic parameters pi of amino-acid side chains from the partitioning of N-acetyl-amino-acid amides. J Eur J Med Chem 18, 369–375.
15. Giangaspero A et al. (2001) Amphipathic alpha helical antimicrobial peptides. Eur J Biochem 268, 5589–5600.
16. Rothemund S et al. (1995) Structure effects of double D-amino acid replacements: a nuclear magnetic resonance and circular dichroism study using amphipathic model helices. Biochemistry 34, 12954–12962.
17. Braunstein A et al. (2004) *In vitro* activity and potency of an intravenously injected antimicrobial peptide and its DL amino acid analog in mice infected with bacteria. Antimicrob Agents Chemother 48, 3127–3129.
18. Tomasinsig L et al. (2008) The human cathelicidin LL-37 modulates the activities of the P2X7 receptor in a structure-dependent manner. J Biol Chem 283, 30471–30481.
19. Vunnam S et al. (1998) Synthesis and study of normal, enantio, retro, and retroenantio isomers of cecropin A-melittin hybrids, their end group effects and selective enzyme inactivation. J Pept Res 51, 38–44.
20. Podda E et al. (2006) Dual mode of action of Bac7, a proline-rich antibacterial peptide. Biochim Biophys Acta 1760, 1732–1740.
21. Fernandez-Reyes M et al. (2010) Lysine N(epsilon)-trimethylation, a tool for improving the selectivity of antimicrobial peptides. J Med Chem 53, 5587–5596.
22. Chan D I et al. (2006) Tryptophan- and arginine-rich antimicrobial peptides: structures and mechanisms of action. Biochim Biophys Acta 1758, 1184–1202.
23. Li S et al. (2010) Water-protein interactions of an arginine-rich membrane peptide in lipid bilayers investigated by solid-state nuclear magnetic resonance spectroscopy. J Phys Chem B 114, 4063–4069.
24. Tang M et al. (2008) Effects of guanidinium-phosphate hydrogen bonding on the membrane-bound structure and activity of an arginine-rich membrane peptide from solid-state NMR spectroscopy. Angew Chem Int Ed Engl 47, 3202–3205.
25. Knappe D et al. (2010) Easy strategy to protect antimicrobial peptides from fast degradation in serum. Antimicrob Agents Chemother 54, 4003–4005.
26. Izdebski J et al. (2007) Synthesis and biological activity of homoarginine-containing opioid peptides. J Pept Sci 13, 27–30.
27. Gobbo M et al. (2009) Substitution of the arginine/leucine residues in apidaecin Ib with peptoid residues: effect on antimicrobial activity, cellular uptake, and proteolytic degradation. J Med Chem 52, 5197–5206.
28. Chongsiriwatana N P et al. (2008) Peptoids that mimic the structure, function, and mechanism of helical antimicrobial peptides. Proc Natl Acad Sci USA 105, 2794–2799.

29. Liu S P et al. (2010) Multivalent antimicrobial peptides as therapeutics: design Principles and structural diversities. Int J Pept Res Ther 16, 199–213.
30. Pini A et al. (2008) Branched peptides as therapeutics. Curr Protein Pept Sci 9, 468–477.
31. Tomasinsig L et al. (2006) Mechanistic and functional studies of the interaction of a proline-rich antimicrobial peptide with mammalian cells. J Biol Chem 281, 383–391.
32. Getahun Z et al. (2003) Using nitrile-derivatized amino acids as infrared probes of local environment. J Am Chem Soc 125, 405–411.
33. Morgera F et al. (2008) Structuring and interactions of human beta-defensins 2 and 3 with model membranes. J Pept Sci 14, 518–523.

Part III

Use of Unnatural Amino Acids in Protein Synthesis

Chapter 11

Experimental Methods for Scanning Unnatural Amino Acid Mutagenesis

Jia Liu and T. Ashton Cropp

Abstract

Site-specific incorporation of unnatural amino acids into proteins *in vivo* relies on the genetic reassignment of nonsense or quadruplet codons. Here, we describe a general procedure for the random introduction of these codons into open reading frames resulting in protein libraries that are scanned with unnatural amino acid residues. These libraries can enable large-scale mutagenesis experiments aimed at understanding and improving protein function.

Key words: Scanning mutagenesis, tRNA, Mu transposon, Codon, Unnatural amino acids

1. Introduction

More than 50 different unnatural amino acids with distinctive properties have been added to the genetic codes of bacteria, yeasts, and mammalian cells (1). Central to this methodology is the reassignment of nonsense or quadruplet codons using orthogonal aminoacyl-tRNA synthetase/tRNA pairs. These chemical tools allow one to control protein functions with light (2), detect transient protein–protein interactions (3), perform protein bioconjugation reactions (4, 5), among other tasks. While the unnatural amino acid mutagenesis methodology is generally efficient, large-scale mutagenesis experiments for scanning unnatural amino acids is hindered by the generation of genes containing random nonsense or quadruplet codons. Indeed, some unnatural amino acids that have been genetically encoded would be even more useful if one could randomly distribute them throughout protein sequence space.

Loredano Pollegioni and Stefano Servi (eds.), *Unnatural Amino Acids: Methods and Protocols*,
Methods in Molecular Biology, vol. 794, DOI 10.1007/978-1-61779-331-8_11, © Springer Science+Business Media, LLC 2012

We and others have described new approaches to generating collections of proteins that contain single amino acid mutations located in random positions (6–8). These libraries are rationally diversified and the mutation in question can be any genetically encoded amino acid. Here, we provide detailed experimental methods for scanning unnatural amino acid mutagenesis that can be applied to any protein of interest. This is a sequential method that consists of creating a small library of open reading frames in which each member contains a random, single and in-frame amber stop codon TAG (or another nonsense codon). The mutants can be separated and expressed individually with desired unnatural amino acids. Alternatively, the mutant library can be expressed with unnatural amino acid as a mixed population, which can be subsequently used for functional screening or genetic selection. The protocol from start to finish can be expected to take approximately 2 weeks to complete.

2. Materials

1. Intein targeting plasmid (pIT) (6).
2. Entranceposon (M1-Cam^R) (Finnzymes, Espoo, Finland) for PCR template (see Fig. 1 and Note 1).
3. DNA containing gene of interest (see Note 2).
4. pInSALect vector (9).
5. Appropriate expression vector (see Note 3).
6. MuA transposase and 10× transposon reaction buffer (Finnzymes, Espoo, Finland), stored at −20°C.
7. 10 mM Tris–HCl buffer, pH 8.5 at 25°C.
8. TE buffer, 10 mM Tris–HCl buffer, 0.1 mM EDTA, pH 8.5 at 25°C.
9. Primers (*Mly* I restriction sites underlined and reverse complement of TAG codon in italic).
   - Forward (see Note 4) and reverse (see Note 5) primers for cloning the gene of interest into pIT vector.
   - Forward and reverse primers for cloning the gene of interest from the pIT vector into a desired expression vector.
   - *Mly*I transposon 5′-gcttagatctgactcggcgcacgaaaaacgcgaaag-3′.
   - Linker FWD 5′Phos-ggatcgactctcctgggtattcgcaataatcttaata ctgag-3′.
   - Linker REV 5′Phos-*cta*gatctgactcaattaccaatgcttaatcagtgag gcacct-3′ (see Note 6).

     Dissolve primer stocks in TE buffer at a final concentration of 100 μM and store at −20°C. Dilute to 10 μM prior to use.

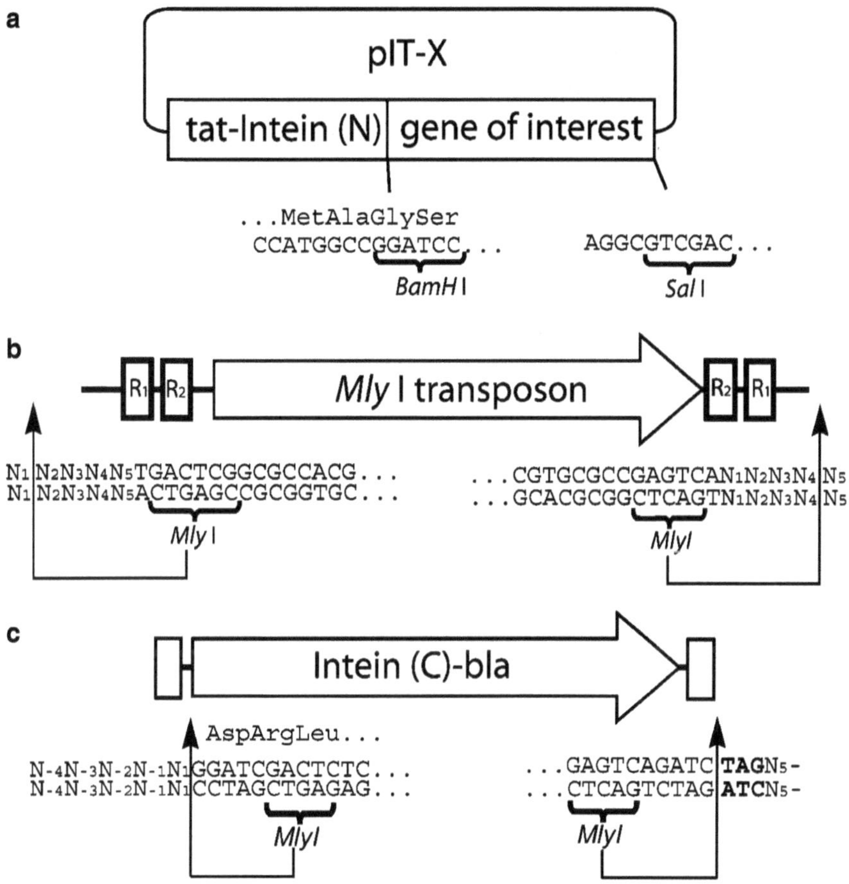

Fig. 1. Schematic diagram showing the DNA components described in this protocol with nucleotides numbered relative to original transposon insertion site. (**a**) The cloning site of pIT-X indicating correct reading frame. Oligonucleotides used for protein of interest must match the correct reading frame. (**b**) *Mly* I transposon showing positioning of restriction sites as it would insert into a segment of DNA. (**c**) Reading-frame selectable linker showing restriction sites, frame, and amber stop codon scar in *bold.*

10. dNTP stock solution, 10 mM total, 2.5 mM each, stored as 20 μL aliquots at −20°C.
11. Phusion DNA polymerase (New England Biolabs, Ipswich, MA), stored at −20°C.
12. Restriction enzymes *Sal* I, *Bam*H I, *Bgl* II, FastDigest *Mly* I and appropriate buffers (Fermentas, Glen Burnie, MD). Store at −20°C.
13. T4 DNA ligase (Fermentas), stored at −20°C.
14. 0.7 and 1% agarose gels.
15. QIAquick PCR purification kit (Qiagen, Valencia, CA).
16. QIAEX II gel extraction kit (Qiagen, Valencia, CA).
17. Chemically or electro-competent *Escherichia coli* cells for cloning, such as DH10B. Store at −80°C.

18. Chemically or electro-competent *E. coli* cells for expression, such as BL21 (DE3). Store at −80°C.
19. 1,000× antibiotic stock solutions: Ampicillin (50 mg/mL), kanamycin (50 mg/mL), and chloramphenicol (35 mg/mL).
20. 1,000× reagents for protein induction, such as 1 M isopropyl β-D-1-thiogalactopyranoside (IPTG) or 20% arabinose.
21. Unnatural amino acid, sufficient to make media at 2 mM final concentration.
22. Luria–Bertani (LB) medium (1 L): 10 g tryptone, 5 g yeast extract, and 10 g sodium chloride.
23. SOC medium (1 L): 20 g tryptone, 5 g yeast extract, 10 mM sodium chloride, and 25 mM potassium chloride. Add magnesium chloride to 10 mM and glucose to 20 mM after autoclaving.
24. 2× YT medium (1 L): 16 g tryptone, 10 g yeast extract, and 5 g sodium chloride, pH 7.2.
25. LB agar plates with appropriate antibiotics.

3. Methods

3.1. Generating a pIT-X Vector Containing Gene of Interest

1. Digest 2 μg of the pIT vector (6) using *BamH* I and *Sal* I. Gel-purify the 3.2 kbp fragment using QIAEX II gel extraction kit (see Note 7).
2. Amplify the gene of interest using PCR to introduce a *BamH* I and *Sal* I at the 5′ and 3′-ends, respectively.
3. Purify the PCR product using with the QIAquick PCR purification kit.
4. Digest the PCR products using *BamH* I and *Sal* I and gel-purify the digestion products using the QIAEX II gel extraction kit.
5. Ligate the purified digest of pIT vector and PCR product using T4 DNA ligase. Name the recombinant plasmid as pIT-X (see Note 8).

3.2. Transposition Reaction

1. Prepare the *Mly* I transposon DNA by PCR as follows: 10 ng Entranceposon (M1-Cam^R) template, 0.2 mM dNTPs, 0.5 μM of primer *Mly* I transposon (which serves as the forward and reverse primer), 1× Phusion HF buffer (NEB) and 0.5 U of Phusion DNA polymerase (NEB) in a 50-μL solution.
2. Incubate the reaction with the following cycle conditions: Initial denaturation at 98°C for 30 s, 30 cycles of 98°C for

10 s, 60°C for 30 s, and 72°C for 1 min and final extension at 72°C for 10 min.

3. Purify the PCR product using with the QIAquick PCR purification kit.
4. Digest the *Mly* I transposon DNA as follows: 2 μg DNA, 10 U of *Bgl* II, and 1× buffer O (Fermentas) in a 50-μL solution (see Note 9).
5. Load the digestion products on a 0.7% agarose gel and purify the 1.3 kbp fragment.
6. Load 1 or 2 μL of the purified transposon DNA on a 1% agarose gel and determine the concentration by comparison with a DNA standard.
7. Store the purified transposon DNA at −20°C prior to use.
8. Perform a 20-μL transposition reaction with the following components: 400 ng of pIT-X plasmid DNA, 1.3 molar excess of *Mly* I transposon DNA (see Note 10), 1× reaction buffer (Finnzymes), and 1 U of MuA transposase (Finnzymes).
9. Incubate the reaction at 30°C for 4 h.
10. Stop the reaction by adding SDS to a final concentration of 0.1% and heating at 75°C for 10 min. Cool the reaction solution on ice.
11. Transform each 1 μL of products into 50 μL of electro-competent *E. coli* cells (see Note 11) and recover in 500 μL SOC at 37°C for 1 h.
12. Plate transformation cells on LB agar supplemented with 50 μg/mL kanamycin and 10 μg/mL chloramphenicol. Grow at 37°C overnight.
13. For a gene of Lbps, collect $9 \times (L + 1{,}500)$ colonies from the transposition reaction (see Note 12). Mix well the collected colonies in LB medium supplemented with 50 μg/mL kanamycin and 10 μg/mL chloramphenicol.
14. Save five tubes of 1-mL stock cells with an OD_{600} of >1.0 in 15% glycerol, store at −80°C (see Note 13). Extract the plasmid DNA from the remaining cells to build the pIT-X-transposon library.

3.3. Isolation of Transposon Insertions Located in the Gene of Interest

1. Digest 2 μg of the transposon library DNA using *Sal* I and *BamH* I.
2. Load the digestion product on a 0.7% agarose gel and gel purify the two fragments using the QIAEX II gel extraction kit corresponding to the pIT vector backbone and the gene of interest with transposon insertions (see Note 14).

3. Ligate the two purified DNA fragments with T4 DNA ligase. Incubate at 16°C for 16 h. Inactivate the ligation product by heating at 70°C for 10 min.
4. Transform the ligation products into electro-competent *E. coli* cells and plate cells on LB agar supplemented with 50 μg/mL kanamycin and 10 μg/mL chloramphenicol.
5. Collect $9 \times L$ colonies (see Note 15) and save cell stocks as above. Extract the plasmid DNA from the remaining cells to obtain the purified transposon library.

3.4. Creation of Random Triplet Nucleotide Deletions

1. Digest the purified transposon library DNA with *Mly* I as follows: 1 μg of DNA, 1 U of FastDigest *Mly* I (Fermentas) (see Note 16), and 1× FastDigest buffer in 50-μL solution.
2. Incubate at 37°C for 1 h (see Note 17).
3. Load the digestion products on a 0.7% agarose gel and purify the larger fragment using the QIAEX II gel extraction kit. This corresponds to a linearized pIT-X library in which each DNA copy contains a random triplet nucleotide deletion in the gene of interest.
4. Load 1 or 2 μL of the purified deletion library DNA on a 0.7% agarose gel and determine the DNA concentration by comparison with DNA standard.
5. Store the purified DNA at −20°C prior to use.

3.5. Linker Ligation and Reading-Frame Selection

1. Amplify the reading-frame selection linker by PCR as follows: 10 ng pInSALect template, 0.2 mM dNTPs, 0.5 μM of each primer LinkerFWD and LinkerFWD, 1× Phusion HF buffer (NEB), and 0.5 U of Phusion DNA polymerase (NEB) in a total 50-μL solution.
2. Incubate the reaction with the following cycle conditions: Initial denaturation at 98°C for 30 s, 30 cycles of 98°C for 10 s, 60°C for 30 s, and 72°C for 1 min and final extension at 72°C for 10 min.
3. Purify the PCR products with QIAquick PCR purification kit. Store at −20°C prior to use (see Note 18).
4. Load 1 or 2 μL of the purified linker DNA on a 1% agarose gel and determine the concentration by comparison with a DNA standard.
5. Perform linker ligation in 100 μL total volume containing: 1 μg linker DNA, 500 ng of linearized pIT-X deletion library DNA, 1 μL of T4 DNA ligase and 1× T4 ligase buffer. This will produce a molar linker:library ratio of approximately 5:1. This can be adjusted appropriately given the size of pIT-X.
6. Incubate reaction at 16°C for 16 h. Heat-inactivate the ligation reaction at 70°C for 10 min.

7. Pool the entire ligation product, ethanol precipitate the DNA and resuspend in 10 μL of sterilized water.
8. Transform 1 μL of the concentrated ligation product into 50-μL electro-competent cells and recover in 500 μL SOC at 37°C for 1 h.
9. Plate cells on LB agar supplemented with 50 μg/mL kanamycin and 40 μg/mL ampicillin (see Note 19). Grow plates at 30°C overnight (see Note 20).
10. Collect 9×*L* colonies and save stock cells as stated above. Extract plasmid DNA from remaining cells to obtain pIT-X-linker library. Store at −20°C prior to use.

3.6. Generation of In-frame TAG Mutations

1. Digest the linker-containing library DNA as follows: 1 μg of DNA, 1 U of FastDigest *Mly*I (see Note 16), and 1× FastDigest buffer in 50-μL solution (see Note 21).
2. Incubate at 37°C for 1 h (see Note 17).
3. Load the digestion product on a 0.7% agarose gel and purify the larger fragment using QIAEX II gel extraction kit. This fragment corresponds to a linearized pIT-X library in which each DNA copy contains a random TAG codon mutation in the gene of interest.
4. Determine the concentration of purified TAG mutation library by comparison with a DNA standard on an agarose gel.
5. Perform a 10-μL intramolecular ligation reaction as follows: ~20–30 ng of linearized TAG mutation library DNA, 1 μL T4 DNA ligase, and 1× T4 ligase buffer.
6. Incubate reaction at 16°C for 16 h. Heat-inactivate ligase at 70°C for 10 min.
7. Transform each 1 μL of the ligation products into electro-competent cells and recover at 37°C for 1 h.
8. Plate cells on LB agar supplemented with 50 μg/mL kanamycin. Grow at 37°C overnight.
9. Collect 9×*L* colonies and save stock cells as stated above. Extract plasmid DNA from remaining cells to obtain pIT-X(TAG) mutation library. Store DNA at −20°C.
10. Select individual colonies to verify mutations by DNA sequencing.

3.7. Cloning the TAG Mutation Library into an Expression Vector and Production of Mutant Proteins

1. PCR amplify the TAG mutation library with primers appropriate for your expression vector and purify the PCR products with QIAquick PCR purification kit.
2. Digest the purified PCR products with desired restriction enzymes and gel-purify the digestion products using QIAEX II gel extraction kit.

3. Digest the desired expression vector using the same restriction enzymes and gel-purify the digestion products using QIAEX II gel extraction kit.
4. Ligate the purified PCR products and vector DNA using T4 DNA ligase as described above.
5. Incubate at 16°C for 16 h. Heat-inactivate the ligase at 70°C for 10 min.
6. Transform the ligation products into electro-competent *E. coli* cells.
7. Plate cells on LB agar supplemented with the antibiotic corresponding to the selection marker in the expression vector.
8. Collect the $9 \times L$ colonies (see Note 15) and save stock cells as stated above. Extract the plasmid DNA from remaining cells (see Note 22).
9. Transform the collected library DNA into competent cells that are suitable for protein expression, such as those containing a pSUP plasmid for incorporation of unnatural amino acids (10) (see Notes 23 and 24).
10. Express the mutant proteins in liquid culture in the presence of desired unnatural amino acid at a final concentration of 2 mM. Depending on the ultimate use of the protein(s), screens or selections can be performed on individual clonal isolates or mixed populations.

4. Notes

1. In this protocol, the *Mly* I transposon DNA is generated by PCR from Entranceposon (M1-Cam^R) DNA. Once constructed, this DNA can be inserted into a cloning vector such that it can be released from the vector by *Bgl* II digestion and contain a 5′-GATC overhang that is important for the transposon reaction (11). This ensures that correctly digested DNA is available for future use (Fig. 1b).
2. The *BamH* I and *Sal* I sites in the genes of interest, if any, should be destroyed by site-directed mutagenesis for downstream manipulations.
3. The expression vector will be double transformed into an expression cell strain containing a plasmid for the incorporation of unnatural amino acids (10). These plasmids contain a chloramphenicol selection marker and p15A origin of replication. Therefore, any expression vector used for the protein of interest should have an alternative selection marker and origin of replication.

4. The forward primer should include a *BamH* I site before the start codon of the gene of interest. The number of oligonucleotides between the start codon and *BamH* I site should be adjusted such that the gene of interest will be in the same reading frame of the N-terminal fusion peptide from the pIT vector (Fig. 1a).
5. The reverse primer should contain a *Sal* I site after the stop codon.
6. It is recommended to PAGE-purify this primer to eliminate truncated products. We have also found that including a phosphate at the 5′-end can increase the efficiency of linker ligation.
7. When gel slices are incubated with solubilization buffer (Qiagen), any temperature above 50°C is not recommended. Higher temperatures denature double-stranded DNA and reduce the yield of gel-purification.
8. When inserted into pIT vector, the gene of interest is in-frame with an N-terminal fusion peptide (Fig. 1a) which contains a Tat signal sequence and an N-terminal region of the VMA *cis*-splicing intein (VMA-N) (9).
9. It is not recommended to use excess DNA (e.g., >2 μg) in *Bgl* II digestion. In our experience, excess DNA can result in incorrect or incomplete digestion products which will reduce the efficiency of transposon integrations.
10. Depending on the size of the target gene (and pIT-X), the amount of transposon DNA should be adjusted. It is important to maintain the molar excess below 2.0 to avoid formation of unstable complexes containing multiple transposon insertions.
11. When using competent cells with a transformation efficiency of 5×10^7 colonies/μg DNA, 20 μL of transposition products typically results in >80,000 colonies. Unused products can be stored at −20°C for at least 1 month without any noticeable decrease of transformation yield.
12. To calculate the number of colonies required for full coverage, the number of allowed sites and site preference of transposon insertions must be considered. The number of insertion sites possible in the pIT vector backbone is 1,500 bps (2,900 of total length minus 600 bps of replication origin and 800 bps of kanamycin resistance gene). Therefore, there are $(L+1{,}500)$ possible insertion sites for the pIT vector carrying a gene of Lbps. Threefold degeneracy $3\times(L+1{,}500)$ of colonies are required to cover 95% of all the possible insertion events, assuming even distribution of insertions (12). To compensate for the site preference of transposon insertions, we assume that additional threefold degeneracy is sufficient to accommodate any intrinsic site preference of Mu transposase (13). Therefore,

$3 \times 3 \times (L + 1{,}500) = 9 \times (L + 1{,}500)$ colonies are sufficient to give a 95% coverage for a gene of Lbps.

13. If needed, the whole tube (1 mL) of stock cells should be used for inoculation to maintain the library diversity.
14. As the transposon can insert both inside and outside of the gene of interest, a *BamH* I/*Sal* I digestion of the transposon library will result in four DNA fragments (1) vector backbone with transposon, (2) vector backbone, (3) gene of interest with transposon insertions, and (4) gene of interest.
15. $9 \times L$ colonies are required for maintaining library diversity because transposon insertions outside of the gene of interest have been purged.
16. Under our experimental conditions, Fermentas FastDigest *Mly* I with FastDigest buffer cleaves DNA more precisely than NEB *Mly* I with NEBuffer 4 does, as judged by sequencing results of obtained mutants.
17. Longer incubation time is not recommended as overdigestion with *Mly* I can remove extra bases.
18. The PCR products of reading-frame selection linker contain the C-terminal region of VMA *cis*-splicing intein (VMA-C) and the β-lactamase gene (BLA), flanked by a *Mly* I site at each end and a TAG codon at the 3′-end (Fig. 1c).
19. When the linker DNA is inserted in-frame with the gene of interest, VMA-N from the pIT vector will be in the same reading frame of VMA-C from the linker sequence. The VMA intein will self-splice and assemble the Tat signal peptide and BLA, which will subsequently confer cells ampicillin resistance.
20. Incubation at a reduced temperature of 30°C is critical for correct intein-mediated splicing (9).
21. This *Mly* I digestion will release the linker sequence from the pIT-X library, leaving an in-frame TAG codon to fill the triplet nucleotide scar.
22. In the pIT-X plasmid, the N-terminal fusion peptide (Tat signal peptide and VMA-N) usually disrupts the native function of the protein of interest. Therefore, activity assay may not be directly performed under the context of pIT vector.
23. We have obtained the best results by the creation of competent cells that already contain a plasmid for expressing an orthogonal aminoacyl-tRNA synthetase/tRNA pair. For example, one can transform BL21(DE3) with a pSUP plasmid and prepare electrocompetent cells using chloramphenicol selection. The TAG mutant library (in the expression vector) can then be transformed into these cells and selected for both plasmids. This approach results in higher protein expression yields and transformation efficiency.

24. We have found acceptable expression results using the pSUP series of plasmids (10) that express orthogonal aminoacyl-tRNA synthetase/tRNA pairs. Recently, alternative plasmid systems (14) have been described that may result in improved yields depending on the protein being expressed. The choice of plasmid does not impact the methodology described here for TAG mutant library construction.

Acknowledgments

The authors thank the National Institutes of Health (GM084396) for financial support. We are also grateful to Prof. Stefan Lutz (Emory University) for the plasmid pInSALect and Prof. Peter G. Schultz for the plasmid pSUP.

References

1. Cropp T A, Schultz P G (2004) An expanding genetic code. Trends Genet 20, 625–630.
2. Wu N et al. (2004) A genetically encoded photocaged amino acid. J Am Chem Soc 126, 14306–14307.
3. Chin J W et al. (2002) Addition of a photocrosslinking amino acid to the genetic code of *Escherichia coli*. Proc Natl Acad Sci USA 99, 11020–11024.
4. Wang L et al. (2003) Addition of the keto functional group to the genetic code of *Escherichia coli*. Proc Natl Acad Sci USA 100, 56–61.
5. Deiters A et al. (2004) Site-specific PEGylation of proteins containing unnatural amino acids. Bioorg Med Chem Lett 14, 5743–5745.
6. Daggett K A, Layer M, Cropp T A (2009) A general method for scanning unnatural amino acid mutagenesis. ACS Chem Biol 4, 109–113.
7. Baldwin A J et al. (2008) Expanded molecular diversity generation during directed evolution by trinucleotide exchange (TriNEx). Nucleic Acids Res 36, e77.
8. Baldwin A J et al. (2009) Expanded chemical diversity sampling through whole protein evolution. Mol Biosyst 5, 764–766.
9. Gerth M L, Patrick W M, Lutz S (2004) A second-generation system for unbiased reading frame selection. Protein Eng Des Sel 17, 595–602.
10. Ryu Y, Schultz P G (2006) Efficient incorporation of unnatural amino acids into proteins in *Escherichia coli*. Nat Methods 3, 263–265.
11. Haapa S et al. (1999) An efficient and accurate integration of mini-Mu transposons *in vitro*: a general methodology for functional genetic analysis and molecular biology applications. Nucleic Acids Res 27, 2777–2784.
12. Patrick W M, Firth A E, Blackburn J M (2003) User-friendly algorithms for estimating completeness and diversity in randomized protein-encoding libraries. Protein Eng 16, 451–457.
13. Poussu E et al. (2004) Probing the alpha-complementing domain of *E. coli* beta-galactosidase with use of an insertional pentapeptide mutagenesis strategy based on Mu *in vitro* DNA transposition. Proteins 54, 681–692.
14. Young T S et al. (2010) An enhanced system for unnatural amino acid mutagenesis in *E. coli*. J Mol Biol 395, 361–374.

Chapter 12

Genetic Incorporation of Unnatural Amino Acids into Proteins in Yeast

Qian Wang and Lei Wang

Abstract

Unnatural amino acids can be genetically incorporated into proteins in live cells by using an orthogonal tRNA/aminoacyl-tRNA synthetase pair. Here we describe a method to efficiently express the orthogonal tRNA and synthetase in *Saccharomyces cerevisiae*, which enables unnatural amino acids to be genetically incorporated into target proteins in yeast with high efficiency. We also describe the use of a yeast strain deficient in the nonsense-mediated mRNA decay, which further increases the unnatural amino acid incorporation efficiency when a stop codon is used to encode the unnatural amino acid. These strategies will facilitate the investigation of proteins and their related biological processes in yeast by exploiting the novel properties afforded by unnatural amino acids.

Key words: Unnatural amino acid, Yeast, Orthogonal tRNA, Orthogonal synthetase, Amber suppression, Polymerase III promoter, Nonsense-mediated mRNA decay, Green fluorescent protein

1. Introduction

To genetically incorporate an unnatural amino acid into proteins in yeast, an orthogonal tRNA/aminoacyl-tRNA synthetase pair needs to be expressed (1, 2). This tRNA/synthetase pair does not cross-talk with endogenous tRNA/synthetase pairs and is engineered to be specific for the desired unnatural amino acid. One challenge is to efficiently express the orthogonal tRNA in yeast (1, 3). Most orthogonal tRNAs used in yeast are derived from bacteria. However, bacteria and yeast differ significantly in tRNA transcription and processing (4) (Fig. 1a). Bacterial tRNAs are transcribed by the sole RNA polymerase (Pol) through promoters upstream of the tRNA gene. The transcription of yeast (and other eukaryotic) tRNAs by Pol III depends principally on promoter elements within

Loredano Pollegioni and Stefano Servi (eds.), *Unnatural Amino Acids: Methods and Protocols*,
Methods in Molecular Biology, vol. 794, DOI 10.1007/978-1-61779-331-8_12, © Springer Science+Business Media, LLC 2012

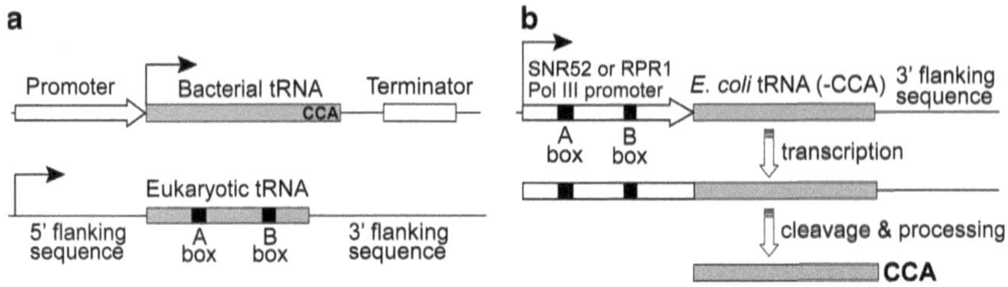

Fig. 1. A general method for efficient expression of orthogonal bacterial tRNAs in yeast. (**a**) Gene elements and organization for tRNA transcription in bacteria (*top*) and in eukaryotic cells (*bottom*) are different. (**b**) Expression of prokaryotic tRNAs in yeast by using an external SNR52 or RPR1 Pol III promoter. These promoters contain the eukaryotic consensus A- and B-box sequences and thus can drive the RNA transcription by Pol III. After the primary RNA is transcribed, the promoter is cleaved to generate the mature tRNA.

the tRNA known as the A- and B-box (4). The A- and B-box identity elements are conserved among eukaryotic tRNAs, but are lacking in many bacterial tRNAs. In addition, bacterial tRNA genes encode full tRNA sequences, whereas eukaryotic tRNAs have the 3′-CCA trinucleotide enzymatically added after transcription (4). Therefore, transplanting bacterial tRNA directly into the tRNA gene cassette in yeast does not generate functional tRNA.

We developed a general method to express bacterial tRNAs in yeast (3), which involves placing an external Pol III promoter containing the consensus eukaryotic A- and B-box sequences upstream of the target bacterial tRNA gene (Fig. 1b). The 3′-CCA trinucleotide of the tRNA is excluded, and the tRNA(−CCA) gene is followed by a 3′-flanking sequence of an endogenous yeast tRNA. A primary RNA consisting of the promoter and the tRNA is transcribed, and the promoter is cleaved posttranscriptionally to yield the mature tRNA. Two yeast Pol III promoters, the RPR1 promoter and the SNR52 promoter, have been shown to efficiently drive the expression of *Escherichia coli* tRNAs in yeast. The expressed *E. coli* tRNA is six- to ninefold more active in translation than the same tRNA transcribed by using the SUP4 5′-flanking sequence. Alternative methods using a strong Pol II promoter with tandem tRNA repeats (5) or the yeast $tRNA^{Arg}$ fused upstream of the target tRNA (6, 7) have also been developed. We will focus on the SNR52/RPR1 promoter method here, as this method works with different *E. coli* tRNAs and has been reproducibly used in different laboratories (8, 9). A similar approach involving the use of type-3 Pol III promoters also works efficiently in mammalian cells (10).

The amber stop codon, UAG, is often introduced into the gene of interest to specify the site at which the unnatural amino acid is to be incorporated. The Nonsense-mediated mRNA decay (NMD) pathway mediates the rapid degradation of mRNAs that contain premature stop codons in yeast (11), whereas no such pathway exists in *E. coli*. When stop codons are used to encode unnatural amino acids, NMD could result in a shorter lifetime for the target mRNA and thus a lower protein yield in yeast. We generated an NMD-deficient yeast strain (LWUPF1Δ) by knocking out the *upf1* gene from the yeast genome, and found that this strain indeed increases the unnatural amino acid incorporation efficiency in comparison to the wild-type (wt) yeast (3).

To demonstrate this method, we describe here the procedures to incorporate a fluorescent unnatural amino acid 2-amino-3-(5-(dimethylamino) napththalene-1-sulfonamido) propanoic acid (abbreviated as DanAla) into the green fluorescent protein (GFP) at position 39. As shown in Fig. 2, the orthogonal *E. coli* leucyl amber suppressor tRNA ($EctRNA^{Leu}_{CUA}$) will be expressed using the SNR52 promoter, and the orthogonal DanAla-specific synthetase (DanAlaRS) (12) will be expressed using the GPD promoter. The target GFP gene with a UAG codon at site 39 will be expressed using the ADH1 promoter in another plasmid. A His6 tag is appended to the C-terminus of GFP for Western detection and affinity purification. The two plasmids will be co-transformed into the wt or LWUPF1Δ yeast strain. The incorporation of DanAla into GFP will be verified with the generation of GFP fluorescence and quantified using flow cytometry or Western blot. DanAla-containing GFP proteins will be extracted from cells and purified using Ni–NTA chromatography.

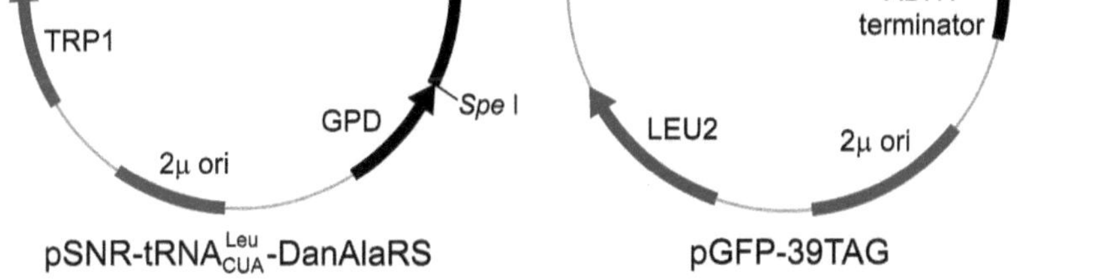

Fig. 2. Schematic illustration of the expression plasmids. (**a**) Expression plasmid for the orthogonal tRNA/synthetase. The orthogonal tRNA (without 3′-CCA trinucleotide) is expressed by the SNR52 promoter; the 3′-flanking sequence of SUP4 is appended at the 3′ end. The orthogonal synthetase gene is expressed by the GPD promoter; mutant synthetase genes specific for different unnatural amino acids can be cloned to replace the DanAlaRS gene by using the unique *Spe*I and *Xho*I sites. (**b**) Expression plasmid for the target gene with the UAG codon introduced at the site of interest. Other genes can be cloned into this vector by using the *Sal*I and *BamH*I sites.

2. Materials

Prepare all solutions with ultrapure water and use analytical grade reagents unless indicated otherwise.

2.1. Stock Solutions

1. 40% D-glucose (1 L): dissolve 400 g of D-glucose in 750 mL water. Adjust the final volume to 1 L. Autoclave at 121°C for 15 min.
2. 50% PEG-3350 (100 mL): add 50 g PEG-3350 in 90 mL water, warm the solution to 50°C to help PEG-3350 to dissolve. Make up the final volume to 100 mL. Autoclave at 121°C for 15 min.
3. 1 M Tris–HCl, pH 8.0 (1 L): dissolve 121.1 g Tris in 800 mL water; adjust pH to 8.0 with HCl. Make up the final volume to 1 L. Autoclave at 121°C for 15 min.
4. 0.5 M EDTA, pH 8.0 (1 L): dissolve 186.1 g Na_2EDTA $2H_2O$ in 800 mL water, adjust pH to 8.0 with NaOH. Make up the final volume to 1 L. Autoclave at 121°C for 15 min.
5. 10× TE (1 L): mix 880 mL water, 100 mL 1 M Tris–HCl, pH 8.0, 20 mL 0.5 M EDTA, pH 8.0 completely. Autoclave at 121°C for 15 min.
6. 10× LiAc (1 M, 100 mL): dissolve 10.2 g lithium acetate dihydrate in 90 mL water; adjust pH to 7.5 with dilute acetic acid. Bring volume up to 100 mL. Autoclave at 121°C for 30 min.
7. 500× DanAla (500 mM, 1 mL): weigh 168.5 mg of DanAla, add 750 μL of water and vortex to mix. Add concentrated HCl dropwise until the solution become clear. Add water to bring to 1 mL (see Note 1).
8. 25× EDTA-free protease inhibitor: dissolve one tablet of EDTA-free protease inhibitor cocktail tablet (Roche) in 1 mL of 1× PBS.

2.2. Media

1. YPD Medium (1 L): 20 g/L Bacto-peptone, 10 g/L yeast extract, 20 g/L D-glucose, pH 6.5. Weigh 20 g bacto-peptone and 10 g yeast extract, transfer to a 1-L graduated cylinder or a glass beaker, and add water to 950 mL. Mix completely and adjust pH to 6.5. Autoclave at 121°C for 15 min. Let the medium cool below 50°C and add 50 mL of sterile 40% D-glucose stock solution.
2. SD/-Leu/-Trp Medium (1 L): dissolve 6.7 g yeast nitrogen base without amino acids and 0.64 g-Leu/-Trp dropout supplement (see Note 2) in 950 mL water; mix to dissolve. Adjust pH to 5.8 if necessary. Autoclave at 121°C for 15 min. Let the medium cool below 50°C and add 50 mL of sterile 40% D-glucose stock solution.

3. SD/-Leu/-Trp/+DanAla Medium (1 L): dissolve 6.7 g yeast nitrogen base without amino acids and 0.64 g-Leu/-Trp dropout supplement (see Note 2) in 948 mL water; mix to dissolve. Adjust pH to 5.8 if necessary. Autoclave at 121°C for 15 min. Let the medium cool below 50°C; add 50 mL of sterile 40% D-glucose stock solution and 2 mL 500× DanAla stock solution.

2.3. Agar Plates

1. YPD Agar medium (1 L): 20 g/L Bacto-peptone, 10 g/L yeast extract, 20 g/L D-glucose, 20 g/L agar, pH 6.5. Weigh 20 g of bacto-peptone, 10 g of yeast extract and 20 g of agar, transfer to a 1-L graduated cylinder or a glass beaker, and add water to 950 mL. Mix completely and adjust pH to 6.5 (see Note 3). Autoclave at 121°C for 15 min. Let the medium cool to ~55°C, and then add 50 mL of sterile 40 % D-glucose stock solution. Mix completely and pour plates. Allow the medium to solidify at room temperature (see Note 4).
2. SD/-Leu/-Trp agar Medium (1 L): dissolve 6.7 g of yeast nitrogen base without amino acids, 0.64 g of Leu/-Trp dropout supplement and 20 g agar in 950 mL of water. Mix completely and adjust pH to 5.8 if necessary. Autoclave at 121°C for 15 min. Let the medium cool to ~55°C, and then add 50 mL of sterile 40% D-glucose stock solution. Mix completely and pour plates. Allow the medium to solidify at room temperature.
3. SD/-Leu/-Trp/+DanAla Agar medium (1 L): dissolve 6.7 g of yeast nitrogen base without amino acids, 0.64 g of Leu/-Trp dropout supplement and 20 g of agar in 948 mL of water. Mix completely and adjust pH to 5.8 if necessary. Autoclave at 121°C for 15 min. Let the medium cool to ~55°C, and then add 50 mL of sterile 40% D-glucose stock solution and 2 mL of 500× DanAla stock solution. Mix completely and pour plates. Allow the medium to solidify at room temperature.

2.4. Transformation Reagents

1. 1× TE buffer (100 mL): add 10 mL 10× TE stock solution into 90 mL sterilized water and mix completely.
2. Carrier DNA (100 mL): dissolve 200 mg of salmon sperm DNA in 100 mL of 1× TE buffer by stirring at 4°C for 1–2 h. Aliquot to 1 mL. Denature it in boiling water bath for 10 min and chill on ice before use (see Note 5).
3. TE/Lithium acetate solution (30 mL): add 3 mL of 10× TE, 3 mL of 10× LiAc into 24 mL of sterilized water, mix completely (see Note 6).
4. PEG/LiAc solution (10 mL): add 1 mL of 10× TE, 1 mL 10× LiAc into 8 mL of 50% PEG-3350 stock solution. Mix completely (see Note 6).

2.5. Other Solutions

1. 1× PBS (pH 7.4): dilute 10× PBS (Roche) to 1× with sterilized water.

2. Washing buffer: 1× PBS, 0.15 mM NaCl, 20 mM imidazole.
3. Elution buffer: 1× PBS, 0.15 mM NaCl, 250 mM imidazole.

3. Methods

3.1. Plasmid Construction

3.1.1. Construct the Expression Plasmid for the Orthogonal tRNA/Synthetase (Fig. 2a)

1. Prepare the gene cassette for expressing the orthogonal $EctRNA^{Leu}_{CUA}$. The final cassette consists of the following: *Pst*I – SNR52 promoter – $EctRNA^{Leu}_{CUA}$ – 3′-flanking sequence – *Sal*I. This cassette can be made using overlapping PCR or ordered directly from companies providing gene synthesis services. See Note 7 for sequence details.
2. Digest the above gene cassette with *Pst*I and *Sal*I, ligate it with the precut plasmid p-TyrRS (3) to make pSNR-$EctRNA^{Leu}_{CUA}$-TyrRS (see Note 8).
3. PCR amplify the DanAlaRS gene using primer FW29 5′-AGC TCG AGT TAG CCA ACG ACC AGA TTG AG-3′ and FW30 5′-AGA CTA GTA TGC AAG AGC AAT ACC GCC CG-3′. Digest the PCR product with *Spe*I and *Xho*I, and ligated into the precut pSNR-$EctRNA^{Leu}_{CUA}$-TyrRS to make pSNR-$EctRNA^{Leu}_{CUA}$-DanAlaRS (see Note 8).

3.1.2. Construct the Expression Plasmid for the Target Gene, pGFP-39TAG (Fig. 2b) and pGFP

A plasmid containing the 2μ ori, *LEU2*, Amp^r, the ColE1 ori, and MCS is used as the backbone to construct the pGFP-39TAG plasmid (3).

1. Use site-directed mutagenesis to introduce Tyr39TAG mutation into the EGFP gene. Amplify the mutant GFP-39TAG gene with primer JT171 5′-TAG TCG GAT CCT CAG TGA TGG TGA TGG TGA TGC TTG TAC AGC TCG TCC ATG CC-3′ and primer JT172 5′-TAG TCG TCG ACA TGG ATT ACA AAG ATG ATG ATG ATA AAG TGA GCA AGG GCG AGG AG-3′to add a His6 tag at the C-terminus.
2. Flank the PCR product with the ADH1 promoter and ADH1 terminator, and clone the whole gene cassette into the backbone plasmid using the *Hind*III and *Eco*RI sites to make pGFP-39TAG (see Note 9).
3. Use similar procedures to make the pGFP plasmid, which expresses the wt EGFP gene without the 39TAG mutation.

3.2. Prepare Yeast Competent Cells

The following procedures are applicable to both wt yeast and the NMD-deficient LWUPF1Δ strain. To determine if the LWUPF1Δ strain would be helpful for your target protein expression, see Note 10.

1. Streak the frozen glycerol stock of the yeast strain onto an YPD agar plate (see Note 11).

2. Incubate the plate at 30°C until yeast colonies are >2 mm in diameter.
3. Scrape the entire colony into 1-mL YPD medium in a Falcon tube, vortex to disperse it (see Note 12).
4. Transfer the medium into a glass tube and add 4-mL fresh YPD medium.
5. Incubate in a shaker at 30°C, 230–270 rpm overnight.
6. Measure OD_{600} of the overnight culture, and inoculate into 50-mL fresh YPD medium to $OD_{600}=0.15$ in a 500-mL flask.
7. Incubate in a shaker at 30°C, 230–270 rpm till $OD_{600}=0.3–0.5$. This will take 3–5 h for a strain with a doubling time of 2 h.
8. Harvest cells in a 50-mL conical tube at 1,500 × *g* for 5 min at room temperature.
9. Resuspend the pellet in 10 mL sterile 1× TE.
10. Pellet the cells at 1,500 × *g* for 5 min at room temperature.
11. Resuspend the pellet gently in 10-mL sterile TE/lithium acetate.
12. Pellet the cells at 1,500 × *g* for 5 min at room temperature.
13. Resuspend the pellet gently in 0.5 mL sterile TE/lithium acetate.

3.3. Transformation with Lithium Acetate

1. Incubate the competent cells at 30°C for 30 min (see Note 13).
2. Boil 10 μg carrier DNA for 10 min, and quickly chill in ice water.
3. Mix the carrier DNA with 2 μg pSNR-$EctRNA^{Leu}_{CUA}$-DanAlaRS and 2 μg pGFP-39TAG in a sterile 1.5-mL Eppendorf tube (see Note 14). Set up a control tube with carrier DNA only.
4. Add 150-μL yeast competent cells to each tube (see Note 15).
5. Incubate at 30°C for 30 min.
6. Add 840 μL PEG/LiAc solution (see Note 16).
7. Incubate at 30°C for 30 min.
8. Heat-shock in a water bath at 42°C for 15 min.
9. Pellet cells at 20,000 × *g* for 5–10 s at room temperature, and remove the PEG/LiAc solution sterilely as quickly as you can.
10. Immediately add 0.2 mL 1× TE to resuspend the pellet by pipetting it up and down gently.
11. Spread on SD/-Leu/-Trp agar plate to select the transformants (see Note 17).
12. Incubate at 30°C until colonies are >1 mm in diameter.

3.4. Re-streak Transformed Colonies

1. Pick ten single colonies and re-streak on SD/-Leu/-Trp agar plate.
2. Incubate at 30°C until colonies are >2 mm in diameter.

3.5. Verify DanAla Incorporation (see Note 18)

1. Re-streak single clones on SD/-Leu/-Trp and SD/-Leu/-Trp/+DanAla agar plates.
2. Incubate at 30°C until colonies are >1 mm in diameter.
3. Excite the plates with 480-nm light.
4. Observe the plates through a viewing filter glass that passes light >500 nm.
5. Green colonies will be seen on SD/-Leu/-Trp/+DanAla plate only. Take a photo with a digital camera if needed.

3.6. Quantify DanAla Incorporation Efficiency (see Note 18)

1. Pick a single colony from the SD/-Leu/-Trp agar plate, suspend in 1 mL SD/-Leu/-Trp medium, and vortex to disperse.
2. Transfer the medium into a glass tube and add 4-mL fresh SD/-Leu/-Trp medium.
3. Incubate in a shaker at 30°C, 230–270 rpm overnight.
4. Measure OD_{600} of the culture, and inoculate into 10 mL fresh SD/-Leu/-Trp/+DanAla medium to $OD_{600}=0.2$ in a 50-mL flask; inoculate into 10 mL fresh SD/-Leu/-Trp without DanAla medium to $OD_{600}=0.2$ in a 50-mL flask as the negative control.
5. Incubate in an orbital shaker at 30°C, 230–270 rpm for 6 h.
6. Pellet cells at 1,500 × g for 5 min at room temperature.
7. Wash cells once with 1× PBS and resuspend in 1× PBS.
8. Perform steps 1–7 for yeast cells transformed with the pSNR-EctRNA$^{Leu}_{CUA}$-DanAlaRS and pGFP in SD/-Leu/-Trp medium as a positive control (see Note 18).
9. Perform steps 1–7 for yeast cells without any plasmids in appropriate medium as a negative control.
10. Perform steps 1–7 for yeast cells transformed with pGFP-39TAG only in SD/-Leu medium as another negative control.
11. Analyze the fluorescence intensity of all cell samples with a flow cytometer (such as the FACScan from Becton and Dickinson). Use 488 nm as the excitation light and 530 nm (bandwidth = 30 nm) as the emission filter. Use the negative and positive control samples to adjust the detection sensitivity of the flow cytometer so that the fluorescence intensity readings of all samples are in a good dynamic range. DanAla is not excited at 488 nm. Determine the mean fluorescence intensity of each sample (Fig. 3a). Repeat steps 1–11 ≥3 times independently.
12. Calculate the DanAla incorporation efficiency using the mean fluorescence intensity of cells (see Note 18).

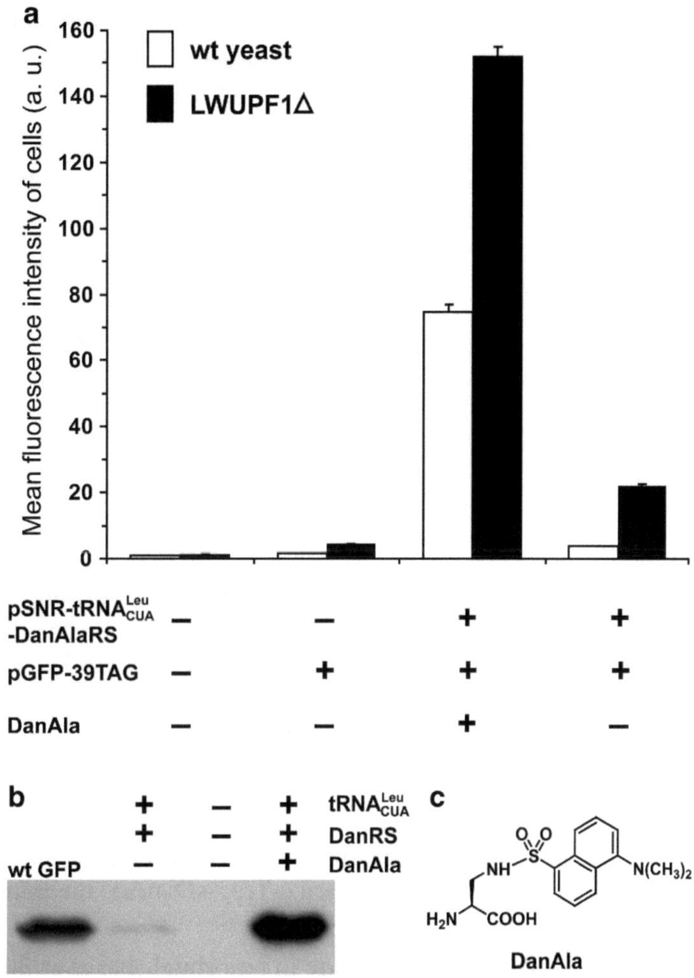

Fig. 3. Analysis of DanAla incorporation into GFP in yeast. (**a**) Flow cytometric analysis of DanAla incorporation into GFP in wt yeast and the NMD-deficient LWUP1Δ strain. Error bars represent S.E.M., n = 3. (**b**) Western blot analysis of the DanAla-incorporated GFP expressed in the LWUPF1Δ strain. Cell lysates were separated by SDS-PAGE and probed with a monoclonal anti-His5 antibody (Invitrogen). A purified wt GFP with a C-terminal His6 tag was used as a positive control. (**c**) The structure of DanAla.

3.7. Western Blot Analysis (see Note 19)

1. Pick a single colony from SD/-Leu/-Trp agar plate, suspend in 1 mL SD/-Leu/-Trp medium, and vortex to disperse.
2. Transfer the medium into a glass tube and add 4 mL fresh SD/-Leu/-Trp medium.
3. Incubate in a shaker at 30°C, 230–270 rpm overnight.
4. Measure OD_{600} of the culture, and inoculate into 5-mL fresh SD/-Leu/-Trp/+DanAla medium to OD_{600} = 0.2 in a glass tube; also inoculate into 5-mL fresh SD/-Leu/-Trp medium without DanAla to OD_{600} = 0.2 in a glass tube as the negative control.
5. Incubate in an orbital shaker at 30°C, 230–270 rpm for 6 h.

6. Pellet cells at 1,500 × *g* for 5 min at room temperature.
7. Resuspend the pellet in 35 μL Y-PER (Pierce), and add 2 μL 25× EDTA stock solution.
8. Agitate at room temperature for 60 min.
9. Centrifuge at 20,000 × *g* for 10 min; keep the supernatant.
10. Load purified GFP-His6 protein as a positive control (see Note 20), 10 μL negative control, and 10 μL sample on 12% SDS-PAGE gel.
11. Transfer the proteins from gel to a PVDF membrane.
12. Detect with the Penta·His mouse monoclonal antibody (Invitrogen) followed by a goat anti mouse IgG-HRP antibody (see Note 21).
13. Develop the film with SuperSignal West Pico Chemiluminescent Substrate (Pierce) (Fig. 3b).
14. Strip the blot and re-probe with an antibody specific for a housekeeping protein for loading control.

3.8. Protein Expression and Purification

1. Pick a single colony from SD/-Leu/-Trp agar plate, suspend in 1-mL SD/-Leu/-Trp medium, and vortex to disperse.
2. Transfer the medium into a glass tube and add 4-mL fresh SD/-Leu/-Trp medium.
3. Incubate in a shaker at 30°C, 230–270 rpm overnight.
4. Measure OD_{600} of the culture, and inoculate into 250 mL fresh SD/-Leu/-Trp/+DanAla medium to OD_{600} = 0.1 in a 1-L flask.
5. Incubate in an orbital shaker at 30°C, 230–270 rpm for 48 h (see Note 22).
6. Pellet cells at 1,500 × *g* for 5 min at room temperature.
7. Resuspend the pellet in 10 mL Y-PER (Pierce), and add 1/2 tablet of EDTA-free protease inhibitor (Roche).
8. Agitate at room temperature for 20 min.
9. Sonicate for 2.5 min (80% power, 3 s on and 10 s off) by using a Sonic Dismembrator (Fisher Scientific).
10. Centrifuge at 20,000 × *g* for 10 min; keep the supernatant.
11. Resuspend the pellet in 10 mL Y-PER (Pierce), add 1/2 tablet of EDTA-free protease inhibitor (Roche).
12. Repeat steps 8–10, combine the supernatant.
13. Add 2 mL Ni–NTA slurry (Qiagen, pre-balanced with Y-PER), and incubate at 4°C for 1 h on a lab rocker.
14. Load the slurry in a column, and wash with 10-bed volumes of 1× PBS buffer, followed by 10-bed volumes of washing buffer.

15. Elute the DanAla-containing GFP with elution buffer, 1 mL, 3×.
16. Exchange the protein into 1× PBS buffer by using Centricon concentrators (Amicon).
17. Run SDS-PAGE to check the purity of the protein.
18. Determine the protein concentration by using the Bradford assay.

4. Notes

1. Some unnatural amino acids may be difficult to dissolve; lower the stock concentration to 200 mM or 100 mM when necessary. Solubility is also dependent on the purity of the unnatural amino acid. If racemic mixture of unnatural amino acids is used, the concentration of the effective L-amino acid will be 50% of the calculated value. Use optical pure L-amino acids whenever possible.
2. Amino acid dropout supplements can be purchased from Clontech (Catalog number 630417). If purchasing from other suppliers, change the amount accordingly by following the product information.
3. The agar will not fully dissolve until it is autoclaved.
4. Prepare agar plates in advance. Unsleeve to dry the plates at room temperature for 1 day prior to plating.
5. Store the boiled carrier DNA at −20°C, which can be reboiled and reused for three times without loss of activity.
6. Prepare these solutions fresh prior to use.
7. The sequence for the SNR52 promoter is the following: tctttgaaaagataatgtatgattatgctttcactcatatttatacagaaacttgatgttttctttcgagtatatacaaggtgattacatgtacgtttgaagtacaactctagattttgtagtgccctcttgggctagcggtaaaggtgcgcatttttcacaccctacaatgttctgttcaaaagattttggtcaaacgctgtagaagtgaaagttggtgcgcatgtttcggcgttcgaaacttctccgcagtgaaagataaatgatc. The sequence for the $EctRNA^{Leu}_{CUA}$ is the following: GCCCGGATGGTGGAATCGGTAGACACAAGGGATTCTAAATCCCTCGGCGTTCGCGCTGTGCGGGTTCAAGTCCCGCTCCGGGTA. Note the underlined anticodon CTA, which recognizes the UAG amber codon. The 3′-CCA trinucleotide of the tRNA is not included in the plasmid. The 3′-flanking sequence of the SUP4 is the following: TTTTTTTGTTTTTTATGTCT. We avoided introducing a restriction enzyme site between the SNR52 promoter and the tRNA because such mutations may impair the promoter strength and/or hamper the generation of the correct 5′ end of the tRNA. If a new tRNA does not show activity after being

expressed using the SNR52 promoter, check if there is any mutation in the promoter and the tRNA.

8. Plasmids pSNR-EctRNA$^{Leu}_{CUA}$-DanAlaRS and pSNR-EctRNA$^{Tyr}_{CUA}$-TyrRS harbor the orthogonal *E. coli* leucyl and tyrosyl amber suppressor tRNA, respectively, and are available from the Wang group (http://wang.salk.edu) upon request. When incorporating unnatural amino acids using orthogonal tRNA/synthetase pairs, the synthetase is evolved to be specific for different unnatural amino acids, but the orthogonal tRNA does not need to be changed and works with all these mutant synthetases. Therefore, mutant synthetases evolved from the *E. coli* TyrRS are all used with the *E. coli* tyrosyl amber suppressor tRNA (EctRNA$^{Tyr}_{CUA}$), and mutant synthetases evolved from the *E. coli* LeuRS are all used with the EctRNA$^{Leu}_{CUA}$. To make an orthogonal tRNA/synthetase expression plasmid for the incorporation of a different unnatural amino acid, one just needs to replace the synthetase gene without recloning the tRNA gene. tRNA genes are generally difficult to be cloned, and the *Pst*I and *Sal*I sites are no longer unique in the final plasmid. The DanAlaRS and TyrRS gene can be swapped for other mutant synthetase genes using the *Spe*I (N-terminus) and *Xho*I (C-terminus) unique sites.
9. To express your gene of interest, any plasmid with *LEU2* and 2μ ori can be used. A strong promoter such as the ADH1 promoter used here is preferred for high expression level. Through the above cloning procedure, we built in a unique *Sal*I site at the N-terminus and a *BamH*I site at the C-terminus of the GFP gene in the pGFP-39TAG. This plasmid is also available through the Wang group (http://wang.salk.edu) as a convenient fluorescent positive control to monitor unnatural amino acid incorporation (see Subheadings 3.5 and 3.6) and to facilitate the cloning of your genes of interest.
10. The NMD-deficient LWUPF1Δ strain is available at the Wang group (http://wang.salk.edu) upon request. To determine whether the LWUPF1Δ strain will help the expression, check the location of the UAG codon in your gene of interest. NMD in yeast shows a polar effect of nonsense codon positions (13). The steady-state mRNA level is reduced by NMD more significantly when the nonsense codon is closer to the 5′ end than to the 3′ end of an mRNA. Consistently, the increase of unnatural amino acid incorporation efficiency in the LWUPF1Δ strain correlates with the position of the UAG codon: more than a twofold increase is measured when the amber codon is within the N-terminal two thirds of the gene, whereas no significant increase is detected when it is within the C-terminal fourth of the coding region.

11. For the LWUPF1Δ strain, add 0.5 mg/mL G418 into the YPD medium to keep the selective pressure. If your yeast strain has additional marker or plasmid, use the appropriate SD/drop out medium to keep selective pressure.
12. Different yeast strains grow at different rates. If colonies are small, or if you are inoculating a larger volume, use several colonies.
13. For the highest transformation efficiency, use competent cells within 1 h of their preparation.
14. The concentration of the plasmid is an important factor for the transformation efficiency. The smaller is the volume ratio of DNA mixture to competent cells, the higher will be the transformation efficiency. The total volume of the DNA mixture less than 10 μL is preferable.
15. The volume of the competent cells should ≥10× volume of the DNA mixture.
16. The PEG/LiAc solution should be freshly prepared before use.
17. In order to get single colonies, spread 190 μL suspension on one plate; then add 180 μL fresh 1× TE to the suspension left and spread on another plate.
18. The pGFP-39TAG can be used as the reporter plasmid to quickly verify the incorporation of an unnatural amino acid by the orthogonal tRNA/synthetase on agar plates, and to quantify the incorporation efficiency of the unnatural amino acid into GFP by using flow cytometry. Green fluorescence of GFP should be detected on cells when the unnatural amino acid is added to the agar plate. In the absence of the unnatural amino acid, no green fluorescence will be detected. To determine the incorporation efficiency, follow the procedures in Subheading 3.6. Measure the mean fluorescence intensity of cells transformed with pSNR- $EctRNA^{Leu}_{CUA}$ -DanAlaRS and pGFP-39TAG grown in 1 mM of DanAla (Int_1), and those grown in the absence of DanAla (Int_2). Also measure the mean intensity of cells transformed with pSNR- $EctRNA^{Leu}_{CUA}$ -LeuRS and pGFP-39TAG (Int_3) and of cells transformed with pGFP-39TAG alone (Int_4). Here Leu is incorporated by the orthogonal *E. coli* $EctRNA^{Leu}_{CUA}$/LeuRS pair through amber suppression. The ratio defined by $(Int_1–Int_2)/(Int_3–Int_4)$ will determine the relative incorporation efficiency of DanAla to Leu. To obtain the net incorporation efficiency of DanAla, measure the mean intensity (Int_5) of cells transformed with pSNR- $EctRNA^{Leu}_{CUA}$ -DanAlaRS and pGFP (a plasmid identical to pGFP-39TAG except that the 39TAG is reverted to wt tyrosine codon TAC). The net DanAla incorporation efficiency is defined by $(Int_1–Int_2)/(Int_5–Int_2)$. Note that this GFP reporter can be generally used to evaluate the incorporation of many unnatural amino acids,

as the 39TAG site is permissive for GFP fluorescence. Also note that the incorporation efficiency of an unnatural amino acid can be protein-dependent and site-dependent.

19. Western blot analysis can also be used to determine the incorporation efficiency of the unnatural amino acid into the target protein. Use densitometry to measure the intensities of target protein bands and use the loading control for sample normalization. If the orthogonal tRNA/synthetase for your unnatural amino acid of interest has not been fully characterized before, it is also necessary to perform mass spectrometric analysis of the purified target protein with the unnatural amino acid incorporated (14). Tandem mass spectrometric analysis of protease digested peptides will determine the identity of the amino acid incorporated at the UAG site. A semiquantitative estimation of the incorporation fidelity can be obtained from the peptide intensities (5). Mass analysis of the intact protein will also reveal if common amino acids are incorporated at the UAG site and if there is any misincorporation at other sites in the target protein.
20. Alternatively, 10 μL of supernatant similarly prepared from cells transformed with pGFP can be used here.
21. If a primary antibody against the target protein is available, it can be used here for Western detection in replacement of the Penta·His antibody.
22. The incubation time for expression is protein dependent. We highly recommend a time course experiment to determine the optimal expression time for your target protein. Western blot analysis of cell lysates can be used to monitor target protein expression level conveniently.

Acknowledgments

We thank Dr. Vicki Lundblad and members of the Lundblad lab for providing reagents and advice on yeast protocols. This work was supported by CIRM (RN1-00577-1) and NIH (1DP2OD004744).

References

1. Wang Q, Parrish A R, Wang L (2009) Expanding the genetic code for biological studies. Chem Biol 16, 323–336.
2. Liu C C, Schultz P G (2010) Adding new chemistries to the genetic code. Annu Rev Biochem 79, 413–444.
3. Wang Q, Wang L (2008) New methods enabling efficient incorporation of unnatural amino acids in yeast. J Am Chem Soc 130, 6066–6067.
4. Sprague K U (1995) Transcription of eukaryotic tRNA genes. In: Soll D, RajBhandary U L (eds), tRNA: Structure, Biosynthesis,

and Function, ASM Press, Washington, DC, pp 31–50.
5. Chen S, Schultz P G, Brock A (2007) An improved system for the generation and analysis of mutant proteins containing unnatural amino acids in *Saccharomyces cerevisiae*. J Mol Biol 371, 112–122.
6. Otter C A, Straby K B (1991) Transcription of eukaryotic genes with impaired internal promoters: the use of a yeast tRNA gene as promoter. J Biotechnol 21, 289–293.
7. Otter C A, Edqvist J, Straby K B (1992) Characterization of transcription and processing from plasmids that use polIII and a yeast tRNA gene as promoter to transcribe promoter-deficient downstream DNA. Biochim Biophys Acta 1131, 62–68.
8. Majmudar C Y et al. (2009) Impact of nonnatural amino acid mutagenesis on the *in vivo* function and binding modes of a transcriptional activator. J Am Chem Soc 131, 14240–14242.
9. Lee H S et al. (2009) Genetic incorporation of a small, environmentally sensitive, fluorescent probe into proteins in *Saccharomyces cerevisiae*. J Am Chem Soc 131, 12921–12923.
10. Wang W et al. (2007) Genetically encoding unnatural amino acids for cellular and neuronal studies. Nat Neurosci 10, 1063–1072.
11. Amrani N, Sachs M S, Jacobson A (2006) Early nonsense: mRNA decay solves a translational problem. Nat Rev Mol Cell Biol 7, 415–425.
12. Summerer D et al. (2006) A genetically encoded fluorescent amino acid. Proc Natl Acad Sci USA 103, 9785–9789.
13. Cao D, Parker R (2003) Computational modeling and experimental analysis of nonsense-mediated decay in yeast. Cell 113, 533–545.
14. Wang L et al. (2001) Expanding the genetic code of *Escherichia coli*. Science 292, 498–500.

Chapter 13

Site-Specific Incorporation of Unnatural Amino Acids into Proteins in Mammalian Cells

Nobumasa Hino, Kensaku Sakamoto, and Shigeyuki Yokoyama

Abstract

Expanding the repertoire of genetically encoded amino acids in cultured mammalian cells requires the expression of the bacterial or archaeal pair of a tRNA and an aminoacyl-tRNA synthetase variant engineered to be specific to the amino acid, along with the supplementation of an unnatural amino acid in the growth medium. The expression of the pair is generally achieved by transfecting the cultured cells with the plasmids bearing the genes encoding the exogenous pair of translation molecules. Here, we provide a description of some of these plasmids and protocols for transfecting cells with the plasmids and preparing growth media supplemented with unnatural amino acids, to facilitate their incorporation into proteins at specific sites.

Key words: Unnatural amino acids, Mammalian cells, Genetic code, tRNA, Aminoacyl-tRNA synthetase, *OriP*, Photo-cross-linker, Azido group, Acetyllysine, Iodotyrosine

1. Introduction

Unnatural amino acids, which are also designated as "nonnatural," have been site-specifically incorporated into proteins in *Escherichia coli* (1), yeasts (2), insect cells (3), and mammalian cells (4). Their site-specific incorporation relies on the expression in these cells of the exogenous pairs of a tRNA and an aminoacyl-tRNA synthetase (aaRS) variant that is engineered to be specific to an unnatural amino acid. These exogenous pairs should not cross-react with any endogenous tRNAs or aaRSs (5, 6). The tRNAs for unnatural amino acids read a stop codon or a quadruplet codon, by which the positions accommodating the amino acids in proteins are defined

Loredano Pollegioni and Stefano Servi (eds.), *Unnatural Amino Acids: Methods and Protocols*,
Methods in Molecular Biology, vol. 794, DOI 10.1007/978-1-61779-331-8_13, © Springer Science+Business Media, LLC 2012

in the coding sequences. The amber stop codon (UAG) has been used to encode unnatural amino acids in most studies with mammalian cells, although the opal and ochre stop codons are also available in the cells (7, 8). The unnatural amino acids are supplemented in the growth media and assimilated by the cells.

The unnatural amino acids, genetically encoded in mammalian cells, have facilitated protein science and technology, by incorporating useful and unique chemical groups or structures into proteins (9–13). Derivatives of the aromatic amino acids are encoded by using engineered variants of *E. coli* tyrosyl-tRNA synthetase (TyrRS) (4, 14) and tryptophanyl-tRNA synthetase (7). L-Lysine and L-leucine derivatives have also become available, by engineering the archaeal pyrrolysyl-tRNA synthetase (PylRS) (10, 12) and *E. coli* leucyl-tRNA synthetase (15), respectively. These aaRS variants have been expressed from the cytomegalovirus (CMV), simian virus 40 (SV40), and thymidine kinase promoters, which are commonly used to express recombinant proteins. The UAG-decoding amber suppressor tRNAs from bacteria and archaea have been expressed from two different types of promoters. The first type is an internal promoter, from which most eukaryotic tRNAs are transcribed. These promoters are utilized when the exogenous tRNA sequence contains the promoter sequence, either naturally or by engineering (4, 5, 7). We have expressed *Bacillus stearothermophilus* $tRNA^{Tyr}$ in this manner, for pairing with *E. coli* TyrRS in mammalian cells, and found that the addition of a 5'-flanking sequence, from the human $tRNA^{Tyr}$ gene, to the *Bacillus* $tRNA^{Tyr}$ gene significantly facilitates its expression. The second type of promoter is an external promoter, such as U6 and H1, which is placed at the 5' side of the exogenous tRNA sequence (10, 12, 15). These promoters have been used to express pyrrolysine tRNA ($tRNA^{Pyl}$), which originally recognizes UAG codons, and amber suppressor tRNAs derived from *E. coli* $tRNA^{Leu}$ and $tRNA^{Tyr}$. The gene encoding an exogenous UAG-decoding tRNA can be tandemly repeated in a plasmid to enhance its expression.

In the following sections, we provide protocols for producing proteins containing derivatives of L-tyrosine/phenylalanine or L-lysine at the amber position, in mammalian cells. The plasmid systems for expressing the aaRS variants specific to these molecules as well as the UAG-decoding tRNAs are described in the second part of Subheading 2. We also provide the recipes for preparing the growth media containing these unnatural amino acids (Subheading 3.1) and protocols for transfecting Chinese hamster ovary (CHO) (Subheadings 3.2 and 3.3) and human embryonic kidney (HEK) 293 (Subheadings 3.4 and 3.5) cells with the plasmids, for the production of recombinant proteins with unnatural amino acids.

2. Materials

2.1. Unnatural Amino Acids

The unnatural amino acids utilized in the protocols are listed below, with their symbols in parentheses. See Table 1 for the names of the specific aaRS variants, along with their references, chemical structures, and recommended concentrations in the growth media, as well as the names of the commercial providers.

1. 3-Iodo-L-tyrosine (iodoTyr).
2. *p*-Benzoyl-L-phenylalanine (*p*Bpa).
3. *p*-Azido-L-phenylalanine (*p*Azpa).
4. *p*-Acetyl-L-phenylalanine (pAcpa).
5. N^{ε}-*tert*-Butyloxycarbonyl-L-lysine (BocLys).
6. N^{ε}-(*o*-Azidobenzyloxycarbonyl)-L-lysine (AzZLys).
7. N^{ε}-Acetyl-L-lysine (AcLys).

2.2. Plasmids

There are no commercially available plasmids for expressing the exogenous pairs of aaRS variants and UAG-reading tRNAs. Users may obtain these plasmids from the laboratories that developed them. The plasmids that we can provide, which are utilized in the protocols, are listed below, and their maps are depicted in Fig. 1.

We used two types of vectors, pcDNA (Invitrogen) and pOriP (10), to construct our plasmids. The pcDNA4/TO vector is used in T-REx CHO cells (Invitrogen), and the recombinant proteins, including aaRS variants, are expressed from the CMV promoter in the presence of tetracycline. This vector and pcDNA3.1 may be used in other cell lines, to allow the constitutive expression of recombinant proteins. The pOriP vector, which also expresses recombinant proteins from the CMV promoter, has the *OriP* sequence from the Epstein–Barr virus. This sequence binds the viral nuclear antigen-1 protein, which is constitutively expressed in HEK293 c-18 cells (American Type Culture Collection or ATCC), and facilitates the transfection of the cells with the pOriP plasmid. The pSuPT vector, derived from pOriP, expresses aaRS variants from the thymidine kinase promoter and also carries a gene encoding a bacterial or archaeal suppressor tRNA.

1. The following eight plasmids, with the prefixes "pc" and "pSuPT," are based on pcDNA4/TO and pSuPT, respectively, and express the TyrRS variants included in the plasmid names:
   - pciodoTyrRS, pcpAzpaRS, pcpAcpaRS, pcpBpaRS,
   - pSuPT-iodoTyrS, pSuPT-pAzpaRS, pSuPT-pAcpaRS, pSuPT-pBpaRS.

Table 1
Unnatural amino acids

Unnatural amino acids (CAS no.) [MW]	Symbol	Chemical structure	aaRS [references]	Working conc. (mM)	Plasmids for incorporation	Providers
Tyrosine/phenylalanine derivatives						
3-Iodo-L-tyrosine (70-78-0) [307.1]	iodoTyr	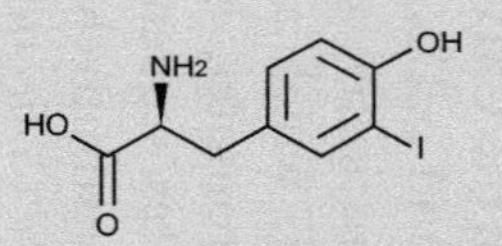	iodoTyrRS [16]	0.3	pciodoTyrRS pSuPT-iodoTyrRS	Sigma Bachem Watanabe Chem[a]
p-Benzoyl-L-phenylalanine (104504-45-2) [269.3]	*p*Bpa	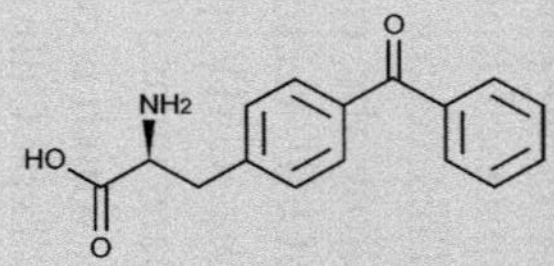	*p*BpaRS [2]	0.5	pcpBpaRS pSuPT-pBpaRS	Bachem Peptech[b] Watanabe Chem[a]
p-Azido-L-phenylalanine (33173-53-4) [206.2]	*p*Azpa	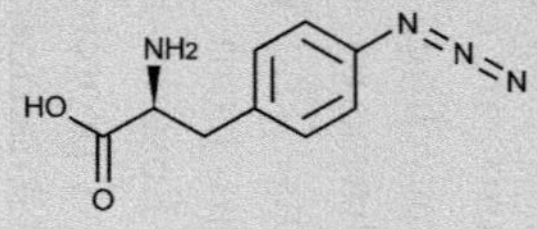	*p*AzpaRS [2]	0.1	pcpAzpaRS pSuPT-pAzpaRS	Bachem Watanabe Chem[a]
p-Acetyl-L-phenylalanine (122555-04-8) [207.2]	*p*Acpa	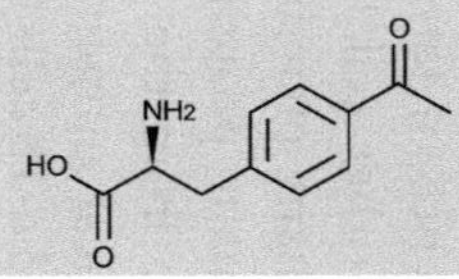	*p*AcpaRS [2]	0.5	pcpAcpaRS pSuPT-pAcpaRS	Peptech[b] Shinsei Chem[c]
Lysine derivatives						
N^{ε}-*tert*-Butyloxycarbonyl-L-lysine (2418-95-3) [246.3]	BocLys	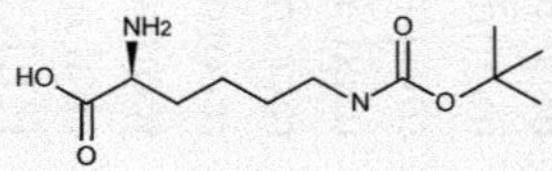	PylRS [17]	1	pcPylRS and pU6tRNAPyl pOriP-PylRS and pOriP-U6tRNAPyl	Sigma Bachem Watanabe Chem[a]
N^{ε}-Acetyl-L-lysine (692-04-6) [188.2]	AcLys	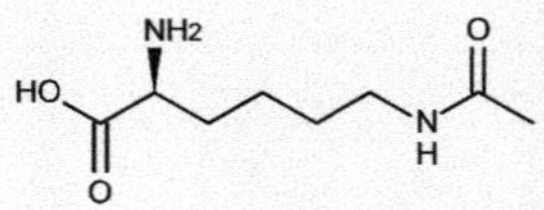	AcLysRS [10]	14	pcAcLysRS and pU6tRNAPyl pOriP-AcLysRS and pOriP-U6tRNAPyl	Sigma Bachem Watanabe Chem[a]
N^{ε}-(*o*-Azidobenzyl-oxycarbonyl)-L-lysine (not assigned) [307.3]	AzZlys	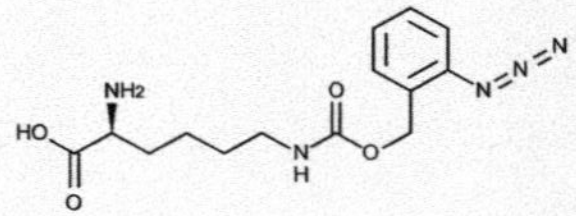	AzZLysRS [17]	0.1	pcAzZLysRS and pU6tRNAPyl pOriP-AzZLysRS and pOriP-U6tRNAPyl	Shinsei Chem[c]

MW molecular weight

[a–c]Watanabe Chemical Co., Ltd., Peptech, and Shinsei Chemical Co., Ltd. are based in Hiroshima (Japan), MA (USA), and Osaka (Japan), respectively

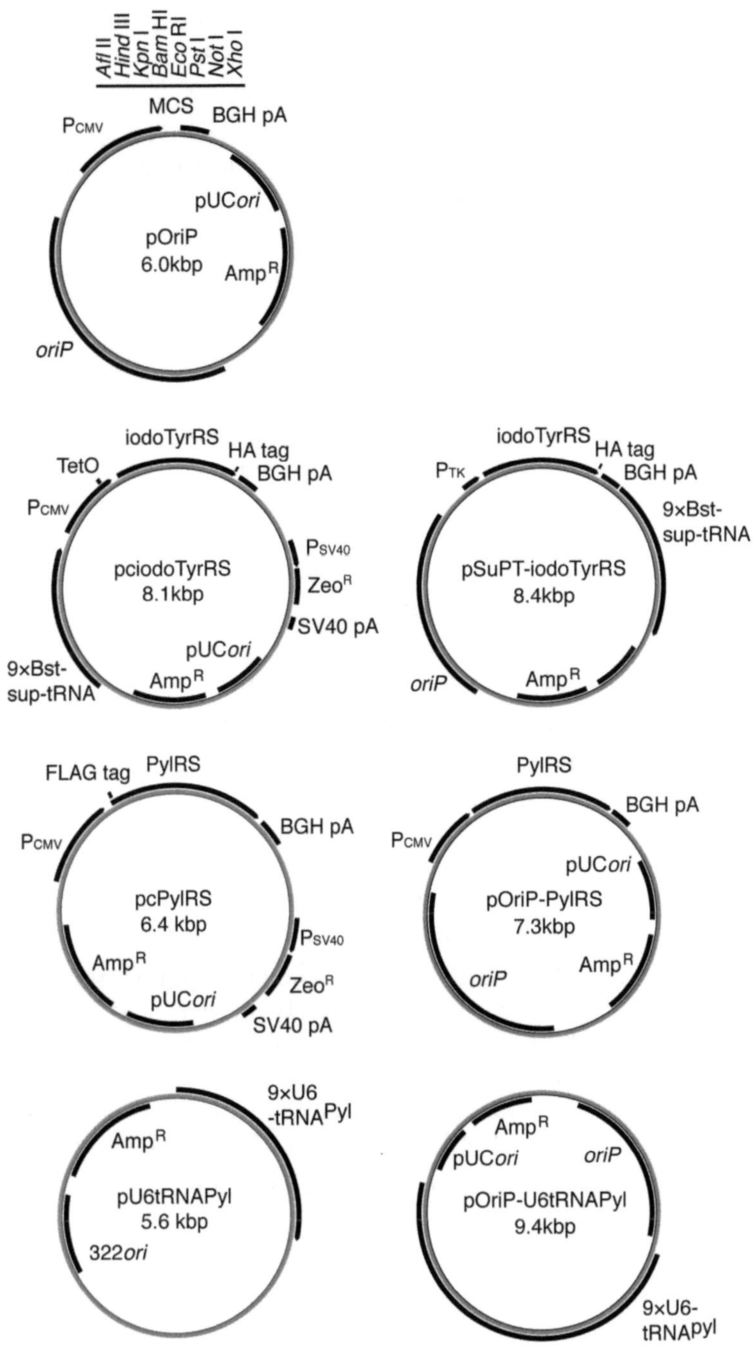

Fig. 1. Plasmid maps. The pc- and pSuPT-plasmids for expressing *p*BpaRS, *p*AzpaRS, and *p*AcpaRS were derived from pciodoTyrRS and pSuPT-iodoTyrRS, by substituting their genes with the gene encoding iodoTyrRS. The pc- and pOriP-plasmids for expressing pAcLysRS and AzZLysRS were derived from pcPylRS and pOriP-PylRS, by substituting their genes with the gene encoding PylRS. P_{CMV} CMV promoter, P_{SV40} SV40 promoter, P_{TK} thymidine kinase promoter, *TetO* tetracycline-responsive sequence, *BGH pA* bovine-growth-hormone polyadenylation sequence, *SV40 pA* SV40 polyadenylation sequence, *HA* hemagglutinin epitope, *oriP oriP* sequence, Zeo^R Zeocin-resistance gene, Amp^R ampicillin-resistance gene, *MCS* multiple cloning site; pUC *ori* and 322 *ori* plasmid replication origins for the pUC and pBR322 plasmids, respectively, functioning in *E. coli*; 9×Bst-sup-tRNA, nine tandem copies of the *Bacillus* UAG-reading suppressor tRNA gene; 9×U6tRNA^{Pyl}, nine tandem copies of the tRNA^{Pyl} gene.

Each plasmid carries nine copies of the gene encoding the *Bacillus* amber suppressor tRNA.

2. The following six plasmids, with the prefixes "pc" and "pOriP," are based on pcDNA3.1 and pOriP, respectively, and express the PylRS variants included in the plasmid names, from the CMV promoter:
   - pcBocLysRS, pcAzZLysRS, pcAcLysRS.
   - pOriP-BocLysRS, pOriP-AzZLysRS, pOriP-AcLysRS.
   - None of the plasmids carries the gene encoding $tRNA^{Pyl}$.
3. The two plasmids: pU6tRNAPyl and pOriP-U6tRNAPyl (based on pBR322 and pOriP, respectively) both carry nine copies of the gene encoding *Methanosarcina mazei* $tRNA^{Pyl}$, each transcribed from the U6 promoter. These plasmids are for use in combination with the plasmids expressing the PylRS variants.

2.3. Cell Lines

1. T-REx CHO (Invitrogen).
2. HEK 293 c-18 (ATCC).

2.4. Cell Culture Media and Serum

1. DMEM medium (Invitrogen).
2. DMEM/F-12 medium (Invitrogen).
3. DMEM medium (1.1× conc.). Dissolve a package of the DMEM powder (Invitrogen) and 3.7 g of $NaHCO_3$ in 800 mL of water. Mix this solution with 20 mL of 1 M HEPES–NaOH buffer (pH 7.2), adjust the volume to 900 mL, and sterilize it by filtration with a 1-L filter system, equipped with a bottle-top filter (0.22 μm) (Millipore).
4. DMEM/F-12 (1.1× conc.) medium. Dissolve a package of the DMEM/F-12 powder (Invitrogen) and 2.4 g of $NaHCO_3$ in 800 mL of water. Mix this solution with 20 mL of 1 M HEPES–NaOH buffer (pH 7.2), adjust the volume to 900 mL, and sterilize it by filtration with a 1-L filter system, equipped with a bottle-top filter (0.22 μm) (Millipore).
5. Opti-MEM (Invitrogen) medium.
6. Fetal bovine serum.

2.5. Other Reagents and Culture Plates

1. QuikChange site-directed mutagenesis kit (Stratagene).
2. Lipofectamine 2000 (Invitrogen).
3. Ethanol solution of 1 mg/mL tetracycline.
4. 24-Well cell culture plates (Falcon).
5. 24-Well cell culture plates, coated with poly-D-lysine (Falcon).

3. Methods

3.1. Preparation of Growth Media Supplemented with Unnatural Amino Acids (see Note 1)

3.1.1. IodoTyr

1. Dissolve 63 mg of iodoTyr in 10 mL of an aqueous acetic acid solution (0.1%, v/v) (see Note 2), by gentle mixing with a rotator at room temperature for 2–3 h, to prepare a stock solution of iodoTyr at a concentration of 20 mM.
2. Sterilize the solution by filtering it through a 0.22-μm filter device connected to a syringe. This solution is stable for at least 2 weeks at 4°C.
3. Dilute 225 μL of the stock solution with 13.5 mL of the DMEM (1.1× conc.) or DMEM/F-12 (1.1× conc.) medium, prewarmed at 37°C, and then adjust the volume to 15 mL with sterile water, to prepare the growth medium with iodoTyr at a final concentration of 0.3 mM.

3.1.2. pBpa

Prepare the *p*Bpa-containing medium just before use (see Note 3).

1. To prepare 15 mL of the medium, dissolve 2.0 mg of *p*Bpa in 300 μL of 1 N HCl in a 15-mL polypropylene conical tube (see Note 4), by mixing vigorously just after the addition of HCl to the amino acid (see Note 3).
2. When the amino acid has dissolved, dilute it immediately with 13.5 mL of DMEM (1.1× conc.) or DMEM/F-12 (1.1× conc.) medium, prewarmed at 37°C.
3. Adjust the pH of the medium to 7.2 with 1 N NaOH, and then adjust the volume to 15 mL with water, to prepare the growth medium with *p*Bpa at a final concentration of 0.5 mM.
4. Finally, sterilize the medium by filtration with a 0.22-μm filter device connected to a syringe.

3.1.3. pAzpa

Prepare the *p*Azpa-containing medium just before use (see Note 3).

1. To prepare 15 mL of the medium, dissolve 0.3 mg of *p*Azpa in 60 μL of 1 N HCl in a 15-mL polypropylene conical tube, by mixing vigorously (see Note 4).
2. When the amino acid has dissolved, dilute it immediately with 13.5 mL of DMEM (1.1× conc.) or DMEM/F-12 (1.1× conc.) medium, prewarmed at 37°C (see Note 5).
3. Adjust the pH of the medium to 7.2 with 1 N NaOH, and adjust the volume to 15 mL with water, to prepare the growth medium with *p*Azpa at a final concentration of 0.1 mM.
4. Finally, sterilize the medium by filtration with a 0.22-μm filter device connected to a syringe.

3.1.4. pAcpa

Prepare the *p*Acpa-containing growth medium just before use (see Note 3).

1. To prepare 15 mL of medium with *p*Acpa at a final concentration of 0.5 mM, dissolve 1.6 mg of *p*Acpa in 15 mL of the DMEM or DMEM/F-12 medium, prewarmed at 37°C.
2. Sterilize the medium with a 0.22-μm filter device connected to a syringe.

3.1.5. BocLys

1. Dissolve 62 mg of BocLys in 10 mL of water by vigorous mixing, to prepare a 25-mM BocLys stock solution.
2. Sterilize the solution by filtering it with a 0.22-μm filter device connected to a syringe. The stock solution is stable for at least 2 weeks at 4°C.
3. To prepare 15 mL of the growth medium with BocLys at a final concentration of 1 mM, dilute 60 μL of the stock solution in 13.5 mL of DMEM (1.1× conc.) or DMEM/F-12 (1.1× conc.) medium, prewarmed at 37°C, and then adjust the volume to 15 mL with sterile water.

3.1.6. AcLys

1. Dissolve 2.6 g of AcLys in 7 mL of water by vigorously mixing, and then adjust the volume to 10 mL, to prepare a stock solution with AcLys at a concentration of 1.4 M.
2. Sterilize the solution by filtering it with a 0.22-μm filter device connected to a syringe. This solution is stable for at least 2 weeks at 4°C.
3. To prepare 15 mL of the growth medium with AcLys at a final concentration of 14 mM, dilute 150 μL of the stock solution with 13.5 mL of DMEM (1.1× conc.) or DMEM/F-12 (1.1× conc.) medium, prewarmed at 37°C, and then adjust the volume to 15 mL with sterile water.

3.1.7. AzZLys

Prepare the AzZLys-containing medium just before use (see Note 3).

1. To prepare 15 mL of the medium, dissolve 0.5 mg of AzZLys in 60 μL of 1 N HCl in a 15-mL polypropylene conical tube (see Note 4), by mixing vigorously just after the HCl is added to the amino acid.
2. When the amino acid has dissolved, dilute it immediately with 13.5 mL of DMEM (1.1× conc.) or DMEM/F-12 (1.1× conc.) medium, prewarmed at 37°C (see Note 5).
3. Adjust the pH of the medium to 7.2 with 1 N NaOH, and adjust the volume to 15 mL with water, to prepare the growth medium with AzZLys at a final concentration of 0.1 mM.
4. Finally, sterilize the medium by filtering it with a 0.22-μm filter device connected to a syringe.

3.2. Creating a Plasmid Carrying an Amber Mutant Gene for Unnatural Amino Acid Incorporation

1. Clone the gene encoding the protein that will contain the unnatural amino acid, downstream of the CMV promoter, in the pcDNA4/TO vector for expressing the recombinant protein in T-REx-CHO cells, or the pOriP vector for expression in HEK 293 c-18 cells (see Note 6).
2. Create an amber codon in the gene, cloned within pcDNA4/TO or pOriP, at the position where the unnatural amino acid will be introduced, by replacing the codon at this position with the amber TAG codon, or inserting TAG before or after the codon (see Notes 7 and 8). A QuikChange site-directed mutagenesis kit (Stratagene) or other conventional methods may be employed for this engineering. If the coding sequence of the gene ends with TAG, then replace this amber stop codon with another stop codon (TAA or TGA), to avoid the incorporation of the unnatural amino acid at the C-terminus.

3.3. Plasmid Transfection for Producing Proteins Containing L-Tyrosine/Phenylalanine Derivatives in T-REx CHO Cells

1. Inoculate 1.0×10^5 of T-REx CHO cells in the well of a 24-well cell culture plate, containing 500 μL of DMEM/F-12 medium, supplemented with fetal bovine serum (FBS) at a 10:1 ratio, in terms of volume.
2. Incubate the cells at 37°C in an incubator with a 5% CO_2 atmosphere. The cell population will become 90–95% confluent after a 16–18 h of incubation.
3. Mix Opti-MEM (50 μL) with 0.4 μg of the pcDNA4/TO plasmid carrying the amber mutant gene (Subheading 3.2), together with 0.4 μg of a "pc-" plasmid for expressing the TyrRS variant and the *Bacillus* suppressor tRNA. Incubate the mixture at room temperature for 5 min. Refer Table 1 or Subheading 2.2 to select the appropriate plasmid for the unnatural amino acid to be incorporated into the recombinant protein.
4. In parallel with step 3, mix 2 μL of Lipofectamine 2000 with 50 μL of Opti-MEM, and then incubate the 52-μL mixture at room temperature for 5 min.
5. Combine these mixtures, prepared at steps 3 and 4, in a tube and mix them gently. Incubate the combined mixture at room temperature for 20 min.
6. Replace the growth medium in the well, containing the cells, with 400 μL of Opti-MEM.
7. Add the mixture, prepared at step 5, to the well and mix gently by rocking the culture plate. Incubate the cells at 37°C in the 5% CO_2 incubator for 4 h.
8. Prepare the unnatural-amino-acid-containing DMEM/F-12 medium, as described in Subheading 3.1. Add 10% (v/v) FBS (see Note 9) and tetracycline to a final concentration of 1 μg/mL.

9. After the 4 h incubation at step 7, replace the Opti-MEM medium with 500 μL of the DMEM/F-12 medium, prepared at step 8.
10. Incubate the cells at 37°C in the CO_2 incubator for 16 h, to allow the cells to produce the recombinant protein containing the unnatural amino acid at the amber position (see Note 10).

3.4. Plasmid Transfection for Producing Proteins Containing L-Lysine Derivatives in T-REx CHO Cells

1. Inoculate 1.0×10^5 of T-REx CHO cells in the well of a 24-well cell culture plate with 500 μL of DMEM/F-12 growth medium, supplemented with FBS at a 10:1 ratio, in terms of volume.
2. Incubate the cells at 37°C in an incubator with a 5% CO_2 atmosphere. The cell population will become 90–95% confluent in the well after a 16–18 h of incubation.
3. Mix Opti-MEM (50 μL) with 0.3 μg of the pcDNA4/TO plasmid carrying the amber mutant gene (Subheading 3.2), together with 0.2 μg of a "pc-" plasmid and 0.3 μg of pU6tRNAPyl. Incubate the mixture at room temperature for 5 min. Refer Table 1 or Subheading 2.2 to select the appropriate plasmid for the unnatural amino acid to be incorporated into the recombinant protein.
4. The succeeding steps are identical to steps 4–10 of Subheading 3.3.

3.5. Plasmid Transfection for Producing Proteins Containing L-Tyrosine/Phenylalanine Derivatives in HEK 293 c-18 Cells

1. Inoculate 2.0×10^5 of HEK 293 c-18 cells in the well of a poly-D-lysine-coated 24-well cell culture plate with 500 μL of DMEM medium, supplemented with FBS at a 10:1 ratio, in terms of volume (see Note 11).
2. Incubate the cells at 37°C in an incubator with a 5% CO_2 atmosphere. The cell population will become 90–95% confluent after a 16–18 h of incubation.
3. Mix Opti-MEM (50 μL) with 0.4 μg of the pOriP plasmid carrying the amber mutant gene (Subheading 3.2), together with 0.4 μg of a "pSuPT-" plasmid for expressing the TyrRS variant and the *Bacillus* suppressor tRNA. Incubate the mixture at room temperature for 5 min. Refer Table 1 or Subheading 2.2 to select the appropriate plasmid for the unnatural amino acid to be incorporated into the recombinant protein.
4. In parallel with step 3, mix 2 μL of Lipofectamine 2000 with 50 μL of Opti-MEM, and then incubate the 52-μL mixture at room temperature for 5 min.
5. Combine these mixtures, prepared at steps 3 and 4, in a tube, and mix them gently. Incubate the combined mixture at room temperature for 20 min.
6. Replace the growth medium of the well, containing the cells, with 400 μL of Opti-MEM.

7. Add the mixture, prepared at step 5, to the well, and mix gently by rocking the culture plate. Incubate the cells at 37°C in the CO_2 incubator for 4 h.
8. Prepare the unnatural-amino-acid-containing DMEM medium, described in Subheading 3.1, and then supplement the medium with 10% (v/v) FBS (see Note 9).
9. After the 4 h of incubation at step 7, replace the Opti-MEM medium with 500 μL of the DMEM medium, prepared at step 8.
10. Incubate the cells at 37°C in the CO_2 incubator for 40 h, to allow the cells to produce the recombinant protein containing the unnatural amino acid at the amber position (see Note 10).

3.6. Plasmid Transfection for Producing Proteins Containing L-Lysine Derivatives in HEK 293 c-18 Cells

1. Inoculate 2.0×10^5 of HEK 293 c-18 cells in the well of a poly-D-lysine-coated 24-well cell culture plate with 500 μL of DMEM medium, supplemented with FBS at a 10:1 ratio, in terms of volume (see Note 11).
2. Incubate the cells at 37°C in an incubator with a 5% CO_2 atmosphere. The cell population will become 90–95% confluent in the well after a 16–18 h of incubation.
3. Mix Opti-MEM (50 μL) with 0.3 μg of the pOriP plasmid carrying the amber mutant gene (Subheading 3.2), together with 0.2 μg of a "pOriP-" plasmid and 0.3 μg of pOriP-U6tRNAPyl. Incubate the mixture at room temperature for 5 min. Refer Table 1 or Subheading 2.2 to select the appropriate plasmid for the unnatural amino acid to be incorporated into the recombinant protein.
4. The succeeding steps are identical to steps 4–10 of Subheading 3.5.

4. Notes

1. The concentrations of the unnatural amino acids in the growth media are critical for achieving their efficient incorporation into proteins. We provide their optimal concentrations for maximizing the yields of the proteins containing the amino acids. For some of these amino acids, the concentrations higher than the recommended ones have adverse effects on the cell growth.
2. Do not use HCl to dissolve iodoTyr, because exposure to a strong acid can remove the iodine atom from the amino acid.
3. Unnatural amino acids, such as *p*Bpa, have reactive groups in their side chains. Note that the amino acids with a benzoyl

or azide group are light-sensitive and labile, and the azide group is vulnerable to reducing reagents. Long-term storage may thus reduce the quality of the stock solution or cause precipitation because of the low solubility of the amino acid.

4. Do not use a small container, such as a 1.5-mL tube, to dissolve the unnatural amino acids. The amino acid may not be mixed well with the solvent in a small tube, and could form aggregates, which cannot be dissolved again completely.
5. Long exposure of an amino acid to strong acid can cause isomerization at the Cα atom, generating the D-isomer. Although this isomer of an amino acid is not recognized by the aaRS, the isomerization reduces the concentration of the available L-isomer.
6. The proteins designed to contain unnatural amino acids may be C-terminally tagged with a FLAG or c-Myc epitope. Such tags are useful for detecting, by using specific antibodies, the full-length products of the amber mutant genes, which should contain unnatural amino acids at the amber positions. A fluorescent protein may be fused at the C-terminus of the protein designed to contain unnatural amino acids. The emission of fluorescence from the products is a convenient indicator of the incorporation of unnatural amino acids.
7. In most of the previous studies, a single amber codon was introduced into a gene, because the yield of the full-length product with the unnatural amino acid at a single site is 10–25%, relative to that of the wild-type product. If two amber codons were introduced into a gene, then the yield would theoretically be 1–6%, which is too low to support many types of experiments. In this regard, note that truncated products, along with the full-length products with unnatural amino acids, are generated, due to the aborted translation at the amber position. If these products are not degraded and are stable in the cells, then they may be functional and should be considered when interpreting the experimental data. Since the truncated products cannot be detected by C-terminal tagging, a peptide tag should be added at the N-terminus, to analyze their expression in the cells.
8. The efficiency of incorporating unnatural amino acids into proteins in response to the amber codon greatly varies, depending on the identity of the amino acid and the position of the amber codon in the gene. The lysine derivatives, iodoTyr, and *p*Acpa are incorporated into proteins more efficiently than *p*Bpa and *p*Azpa, largely because of the higher activities of the specific aaRS variants for the former group. If efficient incorporation of an unnatural amino acid is not achieved at

some position, then change the position of the amber codon in the gene.

9. Adding FBS to the growth medium at a 10:1 ratio reduces the concentration of the unnatural amino acid by a factor of 1.1. The recommended concentrations of the amino acids in the growth media (Subheading 3.1) have been determined with consideration of this dilution.

10. In case the full-length products are not generated, we recommend performing a control experiment, with an amber mutant gene that successfully generated full-length products in the presence of unnatural amino acids (Fig. 2). This experiment will assess whether the unnatural amino acid incorporation system works in your cell lines. We can provide the amber mutant genes of *RAS* and *GRB2*, encoded on pcDNA3.1, pcDNA4/TO, or pOriP, for the control experiment.

11. HEK 293 c-18 cells will easily detach from the bottom of the well of a culture plate upon vigorous shaking or rough treatment. This problem can generally be avoided by using culture plates coated with poly-L/D-lysine, collagen, or other compounds with similar effects.

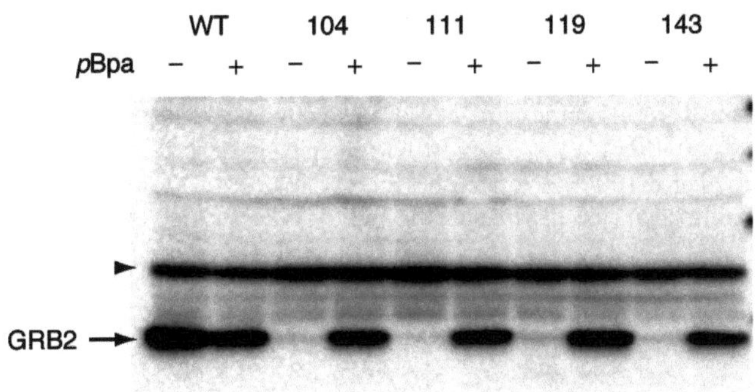

Fig. 2. Western-blot indicating the production of GRB2 proteins with *p*Bpa at different positions in T-Rex CHO cells. The amber codon was separately introduced at positions 104, 111, 119, and 143 of GRB2. The protein was C-terminally tagged with FLAG and expressed from pcDNA4/TO in T-REx CHO cells, expressing *p*BpaRS and the *Bacillus* amber suppressor tRNA. The full-length GRB2 was generated from the wild-type *GRB2* gene and was expressed from the amber mutant genes when *p*Bpa was supplemented in the growth medium. The removal of the amino acid from the medium abolished the expression of the full-length product. The *arrowhead* marks the endogenous protein that reacted with the anti-FLAG antibody. Note here that, if iodoTyr was used instead of *p*Bpa, then the full-length GRB2 protein would be produced even in the absence of iodoTyr, because iodoTyrRS incorporates L-tyrosine in the absence of competing idoTyr in the growth medium. Nevertheless, the supplementation of iodoTyr in the growth medium would increase the yield of the full-length product by a factor of 1.5–2, by expelling L-tyrosine from the amber position almost completely (4).

References

1. Wang L et al. (2001) Expanding the genetic code of *Escherichia coli*. Science 292, 498–500.
2. Chin JW et al. (2003) An expanded eukaryotic genetic code. Science 301, 964–967.
3. Mukai T et al. (2010) Genetic encoding of non-natural amino acids in *Drosophila melanogaster* Schneider 2 cells. Protein Sci 19, 440–448.
4. Sakamoto K et al. (2002) Site-specific incorporation of an unnatural amino acid into proteins in mammalian cells. Nucleic Acids Res 30, 4692–4699.
5. Drabkin HJ, Park HJ, and RajBhandary UL (1996) Amber suppression in mammalian cells dependent upon expression of an *Escherichia coli* aminoacyl-tRNA synthetase gene. Mol Cell Biol 16, 907–913.
6. Furter R (1998) Expansion of the genetic code: site-directed *p*-fluoro-phenylalanine incorporation in *Escherichia coli*. Protein Sci 7, 419–426.
7. Zhang Z et al. (2004) Selective incorporation of 5-hydroxytryptophan into proteins in mammalian cells. Proc Natl Acad Sci USA 101, 8882–8887.
8. Köhrer C, Sullivan EL, and RajBhandary UL (2004) Complete set of orthogonal 21st aminoacyl-tRNA synthetase-amber, ochre and opal suppressor tRNA pairs: concomitant suppression of three different termination codons in an mRNA in mammalian cells. Nucleic Acids Res 32, 6200–6211.
9. Hino N et al. (2005) Protein photo-cross-linking in mammalian cells by site-specific incorporation of a photoreactive amino acid. Nat Methods 2, 201–206.
10. Mukai T et al. (2008) Adding L-lysine derivatives to the genetic code of mammalian cells with engineered pyrrolysyl-tRNA synthetases. Biochem Biophys Res Commun 371, 818–822.
11. Ye S et al. (2010) Tracking G-protein-coupled receptor activation using genetically encoded infrared probes. Nature 464, 1386–1389.
12. Gautier A et al. (2010) Genetically encoded photocontrol of protein localization in mammalian cells. J Am Chem Soc 132, 4086–4088.
13. Hayashi A et al. (2011) Dissecting Cell Signaling Pathways with Genetically Encoded 3-Iodo-L-tyrosine. ChemBioChem 12, 387–389.
14. Liu W et al. (2007) Genetic incorporation of unnatural amino acids into proteins in mammalian cells. Nat Methods 4, 239–244.
15. Wang W et al. (2007) Genetically encoding unnatural amino acids for cellular and neuronal studies. Nat Neurosci 10, 1063–1072.
16. Kiga D et al. (2002) An engineered *Escherichia coli* tyrosyl-tRNA synthetase for site-specific incorporation of an unnatural amino acid into proteins in eukaryotic translation and its application in a wheat germ cell-free system. Proc Natl Acad Sci USA 99, 9715–9720.
17. Yanagisawa T et al. (2008) Multistep engineering of pyrrolysyl-tRNA synthetase to genetically encode N^{ε}-(o-azidobenzyloxycarbonyl) lysine for site-specific protein modification. Chem Biol 15, 1187–1197.

Chapter 14

Incorporation of Unnatural Non-α-Amino Acids into the N-Terminus of Proteins in a Cell-Free Translation System

Takahiro Hohsaka

Abstract

Unnatural amino acid mutagenesis allows us to introduce unnatural α-amino acids into internal positions of proteins in response to expanded codons such as amber and four-base codons. To improve the unnatural amino acid mutagenesis, the incorporation of unnatural α-amino acids and non-α-amino acids into the N-terminus of proteins has been achieved using expanded initiation codons. Here, we describe the method for the incorporation of fluorescent-labeled non-α-amino acids into the N-terminus of proteins in a cell-free translation system.

Key words: Unnatural amino acid, Translation initiation, Cell-free translation, Chemical aminoacylation, N-terminal labeling, Fluorescence labeling

1. Introduction

Incorporation of unnatural amino acids into internal positions of proteins in cell-free and in-cell translation systems is a useful technique for analysis of structure–function relationships of proteins and for design and synthesis of novel functional proteins. The unnatural amino acid mutagenesis allows us to introduce a variety of unnatural amino acids carrying functional groups such as photo-responsible groups, fluorescent groups, biotin, and post-translational modification groups at their side chains.

In contrast, a method to introduce unnatural amino acid derivatives into the N-terminus of proteins has been developed by using initiator Met-tRNAs that contain unnatural moieties at the α-amino group of methionine (1). Various fluorophores (2–4) and biotin (5) were incorporated into the N-terminus of proteins using this method in cell-free translation systems. Amino acids other

Loredano Pollegioni and Stefano Servi (eds.), *Unnatural Amino Acids: Methods and Protocols*,
Methods in Molecular Biology, vol. 794, DOI 10.1007/978-1-61779-331-8_14, © Springer Science+Business Media, LLC 2012

than methionine were also incorporated through the enzymatic misaminoacylation of the initiator tRNA (6–9). Although the N-terminal incorporation in response to an AUG codon competes with the incorporation of methionine by an endogenous initiator tRNA, the use of UAG codon as an initiator codon (6) allows the specific incorporation of unnatural amino acid derivatives (10, 11).

We have revealed that unnatural amino acid derivatives that do not contain amino group at the α-position can also be incorporated into the N-terminus of proteins (Fig. 1) (12, 13). This method allows the incorporation of various compounds having a carboxyl group, and will expand the application fields of the unnatural amino acid mutagenesis technique. Here, we describe the experimental details for the preparation of initiator tRNAs acylated with fluorescent-labeled non-α-amino acids such as β-alanine (β-Ala) (Fig. 2), and the incorporation into the N-terminus of proteins in response the UAG initiation codon in an *Escherichia coli* cell-free translation system.

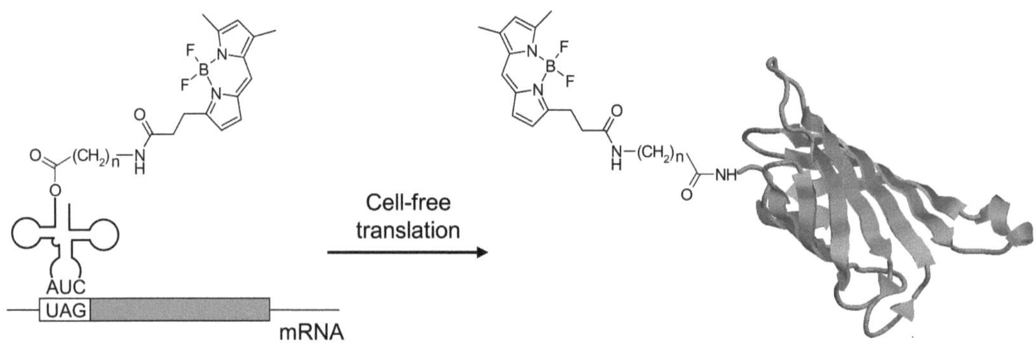

Fig. 1. Illustration of incorporation of fluorescent-labeled non-α-amino acids into the N-terminus of proteins in response to an initiation UAG codon.

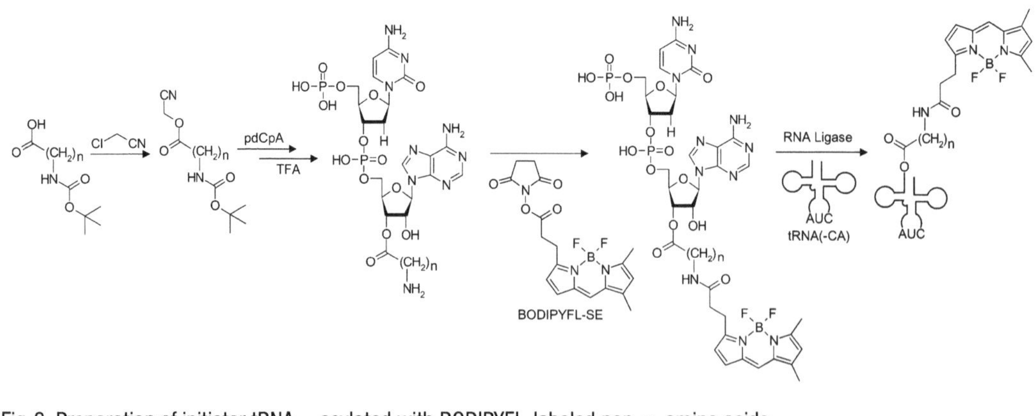

Fig. 2. Preparation of initiator $tRNA_{CUA}$ acylated with BODIPYFL-labeled non-α-amino acids.

2. Materials

2.1. Chemical Synthesis

1. Boc-β-Ala and other Boc-protected non-α-amino acids (Kokusan Chemical, Tokyo, Japan, or Bachem, Bubendorf, Switzerland).
2. 5′-*O*-Phosphoryl-2′-deoxycytidylyl-(3′-5′)-adenosine (pdCpA) tetra-*n*-butyl ammonium salt (14, 15). Dissolve in DMF (dimethylformamide) at 44 mM.
3. BODIPYFL succinimide ester (Molecular Probes, Eugene, OR, USA). Dissolve in DMSO (dimethyl sulfoxide) at 100 mM.
4. XTerra C18 column (Waters, Milford, MA, USA).

2.2. RNA Preparation

1. A vector containing T7 promoter and *E. coli* initiator $tRNA_{CUA}$ gene lacking the 3′ dinucleotide (*TAATACGACT CACTAT ACGC* GGGGTGGAGC AGCCTGGTAG CTCGTCGGGC TCTAAACCCGA AGGTCGTCGG TTCAATCCGG CCCC CGCAAC, in which T7 promoter is italic and the anticodon is underlined).
2. T7 promoter primer CTAATACGAC TCACTATACG (see Note 1).
3. tRNA 3′ primer G(2′-*O*-Me)-U(2′-*O*-Me)-TGCGGGGG CCGGATTTGA, in which the 5′ terminal two nucleotides are 2′-*O*-methylated ribonucleotides (see Note 2).
4. KOD Dash DNA polymerase (2.5 U/μL) (Toyobo, Osaka, Japan).
5. QIAquick PCR purification kit (QIAGEN, Hilden, Germany).
6. 10× Transcription buffer: 400 μL 1 M Tris–HCl (pH 8.0), 200 μL 1 M $MgCl_2$, 50 μL 1 M DTT, 350 μL water (see Note 3).
7. T7 RNA polymerase (50 U/μL) (New England Biolabs, Beverly, MA, USA).
8. RNase inhibitor (40 U/μL) (Takara Bio, Otsu, Japan).
9. Inorganic pyrophosphatase (Sigma, St. Louis, MO, USA). Reconstitute at 0.5 U/μL with 50% glycerol.
10. RNeasy Mini kit (QIAGEN).
11. T4 RNA ligase (25 U/μL) (Takara Bio).
12. 5× Ligation buffer: 500 μL 550 mM HEPES-Na (pH 7.5), 75 μL 1 M $MgCl_2$, 16.5 μL 0.1 M DTT, 50 μL 100 mM ATP, and 358.5 μL water.
13. Phenol/chloroform (for tRNA): saturate phenol by 0.3 M potassium acetate (pH 4.5). Add one volume of chloroform.
14. Poros R2/10 column (Applied Biosystems, Foster, CA, USA).
15. Trithymidine (dT3): Dissolve in water at 70 μM.

16. A vector encoding a protein under the control of T7 promoter, and primers for PCR amplification of the coding region (see Note 4).
17. Vent DNA polymerase (2 U/μL) (New England Biolabs).
18. Phenol/chloroform (for mRNA): Saturate phenol by 10 mM Tris–HCl (pH 7.5), 1 mM EDTA. Add one volume of chloroform.

2.3. Cell-Free Translation

1. LM: 240 μL 2.2 M HEPES-potassium buffer (pH 7.5), 700 μL 2.88 M potassium glutamate, 600 μL 0.42 M phosphoenolpyruvate cyclohexylamine salt, 450 μL 40% (w/v) PEG8000, 150 μL 76 mM ATP potassium salt, 120 μL 2.8 mg/mL folinic acid, 30 μL 88 mM GTP potassium salt, 30 μL 2.2 M ammonium acetate, 30 μL 0.55 M DTT (see Note 5), 20 μL 0.48 M spermidine, and 30 μL water.
2. LM+12: 250 μL LM, 18.2 μL 5.5 mM 20 amino acids, 12 μL 1 M magnesium acetate (see Note 6), and 19.8 μL water.
3. *E. coli* S30 extract for linear template (Promega, Madison, WI, USA).
4. RNase A (Sigma). Reconstitute at 10 mg/mL with water.
5. 2× Sample buffer: 200 μL 0.5 M Tris–HCl (pH 6.8), 200 μL glycerol, 40 μL 2-mercaptoethanol, 40 μL 0.5% BPB, 20 μL 10% SDS, and 500 μL water.

3. Methods

3.1. Synthesis of β-Ala-pdCpA

1. Add 0.1 mg Boc-β-Ala to 0.5 mL triethylamine and 2 mL acetonitrile in a round flask. Add 0.2 mL chloroacetonitrile on ice, and stir at room temperature for 12 h.
2. Acidify the reaction mixture to pH 2–3 with 5% aqueous $KHSO_4$. Extract the product with ethyl acetate. Wash three times with 5% aqueous $KHSO_4$, three times with 4% aqueous $NaHCO_3$, and once with saturated aqueous NaCl. Dry over sodium sulfate. Concentrate by evaporation to afford Boc-β-Ala cyanomethyl ester.
3. Add 2.0 mg Boc-β-Ala cyanomethyl ester (8.8 μmol) to 40 μL 44 mM pdCpA tetra-*n*-butyl ammonium salt (1.76 μmol) in DMF in a 1.5-mL microtube. Incubate the reaction mixture at room temperature for 1 h.
4. Add 1 mL diethyl ether, and centrifuge at 830 × *g* for 1 min. Wash the precipitate twice with 1 mL diethyl ether, and dry under vacuum.

5. Dissolve the pellet in 200-μL trifluoroacetic acid (TFA) in a 1.5-mL microtube. Incubate on ice for 15 min to remove the Boc group. Evaporate TFA by vacuum centrifuge. Wash the pellet twice with 1 mL diethyl ether, and dry under vacuum.
6. Dissolve β-Ala-pdCpA in DMSO (~200 μL). To 1 μL of DMSO solution of β-Ala-pdCpA, add 10 μL of 0.1 M NaOH to hydrolyze the ester linkage between β-Ala and pdCpA. Incubate at room temperature for 2 h (see Note 7). Add 20 μL 0.1 M acetic acid for neutralization. Apply the resulting mixture to a reverse-phase HPLC (XTerra C18, 2.5 μm, 4.6 mm × 20 mm), flow rate 1.5 mL/min with a linear gradient of 0–100% methanol in 0.38% formic acid, over 10 min. Determine the peak area of the released pdCpA. Calculate the concentration of pdCpA by comparing the peak area of the standard pdCpA, whose concentration is determined by UV absorption spectrum using an $\varepsilon_{260} = 23{,}000$/M/cm (14).
7. Adjust the concentration of β-Ala-pdCpA in DMSO to 4.4 mM. Store at −30°C (see Note 8).

3.2. Synthesis of BODIPYFL-β-Ala-pdCpA

1. Combine 200 μL of 4.4 mM β-Ala-pdCpA in DMSO (0.88 μmol), 40 μL of 100 mM BODIPYFL succinimide ester (4 μmol), 160 μL of DMSO, and 400 μL of 0.1 M $NaHCO_3$ on ice in a 1.5-mL microtube. Incubate on ice for 1 h.
2. Add 40 μL 1 M acetic acid for neutralization. Isolate BODIPYFL-β-Ala-pdCpA by a preparative reverse-phase HPLC (XTerra C18, 5 μm, 10 mm × 50 mm), flow rate 3.0 mL/min with a linear gradient of 0–100% methanol in 0.38% formic acid, over 15 min (see Note 9). Remove the solvent by vacuum centrifuge.
3. Determine the concentration of BODIPYFL-β-Ala-pdCpA as described for β-Ala-pdCpA.
4. Adjust the concentration of BODIPYFL-β-Ala-pdCpA in DMSO to 2.2 mM. Store at −30°C (see Notes 10 and 11).

3.3. Preparation of Initiator tRNA Lacking the 3′ Dinucleotide

1. Combine 10 μL of 10× KOD Dash buffer, 10 μL of 2 mM dNTPs, 1 μL of 100 μM T7 promoter primer, 1 μL of 100 μM tRNA 3′ primer, 1 μL of template plasmid encoding *E. coli* initiator $tRNA_{CUA}$, 1 μL of KOD Dash DNA polymerase, and 76 μL of water in a 250-μL microtube. Perform PCR using a thermal cycler (95°C for 3 min; 18 cycles of 98°C for 10 s, 55°C for 2 s, and 74°C for 20 s; 74°C for 3 min).
2. Purify the product using QIAquick PCR purification kit. Elute the PCR product with 50 μL of water.
3. Apply 1 μL eluted solution to 8% PAGE to confirm the amplified DNA (92 bp).

4. Combine 10 μL of 10× Transcription buffer, 16 μL of 25 mM NTPs, 20 μL of 100 mM CMP, 2 μL of 100 mM spermidine, 1 μL of 0.1% BSA, 41 μL of purified DNA solution, 1 μL of RNase inhibitor, 1 μL of inorganic pyrophosphatase, and 8 μL of T7 RNA polymerase. Incubate at 37°C for 15 h (see Note 12).
5. Purify the product using RNeasy Mini kit. Elute the tRNA with water.
6. Determine the concentration of the eluted tRNA by measuring the absorbance at 260 nm for 1/10 diluted tRNA solution (see Note 13). Adjust the concentration of the tRNA to 0.2 mM (0.1 A_{260} U/μL).
7. Apply 1 μL of 1/10 diluted tRNA solution to 8% PAGE containing 7 M urea to confirm the preparation of tRNA.

3.4. Preparation of BODIPYFL-β-Ala-tRNA

1. Combine 10 μL of 5× Ligase buffer, 6.25 μL of 0.2 mM tRNA lacking the 3′ dinucleotide, 1 μL of 0.1% BSA, 5 μL of 2.2 mM BODIPY-β-Ala-pdCpA in DMSO (see Note 14), 6 μL of T4 RNA ligase (25 U/μL), and 21.75 μL of water in a 1.5-mL microtube. Incubate at 4°C for 12 h.
2. Add 50 μL of 0.6 M potassium acetate (pH 4.5). Extract with phenol/chloroform (for tRNA) and chloroform (see Note 15).
3. Add 300 μL of cold ethanol. Precipitate BODIPY-β-Ala-tRNA at −30°C for 30 min.
4. Centrifuge at 16,100×*g* (bench-top microcentrifuge) for 30 min at 4°C. Wash the pellet with cold 70% ethanol and dry under the vacuum.
5. Dissolve the pellet with 5 μL of 1 mM potassium acetate (pH 4.5).
6. Mix 1 μL of the ligation product, 2 μL of 70 μM trithymidine (dT3), and 40 μL of 0.1 M triethylammonium acetate, pH 7.0. Apply to a reverse-phase HPLC (Poros R2/10, 4.6 mm × 100 mm), flow rate 1.0 mL/min with a linear gradient 0–100% of acetonitrile in 0.1 M triethylammonium acetate, pH 7.0, over 20 min. Determine the peak area of BODIPYFL-β-Ala-tRNA and dT3. Calculate the concentration of BODIPYFL-β-Ala-tRNA by comparing the peak area of dT3.
7. Adjust the concentration of BODIPYFL-β-Ala-tRNA to 80 μM (see Note 16).

3.5. Preparation of mRNA

1. Combine 10 μL of 10× ThermoPol buffer, 10 μL of 2 mM dNTPs, 1 μL of 100 μM 5′ primer, 1 μL of 100 μM 3′ primer, 1 μL of template plasmid, 1 μL of Vent DNA polymerase, and 76 μL of water in a 250-μL microtube. Perform PCR using a

thermal cycler (95°C for 3 min; 25 cycles of 95°C for 30 s, 55°C for 30 s, and 72°C for 60 s; 72°C for 3 min).

2. Purify the product using QIAquick PCR purification kit. Elute the PCR product with 50 μL of water.
3. Apply 1 μL of the eluted solution to 4% PAGE to confirm the size of the amplified DNA.
4. Combine 10 μL of 10× Transcription buffer, 16 μL of 25 mM NTPs, 2 μL of 100 mM spermidine, 1 μL of 0.1% BSA, 20 μL of purified DNA solution, 1 μL of RNase inhibitor, 1 μL of inorganic pyrophosphatase, 4 μL of T7 RNA polymerase, and 45 μL of water. Incubate at 37°C for 6 h (see Note 17).
5. Add 100 μL of 5 M ammonium acetate. Incubate on ice for 20 min. Centrifuge at 16,100 × *g* (bench-top microcentrifuge) for 20 min at 4°C.
6. Dissolve the pellet with 200 μL of 10 mM Tris–HCl (pH 7.5), 1 mM EDTA. Extract with phenol/chloroform (for mRNA) and chloroform.
7. Add 20 μL of 3 M potassium acetate (pH 5.2) and 600 μL of ethanol. Precipitate mRNA at −30°C for 30 min.
8. Centrifuge at 16,100 × *g* (bench-top microcentrifuge) for 30 min at 4°C. Wash the pellet with cold 70% ethanol and dry under the vacuum.
9. Dissolve the pellet with 20 μL of water.
10. Determine the concentration of the mRNA by measuring the absorbance at 260 nm for 1/10 diluted mRNA solution. Adjust the concentration to 8 μg/μL (0.2 A_{260} U/μL).
11. Apply 1 μL of 1/10 diluted mRNA solution to 4% PAGE containing 7 M urea to confirm the preparation of mRNA.

3.6. Cell-Free Translation

1. Combine 3 μL of LM + 12, 1 μL of 8 μg/μL mRNA, 1 μL of 80 μM BODIPY-β-Ala-tRNA, 2 μL of S30 extract, and 3 μL of water. Incubate at 37°C for 1 h (see Note 18).
2. Mix 1 μL of the reaction mixture with 1 μL of 10 mg/ml RNase A. Incubate at 37°C for 15 min (see Note 19). Add 10 μL of 2× Sample buffer and 8 μL of water. Heat at 95°C for 5 min. Apply 5 μL of the solution to 15% SDS-PAGE.
3. Visualize the gel by a fluorescence imager scanner with excitation at 488 nm and emission at 520 nm for BODIPYFL (Fig. 3) (see Notes 20 and 21).
4. Visualize by Western blotting using specific antibodies against the expressed proteins or tag peptides fused to the expressed proteins (see Notes 22 and 23).

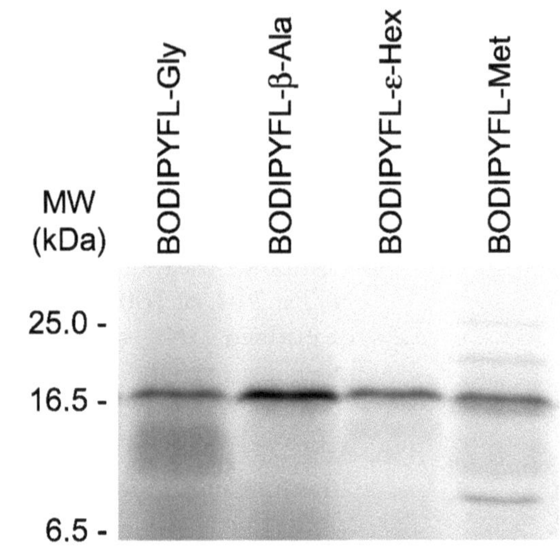

Fig. 3. Fluorescence image of SDS-PAGE gel for the cell-free translation products obtained from a streptavidin gene containing an initiation UAG codon in the presence of initiator $tRNA_{CUA}$ acylated with *N*-BODIPYFL-linked Gly, β-Ala, ε-aminohexanoic acid (ε-Hex), and Met.

4. Notes

1. C is added to the 5′ terminus of the T7 promoter sequence to enhance the transcription by T7 RNA polymerase (16).
2. Two 2′-*O*-methylated ribonucleotides at the 5′ terminus are used to suppress the additional nucleotide extension during the transcription by T7 RNA polymerase (17).
3. Wear disposable gloves while handling reagents for RNA to prevent RNase contamination from the surface of the skin. Reagents, pipette tips, and microtubes should be used exclusively for RNA.
4. We use vectors derived from pGEMEX-1 (Promega), T7 up primer CCCGCGCGTT GGCCGATTCA, and T7 term primer TATTACGCCA GGTTATCCGG.
5. For expression of proteins containing disulfide bonds, DTT is omitted and oxidized glutathione is supplied to the cell-free translation reaction at around 1 mM.
6. The optimized concentration of Mg^{2+} may vary depending on the lot of S30 extracts.
7. Prolonged hydrolysis is required because the ester linkage between pdCpA and non-α-amino acids is more stable against hydrolysis than that of α-amino acids (see Fig. 4). To hydrolyze α-aminoacyl-pdCpA derivatives, incubation with 50 mM NaOH at room temperature for 10 min is sufficient.

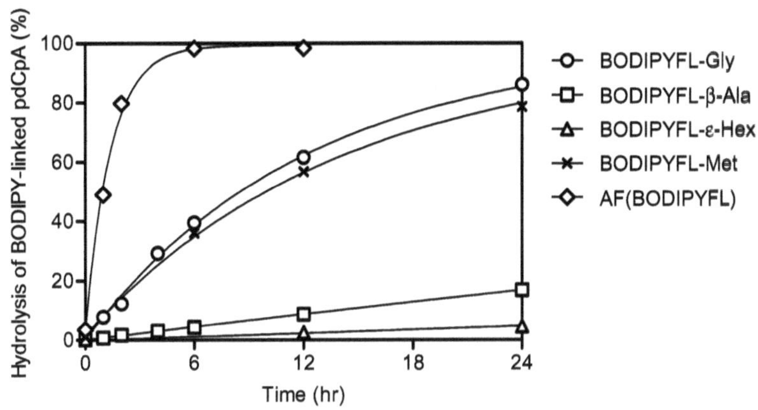

Fig. 4. Spontaneous hydrolysis of the ester linkage between pdCpA and *N*-BODIPYFL-linked Gly, β-Ala, ε-aminohexanoic acid (ε-Hex), and Met in 10 mM Tris–HCl (pH 7.5) at 37°C. AF(BODIPYFL) is α-amino acid (*p*-aminophenylalanine) carrying BODIPYFL at the *p*-amino group.

8. Other non-α-amino acid derivatives can be also synthesized in a similar manner. Derivatives of $NH_2(CH_2)nCOOH$, in which $n = 1–7$ and 11, have been synthesized (13).
9. Fluorophore–pdCpA conjugates may co-elute with fluorophores and succinimide ester derivatives. In that case, other types of reverse-phase columns (e.g., XBridge C18 and Sun Fire C18 columns) may be effective to isolate the desired products.
10. Acyl-pdCpA derivatives can be identified by ESI-MS and MALDI TOF-MS. For ESI-MS, combine 0.2 μL of 2.2 mM acyl-pdCpA in DMSO, 5 μL of 0.38% formic acid, and 5 μL of acetonitrile, and apply to an ESI-MS instrument (Mariner, Applied Biosystems). As calibration standards, trithymidine dT3 (849.1751) and tetrathymidine dT4 (1153.2211) are used.
11. Other fluorophore derivatives can be synthesized in a similar manner. Derivatives of AlexaFluor488, BODIPY558, CR110, Cy3, Cy5, FAM, and TAMRA have been synthesized (13).
12. The incubation can be prolonged to 12–18 h, because tRNAs are relatively resistant against the degradation by contaminating RNase.
13. We use NanoDrop (Thermo Scientific) as a spectrophotometer to measure A_{260} values using 1 μL of tRNA solutions.
14. Warm the DMSO solution at room temperature just before the preparation of the ligation reaction.
15. BODIPYFL-β-Ala-pdCpA and BODIPYFL-β-Ala are extracted to phenol/chloroform phase and BODIPYFL-β-Ala-tRNA remains in aqueous phase. However, some fluorophores may

bring fluorophore–tRNA conjugates to phenol/chloroform phase. In that case, phenol/chloroform and chloroform extraction should be omitted.

16. Other fluorophore–tRNA conjugates can be prepared in a similar manner.
17. Transcription reaction for mRNAs should be terminated within 6 h to avoid the degradation of RNA with contaminating RNase.
18. Reaction mixture of the cell-free translation can be temporarily stored at −80°C.
19. The ester linkage between the tRNA and fluorescent non-α-amino acids is not hydrolyzed during SDS-PAGE, and fluorophore–tRNA conjugates migrate at 15–20 kDa. To avoid the overlapping with the expressed proteins, fluorophore–tRNA conjugates should be digested by RNase before the SDS-PAGE.
20. Clean up the glass plate using Kimwipe moistened with water and with ethanol prior to fluorescence imaging.
21. Fluorescent non-α-amino acids that are not incorporated into proteins migrate at the front of the electrophoresis or at low molecular mass range. If target proteins possess a low molecular mass, a control reaction without the addition of mRNA should be prepared and applied to SDS-PAGE to determine the mobility of the fluorescent non-α-amino acids.
22. We use an anti-His tag antibody (Novagen, La Jolla, CA, USA) to detect His-tagged proteins at the C-terminus.
23. Band intensities on the Western blot are not consistent with those on the fluorescence image because the efficiency of the transfer from gel to membrane varies depending on the protein. Some proteins may be faintly observed on the western blot, whereas distinct fluorescent bands are observed on the fluorescence image.

Acknowledgments

This work was supported by Grants-in-Aid for Exploratory Research (17651122) and Scientific Research on Innovative Areas (20107005) from the Ministry of Education, Culture, Sports, Science, and Technology, Japan.

References

1. Kudlicki W, Odom O W, Kramer G, Hardesty B (1994) Chaperone-dependent folding and activation of ribosome-bound nascent rhodanese: Analysis by fluorescence. J Mol Biol 244, 319–331.
2. Ramachandiran V, Willms C, Kramer G, Hardesty B (2000) Fluorophores at the N terminus of nascent chloramphenicol acetyltransferase peptides affect translation and movement through the ribosome. J Biol Chem 275, 1781–1786.
3. McIntosh B, Ramachandiran V, Kramer G, Hardesty B (2000) Initiation of protein synthesis with fluorophore-Met-tRNA$_f$ and the involvement of IF-2. Biochimie 82, 167–174.
4. Gite S, Mamaev S, Olejnik J, Rothschild K (2000) Ultrasensitive fluorescence-based detection of nascent proteins in gels. Anal Biochem 279, 218–225.
5. Taki M, Sawata S Y, Taira K (2001) Specific N-terminal biotinylation of a protein *in vitro* by a chemically modified tRNA^{fmet} can support the native activity of the translated protein. J Biosci Bioeng 92, 149–153.
6. Varshney U, RajBhandary U L (1990) Initiation of protein synthesis from a termination codon. Proc Natl Acad Sci USA 87, 1586–1590.
7. Drabkin HJ, RajBhandary U L (1998) Initiation of protein synthesis in mammalian cells with codons other than AUG and amino acids other than methionine. Mol Cell Biol 18, 5140–5147.
8. C, Kohrer C, Kenny E et al. (2003) Anticodon sequence mutants of *Escherichia coli* initiator tRNA: Effects of overproduction of aminoacyl-tRNA synthetases, methionyl-tRNA formyltransferase, and initiation factor 2 on activity in initiation. Biochemistry 42, 4787–4799.
9. Chattapadhyay R, Pelka H, Schulman LH (1990) Initiation of *in vivo* protein synthesis with non-methionine amino acids. Biochemistry 29, 4263–4268.
10. Mamaev S, Olejnik J, Olejnik E K, Rothschild KJ (2004) Cell-free N-terminal protein labeling using initiator suppressor tRNA. Anal Biochem 326, 25–32.
11. Olejnik J, Gite S, Mamaev S, Rothschild KJ (2005) N-terminal labeling of proteins using initiator tRNA. Methods 36, 252–260.
12. Muranaka N, Miura M, Taira H, Hohsaka T (2007) Incorporation of unnatural non-α-amino acids into the N terminus of proteins in a cell-free translation system. ChemBioChem 8, 1650–1653.
13. Miura M, Muranaka N, Abe R, Hohsaka T (2010) Incorporation of fluorescent-labeled non-α-amino carboxylic acids into the N-terminus of proteins in response to amber initiation codon. Bull Chem Soc Jpn 83, 546–553.
14. Robertson S A, Ellman J A, Schultz P G (1991) A general and efficient route for chemical aminoacylation of transfer RNAs. J Am Chem Soc 113, 2722–2729.
15. Thorson JS, Cornish VW, Barrett J E et al. (1998) A biosynthetic approach for the incorporation of unnatural amino acids into proteins. Methods Mol Biol 77, 43–73.
16. Baklanov MM, Golikova LN, Malygin E G (1996) Effect on DNA transcription of nucleotide sequences upstream to T7 promoter. Nucleic Acids Res 24, 3659–3660.
17. Kao C, Zheng M, Rudisser S (1999) A simple and efficient method to reduce nontemplated nucleotide addition at the 3 terminus of RNAs transcribed by T7 RNA polymerase. RNA 5, 1268–1272.

Chapter 15

Site-Specific Modification of Proteins by the Staudinger-Phosphite Reaction

Paul Majkut, Verena Böhrsch, Remigiusz Serwa, Michael Gerrits, and Christian P.R. Hackenberger

Abstract

Chemoselective reactions are important tools for the modification of peptides and proteins. Thereby the modification is desired to be site specific and bioorthogonal. Here we describe the site-specific modification of azido-proteins via a Staudinger-type phosphite ligation. The reaction was carried out in aqueous system on proteins containing *p*-azido-phenylalanine in a single position introduced by the amber codon technique. A selective introduction of branched polyethylene scaffolds can be achieved with the application of the methodology reported herein.

Key words: Site-specific protein modification, Azido-proteins, Bioorthogonal, Staudinger-phosphite ligation, Protein PEGylation

1. Introduction

The selective decoration of biomolecules with functional probes or property-modifying residues remains an ongoing challenge for interdisciplinary research. With the possibility to insert site specifically unnatural amino acids by the amber-codon methodology (1–7), the ball was passed to chemists engaged in engineering chemoselective reactions for the application in biocompatible reaction conditions.

Most commonly, azides are applied as chemical reporters in such bioorthogonal reactions (8). The formation of triazoles as a result of a [3 + 2]-cycloaddition between azides and alkynes has proved to be very efficient though requiring the presence of cytotoxic Cu(I)-catalysts limiting the *in vivo* applicability (9, 10). Strain-promoted metal-free alternatives have been developed,

Loredano Pollegioni and Stefano Servi (eds.), *Unnatural Amino Acids: Methods and Protocols*,
Methods in Molecular Biology, vol. 794, DOI 10.1007/978-1-61779-331-8_15, © Springer Science+Business Media, LLC 2012

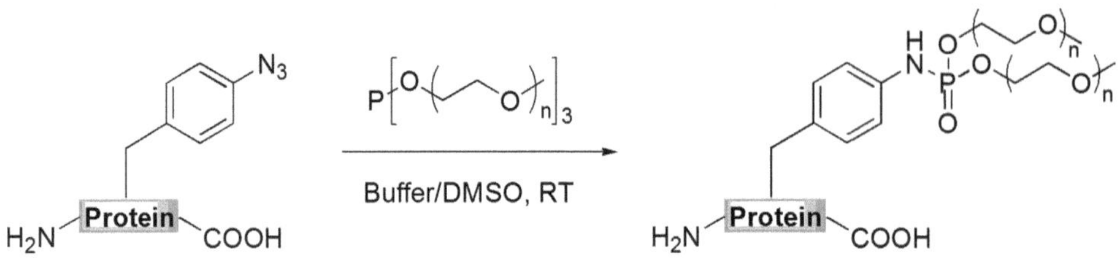

Scheme 1. Site-specific PEGylation of proteins.

which might, however, suffer from the introduction of a large and unnatural moiety between the protein and the probe, could influence biological properties and limit the tunability of the modification (11–13). The Staudinger ligation, a reaction between azides and phosphines has also shown great potential in the labelling of biomolecules *in vivo*, but the oxidation of the phosphines can be limiting to its usability (14–16). The Staudinger-phosphite reaction was introduced as an alternative ligation method that yields a phosphoramidate as the linking unit, which mimics natural-occurring phosphates. The reaction between azido-phenylalanine and air stable phosphites can be conducted in buffers at room temperature requiring no additional components or catalysts (17–19).

Symmetrical phosphites bearing different polyethylene glycol (PEG) chains, which can be synthesized in a straightforward manner as reported in a previous publications (18), can functionalize proteins upon Staudinger reaction with branched PEG-scaffolds (Scheme 1). The elements are expected to have important pharmacological implications when presented on therapeutic polypeptides, since it was shown that branched PEG strands possess reduced immunogenicity and decreased ability to accumulate in animal tissues (20–23). Here, we present the reaction of different proteins with phosphites to yield PEGylated proteins with different chain lengths.

2. Materials

2.1. SDS Polyacrylamide Gel Electrophoresis

1. Resolving gel buffer: 1.5 M Tris–HCl, pH 8.8. Store at 4°C.
2. Stacking gel buffer: 1.0 M Tris–HCl, pH 6.8. Store at 4°C.
3. Thirty percent acrylamide/Bis acrylamide solution (37.5:1 acrylamide:Bis): store at 4°C, in a dark bottle.
4. Ammonium persulfate: 10% Solution in water (see Note 1).
5. *N*,*N*,*N*,*N*′-tetramethyl-ethylenediamine (TEMED) (Garl Roth GmbH®). Store at 4°C (see Note 2).
6. SDS-PAGE Running buffer: 50 mM Tris, 384 mM glycine, and 0.1% SDS (see Note 3).

7. SDS Loading buffer (4×): Roti®-Load1 concentrate (Carl Roth GmbH store at 20°C). For use, dilute with three equivalent volumes of water, then add 1% bromophenol blue.
8. BPB Solution: Dissolve 0.1 g of BPB in 100 mL water.
9. Coomassie staining solution: 0.1% (w/v) Coomassie Brilliant Blue R250, 50% (v/v) MeOH, 10% (v/v) acetic acid (96%) in water.
10. Destaining solution: 20% (v/v) MeOH, 7.5% (v/v) acetic acid (96%) in water.

2.2. TCA Precipitation

1. TCA/Casein solution: 2% (w/v) Casein hydrolysate, 10% (v/v) trichloroacetic acid (TCA) in water.
2. 5% TCA solution: 5% (v/v) TCA in water.

3. Methods

3.1. Synthesis of Azido Proteins SecB and hFABP

For these experiments, two different proteins were synthetized as azidoproteins, namely the *Escherichia coli* protein SecB and the human liver fatty acid-binding protein FABP1. The DNA templates for both proteins comprised the reading frames cloned into the pQE2 vector (Qiagen) containing an amber stop codon for site-specific unnatural amino acid incorporation at the native C-terminus. In the case of SecB, a His-Tag for protein purification followed the amber codon; hFABP1 does not contain a purification tag. Protein synthesis was performed using an *E. coli*-derived cell-free translation system depleted of termination factor RF1 and supplemented with enriched fractions of orthogonal *amber* suppressor tRNA and *p*-azidophenylalanyltRNA synthetase specific for *p*-azidophenylalanine. The system contained *p*-azidophenylalanine (Bachem) in addition to the 20 natural amino acids; the yield of *p*-azidophenylalanine containing SecB was approximately 200 μg per mL reaction, and the yield of hFABP1 was 236 μg per mL reaction. The expressed proteins were expressed in the presence of ^{14}C-labelled leucine to enable radiography.

DNA and amino acid sequence of the *p*-azidophenylalanine containing model proteins FABP and SecB are reported in Fig. 1.

3.2. Determination of Protein Concentration by TCA Precipitation and Scintillation Counting

Two aliquots per reaction (6 μL each) are moved into glass test tubes and mixed with 3.0 mL of TCA/casein solution. For denaturation, the samples are heated in a water bath for 15 min at 85°C, then chilled for 30 min on ice. Afterwards, the samples are vortexed and filtered through a wet (5% TCA) Macherey-Nagel filter paper. The test tubes are washed with 5% TCA and filtered again, then the filters are washed with acetone and dried. Next, the filters are transferred to scintillation tubes and overlaid with 3.0 mL Quicksafe

A scintillation cocktail (Zinsser Analytics). Finally, the samples are incubated by vigorous shaking (300 rpm) for 10 min and analyzed on a liquid scintillation analyzer (Packard Tri-Carb 2900 TR).

3.3. PEGylation of Azido-Proteins by Incubation with Phosphites

To a suspension/solution of an azido-protein (10–20 μM) in phosphate buffer pH 8.0 (20 μL) at 28°C was added a solution of a PEG-phosphite in the same buffer (50 mM), 5–10 μL/24 h, and DMSO, when appropriate (see Note 4). The sample was shaken (2,000 rpm) for up to 48 h at 28°C. Following the incubation, the mixture was analyzed by SDS-PAGE.

a

```
  1 atgtcatttagtggtaaatatcaactgcagagccaggaaaacttt
  1  M  S  F  S  G  K  Y  Q  L  Q  S  Q  E  N  F

 46 gaagccttcatgaaggcaatcggtctgccggaagagctcatccag
 16  E  A  F  M  K  A  I  G  L  P  E  E  L  I  Q

 91 aaggggaaggatatcaagggggtgtcggaaatcgtgcagaatggg
 31  K  G  K  D  I  K  G  V  S  E  I  V  Q  N  G

136 aagcacttcaagttcaccatcaccgctgggtccaaagtgatccaa
 46  K  H  F  K  F  T  I  T  A  G  S  K  V  I  Q

181 aacgaattcacggtgggggaggaatgtgagctggagacaatgaca
 61  N  E  F  T  V  G  E  E  C  E  L  E  T  M  T

226 ggggagaaagtcaagacagtggttcagttggaaggtgacaataaa
 76  G  E  K  V  K  T  V  V  Q  L  E  G  D  N  K

271 ctggtgacaactttcaaaaacatcaagtctgtgaccgaactcaac
 91  L  V  T  T  F  K  N  I  K  S  V  T  E  L  N

316 ggcgacataatcaccaataccatgacattgggtgacattgtcttc
106  G  D  I  I  T  N  T  M  T  L  G  D  I  V  F

361 aagagaatcagcaagagaatttagtaa
121  K  R  I  S  K  R  I AzF *
```

Fig. 1. DNA and amino acid sequence of the *p*-azidophenylalanine containing model protein FABP (**a**) and SecB (**b**). The *p*-azido-phenylalanine (AzF) residue and corresponding amber codon within the SecB amino acid and DNA sequences are *underlined.*

b

```
  1 atgtcagaacaaaacaacactgaaatgactttccagatccaacgt
  1  M  S  E  Q  N  N  T  E  M  T  F  Q  I  Q  R

 46 atttataccaaggatatctctttcgaagcgccgaacgcgccgcac
 16  I  Y  T  K  D  I  S  F  E  A  P  N  A  P  H

 91 gttttccagaaagattggcaaccagaagttaaacttgatctggat
 31  V  F  Q  K  D  W  Q  P  E  V  K  L  D  L  D

136 acggcatcttcccaactggcagatgacgtatacgaagtggtactg
 46  T  A  S  S  Q  L  A  D  D  V  Y  E  V  V  L

181 cgtgttaccgtaacggcctctttgggcgaagaaaccgcgttcctg
 61  R  V  T  V  T  A  S  L  G  E  E  T  A  F  L

226 tgtgaagttcagcagggcggtatttctccatcgcgggtatcgaa
 76  C  E  V  Q  Q  G  G  I  F  S  I  A  G  I  E

271 ggcacccagatggcgcattgcctgggagcatactgcccgaacatt
 91  G  T  Q  M  A  H  C  L  G  A  Y  C  P  N  I

316 ctgttcccgtatgctcgtgagtgcatcaccagcatggtatcccgc
106  L  F  P  Y  A  R  E  C  I  T  S  M  V  S  R

361 ggtacattcccgcaactgaaccttgcgccggttaacttcgatgcg
121  G  T  F  P  Q  L  N  L  A  P  V  N  F  D  A

406 ctgttcatgaactatttgcagcagcaggctggcgaaggtactgaa
136  L  F  M  N  Y  L  Q  Q  Q  A  G  E  G  T  E

451 gaacatcaggatgcctagggtcaccaccatcaccatcactaa
151  E  H  Q  D  A AzF G  H  H  H  H  H  H  *.
```

Fig.1. (continued)

3.4. Acetone Precipitation of Proteins from the Reaction Mixture

To a 10-μL aliquot of the reaction mixture, add 70 μL of water to give a final volume of 80 μL. Gently mix the solution by pipetting up and down. Then add 240 μL (the threefold volume) of ice-cold acetone (stored at −20°C) to the diluted reaction mixture. Gently mix the solution by inverting the reaction vessel 5–6 times. To precipitate the entire amount of protein, keep on ice for 30 min. Collect the precipitated protein by centrifugation at 16,100 × *g* for 5 min at 4°C. Remove the supernatant and let the protein pellets dry in a heating block at 37°C for about 5 min. Then add 16 μL of resolving gel buffer to the pellet. Let the pellet dissolve by shaking for 15 min at 1,000 rpm. When no traces of pellet are detectable (on the vessels bottom), the sample can be analyzed by SDS-PAGE.

3.5. 18% Sodium Dodecyl Sulfate Polyacrylamide Gel Electrophoresis

1. To prepare two gels, mix 2.5 mL of resolving buffer, 6 mL of acrylamide mixture, and 1.44 mL water in a 50-mL conical flask. Add 500 μL of SDS, 4 μL of TEMED and finally 100 μL of ammonium persulfate, and 4 μL of TEMED, and cast gel within an 8.3 cm × 8.2 cm × 0.75 mm gel cassette (Bio-Rad). Allow space for the stacking gel and gently overlay with ethanol.
2. Prepare the stacking gel by mixing 313 μL of resolving buffer, 417 μL of acrylamide mixture, and 1.734 mL water in a 14-mL conical flask. Add 6.25 μL of SDS, 5 μL of TEMED and finally, 40 μL of ammonium persulfate. Insert immediately a 10-well gel comb, without introducing air bubbles.
3. Do not heat the reaction samples mixed with the loading buffer, as this can destroy the phosphoramidates. Experience showed that for a smooth gel-run heating can be skipped. Load same amounts of redissolved reaction mixture (8 μL), as well as the same volume of diluted standard proteins. If empty lanes remain, load the same volume of loading buffer to decrease the probability of smear effects.

   Run the separation at 120 V until the sample has entered the gel and then continue at 150 V until the dye front (from the BPB dye in the samples) has reached the bottom of the gel.
4. Following electrophoresis, open the gel plates with the use of a spatula. The gel remains on one of the glass plates. Rinse the gel with water and transfer carefully to a container with water.

3.6. Coomassie Brilliant Blue Staining and Gel Drying

1. Transfer the gel into a container with Coomassie staining solution. Stain the gel for approximately 10 min under gentle agitation.
2. Remove the staining solution and rinse the gel with destaining solution to remove the excess of stain. Immediately refresh the destaining solution. Let the gel to destain for at least 1 h, if feasible up to 3 h. Refresh the destaining solution every hour. Destaining can also be achieved overnight at 4°C, if necessary.

3. Place the destained gel onto a wet Whatman Paper. Then cover the gel with a plastic sheet and place the obtained sandwich onto a vacuum gel-dryer (Unigeldryer by Uniequip® or similar). Let the gel dry for 50 min at 70°C.

3.7. Filmless Radiography

1. Mark the standard proteins with dots of radioactive dye to indicate the marker on the filmless radiography plot.
2. Erase a filmless radiography screen panel ("Molecular Dynamics Storage Phosphor Screen" by Molecular Dynamics®) by irradiation with visible light on the company's built "Image Eraser". After irradiation, immediately place the screen into a film cassette. Avoid any exposure to sunlight.
3. Place the dried gel onto the bottom of the cassette, then cover with the screen. Close the film cassette and let the gel be exposed to the screen for at least one overnight period (see Note 5).
4. After incubation, transfer the screen to a Typhoon 8600 Variable Mode Imager (Amersham). Develop the screen using the "Typhoon scanner control" software (supplied by the manufacturer) using following settings:
   (a) "Acquisition Mode": Storage phosphor
   (b) "Pixel Size": 200 μm
5. Analyze the obtained figure with the "ImageQuant" software (see Fig. 2).

4. Notes

1. We find that it can be stored at 4°C in the dark and still performs well. Fresh preparation for each casting of gels is not mandatory, but does not hurt too.
2. We find that storing at 4°C reduces its pungent smell.
3. Simple method of preparing running buffer: prepare 5× native buffer (0.25 M Tris, 1.92 M glycine). Weigh 30.3 g of Tris and 144 g of glycine, mix, and bring to 1 L with water.

   Dilute 200 mL of 5× native buffer to 790 mL with water and add 10 mL of 10% SDS. Care should be taken to add SDS solution last, since it makes bubbles.

   Never try to adjust the pH to a correct value. The ratio of Tris and glycine was chosen to yield automatically the right pH. Little variations in the pH will not negatively affect the running performance of the buffer. Adding foreign ions, such as Cl^- to the system, however will!

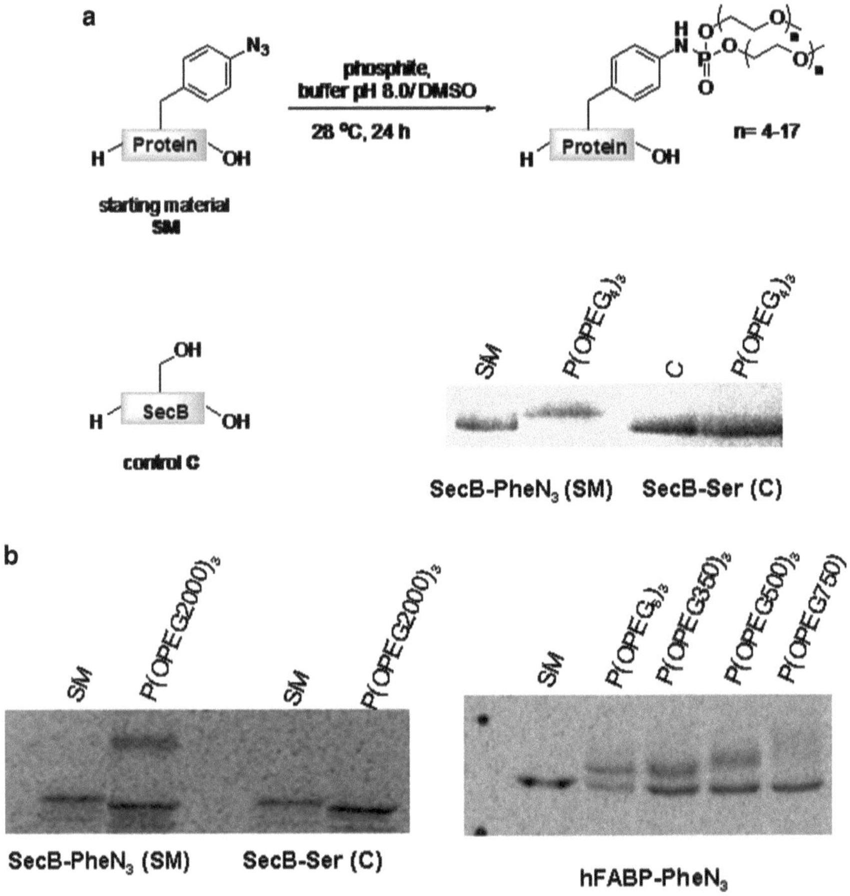

Fig. 2. Gel shift by PEGylation of azido-SecB and azido-hFABP. (**a**) Azido-SecB (purified protein), and (**b**) azido-hFABP (in cell lysate). Upon Staudinger phosphite reaction (*see* also Scheme 1) a phosphoramidate with two PEG chains is formed. The number of the PEG units (PEG350,...) refers to their average molecular mass, i.e. PEG350 means that each residue R corresponds to 350 Da (or approximately 8 units).

4. Protein modification by phosphites proceeds efficiently in plain buffers (Tris or phosphate, pH 7.8–8.2) only if the proteins are sufficiently soluble. For unsoluble proteins, it is recommended to conduct the recation in the presence of GdnCl (up to 6 M) or DMSO (up to 50%) (17, 18).
5. To avoid penetration of sunlight, store the cassette in a drawer or in a dark place.

Acknowledgments

The authors acknowledge financial support from the German Science Foundation (DFG) within the Emmy-Noether program (HA 4468/2-1), the SFB 765, the Fonds der Chemischen Industrie (FCI) and the Böhringer-Ingelheim Foundation ("Plus 3-Perspektiven Programm").

References

1. Wang L, Schultz PG (2005) Expanding the genetic code. Angew Chemie Int Ed 44, 34–66.
2. Wang L, Xie J, Schultz PG (2006) Expanding the genetic code. Ann Rev Biophys Biomol Struc 35, 225–249.
3. Budisa N (2004) Prolegomena to future experimental efforts on genetic code engineering by expanding its amino acid repertoire. Ang Chemie Int Ed 116, 6426–6463.
4. Dougherty DA (2000) Unnatural amino acids as probes of protein structure and function. Curr Opin Chem Biol 4, 645–652.
5. Xie J, Schultz PG (2005) Adding amino acids to the genetic repertoire Curr Opin Chem Biol 9, 548–554.
6. Link AJ, Mock ML, Tirrell DA (2003) Non-canonical amino acids in protein engineering. Curr Opin Biotechn 14, 603–609.
7. Gerrits M et al., (2007) In: Cell-Free Protein Expression, Landes Bioscience, Austin, 2007.
8. Prescher JA, Bertozzi CR (2005) Chemistry in living systems. Nat Chem Biol 1, 13–21.
9. Rostovtsev VV et al. (2002) A stepwise huisgen cycloaddition process: copper(I)-catalyzed regioselective "ligation" of azides and terminal alkynes. Angew Chem Int Ed 2002, 41, 2596–2599.
10. Tornøe CW, Christensen C, Meldal M (2002) Peptidotriazoles on solid phase: [1,2,3]-triazoles by regiospecific copper(i)-catalyzed 1,3-dipolar cycloadditions of terminal alkynes to azides. J Org Chem 67, 3057–3064.
11. Codelli JA et al. (2008) Second-generation difluorinated cyclooctynes for copper-free click chemistry. J Am Chem Soc 130, 11486–11493.
12. Ning X, Guo J, Wolfert MA, Boons G-J (2008) Visualizing metabolically labeled glycoconjugates of living cells by copper-free and fast huisgen cycloadditions. Angew Chem Int Ed 47, 2253–2255.
13. Debets MF et al. (2010) Aza-dibenzocyclooctynes for fast and efficient enzyme PEGylation via copper-free (3+2) cycloaddition. Chem Commun 46, 97–99.
14. Saxon E, Bertozzi CR (2000) Cell surface engineering by a modified Staudinger reaction. Science 287, 2007–2010.
15. Prescher JA, Dube DH, Bertozzi CR (2004) Chemical remodelling of cell surfaces in living animals. Nature 430, 873–877.
16. Agard NJ et al. (2006) A comparative study of bioorthogonal reactions with azides. ACS Chemical Biology 1, 644–648.
17. Böhrsch V et al. (2010) Site-specific functionalisation of proteins by a Staudinger-type reaction using unsymmetrical phosphites. Chem Commun 46, 3176–8.
18. Serwa R et al. (2010) Site-specific PEGylation of proteins by a Staudinger-phosphite reaction. Chemical Science, 596–602.
19. Serwa R et al. (2009) Chemoselective Staudinger-phosphite reaction of azides for the phosphorylation of proteins. Angew Chem Int Ed 48, 8234–8139.
20. Veronese FM (2001) Peptide and protein PEGylation: a review of problems and solutions. Biomaterials 22, 405–417.
21. Veronese FM, Mero A (2008) The impact of PEGylation on biological therapies. Biodrugs 22, 315–329.
22. Roberts MJ, Bentley MD, Harris JM (2002) Chemistry for peptide and protein PEGylation. Adv Drug Deliv Rev 54, 459–476.
23. Caliceti P, Veronese FM (2003) Pharmacokinetic and biodistribution properties of poly(ethylene glycol)-protein conjugates Adv Drug Deliv Rev 55, 1261–1277.

Part IV

D-Amino Acids: Analysis and Applications

Chapter 16

HPLC Methods for Determination of D-Aspartate and N-methyl-D-Aspartate

George H. Fisher and Mara Tsesarskaia

Abstract

D-Amino acids are stereoisomers or optical isomers of naturally occurring L-amino acids and thus possess the same chemical structure, but may differ in their biological/physiological properties. Until a half century ago, D-amino acids had been considered to be unnatural substances found only in microorganisms. However, improvements in analytical instruments and methods have revealed that D-amino acids are present in invertebrates and vertebrates, including humans, and that they possess important physiological functions. D-Aspartate (D-Asp) and its methylated form *N*-methyl-D-aspartate (NMDA) possess neuroendocrine properties in many species. Several methods have been developed for determination of D- and L-enantiomers of amino acids by high performance liquid chromatography (HPLC). We report here improved HPLC methods for the specific determination of D-Asp and NMDA in biological tissues.

Key words: D-aspartate, *N*-methyl-D-aspartate, *N*-methyl-L-aspartate, HPLC

1. Introduction

D-Aspartic acid (D-aspartate, D-Asp) is an endogenous amino acid present in the neuroendocrine tissues of invertebrates and vertebrates. It was first discovered in the nervous system of the mollusks *Octopus vulgaris*, *Sepia officinalis*, and *Loligo vulgaris* (1). Subsequently D-Asp was found in the nervous and endocrine tissues of many other animals (2–4), including rats and humans (5, 6); for a thorough review see ref. 7. D-Asp is also the precursor molecule for the biosynthesis of *N*-methyl-D-aspartate (NMDA) (7), which is a potent neuroexcitatory amino acid (8) that has specific action as an agonist for one of the glutamate sub-type NMDA receptors in the central nervous system of invertebrates and vertebrates (9). NMDA receptors play an important role in learning and memory (10).

Loredano Pollegioni and Stefano Servi (eds.), *Unnatural Amino Acids: Methods and Protocols*,
Methods in Molecular Biology, vol. 794, DOI 10.1007/978-1-61779-331-8_16, © Springer Science+Business Media, LLC 2012

With growing interest in D-Asp and NMDA in biological/physiological tissues, an accurate and quick method for determination and quantification of these amino acids has become of great importance. Several methods have been reported for determination of D- and L-enantiomers of primary amino acids by high performance liquid chromatography (HPLC) after derivatization with chiral fluorogens (11–14). However, very few chiral reagents are known to react with secondary amino acids such as NMDA.

In 1999, Todoroki et al. (15) developed an HPLC method for determination of NMDA in biological tissues using precolumn derivatization with the chiral reagent (+)-1-(9-fluorenyl)ethyl chloroformate (FLEC) followed by isocratic resolution of the two diastereomers formed. This method, however, required the removal of neutral and basic substances by anion-exchange chromatography and removal of primary amino acids by treatment with *o*-phthaldialdehyde (OPA) before derivatization of the NMDA. In 2002, D'Aniello et al. (16) developed an indirect method for determination of NMDA based on the HPLC detection of the methylamine (CH_3NH_2) produced by the oxidation of NMDA in the presence of D-aspartate oxidase (DASPO, EC 1.4.3.1), an enzyme that catalyzes the oxidation of the acidic D-amino acids D-Asp, D-Glu, and NMDA. Also in 2002, the Homma group in Japan (17) developed an automated two-step column-switching HPLC system for resolution of D- and L- *N*-methyl aspartate. In the first step, the sample of amino acids is reacted with an achiral fluorescent reagent 4-fluoro-7-nitro-2,1,3-benzoxadiazole (NBDF) and the NMDA and NMLA derivatives formed are separated from those of primary amino acids by HPLC. In the second step (column switching), the D- and L- *N*-methyl aspartates are resolved on a chiral column. The disadvantages of this method are the high cost and short lifetime of the chiral column and long procedure times for the two-step column-switching HPLC system.

We report here two methods that give better resolution of the D- and L-enantiomers of aspartate and *N*-methyl aspartate with good peak response in minimum retention times.

2. Materials

2.1. Reagents and Buffers

1. Aqueous solutions 150 pmol/μL of D-Asp, L-Asp, and racemic D,L-Asp (Sigma-Aldrich, St. Louis, MO, USA).
2. Aqueous solutions of various concentrations of *N*-methyl(D- and L-)aspartate (NMDA and NMLA), *N*-methyl(D- and L-)glutamate (NMDG and NMLG) (Sigma-Aldrich, St. Louis, MO, USA).
3. 1% Solution in acetone of *N*-α-(2,4-dinitro-5-fluorophenyl)-(D or L)-valine amide (FDNP-Val-NH_2) (NovaBiochem La Jolla, CA, USA).

4. *o*-Phthaldialdehyde-*N*-acetyl-L-cysteine (OPA-NAC reagent): 4 mg OPA and 5 mg NAC dissolved in 0.5 mL of methanol.
5. 0.1 M Sodium borate buffer, pH 9.3–9.5.
6. 30 mM Sodium citrate (8.823 g of $Na_3C_6H_5O_7{\cdot}2H_2O$ in water in a 1-L volumetric flask), adjusted to pH 5.5–6.0 with solid citric acid, to which are then added 111 mL of methanol (final concentration 10% methanol). Alternatively, the HPLC gradient controller can be used to adjust the ratio of sodium citrate to methanol.
7. 0.11% TFA/water (v/v) and 0.11% TFA/acetonitrile (MeCN) (v/v).
8. Water: doubly distilled or deionized through a water purification system such as Millipore® or NANOpure®.
9. D-Aspartate oxidase (DASPO, EC 1.4.3.1) (2 mg/mL) obtained by overexpression.

2.2. HPLC

1. At least a binary gradient pump.
2. In-line solvent degasser.
3. 10-μL Injector loop.
4. ODS-C_{18} pre-column.
5. Hypersil ODS-C_{18} (5 μm) column, 25 cm × 0.46 cm.
6. UV–Vis and fluorescence detectors.
7. Computing integrator data-handling system.

3. Methods

3.1. Determination of D- and L-Asp Alone

3.1.1. Derivatization

1. Based on the method of Aswad (11), the D- and L-enantiomers of Asp react with OPA-NAC reagent in sodium borate buffer to form a pair of fluorescent diastereomeric *N*-alkyl-2-thioalkyl isoindole derivatives (Fig. 1).
2. Borate buffer is added to the Asp standard, followed by OPA-NAC reagent (see Table 1), mixed for 1 min, and then 10 μL are injected into the HPLC column.
3. A similar derivatization procedure can be done with a biological tissue sample extract instead of D,L-Asp standards.

3.1.2. HPLC Conditions

1. The following mobile phases are used to elute the D- and L-Asp diastereomers from the column:

   A: 30 mM sodium citrate, pH 5.5–6.0, to which have been added 111 mL of methanol (final concentration 10% methanol). Alternatively, the HPLC gradient controller can be used to adjust the ratio of sodium citrate to methanol; B: 100% methanol (MeOH); and C: 100% acetonitrile (AcCN).

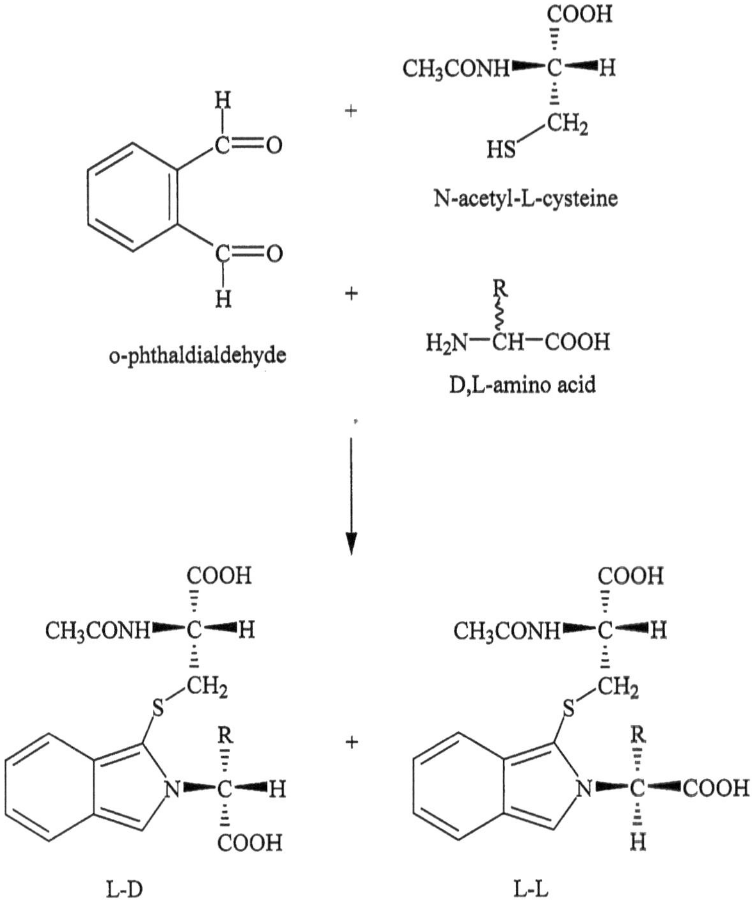

Fig. 1. Derivatization of D- and L-Asp with OPA-NAC to form a pair of fluorescent diastereomeric *N*-alkyl-2-thioalkyl isoindole derivatives.

Table 1
D,L-Asp calibration

D,L-Asp std (pmol/μL)	Std (μL)	OPA-NAC (μL)	Borate (μL)	Total (μL)	Injected (μL)	Total Asp (pmol)	D-Asp (pmol)	L-Asp (pmol)
150	5	5	365	375	10	20	10	10
150	10	5	285	300	10	50	25	25
150	15	5	205	225	10	100	50	50
150	30	10	185	225	10	200	100	100
150	50	10	90	150	10	500	250	250
150	100	20	30	150	10	1000	500	500

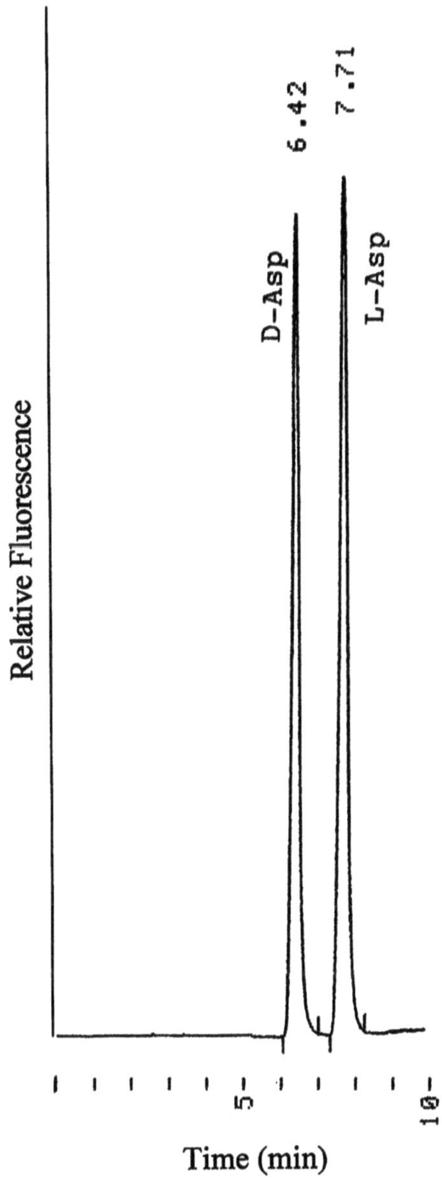

Fig. 2. HPLC chromatogram of resolution of D- and L-Asp derivatives.

2. The following gradient is used: 0–6 min, 100% A; 6–10 min linear gradient to 25% A, 40% B, and 35% C (to wash the column); returning to 100% A and equilibrating at 100% A over the next 5–8 min. Flow rates are 1.0 mL/min.
3. The fluorescence of the derivatives was measured using a fluorescence detector at 325-nm excitation and 415-nm emission.
4. Under these conditions, D-Asp is well separated by 1.2–1.3 min from L-Asp (D-Asp elutes at ~6.3–6.5 min and L-Asp at ~7.6–7.8 min) (Fig. 2).

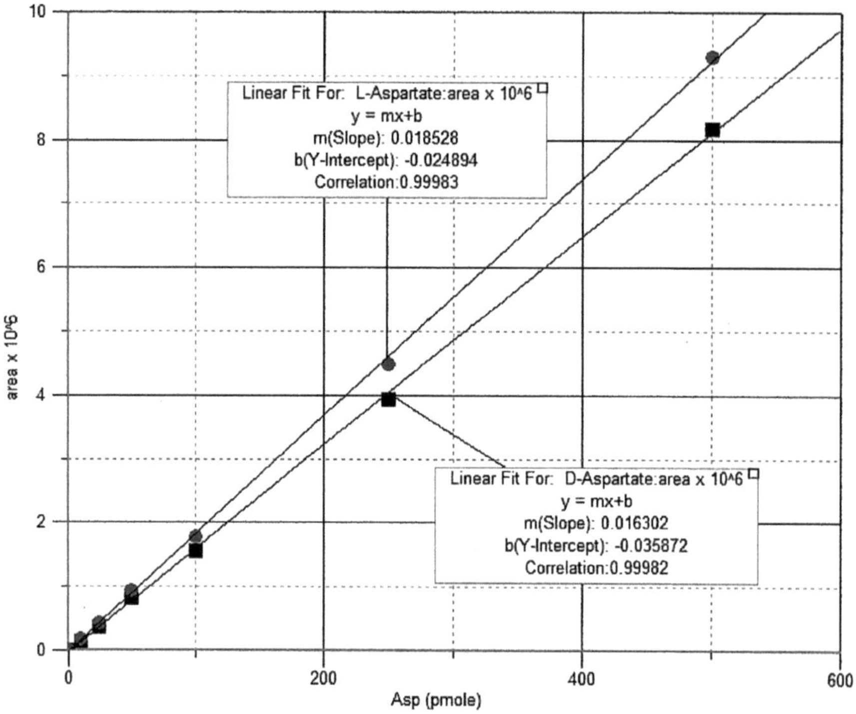

Fig. 3. Fluorescence responses of D- and L-Asp. Calibration showing a linear correlation of peak areas with picomoles of L-Asp (*filled circle*) and D-Asp (*filled square*), from chromatograms as in Fig. 2.

3.1.3. HPLC Calibration

The HPLC is calibrated with racemic D,L-Asp standards, using the ratios shown in Table 1, borate buffer, and OPA-NAC to inject precise amounts of D- and L-Asp onto the column.

3.1.4. Detection and Quantification

A plot of peak area vs. nanomoles of Asp is made from the D,L-Asp standard (Fig. 3). The areas under the peaks are used to quantify the picomoles of D- and L-Asp of the standards, which can then be used to quantify the picomoles of D- and L-Asp in biological samples.

4. Results

1. Fluorescence responses (peak areas) of both the D- and L-Asp derivatives are linear over the range of 10–500 pmol injected (correlation $r \geq 0.999$) (Fig. 3).
2. Since the derivatized aspartates are diastereomers the specific fluorescence of the D-Asp derivative is always about 12 ± 1% lower than that of L-Asp.
3. The D-Asp peak disappears when incubated with DASPO, which oxidizes only D-aspartate (15), giving further proof that the first peak eluted at ~6.4 min is actually D-Asp.

4. Sodium citrate-methanol (NaCit-MeOH) gives better resolution in shorter times than previously reported HPLC methods using sodium acetate-methanol gradients. In general, citrate also tends to give better separation (resolution) than acetate. Since the molecular weight of citrate is much greater than that of acetate, an equimolar solution of citrate would tend to have a greater effect in its interaction with the OPA-NAC amino acid derivatives than acetate has. Methanol or acetonitrile tend to act as additional organic modifiers and decrease retention times as their concentration increases. Thus, this improved method is suitable for routine determination of small concentrations of D-Asp in multiple samples of biological tissues (see Notes 1–4).

4.1. Simultaneous Determination of NMDA, NMLA, NMDG, NMLG, and D- and L-Asp in Combination

We have developed a sensitive, one-step derivatization and HPLC method for simultaneous determination and resolution of *N*-methyl (D- and L-)aspartate (NMDA, NMLA), *N*-methyl(D- and L-)glutamate (NMDG, NMLG), and D- and L-Asp in combination with each other. This method is based on a derivatization procedure with the chiral reagent *N*-α-(2,4-dinitro-5-fluorophenyl)-(D- or L-)valine amide (FDNP-Val-NH_2), a close analog of Marfey's reagent 1-fluoro-2,4-dinitrophenyl-5-L-alanine amide (18). The diastereomers formed (Fig. 4) are separated on an inexpensive

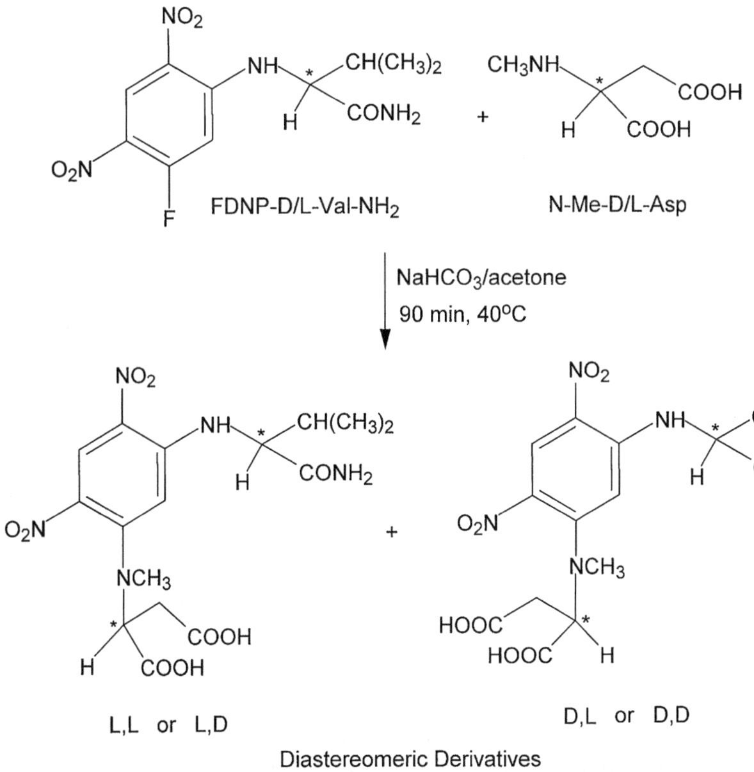

Fig. 4. Derivatization scheme with the chiral reagent *N*-α-(2,4-dinitro-5-fluorophenyl)-(D- or L-) valine amide (FDNP-Val-NH_2) to form a pair of diastereomers.

reversed phase ODS-Hypersil column with elution by a 0.11% TFA/water – 0.11% TFA/MeCN gradient. UV absorption at 340 nm permits detection levels in the range of 5–10 pmol.

4.1.1. Derivatization

1. D,L-Amino acids or tissue samples are derivatized with D- or L-FDNP-Val-NH_2, according to the procedure of Szókán et al. (19) (Fig. 4).
2. An amount of 2.5 μmol of amino acid or a tissue homogenate from a biological sample are dissolved in 100 μL of 0.5 M $NaHCO_3$ to which 400 μL of a 1% solution of FDNP-Val-NH_2 in acetone are added.
3. The mixture is incubated for 90 min at 40°C in a Thermomixer, and after cooling acidified with 2 M HCl to pH ~4.
4. After 100- to 150-fold dilution with MeOH, 10 μL are injected into the HPLC.

4.1.2. HPLC Analyses

1. The derivatized diastereomers formed are separated by HPLC on a 5-μm ODS-Hypersil reversed phase column (25 cm × 0.46 cm) using a gradient pump, with UV detection at 340 nm, connected to a computing integrator.
2. Two solvent mixtures are used: (A) 0.11% TFA in water (v/v) and (B) 0.11% TFA in MeCN (v/v).
3. The following HPLC program is used: isocratic for 2 min with 80% A:20% B, then a linear gradient to 62% A:38% B over the next 46 min, followed by a 4-min wash with 5% A:95% B, then back to 80% A:20% B. All flow rates are 1.1 mL/min.

5. Results

1. Figure 5 shows HPLC chromatograms for standards of (a) NMLA and NMDA, (b) D-Asp and L-Asp, and (c) NMDG and NMLG, derivatized with FDNP-D-Val-NH_2. The diastereomers of each pair are well resolved. The order of elution of the enantiomers is reversed when derivatized with FDNP-L-Val-NH_2.
2. Figure 6 shows an HPLC chromatogram of a standard mixture of all the amino acids that elute in the region from 15 to 45 min. Other amino acids eluting in this region do not interfere with NMDA, NM(D/L)G or D/L-Asp. However, D-Thr coelutes with NMLA. This should not be a problem since most biological samples probably do not contain NMLA or D-Thr. Also, when the sample is derivatized with FDNP-L-Val-NH_2, then L-Thr and NMDA coelute and NMLA is resolved. Thus, NMLA can be identified.

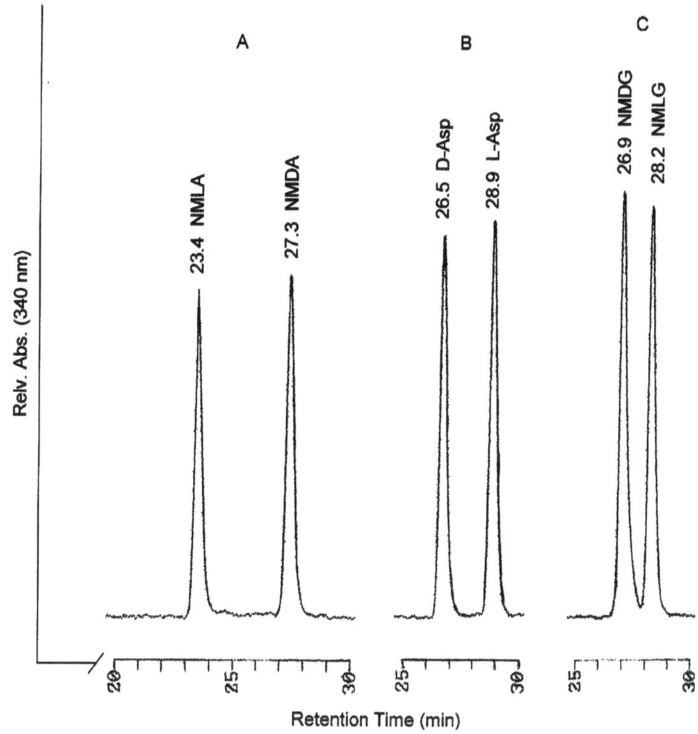

Fig. 5. HPLC chromatograms for standards of (**A**) NMLA and NMDA, (**B**) D-Asp and L-Asp, and (**C**) NMDG and NMLG, each derivatized with FDNP-D-Val-NH_2.

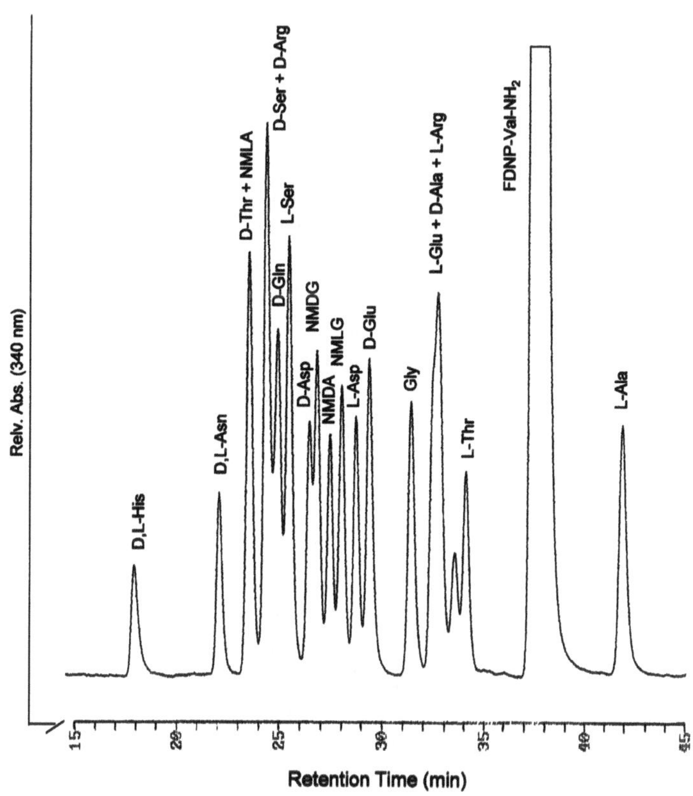

Fig. 6. HPLC chromatogram of a standard mixture of amino acids derivatized with FDNP-D-Val-NH_2.

3. Peak areas as a function of concentration are linear in the range down to 5–10 pmol, thus allowing calculation of the concentrations of these amino acids in biological tissue samples (see Notes 5–7).

6. Notes

1. The OPA-NAC reagent is light sensitive and should be stored in a dark (light resistant) vial.
2. The HPLC should be calibrated with the D- and L-Asp standards each time a new batch of OPA-NAC reagent is prepared.
3. The picomoles of D- and L-Asp in 10 μL of a biological sample injected into the HPLC can be calculated by dividing the areas of D- and L-Asp peaks of the sample by the slope of the respective D- and L-Asp calibrations (as determined by calibration standards, Fig. 3).
4. If the D- and L-Asp are obtained by hydrolysis of a biological sample with 6 M HCl for 6 h at 100°C, then the picomoles obtained from the HPLC should be multiplied by 0.995 to account for 0.5% hydrolysis-induced racemization.
5. This method is highly reliable and relatively fast (approximately 50-min HPLC run time). With this method, the D- and L-enantiomers of *N*-methyl aspartic acid, *N*-methyl glutamic acid, and aspartic acid are well resolved. After extensive studies with all 20 natural amino acids plus GABA and taurine, only D-Thr partially overlaps with NMLA while NMDA stays well separated from any other amino acids. However, if one is interested in detecting specifically NMLA, this problem can be easily overcome by using the other enantiomer of the derivatizing agent (FDNP-L-Valine-NH_2); then the order of elution will be reversed, and now NMLA will be resolved and NMDA will be obscured by L-Thr. A large peak at 37.4 min is unreacted FDNP-Val-NH_2.
6. A definitive advantage to this method is that pretreatment of tissue samples with OPA to remove primary amino acids is not necessary since none of the diastereomers of these other amino acids overlap with those of D-Asp, NMDA, or NMDG.
7. The order of elution of the D- and L- enantiomers of the derivatized amino acid can be proven by any of the following methods (a) derivatize and run each pure enantiomer and note its retention time; (b) spike a racemic mixture of a D,L-amino acid with one of its pure enantiomers and note which HPLC peak increases; (c) treat the amino acid with DASPO that specifically catalyzes oxidization of the D-amino acids NMDA, NMDG, and D-Asp such that the peak corresponding to the D-enantiomer is reduced or disappears.

Acknowledgments

We wish to acknowledge the Barry University students who helped develop the experimental procedures described here: Erika Galindo, Susana Lopez, and Collins Boston. Separate reports of these methods have previously been published in references (19–21).

References

1. D'Aniello A, Giuditta A (1977) Identification of D-aspartic acid in the brain of *Octopus vulgaris*. J Neurochem 29, 1053–1057.
2. Hashimoto A, Oka T (1997) Free D-aspartate and D-serine in the mammalian brain and periphery. Prog Neurobiol 52, 325–353.
3. D'Aniello A, DiFiore M M, Fisher G (1998) Occurrence of D-aspartic acid in animal tissues and its role in the nervous and endocrine systems. Trends Comp Biochem Physiol 4, 1–24.
4. Spinelli P, Brown E et al. (2006) D-Aspartic acid in the nervous system of *Aplysia limacina*: Possible role in neurotransmission. J Cell Physiol 206, 672–681.
5. D'Aniello A, DiFiore M M, Fisher G H et al. (2000) Occurrence of D-aspartic acid and N-methyl-D-aspartic acid in rat neuroendocrine tissues and their role in the modulation of luteinizing hormone and growth hormone release. FASEB J 14, 699–714.
6. Fisher G H, D'Aniello A et al. (1991) Free D-aspartate and D-alanine in normal and Alzheimer brain. Brain Res Bull 26, 983–985.
7. D'Aniello A (2007) D-Aspartic acid: An endogenous amino acid with an important neuroendocrine role. Brain Res Rev 53, 215–234.
8. Watkins J C, Evans R H (1981) Excitatory amino acid transmitters. Ann Rev Pharmacol Toxicol 21, 165–204.
9. Monaghan D T, Cotman C W (1986) Identification and properties of *N*-methyl-D-aspartate receptors in rat brain synaptic plasma membranes. Proc Natl Acad Sci USA 83, 176–179.
10. Mondadori C, Weiskrantz L et al. (1989) NMDA receptor antagonists can enhance or impair learning performance in animals. Exp Brain Res 75, 449–456.
11. Aswad D W (1984) Determination of D- and L-aspartate in amino acid mixtures by high performance liquid chromatography after derivatization with a chiral adduct of *o*-phthaldialdehyde. Anal Biochem 137, 405–407.
12. Nimura N, Kinoshita T (1986) *o*-Phthaldialdehyde-N-acetyl-L-cysteine as a chiral derivatization reagent for liquid chromatographic optical resolution of amino acid enantiomers and its application to conventional amino acid analysis. J Chromatog 352, 169–177.
13. Nishikawa T, Oka T et al. (1992) Determination of free amino acid enantiomers in rat brain and serum by HPLC after derivatization with N-*tert.*-butoxycarbonyl-L-cysteine and *o*-phthaldialdehyde. J Chromatog 582, 41–48.
14. Brüchner H, Haasmann S et al. (1994) Liquid chromatographic determination of D- and L-amino acids by derivatization with *o*-phthaldialdehyde and chiral thiols. J Chromatog 666, 259–273.
15. Todoroki N, Shibata K et al. (1999) Determination of *N*-methyl-D-aspartic acid in tissues of bivalves by HPLC. J Chromatog B 728, 41–47.
16. D'Aniello A, De Simone A et al. (2002) A specific high-performance liquid chromatography method to determine *N*-methyl-D-aspartic acid in biological tissues. Anal Biochem 308, 42–51.
17. Skine M, Fukuda H, Nimura N et al. (2002). Automated column-switching high-performance liquid chromatography system for quantifying *N*-methyl-D- and -L-aspartate. Anal Biochem 310, 114–121.
18. Szókán G, Mezö G, Hudecz F (1988) Application of Marfey's reagent in racemization studies of amino acids and peptides. J Chromatog 444, 115–122.
19. Galindo E, Tsesarskaia M, Fisher G et al. (2009) An Improved HPLC Method for Determination and Quantification of D- and L-Aspartic Acid. In: Konno, Brueckner, D'Aniello, Fisher, Fujii, Homma (eds), D-Amino acids: practical methods and protocols, Volume 1: Analytical methods for D-amino acids. Nova Science Publishers, New York. pp. 43–48.
20. Tsesarskaia M, Galindo E, Fisher G, Szókán G (2009) A Sensitive One Step HPLC Method for

Simultaneous Determination of N-Methyl-(D and L)-Aspartate, N-Methyl-(D and L)-Glutamate and (D and L)-Aspartate in Biological Tissues. In: Konno, Brueckner, D'Aniello, Fisher, Fujii, Homma (eds), D-Amino acids: practical methods and protocols, Analytical methods for D-amino acids. Nova Science Publishers, New York, pp 25–31.

21. Tsesarskaia M, Galindo E, Szókán G, Fisher G (2009) HPLC determination of acidic D-amino acids and their *N*-methyl derivatives in biological tissues. Biomed Chromatog 23, 581–587.

Chapter 17

Estimation of Chronological Age from the Racemization Rate of L- and D-Aspartic Acid: How to Completely Separate Enantiomers from Dentin

Toshiharu Yamamoto and Susumu Ohtani

Abstract

Estimation of chronological age is essential in forensic and archeological science. The racemization method is one of the best methods to meet the demands of these scientific fields, providing both accuracy of the estimated age and simplicity of technique. In general, living organs are composed of L-form amino acids. Conversion from L-form to D-form amino acids is a first-order chemical reaction. Thus, the quantity of D-form amino acids in an organ is proportional to the passed time (age) after organ completion if no protein turnover occurs after organization. However, every living organ undergoes some degree of protein turnover. Therefore, organs with low metabolic rates, such as teeth and bone, should be targeted for the racemization method. The most critical point of the technique may be the complete separation of D- and L-forms by gas chromatography because of the very small amounts of D-form amino acids present. We describe the detailed procedures and the critical points for obtaining reliable estimated ages using the racemization method.

Key words: Racemization, D-aspartic acid, Age estimation, Dentin, Gas chromatography

1. Introduction

Various methods exist for estimating chronological age, involving evaluation of morphological or biochemical changes with age (1). The racemization method for age estimation is based on the detection of biochemical changes. Biological proteins are usually synthesized using L-form amino acids. Racemization is the first-order chemical reaction converting L-form amino acids to D-form amino acids. Therefore, theoretically, if there is no turnover of proteins following complete organization of an organ, the amount of D-form amino

Loredano Pollegioni and Stefano Servi (eds.), *Unnatural Amino Acids: Methods and Protocols*,
Methods in Molecular Biology, vol. 794, DOI 10.1007/978-1-61779-331-8_17, © Springer Science+Business Media, LLC 2012

acids present is proportional to the time since complete organization. Therefore, organs with low metabolic rates, such as teeth and bone, are the most suitable targets for age estimation (2). However, every organ has some degree of protein turnover, even teeth and bone. Many kinds of proteins present in a sample are related to the time since complete organization of the organ. Thus, by examining a large number of standard samples and equalizing the tissue components it is possible to envisage reliable estimated ages. The racemization method is also applicable to a purified single protein, for example, osteocalcin (3), although a high degree of purification is required, together with an accurate time for protein completion. Among the amino acids, aspartic acid (Asp) shows the largest rate constant for racemization (4), although the racemization rate differs depending on the position of L-Asp within the protein and the protein's higher structures (5–7). We believe that the technical simplicity of the racemization method using dentin makes it one of the best methods to satisfy the demand for accuracy in age estimation in the forensic and archeological sciences (2). Although the racemization method is a reliable and simple technique, there are many factors that may influence racemization rates (2). Here we focus on the procedure of the racemization method for dentin and note areas that require special attention to obtain reliable estimated ages.

2. Materials

2.1. Instruments

1. Isomet low speed saw (Buehler, Chicago, USA), diamond wafering blade, Series 15 LC diamond (102 mm × 0.3 mm).
2. Pulverizer, vibratory sieve shaker (Fritsch, Idar-Oberstein, Germany).
3. Gas chromatography, GC-17A (Shimazu, Kyoto, Japan).
4. Capillary column, length 15 m, inner diameter 0.25 mm, coated with Chirasil-Val (8).
5. Heating chamber for hydrolysis, heating block, HE-21 (Yamato, Tokyo, Japan).
6. Another heating chamber for amino acid conversion, Reacti-Therm heating/stirring module (Pierce Inc., Rockford, USA).
7. Columns to collect amino acids, one column (length 260 mm, diameter 35 mm) connected with a second column (length 250 mm, diameter 15 mm) (Fig. 1).
8. Software, Chrom Pack (Shimazu, Kyoto, Japan) to calculate D/L ratio and regression lines.
9. Rotary evaporator, Eyela N-1100 (Tokyo Rikakikai Co. Ltd., Tokyo, Japan).

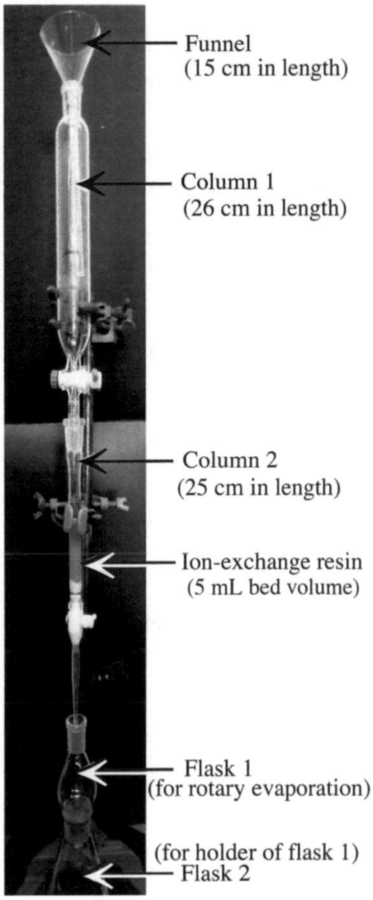

Fig. 1. Apparatus for collection of hydrolyzed amino acids. A stopcock is inserted between *columns 1* and *2* to avoid disturbing the resin surface. *Flask 1* is round-bottomed to facilitate rotary evaporation. *Flask 2* is flat-bottomed.

2.2. Reagents

Deionized water and analytical grade reagents are recommended.

1. 0.2 N HCl.
2. Ethanol.
3. Ether.
4. 6 N HCl.
5. Ion-exchange resin, Dowex, 50W-X8, bead size 50–100 mesh (Dow Chemical Company, Midland, USA).
6. 2 N (3.4%) NH_4OH: Mix 28 mL of 12% ammonia solution with 72 mL of deionized water.
7. A mixture of isopropyl alcohol and acetyl chloride, 8:2, v/v. Note that the acetyl chloride should be added to the isopropyl alcohol in small aliquots to mitigate the heat of fusion.

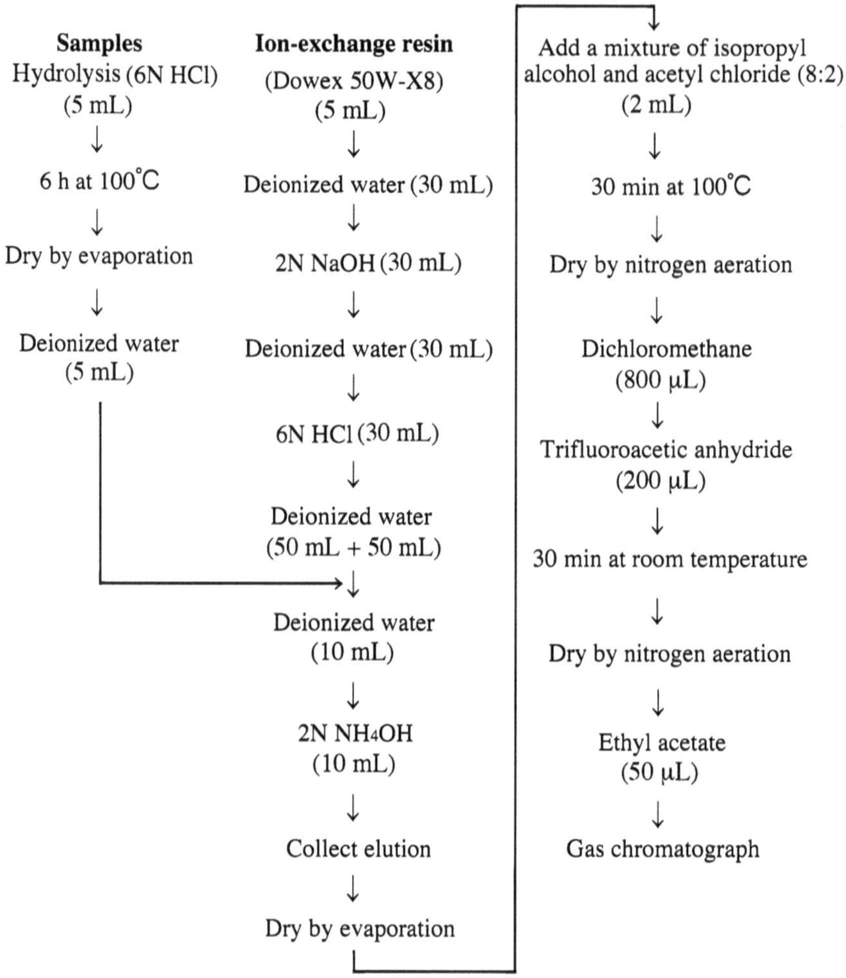

Fig. 2. Flowchart of the procedure.

8. Nitrogen gas.
9. Dichloromethane.
10. Trifluoroacetic anhydride.
11. Ethyl acetate.

3. Methods

A flowchart of the method is set out in Fig. 2.

1. Collect test samples for age estimation and select appropriate standard samples (see Notes 1 and 2) to deduce a regression line. In the case of teeth, prepare test teeth and four or more standard teeth of known ages. Cut into sagittal sections (labial-lingual

or buccal-lingual direction) of 1-mm thickness with a low speed cutter (see Note 3).

2. Equalize standard and test samples in view of the tissue composition (see Note 4). In the case of teeth, completely remove enamel, cementum, and root soft tissues with the aid of a stereoscopic microscope. If necessary, stain with van Gieson solution (a mixture of 100 mL of saturated picric acid solution and 10 mL of 1% acid fuchsine) to demarcate these structures.
3. Wash samples with 0.2 M HCl for 5 min.
4. Wash samples three times with deionized water for 5 min each.
5. Wash samples with ethanol for 5 min.
6. Wash samples with ether for 5 min.
7. Pulverize samples with a pulverizer (see Note 5).
8. Add 10 mg of each pulverized sample into a temperature-resistant screw cap test tube.
9. Add 5 mL of 6 N HCl into the test tube.
10. Close the test tube and heat at 100°C for 6 h in a heating chamber to hydrolyze proteins into amino acids. In general, this step liquefies the powder. If sediment is still present, spin down.
11. Leave until sample returns to room temperature.
12. Dry the sample solution with a rotary evaporator.
13. Add 5 mL of deionized water to the test tube and agitate for 10–20 s.
14. Apply the sample solution to 5 mL of an ion-exchange resin to collect amino acids (Fig. 1) (see Note 6).
15. Wash the resin with 10 mL of deionized water.
16. Elute the amino acids from the resin into a flask with 10 mL of 2 N NH_4OH.
17. Dry the eluted fraction with a rotary evaporator.
18. Add 2 mL of a mixture of isopropyl alcohol and acetyl chloride (8:2 v/v) to the flask and sonicate for 1 min.
19. Put the solution into a screw cap test tube and seal it.
20. Let sample stand for 30 min at 100°C in a heating chamber to convert amino acids into isopropyl ester forms (see Note 7).
21. Allow sample to cool, then dry using nitrogen gas aeration.
22. Add 800 μL of dichloromethane to the test tube, seal it, and sonicate for 1 min.
23. Open the test tube, add 200 μL of trifluoroacetic anhydride, seal it, and sonicate again for 1 min.

24. Let the sample stand for 30 min at room temperature.
25. Dry the sample using nitrogen gas aeration.
26. Add 50 μL of ethyl acetate to the test tube.
27. Apply 1–2 μL of the sample to a gas chromatography column to separate and measure L- and D-aspartic acids (see Note 8).
28. Calculate D/L ratios of the standard samples and build a regression line.
29. Substitute the D/L ratios of the test sample into the regression line and calculate estimated ages.

4. Notes

1. Selection of the standard samples of known age is important. In the case of teeth, four or more homonymous teeth from the same jaw as the test samples should be selected because these teeth develop in an individual at approximately the same time, resulting in a similar degree of racemization (9, 10).
2. In general, we keep standard and test teeth in a desiccator. However, fixed teeth may also be used if they are stored in a closed container at temperatures below 15°C. Teeth stored in a fixative of 10% neutral formalin had the highest degree of racemization, followed by those stored in 10% formalin, while those stored in 95% ethanol had the lowest degree of racemization. However, the racemization did not proceed significantly and the degree of racemization was almost the same as at the time of extraction even 10 years later when teeth had been stored at 15°C (11).
3. Enamel, cementum, and dentin all have different racemization rates (12–14). Even different parts of the dentin have different D/L ratios (15) because dentin formation starts at the boundary between the enamel and dentin and gradually shifts toward the dental pulp and the root apex region. Therefore, consistent dissection of the standard and test samples is very important to equalize the tissue components and tissue ages between the standard and test samples.
4. For example, note that compact bone is better for age estimation than spongy bone because the latter contains bone marrow that has higher metabolic rates than bone itself and may disturb the measuring of real D/L ratios of bone. Complete elimination of bone marrow from spongy bone is difficult (because of easier homogenization of the tissue composition).
5. The degree (time) of pulverization should be constant because different degree of pulverization may result in different D/L ratios (16).

6. The ion-exchange resin should be prepared according to the manufacturer's instructions. Briefly soak the resin in 30 mL of deionized water, decant, and change the solution to 30 mL of 2 N NaOH. Decant again and change the solution to 30 mL of deionized water, decant and change solution again to 30 mL of 6 N HCl, then rinse twice with 50 mL of deionized water (Fig. 2).
7. Conversion of amino acids into isopropyl esters has several merits as a means to separate enantiomers. Firstly, nonvolatile amino acids are changed into a volatile form. Secondly, the amino acids become more stable to temperature preventing further decomposition. Thirdly, the amino acids become most easy (sensitive) to detect. Fourthly, the process makes it easier to separate enantiomers.
8. Procedure for gas chromatography: samples of 1 μL were injected into the chiral phase column which was temperature controlled. The injector temperature was 250°C; and the detector temperature was 200°C. The initial column temperature was 104°C; which was then increased to 180°C by 1°C/min. Larger injection volumes may cause earlier contamination of the capillary column resulting in incomplete separation of

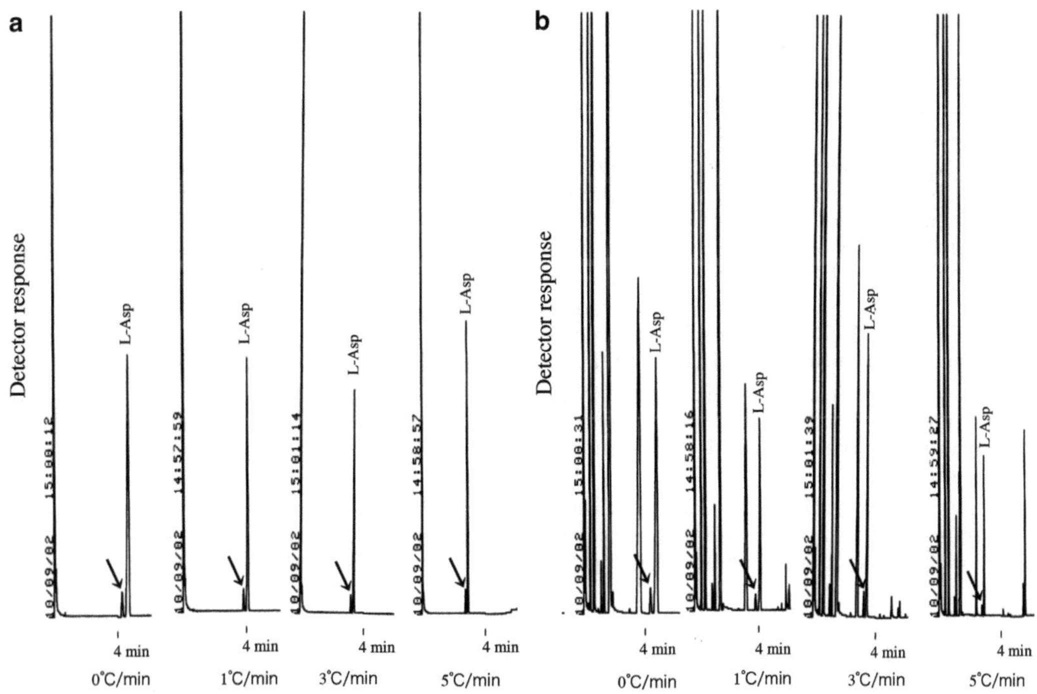

Fig. 3. Representative gas chromatograms showing the effects of increasing temperature rates per min on the capillary column. Sample in a): An artificial mixture of L- and D-Asp (19:1). *Arrows* indicate the D-Asp level. Sample in b): Dentin. *Arrows* indicate the D-Asp level. Note that at a higher rate of temperature increase, (°C/min) the D- and L-Asp peaks are closer.

D- and L-form amino acids and a curved baseline. Separation of the D- and L-forms of aspartic acid depends on various conditions. Initially, it is recommended to carry out a test run using an artificial mixture of D- and L-aspartic acid. In general, the longer the column, the longer detection takes, and a column with a smaller caliber results in more complete separation of the enantiomers. The initial and final injector, detector and column temperatures, and the rate of temperature increase are also factors affecting complete separation. Figure 3 shows how increasing the rate of temperature increase in the capillary column affects the separation.

References

1. Ritz-Timme S, Cattaneo C, Collins MJ et al. (2000) Age estimation: The state of the art in relation to the specific demands of forensic practise. Int J Legal Med 113, 129–136.
2. Ohtani S, Yamamoto T (2005) Strategy for the estimation of chronological age using the aspartic acid racemization method with special reference to coefficient of correlation between D/L ratios and ages. J Forensic Sci 50, 1020–1027.
3. Ritz S, Turzynski A, Schutz HW et al. (1996) Identification of osteocalcin as a permanent aging constituent of the bone matrix: basis for an accurate age at death determination. Forensic Sci Int 770, 13–26.
4. Helfman P M, Bada J L (1975) Aspartic acid racemisation in tooth enamel from living humans. Proc Nat Acad Sci USA 72, 2891–2894.
5. Fujii N, Momose Y, Ishii N et al. (1999) The mechanism of simultaneous stereoinversion, racemization, and isomerization at specific aspartyl residues of aged lens protein. Mech Ageing Develop 107, 347–358.
6. Groenen P J T A, van den Ijssel P R L A, Voorter C E M et al. (1990) Site-specific racemization in aging αA-crystallin. FEBS J 269, 109–112.
7. Shapira R, Wilkinson KD, Shapira G (1988) Racemization of individual aspartate residues in human myelin basic protein. J Neurochem 50, 649–654.
8. Abe I, Ohtani S (2006) Novel chiral selectors anchored on polydimethylsiloxane as stationary phases for separation of derivatized amino acid enantiomers by capillary gas chromatography. J Sep Sci 29, 319–324.
9. Logan W H G, Kronfeld R (1933) Development of the human jaws and surrounding structures from birth to the age of fifteen years. J Am Dent Accoc 20, 379–427.
10. Ohtani S, Ito R, Yamamoto T (2003) Differences in the D/L aspartic acid ratios in dentin among different types of teeth from the same individual and estimated age. Int J Legal Med 117, 149–152.
11. Ohtani S, Ohhira H, Watanabe A et al. (1997) Estimation of age from teeth by amino acid racemization: influence of fixative. J Forensic Sci 42, 137–139.
12. Ohtani S, Yamamoto K (1992) Estimation of age from a tooth by means of racemization of an amino acid, especially aspartic acid-comparison of enamel and dentin. J Forensic Sci 37, 1061–1067.
13. Ohtani S (1995) Studies on age estimation using racemization of aspartic acid in cementum. J Forensic Sci 40, 805–807.
14. Ohtani S, Sugimoto H, Sugeno H et al. (1995) Racemization of aspartic acid in human cementum with age. Arch Oral Biol 40, 91–95.
15. Ohtani S (1997) Different racemization ratios in dentin from different locations within a tooth. Growth Develp Aging 61, 93–99.
16. Ohtani S, Matsushima Y, Kobayashi Y, Kishi K (1998) Evaluation of aspartic acid racemization ratios in the human femur for age estimation. J Forensic Sci 43, 949–953.

Chapter 18

Enzymatic Detection of D-Amino Acids

Gianluca Molla, Luciano Piubelli, Federica Volontè, and Mirella S. Pilone

Abstract

D-Amino acids play several key roles and are widely diffused in living organisms, from bacteria (in which D-alanine is a component of the cell wall) to mammals (where D-serine is involved in glutamatergic neurotransmission in the central nervous system). The study of the biological processes involving D-amino acids and their use as clinical or biotechnological biomarkers requires reliable methods of quantifying them. Although "traditional" analytical techniques have been (and still are) employed for such tasks, enzymatic assays based on enzymes which possess a strict stereospecificity (i.e., that are only active on the D-enantiomers of amino acids) allowed the set-up of low-cost protocols with a high sensitivity and selectivity and suitable for determining the D-amino acid content of complex biological samples. The most exploited enzyme in these assays is D-amino acid oxidase, a flavoenzyme that exclusively oxidizes D-amino acids and possesses with a broad substrate specificity and a high kinetic efficiency.

Key words: Enzymatic assay, D-Amino acid oxidase, Coupled assay, Spectrophotometry, Analytical detection

1. Introduction

Our understanding of the function of D-amino acids and distribution in animals has increased remarkably in the last 20 years mainly as the result of the development of sensitive enantioselective separation strategies by which D-amino acids can be quantified in complex biological samples. The presence of D-amino acids in animals was first reported in 1950 by Auclair and Patton (1) and later confirmed by Corrigan (2): they were assumed to arise from endogenous microbial flora, from ingestion with the diet, or from spontaneous racemization of L-amino acids incorporated in polypeptides during aging (3). Consequently, it was assumed that free D-amino acids do

Loredano Pollegioni and Stefano Servi (eds.), *Unnatural Amino Acids: Methods and Protocols*,
Methods in Molecular Biology, vol. 794, DOI 10.1007/978-1-61779-331-8_18, © Springer Science+Business Media, LLC 2012

not possess a specific biological function in mammals. In 1992, Hashimoto and colleagues (4) finally established that relevant concentrations of D-aspartate and D-serine were present in the central nervous system (CNS) of mammals by using a highly specific chiral HPLC apparatus coupled with an amino acid derivatization procedure and a sensitive fluorescent detection technique. This study paved the way for identifying specific and relevant roles of D-amino acids in the CNS: D-aspartate was proposed to play a role in the development of some brain areas whereas D-serine was discovered to be an endogenous ligand for the NMDA receptors required, together with L-glutamate, for glutamatergic neurotransmission (5). Recently, the D-serine concentration has been proposed as a biomarker for several relevant mental disorders (e.g., schizophrenia or bipolar disorder) (6, 7). In addition, it has been demonstrated that an increased D-amino acid concentration in blood plasma is associated with aging or kidney disorder. For a summary of the beneficial and harmful activity of D-amino acids in mammals, see (8).

In the field of biotechnology, determining D-amino acid concentrations in food samples represents a highly important task since this parameter is considered an index for food quality assessment and a molecular marker for food ripening and/or microbiological contamination (9, 10). Moreover, unnatural D-amino acids are of great value since they are valuable pharmaceuticals in their own right and also components of numerous therapeutically relevant compounds (e.g., D-2-naphthylalanine is a component of the peptide drug Nafarelin) (11).

Over the years, researchers have developed protocols for high-resolution separation of D-amino acids by chiral gas chromatography, HPLC, and capillary electrophoresis coupled with detection systems based on mass spectroscopy or fluorescence. Unfortunately, these techniques are time-consuming and require very expensive instruments and highly qualified and trained operators. In addition, they cannot be employed to directly analyze complex samples (e.g., food samples and tissue extracts). In contrast, techniques based on enzymatic assays exploit the strict enantioselectivity of enzymes combined with fast, easy, and inexpensive procedures that do not involve preliminary and time-consuming cleaning steps; thus, even complex samples can be processed by using these techniques.

Due to the predominance of L-amino acids in nature, only few enzymes showing an absolute specificity for D-amino acids have been identified. Among them, the flavoprotein D-amino acid oxidase (DAAO, EC 1.4.3.3.) purified from the yeast *Rhodotorula gracilis* (RgDAAO) is the most suitable biotool for use in *in vitro* enzymatic assays. RgDAAO is able to oxidize D-amino acids with a strict stereospecificity and a highly specific activity to produce the corresponding α-keto acids and ammonia. The FAD cofactor of the enzyme is then re-oxidized by molecular oxygen, producing H_2O_2 (Scheme 1) (12). The substrate specificity of wild-type RgDAAO encompasses almost all of the 20 natural amino acids; only acidic amino acids (i.e., D-aspartate and D-glutamate) are poor substrates (Fig. 1). These latter D-amino acids are substrates of a different flavoprotein, namely, D-aspartate oxidase (DASPO, EC 1.4.3.1). Recently, RgDAAO variants possessing an altered substrate specificity were generated by protein engineering (Fig. 1) (13, 14).

Because of the simple stoichiometry of the reaction catalyzed by RgDAAO, D-amino acids can be detected by determining the oxygen consumed or by quantifying the reaction products with various spectrophotometric techniques or through coupled enzymatic assays. Moreover, in the analytical field, RgDAAO is also

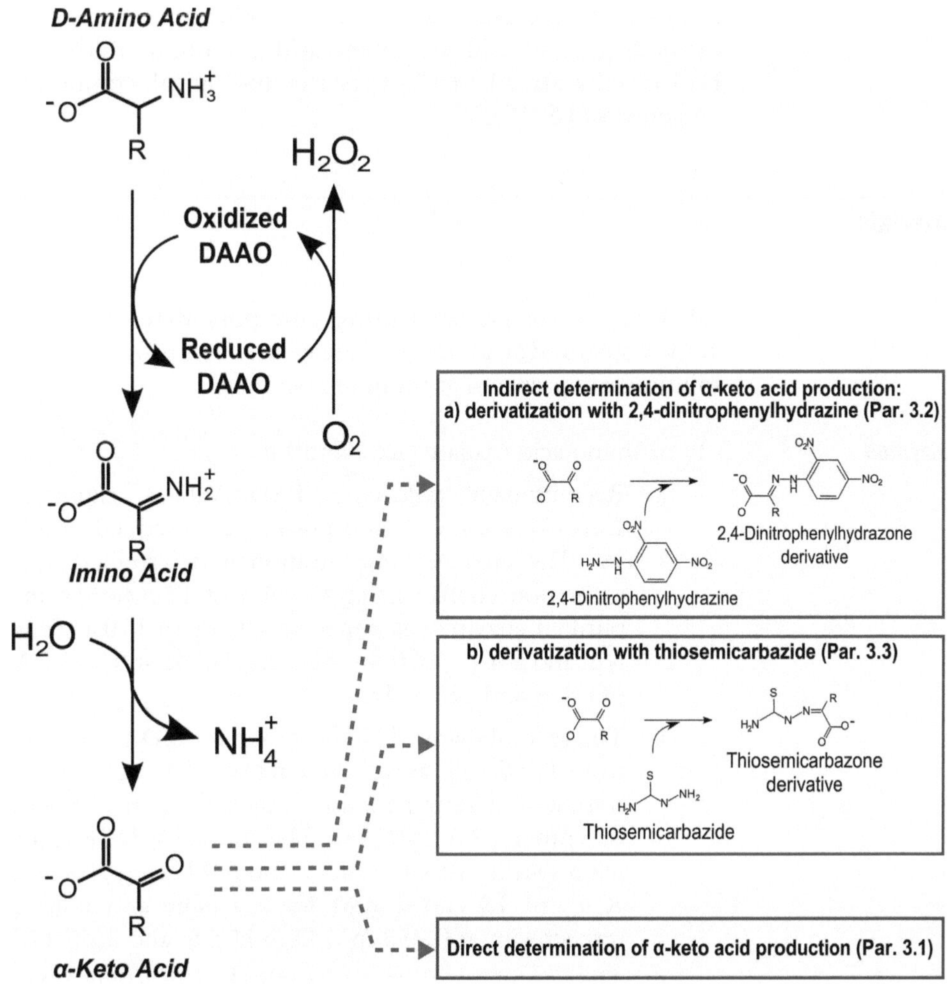

Scheme 1. Reaction catalyzed by DAAO and spectrophotometric assays for analytical determination of α-keto acids produced during the reaction.

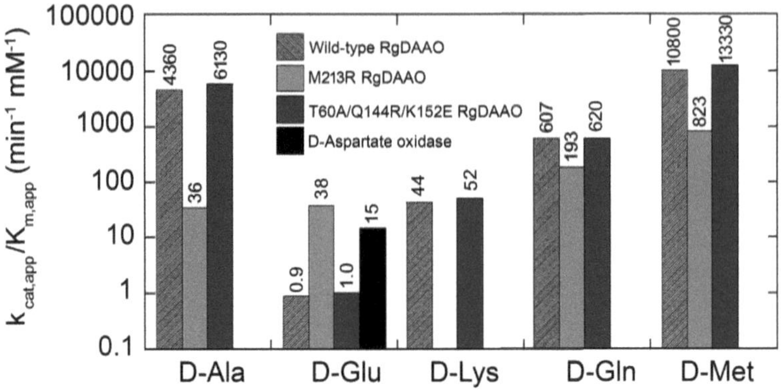

Fig. 1. Substrate specificity of different RgDAAO variants and D-aspartate oxidase represented as catalytic efficiency (k_{cat}/K_m ratio). Values were measured using the oxygen consumption assay at 25°C and pH 8.5 (13, 14, 16, 20, 26).

employed in the preliminary steps of other analytical separation methods (e.g., to oxidize D-amino acids prior to derivatization and HPLC separation) (15) or as the biological component of biosensors (16, 17).

2. Materials

All solutions are prepared using ultra-pure water (e.g., Milli-Q grade water) and analytical grade reagents. Unless otherwise stated, solutions are prepared at room temperature.

2.1. Enzymes

1. D-Amino acid oxidase (see Note 1):
   (a) Recombinant wild-type RgDAAO is expressed in *Escherichia coli* cells and purified as described in refs. 18, 19. The enzyme concentration is determined using an extinction coefficient at 455 nm of 12.6/mM/cm. The purified enzyme has a specific activity of 110 U/mg protein and a $K_{m,app}$ of 0.8 mM on D-alanine at 0.253 mM O_2, pH 8.5, and 25°C (18).
   (b) The recombinant M213R variant of RgDAAO is purified from *E. coli* cells as reported in (14). The enzyme concentration is determined using an extinction coefficient at 455 nm of 13.5/mM/cm. The purified M213R RgDAAO has a specific activity of 15.7 and 5.9 U/mg protein and a $K_{m,app}$ of 18 and 2 mM for D-alanine and D-aspartate, respectively, at 0.253 mM O_2, pH 8.5, and 25°C (13).
   (c) The recombinant T60A/Q144R/K152E RgDAAO variant was purified from *E. coli* as reported in (16). The enzyme concentration is determined using the extinction coefficient

of the wild-type enzyme (12.6/mM/cm). The purified T60A/Q144R/K152E RgDAAO has a specific activity of 140 U/mg protein and a $K_{m,app}$ of 0.4 mM on D-alanine (16).

2. D-Aspartate oxidase (DASPO):

Bovine DASPO is purified from the natural source (porcine kidney) as reported in ref. 20 or produced as recombinant protein in *E. coli* and purified as reported in ref. 21. The enzyme concentration is determined using the extinction coefficient at 454 nm of 11.8/mM/cm (20). The specific activity of the purified enzyme on D-aspartate is 30 U/mg protein and it is similar for both native and recombinant enzyme.

One DAAO/DASPO enzymatic unit corresponds to the amount of enzyme that converts 1 μmol of D-alanine/D-aspartate per minute at 25°C (11, 18, 20).

2.2. Direct Determination of α-Keto Acid Production

1. UV/Vis spectrophotometer and 1 mL quartz semi-micro cuvettes.
2. 150 mM Disodium pyrophosphate buffer, pH 8.5.
3. 1 U/μL RgDAAO solution in 75 mM disodium pyrophosphate buffer, pH 8.5.
4. 50 mM Pyruvate (or the desired α-keto acid or amino acid) in 75 mM disodium pyrophosphate, pH 8.5 (for the calibration curve, see Note 2).

2.3. Indirect Determination of α-Keto Acid Production: Derivatization with 2,4-Dinitrophenylhydrazine

1. Visible spectrophotometer and 1 mL glass or plastic semi-micro cuvettes (see Note 3).
2. 150 mM Disodium pyrophosphate buffer, pH 8.5.
3. 1 U/μL RgDAAO solution in 75 mM disodium pyrophosphate buffer, pH 8.5.
4. 1 mM 2,4-Dinitrophenylhydrazine solution in 1 N HCl (see Note 4).
5. 0.6 N NaOH.
6. 200 μM Pyruvate solution (or the desired α-keto acid) in 75 mM disodium pyrophosphate, pH 8.5 (for the calibration curve, see Note 5).

2.4. Indirect Determination of α-Keto Acid Production: Derivatization with Thiosemicarbazide

1. UV/Vis spectrophotometer and 1 mL quartz semi-micro cuvettes.
2. 150 mM Disodium pyrophosphate buffer, pH 8.5.
3. 1 U/μL RgDAAO solution in 75 mM disodium pyrophosphate buffer, pH 8.5.
4. 10 mM Thiosemicarbazide dissolved in 0.1 M sodium acetate, pH 5.4.

5. 1 mM Pyruvate (or the desired α-keto acid or amino acid) solution in 75 mM disodium pyrophosphate, pH 8.5 (for the calibration curve, see Note 6).

2.5. Determination of Hydrogen Peroxide Production: o-Dianisidine Method

1. Visible spectrophotometer and 1 mL glass or plastic semi-micro cuvettes.
2. 150 mM Disodium pyrophosphate buffer, pH 8.5.
3. 10 mM *o*-Dianisidine dihydrochloride in H_2O (see Note 7).
4. 100 U/mL Horseradish peroxidase (HRP, EC 1.11.1.7) in 75 mM disodium pyrophosphate, pH 8.5 (see Note 8).

2.6. Determination of Hydrogen Peroxide Production: 4-Aminoantipyrine Method

1. Visible spectrophotometer and 1 mL glass or plastic semi-micro cuvettes.
2. 150 mM Disodium pyrophosphate buffer, pH 8.5.
3. 15 mM 4-Aminoantipyrine (4-AAP) in H_2O (see Note 9).
4. 200 mM Phenol.
5. 250 U/mL HRP in 75 mM disodium pyrophosphate, pH 8.5 (see Note 8).

2.7. Determination of Hydrogen Peroxide Production: Amplex® Red/UltraRed Assay (Molecular Probes Inc., USA)

1. Fluorescent microplate reader (e.g., Infinite 200, TECAN, Switzerland).
2. 96-Well microplates.
3. 300 mM Disodium pyrophosphate buffer, pH 8.5.
4. 10 U/mL HRP in 75 mM disodium pyrophosphate, pH 8.5 (see Note 8).
5. Amplex® Red/UltraRed stock solution: open the vial under nitrogen and add 340 μL of DMSO. The final concentration is 10 mM. Prepare 25 μL aliquots and store at −20°C, protecting from light.
6. Amplex® Red/UltraRed stop solution: add 1.45 mL of ethanol to the stop solution reagent vial. Mix 1.0 mL of this solution with 1.0 mL of milliQ-grade H_2O. This stop solution is stable for 1 month at 4°C in the dark.
7. 0.1 mM D-Alanine in 75 mM disodium pyrophosphate buffer, pH 8.5 (for the calibration curve, see Note 10).

2.8. Determination of Ammonia Production

1. 150 mM Disodium pyrophosphate, pH 8.5.
2. 1,000 U/mL Glutamate dehydrogenase (GldH, Roche, Basel, Switzerland) solution in 75 mM disodium pyrophosphate, pH 8.5.
3. 25 mM Nicotinamide adenine dinucleotide reduced form (NADH) in 75 mM disodium pyrophosphate, pH 8.5.
4. 100 mM α-Ketoglutaric acid in H_2O.

2.9. Determination of O_2 Consumption

1. Hansatech oxygen electrode connected to a PC for digital data acquisition (Oxygraph, Hansatech Instruments Ltd., UK) (see Note 11).
2. 150 mM Disodium pyrophosphate, pH 8.5.
3. D-Alanine solution at different concentrations in 75 mM disodium pyrophosphate, pH 8.5 (as standard for the set up of the assay).

3. Methods

For all the spectrophotometric assays, data are analyzed using the Lambert–Beer law: the concentration of the desired compound is calculated according to the following equation:

$$[\text{Compound}] = \Delta\text{Abs} / [1\ (\text{cm}) \times \text{extinction coefficient}\ (\text{mM}^{-1}\text{cm}^{-1})] = \text{mM} \quad (1)$$

where: $\Delta\text{Abs} = \text{Abs}_{\text{final}} - \text{Abs}_{\text{initial}}$ or $\Delta\text{Abs} = \text{Abs}_{\text{sample}} - \text{Abs}_{\text{blank}}$; l = cuvette path (usually 1 cm); ε = extinction coefficient (usually expressed as /mM/cm).

As a general rule, spectrophotometric assays are best suited for analyzing samples containing only, or prevalently, a single amino acid. In fact, these methods are based on specific calibration curves prepared using a specific α-keto acid as standard (see Scheme 1 and Note 12). Samples in which the D-amino acid composition is more complex are better analyzed by applying coupled enzymatic methods and calculating the total amount of D-amino acids according to H_2O_2 (see Subheadings 3.4–3.6) or NH_4^+ (see Subheading 3.7) production (Scheme 2). For samples containing D-amino acids that are not or poor substrates of wild-type RgDAAO, engineered variants of RgDAAO (13, 14, 16) or DASPO (20, 21) and/or a mixture of these enzymes can be employed (see Fig. 1 and Subheading 2.1).

All assays are carried out at 25°C using a thermostated system (circulating water bath or Peltier control).

3.1. Direct Determination of α-Keto Acid Production

α-Keto acids, the main products of D-amino acid oxidation by DAAO, show strong absorption in the UV region (see Note 13): e.g., pyruvate, the product of the oxidation of D-alanine, possesses a maximum absorbance peak at 220 nm and an additional peak at 320 nm (of lower intensity). Therefore, α-keto acid concentrations can be rapidly and easily determined by using spectrophotometry.

Since different α-keto acids possess slightly different extinction coefficients (both for wavelength and intensity), the highest accuracy is achieved with a specific calibration curve generated for each D-amino acid under investigation. The protocol reported below refers to the assay of D-alanine.

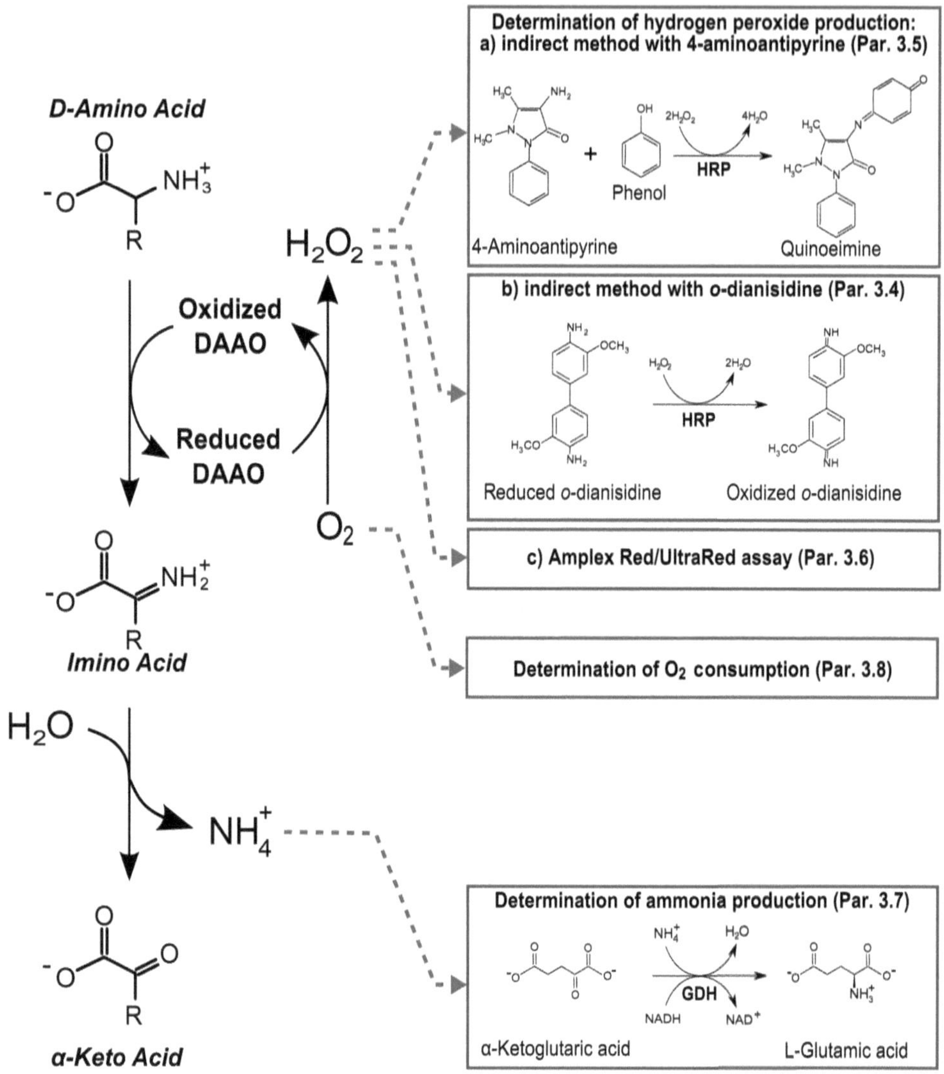

Scheme 2. Methods for analytical determination of H_2O_2 and NH_4^+ (using coupled enzymatic assays) and method for the determination of O_2 consumption (using a Clark-type oxygen electrode).

Assay protocol:

1. Prepare the reaction mixture directly in a quartz semi-micro cuvette, adding the sample containing D-alanine at unknown concentration (usually 100 μL) to disodium pyrophosphate buffer, pH 8.5 (final concentration: 75 mM).
2. Record the initial absorbance at 320 nm ($Abs_{initial}$).
3. Add 10 μL of 1 U/μL DAAO solution (in 75 mM disodium pyrophosphate buffer, pH 8.5) to the mixture (final volume: 1 mL).

4. The reaction is incubated at 25°C for ~20 min, or until the absorbance reaches a plateau (see Note 14). At this point, the final absorbance is recorded (Abs_{final}).
5. The concentration of pyruvate (and consequently, the concentration of D-alanine in the starting sample) is determined using Eq. 1 and the extinction coefficient at 320 nm for pyruvate (21/M/cm) as obtained from the corresponding calibration curve (see Notes 15 and 16).

3.2. Indirect Determination of α-Keto Acid Production: Derivatization with 2,4-Dinitro phenylhydrazine

In this assay, α-keto acids produced by D-amino acid oxidation by DAAO are reacted with 2,4-dinitrophenylhydrazine to produce the corresponding 2,4-dinitrophenylhydrazone derivatives that strongly absorb in the visible region at 445 nm (22) (see Scheme 1 and Note 17).

Assay protocol:

1. The reaction mixture is prepared in a microcentrifuge tube: 100 μL of the sample containing D-alanine at unknown concentration are added to 395 μL of disodium pyrophosphate buffer, pH 8.5 (buffer final concentration: 75 mM).
2. Add 5 μL of 1 U/μL DAAO solution (in 75 mM disodium pyrophosphate buffer, pH 8.5).
3. The reaction is incubated at 25°C for ~20 min.
4. Add 150 μL of 1 mM 2,4-dinitrophenylhydrazine dissolved in 1 N HCl.
5. Incubate for 10 min at 37°C to allow complete reaction of the produced α-keto acids with 2,4-dinitrophenylhydrazine.
6. Add 1,050 μL of 0.6 N NaOH to stop the derivatization reaction.
7. Incubate at room temperature for 5 min.
8. Measure the absorbance at 445 nm (Abs_{sample}).
9. The pyruvate concentration is calculated using Eq. 1, where ΔAbs is the Abs_{sample} subtracted by the absorbance of the blank (Abs_{blank}), which is prepared following the same procedure but 100 μL of sample are replaced with 100 μL of water. The extinction coefficient of the pyruvate derivative is calculated from a calibration curve (see Fig. 2a and Note 5).

3.3. Indirect Determination of α-Keto Acid Production: Derivatization with Thiosemicarbazide

This spectrophotometric assay is similar to the 2,4-dinitrophenylhydrazine one with the sole difference that, in this case, the concentration of the α-keto acid is determined after it has reacted with thiosemicarbazide and the corresponding thiocarbazone derivative has been produced, Scheme 1 (23).

Assay protocol:

1. Add 100 μL of the products of the oxidation reaction of D-amino acids by DAAO (see above, Subheading 3.2 for setting

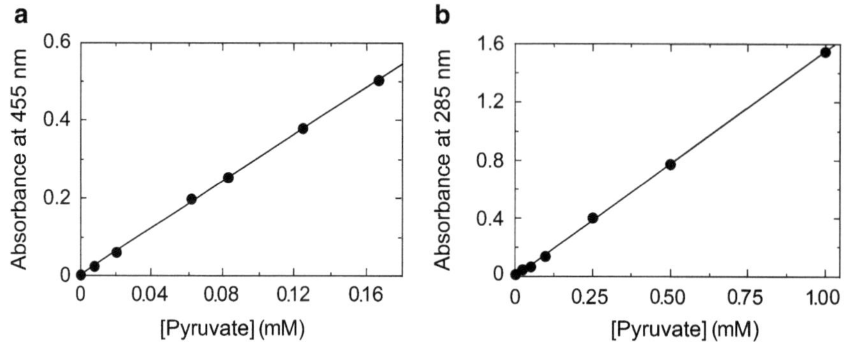

Fig. 2. Calibration curves for the spectrophotometric detection of pyruvate following derivatization with 2,4-dinitrophenylhydrazine (**a**) or with thiosemicarbazide (**b**). The slope of the lines represents the millimolar extinction coefficient of each compound (determined at 25°C, pH 8.5).

up the reaction) to 1 mL of 10 mM thiosemicarbazide dissolved in 0.1 M sodium acetate, pH 5.4 (see Note 18).

2. Incubate the derivatization reaction for 10 min at room temperature.
3. When formation of the thiosemicarbazone derivative is complete, the absorbance at 285 nm (corresponding to the wavelength of maximum absorbance) is recorded (Abs_{sample}).
4. The concentration of pyruvate is calculated using Eq. 1, where ΔAbs is the Abs_{sample} subtracted by the absorbance of the blank (Abs_{blank}), which is prepared using the same procedure but the sample is replaced with 100 μL of water. The extinction coefficient of the pyruvate derivative is calculated from a calibration curve (see Fig. 2b and Note 6).

3.4. Determination of Hydrogen Peroxide Production: o-Dianisidine Method

The H_2O_2 produced by the DAAO reaction is reduced to H_2O by HRP, whereby the electron donor *o*-dianisidine is simultaneously oxidized (see Scheme 2). Its oxidized species is brownish-red in color, showing an absorption maximum at about 440 nm with an extinction coefficient of 13/mM/cm (24) (see Note 3).

Assay protocol:

1. The reaction mixture is prepared directly in a disposable plastic semi-micro cuvette, mixing the amino acid solution at unknown concentration (usually 100 μL) with 1 mM *o*-dianisidine and 1 U of HRP in 75 mM disodium pyrophosphate, pH 8.5 (all final concentrations).
2. Record the initial absorbance ($Abs_{initial}$) of the mixture at 340 nm.
3. Add 10 μL of 0.1 U/μL DAAO solution (final volume: 1 mL) and monitor the absorbance at 440 nm until the value is stable (~10 min, Abs_{final}).

4. The concentration of oxidized *o*-dianisidine produced by the reaction, corresponding to the concentration of D-amino acid(s) in the original sample, is determined by measuring the increase in absorbance at 440 nm ($Abs_{final} - Abs_{initial}$) using Eq. 1 and an extinction coefficient of 13/mM/cm (see Notes 19 and 20).

3.5. Determination of Hydrogen Peroxide Production: 4-Aminoantipyrine Method

This assay is based on the same principle as the previous one (see Subheading 3.4), but 4-aminoantipyrine (instead of *o*-dianisidine) is oxidized by HRP in the presence of phenol (see Scheme 2). The resulting quinoeimine product shows an extinction coefficient of 6.58/mM/cm at 505 nm (25). See Note 3.

Assay protocol:

1. The reaction mixture is prepared directly in a disposable plastic semi-micro cuvette, mixing the amino acid solution at unknown concentration (usually 100 μL) with 1.5 mM 4-AAP, 2 mM phenol, and 2.5 U of HRP in 75 mM disodium pyrophosphate, pH 8.5 (all final concentrations).
2. Record the initial absorbance ($Abs_{initial}$) of the mixture at 505 nm.
3. Add 10 μL of 0.1 U/μL DAAO solution (final volume: 1 mL) and monitor the absorbance at 505 nm until the value is stable (~10 min, Abs_{final}).
4. The concentration of quinoneimine produced by the reaction is determined by measuring the increase in absorbance at 505 nm using Eq. 1 ($\varepsilon = 6.58$/mM/cm). Because of the stoichiometry of the reaction, in which two molecules of hydrogen peroxide react with one molecule of 4-AAP (Scheme 1), the concentration of H_2O_2 (corresponding to the concentration of D-amino acid(s) in the original sample) is twice the concentration of quinoneimine (see Notes 20 and 21).

3.6. Determination of Hydrogen Peroxide Production: Amplex® Red/UltraRed Assay

This assay is based on a fluorogenic substrate for HRP (Amplex® Red/UltraRed) that reacts with hydrogen peroxide at a 1:1 stoichiometric ratio to produce a brightly fluorescent product: this assay is highly sensitive. Since the protocol is set up for 96-well microplates, this assay is particularly useful to test several samples simultaneously.

Assay protocol (see Note 22):

1. Prepare the working solution by mixing 100 μL of 10 U/mL HRP and 50 μL of 10 mM Amplex® Red/UltraRed reagent in 150 mM disodium pyrophosphate, pH 8.5 (buffer final concentration). This reagent must be prepared fresh daily and kept at 4°C. Final volume: 5 mL.
2. In each well, mix 50 μL of working solution and 45 μL of sample and add 1 U of DAAO in a volume of 5 μL; maintain

the plate under nitrogen. Incubate on a shaker for 1 h in the dark.

3. Add 20 μL of stop solution to each well and incubate for 5 min in the dark.
4. Read the fluorescence intensity using a fluorimeter microplate reader: exc. = 535 nm (slit = 25 nm); em. = 590 nm (slit = 20 nm).
5. The D-amino acid concentration is calculated on the basis of a calibration curve obtained using known concentrations of D-alanine, ranging from 1 to 5 μM (final concentration in the well) (see Notes 10, 20 and 23).

3.7. Determination of Ammonia Production

In the presence of NH_3, produced by deamination of D-amino acids by DAAO, the enzyme glutamate dehydrogenase catalyzes the reductive amination of α-ketoglutaric acid to L-glutamic acid. During this reaction, the cofactor NADH is oxidized to NAD^+ (Scheme 1).

Assay protocol:

1. The reaction mixture is prepared in a quartz semi-micro cuvette, mixing the amino acid solution at unknown concentration (usually 100 μL) with 5 mM α-ketoglutaric acid, 2.5 mM NADH, and 20 U of glutamate dehydrogenase in 75 mM disodium pyrophosphate, pH 8.5 (all final concentrations).
2. Record the initial absorbance ($Abs_{initial}$) of the mixture at 340 nm.
3. Add 10 μL of 0.1 U/μL DAAO solution (final volume: 1 mL) and monitor the absorbance decrease at 340 nm for ~20 min until the value is stable (Abs_{final}).
4. The concentration of NAD^+ produced by the reaction, corresponding to the converted D-amino acid, is determined from the decrease in absorbance at 340 nm using Eq. 1 and an extinction coefficient of 6.22/mM/cm (see Note 24).

3.8. Determination of O_2 Consumption

The concentration of D-amino acids in solution can be correlated to the rate of the O_2 consumption during the reaction catalyzed by DAAO (Scheme 1). The O_2 concentration is recorded in real-time during the reaction using a Hansatech Clark-type oxygen electrode, with the acquisition device connected to a PC. The reaction takes place in a thermostated chamber mixed by a magnetic stirrer (see Note 25).

1. The reaction mixture is prepared directly in the oxymeter chamber, adding the sample containing the D-amino acid at unknown concentration (usually 100 μL) to disodium pyrophosphate buffer, pH 8.5 (final concentration: 75 mM).
2. Once the signal is stable, add 10 μL of RgDAAO (0.01 μg, ~0.01 U); final volume: 1 mL.

3. Measure the initial O_2 consumption rate (ΔpO_2/min) and calculate the enzymatic activity using the following equation:

$$\text{enzymatic activity} = \text{U/mg protein} = \frac{\Delta pO_2/\text{min} \times 0.253\ \text{mM}\ (O_2\ \text{solubility at}\ 25^{\circ}\text{C})}{\text{mg of added DAAO}} \times \text{assay volume (1 mL)}. \quad (2)$$

4. The concentration of D-amino acid is calculated using the Michaelis–Menten equation and the appropriate kinetic parameters (see Note 26):

$$[\text{D-amino acid}] = \frac{v_0 \times K_m}{V_{max} - v_0} = \text{mM}. \quad (3)$$

4. Notes

1. DAAO can also be purchased from commercial suppliers: e.g., from CPC Biotech (Italy, microbial origin), Calzyme Laboratories (USA, porcine), Affymetrics (USA, porcine), and Sigma-Aldrich (USA, porcine). Caution must be taken if using DAAOs from different sources since they possess different kinetic parameters and different optimal experimental conditions (26–28).
2. If the desired α-keto acid is not available, the corresponding D-amino acid can be used. In this case, the D-amino acid must be quantitatively converted (using DAAO) to the corresponding α-keto acid before recording the absorbance value for each calibration data point.
3. This assay does not require quartz cuvettes and a UV spectrophotometer since the absorbance is measured in the visible region of the spectrum.
4. 1 mM 2,4-Dinitrophenylhydrazine is not entirely dissolved in 1 N HCl: filter the solution before use to remove particulates. The consequent, slight decrease in the 2,4-dinitrophenylhydrazine concentration does not pose a problem since this reagent is still in large excess compared to the amino acid/α-keto acid.
5. To generate the calibration curve, a set of standard solutions of α-keto acid (from 10 to 200 μM) in 75 mM disodium pyrophosphate is used. The absorbance values at 445 nm are then plotted as a function of the concentration of pyruvate and the millimolar extinction coefficient of the 2,4-dinitrophenylhydrazone derivative is determined from the slope of the linear fitting of the data: e.g., 3.5/mM/cm for pyruvate (Fig. 2a) or 2.3/mM/cm for the 2,4-dinitrophenylhydrazone derivatives of

benzoylformic acid (the product of oxidation of D-phenylglycine) and 2-ketoglutaric acid (the product of oxidation of D-glutamate).

6. To generate the calibration curve, prepare a set of standard solutions of pyruvate (or the desired α-keto acid, from 25 to 1,000 μM) in 75 mM disodium pyrophosphate. The absorbance values at 285 nm are plotted as a function of the concentration of pyruvate and the millimolar extinction coefficient of the pyruvate-thiocarbazone derivative is determined from the slope of the linear fitting of the data (Fig. 2b).
7. The *o*-dianisidine solution in water must be freshly prepared (i.e. on the day of the experiment) and kept on ice.
8. Horseradish peroxidase is available as lyophilized powder (e.g., grade I, Roche, Switzerland). It is usually resuspended in disodium pyrophosphate buffer, pH 8.5, to prepare a concentrated stock solution (e.g., 25,000 U/mL) that can be stored at 4°C for several months. From this stock solution, appropriately diluted working solutions are prepared freshly every day.
9. The 4-aminoantipyrine solution in water must be freshly prepared (i.e. on the day of the experiment) and kept on ice.
10. To generate the calibration curve, follow the assay protocol reported in Subheading 3.6 with the following changes: (a) prepare the working solution in 75 mM disodium pyrophosphate, pH 8.5 (instead of 150 mM); (b) mix the working solution with D-alanine standard solutions ranging from 2 to 10 μM (1–5 μM final concentration in the well).
11. Oxymeter requires calibration before use: the amperometric value corresponding to 100% of O_2 relative concentration is measured when air-saturated pure water is present in the oxymeter chamber (the dissolved O_2 concentration is 0.253 mM at 25°C). The value corresponding to 0% of O_2 relative concentration is measured after adding few crystals of sodium dithionite. Setting up and maintaining the oxygen electrode is a delicate task and specific training is required.
12. All the assays, with the sole exception of the one based on O_2 consumption (i.e. all the end-point assays), allow the detection of a maximum D-amino acid concentration equal to the concentration of O_2 (e.g., 0.253 mM at 25°C). To determine higher D-amino acid concentrations, O_2 must be pumped into the reaction mixture.
13. The presence of contaminants in the sample might increase the absorbance value at the desired wavelength to values above reliable spectrophotometer limits. This is particularly relevant in the UV region (i.e. for the direct determination of α-keto acids, Subheading 3.1, or for the assay based on thiosemicarbazide,

Subheading 3.3): the presence of nucleic acids and/or proteins increases the absorbance intensity in the 260–280 nm range.

14. If necessary, the reaction rate can be increased by incubating the reaction at 37°C.
15. Because of the low extinction coefficient of pyruvate at 320 nm (21/M/cm), the lowest concentration of D-alanine that can be determined in the assay mixture is 0.2 mM (resulting in an absorbance of ~0.005 AU), corresponding to 2 mM in the original sample; the highest concentration is ~75 mM (corresponding to an absorbance of ≥1.5 AU). Sensitivity can be increased by measuring the absorbance at a lower wavelength (i.e. 220 nm for pyruvate).
16. The calibration curve is prepared using pyruvate solutions at different concentrations (ranging from 0.2 to 50 mM) prepared starting from a 50 mM pyruvate stock solution (in 75 mM disodium pyrophosphate, pH 8.5). The absorbance values at 320 nm are then plotted as a function of pyruvate concentration: the extinction coefficient of pyruvate is determined from the slope of the linear fitting of the experimental values.
17. Since the 2,4-dinitrophenylhydrazone derivatives of the α-keto acid strongly absorb in the visible range of the spectrum, the sensitivity of this assay is at least 20-fold higher than direct measurement of pyruvate absorption (at 320 nm).
18. This assay requires only a small volume of sample (≤100 μL) compared to the previous ones (Subheadings 3.1 and 3.2).
19. For this assay, the D-amino acid concentration in the reaction mixture must be in the 0.02–0.1 mM range.
20. Since this assay is based on determining the concentration of H_2O_2, the calibration curve does not depend on the D-amino acid used for the calibration. As a consequence, this method is suitable for analyzing complex samples that contain even more than one D-amino acid species.
21. For this assay, the concentration of D-amino acids in the reaction mixture must be in the 0.02–0.2 mM range.
22. Detailed protocols for using Amplex® Red/UltraRed Reagent and Stop Reagent are available at the Invitrogen/Molecular Probe web site. We strongly recommend reading these protocols carefully prior to performing the assay.
23. Due to the response variability of the fluorescence signal, calibration curves with known D-alanine concentrations must be repeated on each 96-well microplate. Furthermore, triplicates for each standard and sample are recommended.
24. Also in this case the calculated calibration curve is not dependent on the D-amino acid used for the calibration since the assay is based on determining the concentration of NADH.

It is thus suitable for analyzing complex samples that contain even more than one D-amino acid species.

25. This assay is not an end-point method and relies on correct determination of the initial velocity of the reaction that, in turn, is potentially subject to larger errors than the previous assays. For this reason, this assay is very sensitive to chemical (e.g., small pH changes or the presence of inhibitors) and/or physical (e.g., the reaction temperature) factors. Small changes in these parameters might result in large alterations of the experimental values.
26. The operative range of this assay is from 0.1 to 4 mM D-alanine. This range might be different if different D-amino acids are assayed; in this case, a preliminary calibration curve is required to correctly determine the linear region of this assay. For a number of D-amino acids, the kinetic properties (K_m and V_{max} parameters) of wild-type and variants of RgDAAO are known (12–14, 16, 26).

Acknowledgments

This work was supported by grants from Fondo di Ateneo per la Ricerca (University of Insubria) to G. Molla and L. Piubelli.

References

1. Auclair J L, Patton R L (1950) On the occurrence of D-alanine in the haemolymph of the milkweed bug *Oncopeltus fasciatus*. Rev Can Biol 9, 3–8.
2. Corrigan J J (1969) D-amino acids in animals. Science 164, 142–149.
3. Helfman PM, Bada J L, Shou M Y (1977) Considerations on the role of aspartic acid racemization in the aging process. Gerontology 23, 419–425.
4. Hashimoto A, Nishikawa T, Oka T et al. (1992) Determination of free amino acid enantiomers in rat brain and serum by high-performance liquid chromatography after derivatization with N-tert-butyloxycarbonyl-L-cysteine and *o*-phthaldialdehyde. J Chromatogr 582, 41–48.
5. Snyder S H, Kim P M (2000) D-amino acids as putative neurotransmitters: focus on D-serine. Neurochem Res 25, 553–560.
6. Oliet S H, Mothet J P (2006) Molecular determinants of D-serine-mediated gliotransmission: from release to function. Glia 54, 726–737.
7. Pollegioni L, Sacchi S (2010) Metabolism of the neuromodulator D-serine. Cell Mol Life Sci 67, 2387–2404.
8. Rubio-Barroso S, Santos-Delgado M J, Martin-Olivar C, Polo-Diez L M (2006) Indirect chiral HPLC determination and fluorimetric detection of D-amino acids in milk and oyster samples. J Dairy Sci 89, 82–89.
9. Friedman M J (1999) Chemistry, nutrition, and microbiology of D-amino acids. Agric Food Chem 47, 3457–3479.
10. Marchelli R, Palla G, Dossena A et al. (1997) D-ammino acidi: marker molecolari di stagionatura e di tipicità per il Parmigiano-Reggiano e il Grana Padano. Scienza Tecnica Lattiero-Casearia 48, 21–32.
11. Caligiuri, A., D'Arrigo, P., Rosini, E. et al. (2006) Enzymatic conversion of unnatural amino acids by yeast D-amino acid oxidase. Adv Synth Catal 348, 2183–2190.
12. Pollegioni L., Piubelli L., Sacchi S et al. (2007) Physiological functions of D-amino acid oxidases: from yeast to humans. Cell Mol Life Sci 64, 1373–1394.
13. Sacchi S, Lorenzi S, Molla G et al. (2002) Engineering the substrate specificity of D-amino-acid oxidase. J Biol Chem 277, 27510–27516.

14. Sacchi S, Rosini E, Molla G et al. (2004) Modulating D-amino acid oxidase substrate specificity: production of an enzyme for analytical determination of all D-amino acids by directed evolution. Protein Eng Des Sel 17, 517–25.
15. Oguri S, Nomura M, Fujita Y (2005) A new strategy for the selective determination of D-amino acids: enzymatic and chemical modifications for pre-column derivatization. J Chromatogr A 1078, 51–58.
16. Rosini E, Molla G, Rossetti C et al. (2008) A biosensor for all D-amino acids using evolved D-amino acid oxidase. J Biotechnol 135, 377–384.
17. Pernot P, Mothet J P, Schuvailo O et al. (2008) Characterization of a yeast D-amino acid oxidase microbiosensor for D-serine detection in the central nervous system. Anal Chem 80, 1589–1597.
18. Molla G, Vegezzi C, Pilone MS, Pollegioni L (1998) Overexpression in *Escherichia coli* of a recombinant chimeric *Rhodotorula gracilis* D-amino acid oxidase. Protein Expr Purif 14, 289–294.
19. Fantinato S, Pollegioni L, Pilone M S (2001) Engineering, expression and purification of a His-tagged chimeric D-amino acid oxidase from *Rhodotorula gracilis*. Enzyme Microb Technol 29, 407–412.
20. Negri A, Massey V, Williams C H Jr (1987) D-Aspartate oxidase from beef kidney: purification and properties. J Biol Chem 262, 10026–10034.
21. Simonic T, Duga S, Negri A et al. (1997) cDNA cloning and expression of the flavoprotein D-aspartate oxidase from bovine kidney cortex. Bioch J 322, 729–735.
22. Bohme A, Winkler O (1954) Zur Bestimmung geringer Mengen Acetaldehyd. Z Anal Chem 412, 1–5.
23. Walsh C T, Schonbrunn A, Abeles R H (1971) Studies on the mechanism of action of D-amino acid oxidase: evidence for removal of substrate α-hydrogen as a proton. J Biol Chem 216, 6855–6866.
24. Chlumsky L J, Zhang L, Rhamsey A J, Schuman Jorns M (1993) Preparation and properties of recombinant corynebacterial sarcosine oxidase: evidence for posttranslational modification during turnover with sarcosine. Biochemistry 32, 11132–11142.
25. Mori N, Sano M, Tani Y, Yamada H (1980) Purification and properties of sarcosine oxidase from *Cylindrocarpon didymum* M-1. Agric Biol Chem 44, 1391–1397.
26. Pollegioni L, Sacchi S, Caldinelli L et al. (2007) Engineering the properties of D-amino acid oxidases by a rational and a directed evolution approach. Curr Protein Pept Sci 8, 600–618.
27. Molla G, Sacchi S, Bernasconi M et al. (2006) Characterization of human D-amino acid oxidase. FEBS Lett 580, 2358–2364.
28. Pollegioni L, Caldinelli L, Molla G et al. (2004) Catalytic properties of D-amino acid oxidase in cephalosporin C bioconversion: a comparison between proteins from different sources. Biotechnol Prog 20, 467–473.

Chapter 19

An Enzymatic-HPLC Assay to Monitor Endogenous D-Serine Release from Neuronal Cultures

Inna Radzishevsky and Herman Wolosker

Abstract

D-Serine is a transmitter-like molecule that physiologically activates NMDA receptors in the brain. Although D-serine was thought to be exclusively released by astrocytes, we recently demonstrated endogenous D-serine release from neurons in cultures and slices. So far high-performance liquid chromatography (HPLC) has been the standard technique to monitor D-serine and other amino acids. This method employs pre-column derivatization with a chiral reagent to produce fluorescence derivatives that can be further separated on a reversed-phase column. Due to the close retention times of L-serine, L-glutamine, and D-serine, the quantification of low levels of endogenous D-serine synthesis and release from cell cultures and tissues can be challenging. We here describe an enzymatic treatment method to specifically destroy L-glutamine and L-serine by glutaminase and L-serine dehydratase enzymes, respectively, allowing accurate determination of nanomolar D-serine concentrations by subsequent HPLC analysis.

Key words: D-Serine, High-performance liquid chromatography, *N*-tert-butyloxycarbonyl-L-cysteine, *o*-Phthaldialdehyde, L-Serine dehydratase, Glutaminase, D-Serine deaminase, Serine racemase

1. Introduction

Studies from several laboratories demonstrate that D-serine is a transmitter/modulator in the brain that binds to the co-agonist site at the N-methyl-D-aspartate (NMDA) receptors and, along with glutamate, mediates several important physiological processes (1–3). Dysfunction of NMDA receptors is involved in a variety of neurodegenerative disorders, like Alzheimer's and Huntington's diseases (4, 5).

D-Serine is synthesized from L-serine by the serine racemase enzyme (6). Detection of D-serine was made possible by a high-performance liquid chromatography (HPLC) method developed in

Loredano Pollegioni and Stefano Servi (eds.), *Unnatural Amino Acids: Methods and Protocols*,
Methods in Molecular Biology, vol. 794, DOI 10.1007/978-1-61779-331-8_19,

the early 1990s (7). This method utilizes pre-column derivatization with *o*-phthaldialdehyde (OPA) and an optically active thiol, such as *N*-tert-butyloxycarbonyl-L-cysteine (Boc-L-Cys). The resultant diastereoisomeric derivatives are injected into the HPLC apparatus for subsequent separation on a reverse-phase column and detection. Nevertheless, this method is of limited use to monitor endogenous D-serine synthesis and release from primary cultures. The high levels of L-glutamine and L-serine present in the cells and culture medium prevent the accurate determination of D-serine because their retention times are very close. Furthermore, L-serine used to stimulate D-serine synthesis in cultures commonly contains contaminant D-serine (up to 1%) which exceeds the endogenous levels of D-serine. We now describe methods to selectively remove L-serine and L-glutamine from samples, which improve the HPLC separation of D-serine and increase the sensitivity of D-serine detection to nanomolar concentrations. Additionally, we provide a simple method to remove contaminant D-serine from commercial L-serine to allow accurate detection of D-serine synthesis.

2. Materials

1. Deionized water (18 mΩ) is prepared by a Milli-Q water purification system (Millipore Corporation).
2. L-serine (Bachem) is cleaned from D-serine impurities by D-serine deaminase (yDsdA) isolated from *Saccharomyces cerevisiae.*
3. All solvents are of HPLC-grade. Reagents for neuronal culture are of cell culture grade. All other reagents are of analytical grade.

2.1. Enzymes

1. Recombinant D-serine deaminase from *S. cerevisiae* (yDsdA) with an N-terminal histidine tag is kept frozen in a −70°C freezer at 6 mg/mL. The enzyme was cloned by PCR (Biometra) from yeast genomic DNA using primers 5′ACGCGCTA GCATGAGCGATGTTCTATCTCAA3′ (forward) and 5′ACGCCTCGAGTTACCATTTCTGAAAGGTAA3′ (reverse). Subsequently, the PCR product was sub-cloned into pET28c+ vector (Novagen) at NheI and XhoI sites. The yDsdA enzyme is expressed and purified as described (8). The purity of the enzyme is always verified by SDS-PAGE.
2. Purified glutaminase from *Escherichia coli* is obtained from Sigma (grade VII). The enzyme is resuspended in 5 mM sodium acetate buffer (pH 6.2) and divided in 10 μL aliquots (0.25 U/μL) that are kept frozen at −70°C.
3. Recombinant human L-serine dehydratase (hSDH) clone was obtained from Dr. H. Ogawa (Toyama Medical and

Pharmaceutical University, Japan). The hSDH was sub-cloned by PCR (Biometra) into pET28c+ at SalI and NotI sites using primers 5′ACGCGTCGACAATGATGTCTGGAGAA C3′ (forward) and 5′ATAAGAATGCGGCCGCTCACTTGG GCAACCT3′ (reverse). The enzyme is expressed and purified as described (8).

2.2. HPLC

2.2.1. HPLC Solvents

1. Preparation of the buffers should be carried out in a chemical hood.
2. Make a solution of 0.2 M acetate buffer, pH 6.2 by adding 11.5-mL glacial acetic acid in 900-mL HPLC-grade water.
3. The pH is adjusted to 6.2 by adding about 7 g NaOH (Aldrich; semiconductor grade) pellets and the final volume is adjusted to 1,000 mL.
4. Eluent A: 905 mL of 0.1 M acetate buffer, pH 6.2, 65 mL of acetonitrile (J.T. Baker), and 30 mL of tetrahydrofurane (THF, without stabilizer; Sigma).
5. Eluent B: 500 mL of 0.1 M acetate buffer, pH 6.2, 470 mL acetonitrile, and 30 mL THF.
6. Eluent C: 80% Methanol (JT Baker) in water.
7. Eluent D: 80% Acetonitrile in water.

2.2.2. Derivatizer Solution

Dissolve 1 mg/mL *N*-tert-butyloxycarbonyl-L-cystein (Nova-biochem) and 1 mg/mL *o*-phthaldialdehyde (Sigma) in methanol. Use fresh solution.

2.2.3. Analytical HPLC Columns and Instrumentation

1. The HPLC analysis can be carried with any basic analytical HPLC system equipped with a fluorescent detector.
2. We use a Merck-Hitachi LaChrom liquid chromatograph equipped with autosampler L-7250, a quaternary gradient-pump L-7100 (Merck-Hitachi) and an L-7350 (Merck) column oven adjusted to 30°C. The solvents are continuously degassed with degasser unit L-7614 (Merck). The instrument and chromatographic data are managed by the D-7000 HPLC System Manager Software (Merck-Hitachi).
3. Fluorescent amino acid derivatives are detected using a FL-7485 detector (Merck-Hitachi). Excitation and emission wavelengths are 344 and 443 nm, respectively.
4. The separation is carried out on a Spheri-5C18 column (220 mm × 4.6 mm i.d., 5-μm particle size; 300-Å pore size) from Alltech (Deerfield), fitted with precolumn NewGuard RP-18, 7 μm, 15 mm × 3.2 mm by Grace (Deerfield).
5. The rack with the samples and derivatizer is maintained at room temperature.

3. Methods

3.1. L-Serine Clean-up

The presence of up to 1% contaminant D-serine in commercial L-serine is frequently overlooked and will give wrong values for D-serine synthesis and release, as the exogenous D-serine will exceed the endogenous D-serine produced by cells. In our experience, it is crucial to completely remove any contaminant D-serine from L-serine. This includes L-serine that is present in most commercial culture media; otherwise the exogenous D-serine will be incorporated into the cells and may be mistaken with the endogenous pool.

1. To remove contaminant D-serine, dissolve 10 mM L-serine (Bachem) in 3 mM Tris–HCl (pH 8 ± 0.4).
2. Add purified recombinant yDsdA at a final concentration of 10 μg/mL for 2 h at 37°C.
3. To stop the reaction, heat-inactivate yDsdA by incubating the solution at 100°C in a heat block for 5 min.
4. Denaturated yDsdA is removed by a 5-min centrifugation at 16,000 × *g*.
5. Purified L-serine solution can be directly used or concentrated by lyophilization to a 200- to 300-mM stock.
6. The absence of contaminant D-serine and the final stock L-serine concentration are monitored by HPLC (see below).

3.2. D-Serine-Free Medium

In experiments of D-serine production or release from culture cells, it is important to employ culture media that lack L-serine and consequently are devoid of D-serine as well. Some commercial media, like Minimal Essential Medium and Basal Medium Eagle, lack both L- and D-serine, but the supplemented serum has about 1–2 μM D-serine. In this context, we suggest the use of defined-serum-free medium prepared *in-house*. For neuronal cell culture, we prepare a modified neurobasal (MNB) medium by mixing all components according to (9), but omitting L-serine.

3.3. Primary Neuronal Culture

1. Cortical neuronal cultures are prepared from Sprague–Dawley rat embryos (embryonic days 16–18) and maintained in 6-well culture plates (NUNC) pre-coated with poly-D-lysine, as previously described (10).
2. Two million cells per well are seeded in L-serine free MNB medium supplemented with penicillin/streptomycin and B27 (9).
3. Half of the medium is replaced every 2–3 days with fresh medium.

3.4. D-Serine Synthesis and Release

1. The experiments of D-serine synthesis or release are carried out at 12–14 days after seeding the neurons.
2. D-Serine synthesis/release can be monitored by adding different amounts of L-serine to stimulate D-serine production, albeit some D-serine synthesis/release is also observed without adding L-serine (11).
3. For HPLC analysis of D-serine, cell culture medium (20–200 μL), is collected and treated with purified glutaminase enzyme.
4. For this, the pH is first adjusted to 4.9 by addition of sodium acetate buffer (pH 4.5) to a final concentration of 100 mM.
5. Then glutaminase is added to a final concentration of 0.6 U/mL, and the samples are incubated overnight at 37°C.
6. The samples are deproteinized by addition of trichloroacetic acid (TCA) at 5% final concentration followed by centrifugation at 16,000 × *g* for 5 min.
7. The supernatant is collected and the TCA is removed by four extractions with 10 volumes of water-saturated diethyl ether. Excess of ether is removed by SpeedVac system (Savant) for 5 min.
8. As internal standard, we add 50–500 pmol L-homocysteic acid (L-HCA). The samples are alkalinized by addition of 10–100 μL of 0.4 M borate buffer (pH 9) and derivatized for HPLC analysis (see next section).
9. The results of glutaminase treatment can be seen in Fig. 1, in which removal of glutamine disclosed a discrete D-serine peak that was previously obscured by the large glutamine peak present in the culture media. This allowed us to quantify the amount of endogenous D-serine release from neuronal cell cultures elicited by AMPA receptor stimulation or cell depolarization (11).
10. In some experiments, we add high concentrations of L-serine (1–10 mM) in order to stimulate D-serine synthesis, either from cell cultures (12) or *in vitro* by the purified recombinant serine racemase (13). In these experiments, high L-serine amounts may interfere with the HPLC analysis of D-serine concentration, since the very large L-serine peak will prevent optimal separation of the L- and D-enantiomers.
11. To overcome this problem, L-serine can be removed from the samples by treatment with 250 μg/mL serine dehydratase (hSDH) for 3 h at 37°C (14) (see Note 1). Subsequently, the samples can be deproteinized and processed as described above for HPLC analysis.

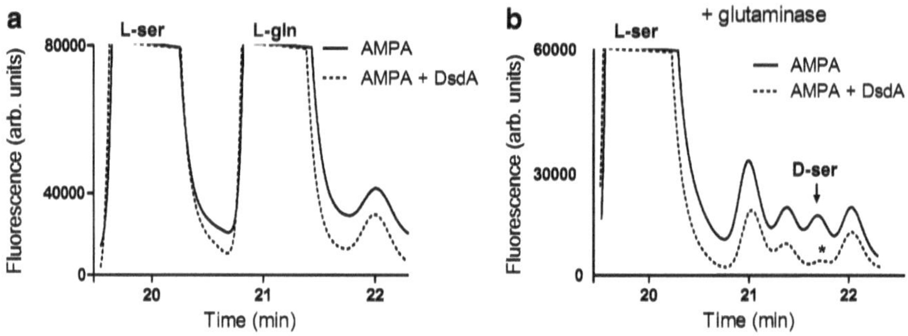

Fig. 1. Representative HPLC chromatogram of endogenous D-serine release from primary neuronal cultures. D-serine release is induced by α-amino-3-hydroxyl-5-methyl-4-isoxazolepropionate (AMPA), a selective agonist of one of the three groups of ionotropic glutamate receptors. (**a**) Conventional HPLC analysis for D-serine does not reveal any endogenous D-serine release from neuronal cultures, as the D-serine peak is obscured by the large L-glutamine (L-gln) peak. (**b**) Incubation of culture medium with purified glutaminase discloses a peak of released D-serine that was previously fused to L-glutamine peak (*arrow*). The identity of D-serine peak was confirmed by treatment of the medium with D-serine deaminase (DsdA) (see Note 3), denoted by an *asterisk*. Baseline of the sample treated with DsdA (*dash curve*) was offset for clarity. Adapted from ref. 11, with permission from Faseb J.

3.5. HPLC

The simultaneous determination of free amino acids in the samples is accomplished by HPLC with fluorometric detection as describe previously (7), but with some modifications.

3.5.1. Derivatization

1. Mix 1–200 μL sample with 10 μL of 50 μM L-HCA (internal standard), 60–100 μL l of 0.4 M borate buffer (pH 9), and complete the final volume to 1 mL with HPLC-grade water.
2. The derivatization is carried out by mixing 40 μL of derivatizer solution (see Note 2) with the sample.
3. After 1 min, 100 μL is injected to the HPLC apparatus.

3.5.2. HPLC Separation Conditions

1. The amino acids are eluted with a flow rate of 1 mL/min and a linear gradient between solvent A and solvent B.
2. The gradient is programmed as follows: 0 min, 100% A; 0–40 min, linear gradient to 33% B; 41–50 min, 100% C (washing step); 51–60 min, 100% D (washing step); 61–70 min, 100% A to equilibrate the column before a new injection.

4. Notes

1. This enzyme is specific for L-serine/L-threonine and does not recognize D-serine as substrate (14).
2. The derivatizer chemicals are used at a tenfold lower concentration than previously reported (7). This helps to decrease background levels.

3. To identify the D-serine peak, samples can be treated with purified yDsdA (10–20 μg/mL) prior to HPLC analysis. This enzyme was originally cloned by Yoshimura and colleagues (15). Since this enzyme is most active at neutral/alkaline pH, TCA-treated samples should be neutralized with a suitable buffer (e.g., 0.4 M borate buffer, pH 9) before the addition of the enzyme. The final pH should be between 7 and 8.5. Alternatively, yDsdA treatment can be done in the culture medium before glutaminase or TCA addition.

References

1. Basu A C et al. (2009) Targeted disruption of serine racemase affects glutamatergic neurotransmission and behavior. Mol Psychiatry 14, 719–727.
2. Mothet J P et al. (2000) D-serine is an endogenous ligand for the glycine site of the N-methyl-D- aspartate receptor. Proc Natl Acad Sci USA 97, 4926–4931.
3. Wolosker H et al. (2008) D-amino acids in the brain: D-serine in neurotransmission and neurodegeneration. FEBS J 275, 3514–3526.
4. Lipton S A (2006) Paradigm shift in neuroprotection by NMDA receptor blockade: memantine and beyond. Nature Reviews 5, 160–170.
5. Shleper M, Kartvelishvily E, Wolosker H (2005) D-serine is the dominant endogenous coagonist for NMDA receptor neurotoxicity in organotypic hippocampal slices. J Neurosci 25, 9413–9417.
6. Wolosker H, Blackshaw S, Snyder S H (1999) Serine racemase: a glial enzyme synthesizing D-serine to regulate glutamate-N-methyl-D-aspartate neurotransmission. Proc Natl Acad Sci USA 96, 13409–13414.
7. Hashimoto A et al. (1992) Determination of free amino acid enantiomers in rat brain and serum by high-performance liquid chromatography afterderivatizationwithN-tert.-butyloxycarbonyl-L-cysteine and o-phthaldialdehyde. J Chromatogr 582, 41–48.
8. Foltyn V N, Wolosker H (2006) Recombinant serine racemase preparation. In: Fisher G H, Brueckner N, Fujii H, Homma H, Konno R (eds), D-Amino acids: A new frontier in amino acid and protein research, Nova Science Publishers Inc., pp 463–466.
9. Price P J, Brewer G J (2001) Serum-free media for neural cell cultures. In: Protocols for neural cell culture, Human Press, Totowa, New Jersey, pp 255–264.
10. Kartvelishvily E et al. (2006) Neuron-derived D-serine release provides a novel means to activate N-methyl-D-aspartate receptors. J Biol Chem 281, 14151–14162.
11. Rosenberg D et al. (2010) Neuronal release of D-serine: a physiological pathway controlling extracellular D-serine concentration. FASEB J 24, 2951–2961.
12. Balan L et al. (2009) Feedback inactivation of D-serine synthesis by NMDA receptor-elicited translocation of serine racemase to the membrane. Proc Natl Acad Sci USA 106, 7589–7594.
13. Foltyn V N et al. (2005) Serine racemase modulates intracellular D-serine levels through an alpha,beta-elimination activity. J Biol Chem 280, 1754–1763.
14. Ogawa H et al. (2006) Enzymatic and biochemical properties of a novel human serine dehydratase isoform. Biochim Biophys Acta 1764, 961–971.
15. Ito T et al. (2008) A novel zinc-dependent D-serine dehydratase from *Saccharomyces cerevisiae.* Biochem J 409, 399–406.

Chapter 20

Electrophysiological Analysis of the Modulation of NMDA-Receptors Function by D-Serine and Glycine in the Central Nervous System

Fabrice Turpin, Glenn Dallérac, and Jean-Pierre Mothet

Abstract

The NMDA subtypes of glutamatergic receptors (NMDARs) are unusual in that their activation requires the binding of both glutamate and a co-agonist glycine or D-serine. Whereas glycine was first suggested to play such a role, it was later established that D-serine could serve as an endogenous co-agonist at different central synapses. We still do not know the exact nature of the endogenous co-agonist(s) of NMDARs and the function of the so-called glycine B site in many brain structures. We introduced few years ago the use of enzymes that specifically degrade either D-serine or glycine to decipher the influence of these amino acids on NMDA receptors function. The use of these enzymatic scavengers represents an invaluable technique for neurophysiologists investigating the neuromodulation of the glycine B site in the CNS. Here, we describe the proper way to manipulate these enzymes during electrophysiological recordings in acute brain slices and highlight the experimental tricks.

Key words: D-Serine, Glycine, NMDA receptors, Co-agonist, Synapse, Neurons, Astrocytes, Recombinant oxidases, Long term potentiation, Glutamate

1. Introduction

It has long been thought that D-amino acids are unnatural and that only L-amino acid occur in mammals because of the stereospecificity of biological protein synthesis (1). Two decades ago, the development of techniques aiming at distinguishing isomers of amino acids in biological samples allowed the identification of substantial amounts of two D-amino acids, namely D-serine and D-aspartate in the central nervous system (CNS) of mammals (1).

Loredano Pollegioni and Stefano Servi (eds.), *Unnatural Amino Acids: Methods and Protocols*,
Methods in Molecular Biology, vol. 794, DOI 10.1007/978-1-61779-331-8_20, © Springer Science+Business Media, LLC 2012

The discovery of D-serine in the CNS had a profound impact on our understanding of how the transfer of information in the brain is operating. Indeed, this atypical amino acid is now recognized to play a major role in the brain by serving as a key neuromodulator at the *N*-methyl-D-aspartate receptors (NMDAR), a subtype of ionotropic receptor to glutamate, in many brain regions (2, 3). Pioneer studies in the late 1980s have shown that activation of NMDAR not only requires the binding of glutamate both also of an endogenous coagonist. Because exogenous glycine and inhibi-tors of glycine transporters increases NMDA synaptic responses, it has long been assumed that glycine is the endogenous coagonist. However, several lines of evidence accumulated over the last decade have led to the hypothesis that D-serine rather than glycine is the endogenous coagonist for the so-called glycine site (or glycine B site) of NMDAR (2, 3). D-Serine and its biosynthetic enzyme serine racemase (SR) occur primarily in the brain, with highest concentrations in regions enriched in *N*-methyl-D-aspartate receptors (NMDAR). Treatment of cell culture and brain slices with D-amino acid oxidase (DAAO) reduces synaptic NMDA currents in the hippocampus (4, 5), the hypothalamus (6) or the retina (7). Invalidation of serine racemase in mice leads to reduction of NMDAR activity and long-term potentiation in the hippocampus (8). Compelling evidence suggests that besides its physiological functions, D-serine may also participate in excitotoxic events when released in excess (2). Indeed, SR-deletion confers neuronal protection to cerebral ischemia and excitotoxicity (9) as well as β-amyloid 1–42 peptide injury in the forebrain (10). Moreover, loss of function in DAAO is a fundamental component of sporadic amyotrophic lateral sclerosis (11). In contrast, aging and schizophrenia are associated with decreased D-serine levels (12, 13).

Whether endogenous glycine is a true co-agonist of NMDARs requires further investigations and the relative contribution of the two co-agonists needs to be elucidated. Here, we describe the use of recombinant glycine oxidase (GO) and DAAO as enzymatic scavengers to probe the relative functions of D-serine and glycine at glycine B site in the central nervous system.

2. Materials

2.1. Solutions and Drugs

All solutions are prepared with ultrapure water (prepared with purifying deionized water reaching a sensitivity of 18.2 MΩ cm at 25°C). Artificial cerebrospinal fluid (ACSF) should be freshly prepared every day, prior to experimentation. All drugs are prepared as stock solutions and stored at −20°C before use.

2.1.1. Extracellular Solutions

1. ACSF is composed of (mM): NaCl 123, KCl 2.5, Na_2HPO4 1, $NaHCO_3$ 26, and glucose 10. Add ~600 mL water into a 1 L glass beaker. Add (in g) NaCl 7.134, KCl 0.186, Na_2HPO_4 0.156, $NaHCO_3$ 2.184, and glucose 1.8. Make up to 1 L.
2. Three different ACSF are used for dissection (ACSF1), slice recovery (ACSF2) and recording (ACSF3). Their difference lies on the concentrations of $MgCl_2$ and $CaCl_2$ as described in Table 1.
3. $MgCl_2$ and $CaCl_2$, stock solutions should be renewed every week at a concentration of 1 M with ultrapure water. Add 40 mL of water in to a 50 mL Falcon tube. Add (in g) $MgCl_2$ 5.082, $CaCl_2$ 3.675 and make up to 50 mL. Store at room temperature.
4. ACSF is oxygenated (95% O_2/5% CO_2) for at least 15 min before pH is adjusted to 7.2–7.4 with KOH (1 M) and stored at +4°C until use.
5. The osmolarity of the extracellular solution is determined by the freezing point depression method using an automatic osmometer (osmometer 3320, Advanced Instruments, France) and set at 295–300 mosmol/L with mannitol (Sigma, France); 1 mM of mannitol corresponds to 1 mosmol.

2.1.2. Intracellular Solution

The composition of the intracellular solution used is described in Table 2. Using a caesium-based intrapipette solution enables to block most of the potassium currents. Lidoicaine *N*-ethyl chloride (QX314 Cl, Tocris, Cat. No. 2313) is also added to the intracellular solution in order to block voltage-activated Na^+ channels. Inactivation of both Na^+ and K^+ voltage-sensitive channels will prevent triggering

Table 1
Composition of extracellular solutions

Compounds	ACSF 1 (mM)	Volume (μL) of stock solution for 200 mL ACSF 1	ACSF 2 (mM)	Volume (μL) of stock solution for 200 mL ACSF 2	ACSF 3 (mM)	Volume (μL) of stock solution for 200 mL ACSF 3
NaCl	123		123		123	
$NaHCO_3$	26		26		26	
Glucose	10		10		10	
KCl	2.5		2.5		2.5	
Na_2HPO_4	1		1		1	
$MgCl_2$	4	800	2.5	500	1.3	260
$CaCl_2$	0.5	100	1	200	2.5	500

Table 2
Composition of intracellular solution in mM (for 50 mL)

Compounds	Compounds (mM)	Quantity (mg)
CsCl	123	1,092
NaCl	26	76
HEPES	10	119
QX-314-Cl	2.5	37
EGTA	1	500 μL of 0.1 M solution
$CaCl_2$	0.1	5 μL of 1 M solution

of action potential and therefore allow, in the voltage-clamp mode, to bring the membrane potential at +40 mV. Greater NMDA currents are indeed recorded at positive membrane potential values as the magnesium block of the NMDA-receptor is relieved, the magnesium ion being a cation. The pH of the solutions adjusted to 7.3 with CsOH and the osmolarity set at ~290 mosm/L using mannitol.

2.1.3. Drugs

1. AMPA antagonist NBQX disodium salt (Sigma, France): stock solutions are prepared at 10 mM; i.e. 10 mg in 2.629 mL of water. Store at −20°C.
2. $GABA_A$ antagonist Picrotoxin (Sigma, France): stock solutions are prepared at 50 mM; i.e. 100 mg in 3.319 mL in pure ethanol. Store at −20°C.
3. NMDA antagonist AP5 (Sigma, France): stock solutions are prepared at 50 mM; i.e. 10 mg in 2.627 mL of water.

2.1.4. Enzymes

Recombinant *Rhodotorula gracilis* D-amino acid oxidase (*Rg*DAAO, EC 1.4.3.3) is produced by overexpression in *Escherichia coli* cells and purification as reported earlier (14); the final enzyme preparation has a specific activity of 100 ± 15 U/mg protein on D-serine as substrate. Recombinant *Bacillus subtilis* glycine oxidase (*Bs*GO, EC 1.4.3.19) is also generated by overexpression in *E. coli* cells and subsequent purification (15); the final enzyme preparation has a specific activity of 0.9 ± 0.2 U/mg protein on glycine as substrate. These flavoenzymes specifically degrade D-ser (*Rg*DAAO) and glycine (*Bs*GO), as demonstrated by the corresponding apparent catalytic efficiency (k_{cat}/K_m ratio) values: k_{cat}/K_m ratios of 3.0 and 0.058/mM/s were determined for *Rg*DAAO on D-serine and glycine, respectively (14), while the k_{cat}/K_m ratios determined for *Bs*GO were 0.00025 and 0.867/mM/s on D-ser and glycine, respectively (15). In order to degrade D-serine or glycine, slices must be incubated for at least 45 min and then continuously perfused with ACSF containing *Rg*DAAO (0.2–0.4 U/mL) or *Bs*GO (0.1–0.2 U/mL), respectively (see Note 1).

2.2. Animals

Male Wistar rats (Charles River, France) are used for these experiments. Rats are housed individually with food and water ad libitum in a temperature-controlled room and on a 12 h light/dark cycle.

2.3. Dissection Material

1. Vibratome (VT1000S, Leica, France).
2. Brush size 4.
3. 2 Scalpels (FST, Phymep, France).
4. Thin spatula (FST, Phymep).
5. Large spatula (FST, Phymep).
6. Curved forceps (FST, Phymep).
7. Friedman bone-cutting forceps (FST, Phymep).
8. Large scissors (FST, Phymep).
9. Bechers (VWR, France).
10. 2 Petri dishes (VWR).
11. Pasteur pipette (VWR).
12. Carbogen (95% O_2, 5% CO_2, Air liquide, France).

2.4. Recording Set-up

1. Microscope with ×10 and ×40 optics (Olympus BX61, France).
2. Recording chamber (Luigs&Neumann, Germany).
3. Oscilloscope (Teknotronix, France).
4. Multiclamp 700B amplifier (Axon Instruments, USA).
5. Digidata 1440A interface (Axon Instruments).
6. Anti-vibration table (TMC, France).
7. Peristaltic pump (Gilson, France).
8. Camera (Hamamatsu, France).
9. Temperature controller (Luigs&Neumann).
10. 2 Motorized micromanipulators (Luigs&Neumann).
11. CV7B headstage (Axon Instruments).
12. Electrode holder (Axon Instruments).
13. Stimulator (Master 8, AMPI, Israel).
14. Stimulating electrode (Phymep).
15. 15, Electrode Puller (P-97, Sutter Instrument, USA).
16. Thin-walled filamented borosilicate glass (OD = 1.5 mm, ID = 1.17 mm; Harvard Apparatus, France).
17. The recording chamber must be adapted under the microscope optics. Recording chambers are usually of circular or ovoid shape, made out of Plexiglas and their floor consists in a glass

coverslip. Oxygenated ACSF3 should be running through the recoding chamber at a rate of 2–3 mL/min by means of the peristaltic pump.

18. For visualization of the tissue sample, the camera must be fitted on the Olympus microscope. The Hamamatsu camera (Orca R^2) used herein runs with a control panel (HC image) program installed on a Dell computer.
19. Electrophysiological experiments are conducted under voltage-clamp conditions, using a computer-controlled MultiClamp 700B microelectrode patch-clamp amplifier (Axon Instruments) in the whole-cell recording mode. The amplifier is operated using a control panel program (MultiClamp Commander host software, Axon Instruments) which runs on a Dell personal computer and communicates with the amplifier via serial cables (Fig. 1).
20. Recording electrodes are pulled on a horizontal flaming/Brown micropipette puller (Sutter Instrument Model P-97) using thin-walled filamented borosilicate glass. The electrode when filled with the internal pipette solution has a 2–5 MΩ nominal resistance. The electrode puller program is set as follow: heat = RAMP, time = 200, velocity should be adjusted to find the region where three loops are needed to separate the glass.

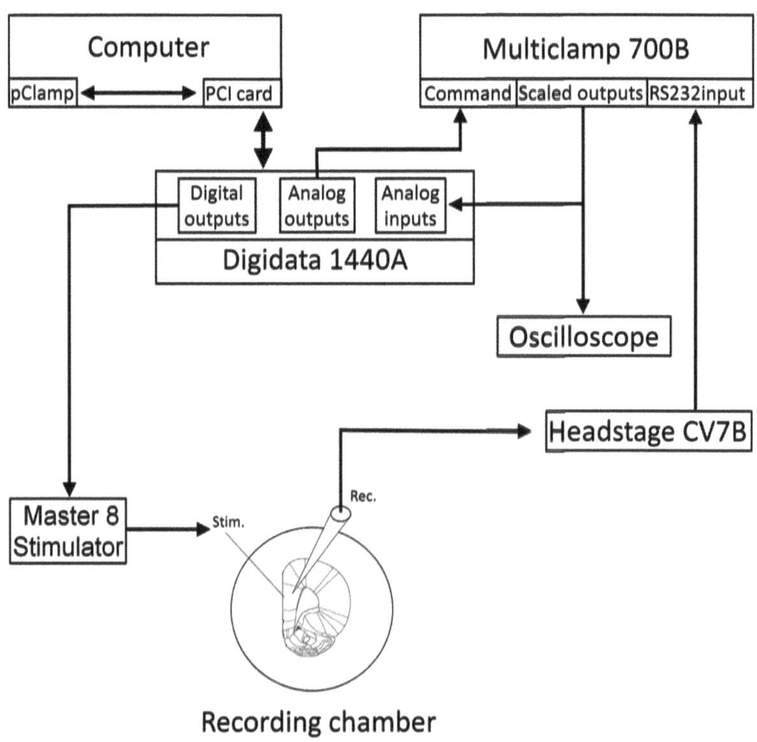

Fig. 1. Schematic drawing of the recording whole-cell patch clamp recording set-up.

Adjust the velocity until the correct pipette shape and resistance is reached.

21. For recording, patch electrodes are inserted in a pipette holder (connected to the CV7B headstage) such that a silver/silver chloride (Ag/AgCl) wire, which runs through the holder, is in contact with the internal solution. The Ag/AgCl wire is constructed from a silver wire (Phymep) onto which is coated a layer of silver chloride (AgCl; Prolabo, France). The pipette holder also exhibits a side access onto which tubing (OD = 2.4 mm, ID = 0.8 mm; Tygon, France) can be fitted and connected to an empty syringe, which enables to apply positive or negative air pressure within the electrode.
22. The series resistance is compensated prior to recording. Whole-cell currents are amplified (×2), filtered at 2 kHz, and sampled at 5 kHz using a Digidata 1440A interface (AD/DA converter) driven by the pClamp 10 software package (Axon Instruments) installed on a Dell personal computer.
23. The master 8 stimulator is used to supply voltage pulses to the stimulating electrode fitted on the micromanipulator.
24. All recordings are carried out at 30°C on the stage of an Olympus BX51 microscope (Fig. 2).

3. Methods

3.1. Dissection

1. Postnatal day 45–60 male Wistar rats may be lightly anesthetized with Isofluorane prior to decapitation. To this end, the animal should be placed in a transparent and hermetic box containing a tissue soaked with liquid isofluorane. Once passed out the animal should be taken out of the box before death occurs.
2. Use a guillotine to cut off the head.
3. With a scalpel cut the skin and conjunctive tissue along the midline of the skull and pull it apart in order to expose the skull.
4. Using the bone-cutting forceps, remove all bones encapsulating the cerebellum.
5. Make a small cut along the midline of the skull using large scissors and with Friedman or curved forceps sequentially remove both parietal bones.
6. Delicately remove the frontal bones with the Friedman forceps.
7. Using fine scissors or curved forceps ensure that the pia is removed and scoop out the brain with a small spatula.
8. Place the brain in oxygenated (95% O_2/5% CO_2) ice-cold ACSF1 within 60 s after decapitation. see Note 2.

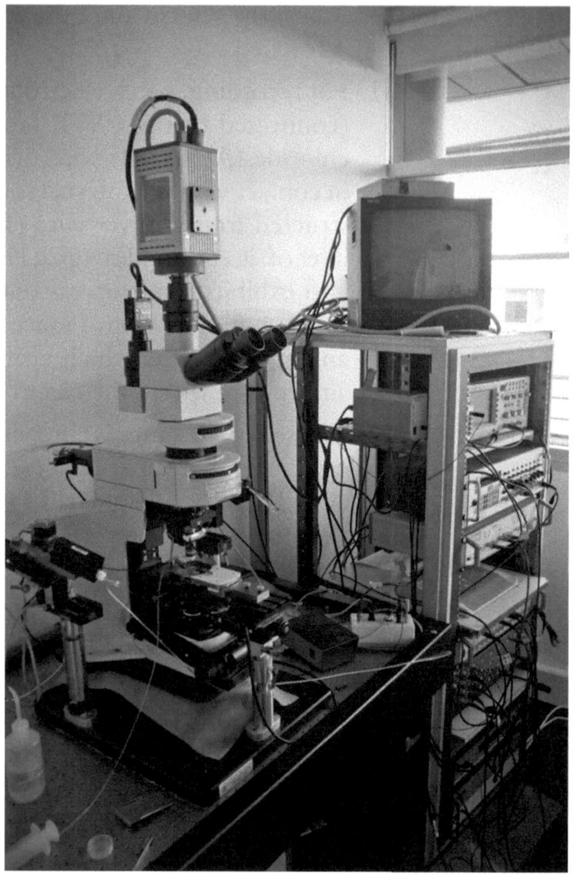

Fig. 2. Picture of an electrophysiological set-up.

9. While the tissue cools down (~3 min), apply a thin layer of cyanoacrylate glue (Super-glue) to the stage of the tissue slicer.
10. Take out the brain with the large spatula, and transfer it onto a Petri dish placed on ice. Let the brain lie on its ventral aspect and make a coronal cut through the middle of the cerebral cortex (approximately at the level of dorsal hippocampi), with a scalpel or a vibroslice blade. See Note 3.
11. Gently glue the tissue by its caudal aspect in the correct orientation (see Figs. 1 and 3) and immediately submerge the tissue with ice cold ACSF1.
12. Cut 500 μm slices until the forceps minor of the corpus callosum can clearly be distinguished. From that point, cut and collect 300 μm slices. Using a fire-polished Pasteur pipette or a size 4 brush, gently transfer each slice in a Petri dish filled with oxygenated ACSF1 and stop collecting when the rostral caudate-putamen appears. The forward speed of the vibratome should be slow while the vibration of the blade should be set as fast as possible.

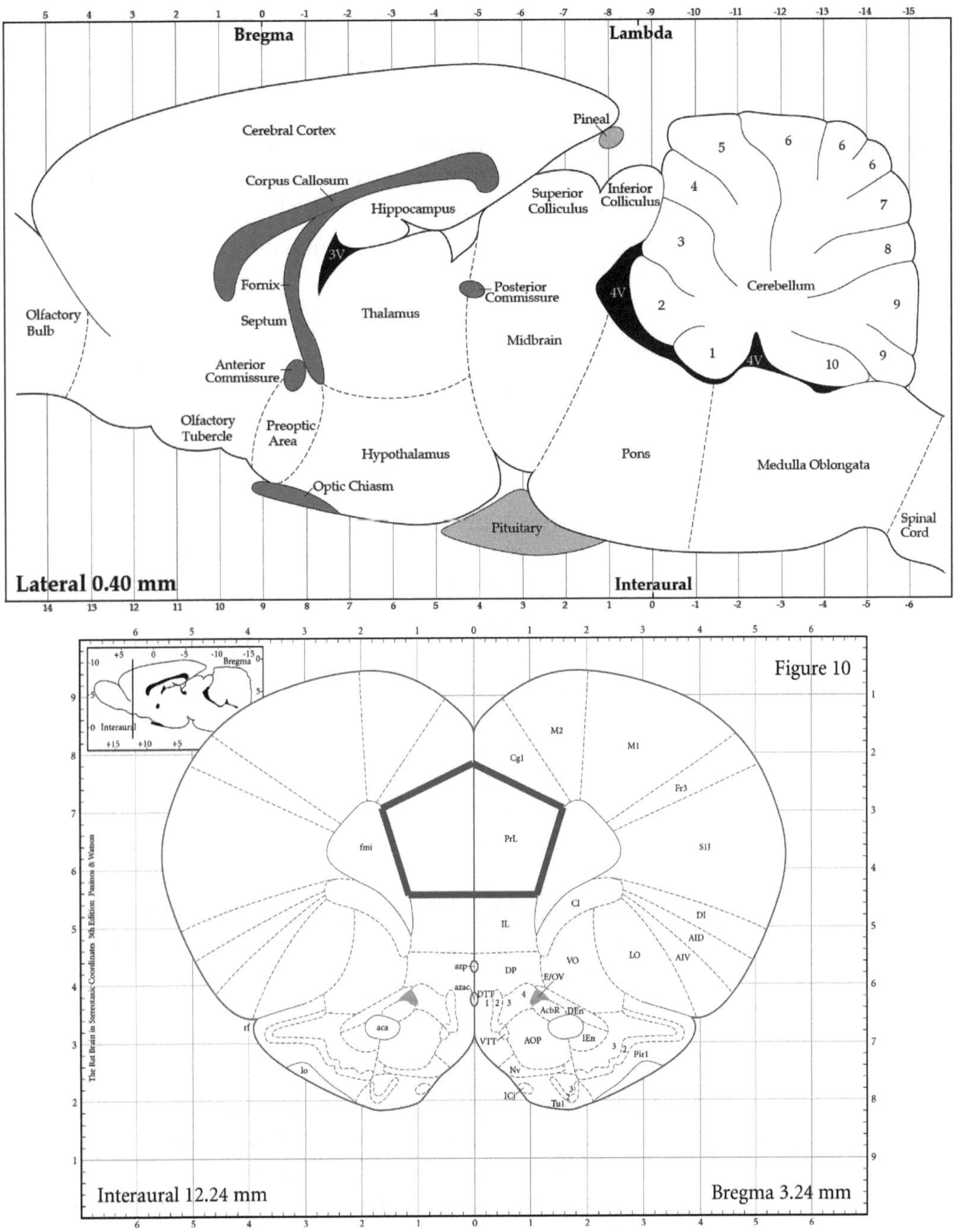

Fig. 3. Schematic representation of longitudinal (left, 0.4 lateral) and coronal (right, −3.2 from Bregma) section of the rat brain (16).

13. Cut slices in halves through the midline with a scalpel and transfer them by mean of a fire polished Pasteur pipette in a holding chamber containing oxygenated ACSF2 warmed at 31°C in a water bath.
14. Incubate the slices at 31°C for 40 min and at least 1 h at room temperature prior to electrophysiological recordings.

15. Finally, using a Pasteur pipette cut and fire-polished to an opening of 3–5 mm across, transfer each slice to a holding chamber containing oxygenated ACSF3 at room temperature.

3.2. Patch-Clamp Recording

1. Place a slice in the submerged chamber. The slice may be held down with a platinium wire as this metal in unalterable and does not release particles into ACSF.
2. Place a bipolar or monopolar stimulating electrode in layer I/II of the prelimbic cortex (PL).
3. Place the field of view of the microscope at ×40 in layer V of the PL.
4. Fill up a patch pipette with internal solution, place it on the headstage, in the ACSF3, put on positive pressure (using the syringe connected to the electrode holder) and offset pipette in the voltage clamp mode.
5. Target a layer V pyramidal cell that is preferentially not close to the surface. Healthy cells are typically smooth and exhibit a regular shape. See Note 4.
6. Once the cell has been reached, the positive pressure will result in the formation of a halo by the cell membrane surrounding the pipette. Upon formation of this halo, suddenly remove the positive pressure, immediately apply a gentle suction to the pipette, and clamp the voltage to −70 mV until formation of a "giga-seal". This occurs when the access resistance, which can be measured online using the pClamp software, goes up to several GΩ. Such configuration is called "cell-attached". See Note 5.
7. At that point, the whole-cell configuration can be obtained by briefly applying a strong suction to break the cell membrane. When membrane rupture is successful, the input resistance (resistance of the couple electrode + cell) should reduce drastically. For pyramidal neurons, input resistance lies usually between 100 and 150 MΩ. See Notes 6 and 7.
8. As the pipette solution is now in contact with the intracellular content, the neuron will dialyze. Since soluble endogenous components may be lost during this process, in order to enable full equilibration of the intracellular content, 10–15 min should be allowed prior to recording. See Note 8.

3.3. Drug Application and Recording Protocol

1. For whole-cell recording of NMDA currents, NBQX disodium salt and picrotoxin are added to the ACSF3 (1/1,000 of the stock solutions). For AMPA currents recording, AP5 and picrotoxin are added to the recording medium (1/1,000 of stock solutions).
2. While AMPA currents are recorded at −70 mV, recording of NMDA EPSC are performed at +40 mV. The layer V pyramidal

neuron should still be patched and dialyzed at −70 mV, once this has been completed, the membrane potential may slowly (i.e. 5–10 min) be raised from −70 to +40 mV.

3. Upon stimulation of the local afferent (layer I/II) AMPA EPSC should be recorded as fast inward current at −70 mV while NMDA current are much slower and outwardly-going at +40 mV (Fig. 4).
4. For *Rg*DAAO and *Bs*GO experiments, prior to recording, slices must incubated during 45 min in ACSF3 supplemented with 0.3 U/mL of *Rg*DAAO or 0.2 U/mL of *Bs*GO in order to degrade specifically D-serine and glycine, respectively. Slices are then transferred in the same medium running through the recording chamber.
5. In all experiments, a stable baseline of at least 5 min should be collected. All drugs are bath applied for 10 min and subsequently washed for a minimum of 10 min. Amplitude of AMPA and NMDA currents are respectively measured every 15 and 30 s for the whole time of experimentation (Fig. 4).

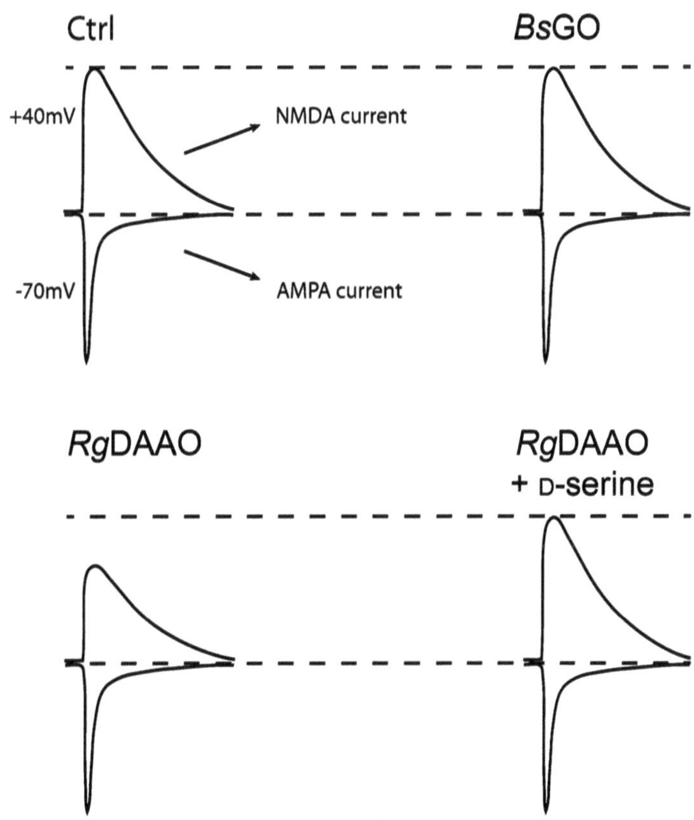

Fig. 4. Illustration of theoretical AMPA and NMDA currents in control situation and under *Rg*DAAO or *Bs*GO. Note that only *Rg*DAAO decreases NMDA currents which can be restored by application of exogenous D-serine.

6. For data analysis, in order to control for variations in the stimulating electrode placement and afferent activation, NMDA EPSC amplitudes are normalized to AMPA currents. Hence, the NMDA/AMPA ratios rather than the amplitude of NMDA currents are to be compared between treatments.

4. Notes

1. In our initial work, we purchased DAAO from Boehringer Mannheim (4). Different sources of DAAO can be used as substitutes provided enough purity and high activity. We and colleagues have tested the effects of the enzyme from Worthington Biochemical Corporation (http://www.worthington-biochem.com) or from Sigma and found that these enzymes worked well if accurate controls are made. Other commercial DAAOs can be purchased from USB Corporation (http://www.usbweb.com, OH, USA) or ABCR (http://www.abcr.de, Karlsruhe, Germany) but we have no experience with these enzymes. All commercial DAAOs are partially purified enzyme provided as a lyophilized powder or as a suspension in 3.2 M ammonium sulfate solution. Each batch of enzyme should be dialyzed (after re-suspension when starting with lyophilized powder) for 8–10 h at 4°C against 20 mM phosphate buffer (pH 7.4) containing flavin adenine dinucleotide (FAD) in order to remove contaminants and ammonium sulphate. The dialyzed enzyme is stored at −20°C until use. It is noteworthy to indicate that we have found that FAD present in the batch may have its own effect on NMDA receptor function just mimicking the inhibitory effect of commercial DAAO. Thus, baseline should be recorded in the presence of FAD alone before applying commercial DAAO.
2. The solution should be prepared in advance and placed for 20–30 min in the freezer.
3. Avoid squeezing or otherwise deforming the tissue at this stage. Regularly wet the brain with oxygenated ACSF using a Pasteur pipette.
4. The positive pressure enables to push away cell debris while the pipette is approaching the targeted neuron. One should not apply an excessive amount of positive pressure as this will almost certainly damage the slice.
5. Visually, the seal is typically assessed by applying small (e.g. 20 mV) voltage steps (in the voltage-clamp mode) of which the amplitude will decrease as the resistance increases.

6. The drop in input resistance is visually obvious on the oscilloscope trace while applying small voltage steps. It will indeed augment the size of the observed steps from virtually zero to several millivolts. Furthermore, in a neuron the steps shall not appear linear any longer but will exhibit a charging phase.
7. For accurate current measures, the access resistance (resistance of the pipette + resistance of the junction) should remain under 15 MΩ. Access resistance can be measured online with the pClamp software.
8. When using cesium based intracellular solution, always stay in the voltage clamp mode (i.e. do not switch to current clamp).

Acknowledgments

This work was supported by ANR grant (ANR-09-MNPS-022-01) and SERVIER grant to JPM. The authors are grateful to their collaborators for their support and notably to Dr Silvia Sacchi and Prof. Loredano Pollegioni for graciously providing us with the enzymes.

References

1. Hashimoto A, Oka T (1997) Free D-aspartate and D-serine in the mammalian brain and periphery. Prog Neurobiol 52, 325–353.
2. Martineau M, Baux G, Mothet J P (2006) D-Serine signalling in the brain: friend and foe. Trends in Neurosci. 29, 481–491.
3. Wolosker H (2007) NMDA receptor regulation by D-serine: new findings and perspectives. Mol Neurobiol 36, 152–164.
4. Mothet J P, Parent A T, Wolosker H et al. (2000) D-serine is an endogenous ligand for the glycine site of the N-methyl-D-aspartate receptor. Proc Natl Acad Sci USA 97, 4926–4931.
5. Yang, Y, Ge W, Chen Y et al. (2003) Contribution of astrocytes to hippocampal long-term potentiation through release of D-serine. Proc Natl Acad Sci USA 100, 15194–15199.
6. Panatier A, Theodosis D T, Mothet J P et al. (2006) Glia-derived D-serine controls NMDA receptor activity and synaptic memory. Cell 125, 775–784.
7. Stevens E R, Esguerra M, Kim P M et al. (2003) D-Serine and serine racemase are present in the vertebrate retina and contribute to the physiological activation of NMDA receptors. Proc Natl Acad Sci USA 100, 6789–6794.
8. Basu A C, Tsai G E, Ma C L et al. (2009) Targeted disruption of serine racemase affects glutamatergic neurotransmission and behavior. Mol Psychiatry 14, 719–727.
9. Mustafa A K, Ahmad A S, Zeynalov E et al. (2010) Serine racemase deletion protects against cerebral ischemia and excitotoxicity. J Neurosci 30, 1413–1416.
10. Inoue R, Hashimoto K, Harai T, Mori H (2008) NMDA- and beta-amyloid1-42-induced neurotoxicity is attenuated in serine racemase knockout mice. J Neurosci 28, 14486–14491.
11. Mitchell J, Paul P, Chen H J et al. (2010). Familial amyotrophic lateral sclerosis is associated with a mutation in D-amino acid oxidase. Proc Natl Acad Sci USA 20, 7556–7561.
12. Sacchi S, Bernasconi M, Martineau M et al. (2008) pLG72 modulates intracellular D-serine levels through its interaction with D-amino acid oxidase: effect on schizophrenia susceptibility. J Biol Chem 283, 22244–22256.

13. Turpin F R, Potier B, Dulong J R et al. (2011) Reduced serine racemase expression contributes to age-related deficits in hippocampal cognitive function. Neurobiol Aging 32, 1495–1504.
14. Molla G, Vegezzi C, Pilone M S, Pollegioni L (1998) Overexpression in *Escherichia coli* of a recombinant chimeric *Rhodotorula gracilis* D-amino acid oxidase. Protein Expr Purif 14, 289–294.
15. Job V, Marcone G L, Pilone M S, Pollegioni L (2002) Glycine oxidase from *Bacillus subtilis*. Characterization of a new flavoprotein. J Biol Chem 277, 6985–6993.
16. Paxinos G, Watson C (1997) The rat brain in stereotaxic coordinates, Compact 3rd Edition CD-Room. Academic Press, San Diego.

Chapter 21

Biosensors for D-Amino Acid Detection

Silvia Sacchi, Elena Rosini, Laura Caldinelli, and Loredano Pollegioni

Abstract

The presence of D-amino acids in foods is promoted by harsh technological processes (e.g., high temperature or extreme pH values) or can be the consequence of adulteration or microbial contamination (D-amino acids are major components of the bacterial cell wall). For this reason, quality control is becoming more and more important both for the industry (as a cost factor) and for consumer protection. For routine food analysis and quality control, simple and easily applicable analytical methods are needed: biosensors can often satisfy these requirements. The use of an enzymatic, stereospecific reaction could confer selectivity to a biosensor for detecting and quantifying D-amino acids in foodstuffs. The flavoenzyme D-amino acid oxidase from the yeast *Rhodotorula gracilis* is an ideal biocatalyst for this kind of application because of its absolute stereospecificity, very high turnover number with various substrates, tight binding with the FAD cofactor, and broad substrate specificity.

Furthermore, alterations in the local brain concentrations of D-serine (predominantly D-amino acid in the mammalian central nervous system) have been related to several neurological and psychiatric diseases. Therefore, quantifying this neuromodulator represents an important task in biological, medical, and pharmaceutical research. Recently, an enzymatic microbiosensor, also using *R. gracilis* D-amino acid oxidase as biocatalyst, was developed for detecting D-serine *in vivo*.

Key words: Amperometric detection, D-Amino acid oxidase, D-Serine, Food quality, Analytical detection

1. Introduction

Microorganism-synthesized D-amino acids are estimated to constitute approximately one third of the D-amino acid burden in humans (1). Moreover, naturally occurring L-amino acids may be transformed to their mirror image configuration isomers through exposure of food proteins to processing conditions such as high pH, heat, and acids. The presence of D-amino acids in food is normally associated with a decrease in protein digestibility, thus affecting the bioavailability of essential amino acids and ultimately impairing the

Loredano Pollegioni and Stefano Servi (eds.), *Unnatural Amino Acids: Methods and Protocols*,
Methods in Molecular Biology, vol. 794, DOI 10.1007/978-1-61779-331-8_21, © Springer Science+Business Media, LLC 2012

nutritional value of food (2). Both the concentration of D-amino acids and the D/(D+L) ratio have been proposed for assessing food quality and identifying food adulteration (2, 3). Indeed, D-amino acid concentrations in food often depend on the original content of these "atypical compounds" in the raw material: D-Ala, D-Glu, D-Asp, D-Lys, D-Ser, D-Pro, D-Phe, D-Arg, and D-Leu have been identified as natural components of various foods (2). Significant levels of these D-amino acids were measured in bread (4), coffee (5), fruit and vegetable juices (6), honey, chocolate (7), vinegars (8), wine (9), cheese (10), and dairy products (11, 12). In particular, raw milk from ruminants contains D-Ala, D-Asp, D-Glu, D-Lys, and D-Ser. Relatively high concentrations of these D-amino acids are also present in widely consumed fermented milk products, among which ripened cheeses are the richest in D-amino acids (10–12). The D-amino acid content differs in different cheeses and also changes according to means of production and storage. For example, Parmigiano Reggiano and Grana Padano cheeses can be distinguished from each other by the relative amounts of D-Ala, D-Asp, and D-Glu, while the racemization degree, the D/(D+L) isomer ratio, can be used to estimate the sample age (11). Furthermore, all D-amino acids can be found in foods as a consequence of adulteration, e.g., when hydrolyzed proteins are added to mask low nutrient content (11). For these reasons, determination of the overall D-amino acid content in food specimens constitutes an adequate analytical parameter.

Generally, efficient enantiomeric separation and analysis procedures using high performance liquid chromatography, gas chromatography, or capillary electrophoresis are performed to detect the presence of the different amino acids in foodstuffs, beverages, and biological samples (2, 12). Although these analytical methods are very sensitive, reliable, and reproducible, they are time-consuming and expensive (requiring specialized personnel) and often too slow for the food industry. For routine and quality control measurements, simple and easily applicable analytical methods are needed: biosensors can frequently satisfy these requirements.

During the last decade, many research groups have developed enzyme sensors that specifically detect D-amino acids based on the stereospecific, oxidative deamination catalyzed by the FAD-containing flavoenzyme D-amino acid oxidase (EC 1.4.3.3, DAAO): DAAO converts D-amino acids into the corresponding α-keto acids and ammonia using molecular dioxygen and producing hydrogen peroxide; for a review, see refs. 13, 14. Currently, the most efficient biosensors use the enzyme from the yeast *Rhodotorula gracilis* (RgDAAO), which exhibits a very high turnover number, a broad substrate specificity, and a tight binding with the FAD cofactor (13, 14). Moreover, the possibility of employing engineered RgDAAO variants with an enlarged substrate specificity and, in particular, with a remarkably improved catalytic efficiency on acidic and basic substrates (15–17) renders "the latest-generation biosensors" powerful analytical tools for determining the total D-amino acid content, even on complex matrices (17, 18).

2. Materials

2.1. Enzymes

1. Recombinant wild-type (19) and M213R, M213G, and T60A/Q144R/K152E variants of RgDAAO were overexpressed in *Escherichia coli* BL21(DE3)pLysS cells using the pT7-HisDAAO expression vector and purified by Hitrap Chelating chromatography because of the addition of an His-tag at the N-terminus, as detailed in refs. 15–17. The enzyme concentration is determined using an extinction coefficient of ~12.6/mM/cm at 455 nm. The kinetic parameters of different purified DAAO variants were determined at pH 8.5 and 25°C. The specific activity on D-Ala as a substrate is: 98 U/mg protein (wild-type RgDAAO), 6.6 U/mg protein (M213R), 19.2 U/mg protein (M213G), and 92 U/mg protein (T60A/Q144R/K152E). One DAAO unit corresponds to the amount of enzyme that converts 1 μmol of D-amino acid per minute at 25°C.
2. The DAAO variants used with the Specialities biosensor were covalently immobilized on an Amberzyme Oxirane support (Rohm and Haas Advanced Biosciences, USA) by a coupling procedure involving protein-free amino groups (17). The activity of the immobilized enzyme was determined by an amperometric assay (oxygen electrode), using a weighted amount of matrix (100 mg/mL) in 100 mM potassium phosphate, pH 8.0.
3. The final RgDAAO preparation used for the microbiosensor was concentrated up to 58 mg/mL in a solution containing 25 mg/mL bovine serum albumin and 1% glycerol in 20 mM phosphate buffer, pH 8.5. The enzyme was immobilized under saturated glutaraldehyde vapors (20).

2.2. Midaspro Amperometric Biosensor (15, 18)

1. Midaspro biosensor (Sartorious A.G., Germany).
2. Calibration solution: 0.2 mg/mL hydroquinone.
3. Working buffer solution: 100 mM disodium pyrophosphate, pH 8.5.
4. Substrate standard solutions: 0.15–3 mM D-Ala or 0.5–20 mM D-Asp in working buffer.
5. Washing solution: 1% sodium hypochlorite.

2.3. Specialities Amperometric Biosensor (17)

1. Electrochemical cells (Specialities s.r.l., Italy).
2. Data acquisition and processing system.
3. Working buffer solution: 100 mM potassium phosphate, pH 7.0.

4. D-Ala standard solutions: 0.25–5 mM in 100 mM potassium phosphate, pH 7.0.

2.4. Microbiosensor (20)

1. The biosensor is constituted of a 90% Pt/10% Ir wire of 25 μm in diameter (Goodfellow, UK) glued to a 0.3-mm copper wire using electroconductive silver paint (Radiospares, Beauvais, France) and inserted into a pulled glass capillary (Harvard Apparatus, UK).
2. Patch-clamp amplifier (Geneclamp GC500, Molecular Devices, USA) or electrochemistry amplifier (VA-10, NPI Electronics, Germany) used with a two-electrode potentiostat.
3. Acquisition board (ITC-18, Instrutech, USA).
4. Reference electrode: chlorided silver wire electrode placed directly in the recording chamber.
5. D-Ser Standard solutions: 0.5–10 μM in 10 mM PBS (phosphate-buffered saline constituted by 8 g/L NaCl, 0.2 g/L KCl, 1.44 g/L Na_2HPO_4, and 0.24 g/L KH_2PO_4), pH 7.4.
6. Artificial extracellular medium: 126 mM NaCl, 1.5 mM KCl, 1.25 mM KH_2PO_4, 1.5 mM $MgSO_4$, 2 mM $CaCl_2$, and 10 mM HEPES, pH 7.4.

3. Methods

3.1. Determination of Neutral and Acidic D-Amino Acid Content (Midaspro Biosensor)

The electrochemical measurements are carried out at room temperature in a 5-mL amperometric cell consisting of a plastic flow cell that contains a working and a reference electrode at an applied fixed voltage of +400 mV vs. Ag/AgCl (15, 18). The enzyme (approximately 8–25 μg of protein) is immobilized by simple absorption on the surface of the working electrode (a graphite disk 20 mm in diameter and 2 mm thick, Fig. 1).

The assays employ two RgDAAO variants displaying a different substrate specificity: (a) the wild-type enzyme, which is active on neutral and polar D-amino acids but inactive on the acidic ones (the catalytic efficiency, expressed as the k_{cat}/K_m ratio, is ~4,000-fold higher on D-Ala than on D-Asp as a substrate) (13, 14) and (b) the M213R variant, which is active on D-Ala as well as on D-Asp and D-Glu (15).

The detection procedure is performed as follows:

1. Calibrate the electrochemical cell that was previously filled with 5 mL of the working buffer solution: when the anodic signal is stable, inject 20 μL of the calibration solution and measure the corresponding current intensity (calibration peak).
2. Wash the calibrated electrode with 5 mL of the washing solution and 20 mL distilled water.

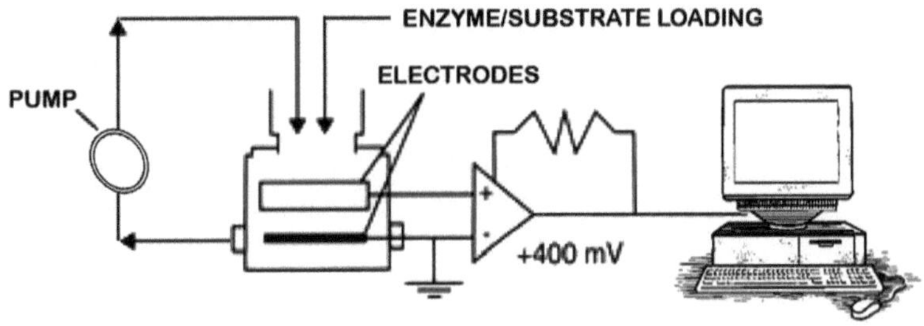

Fig. 1. Scheme of the amperometric biosensor Midaspro (15, 18). By using the peristaltic pump, the substrate/sample solutions can be efficiently mixed through the electrochemical cell. The signal is recorded at a fixed potential of +400 mV.

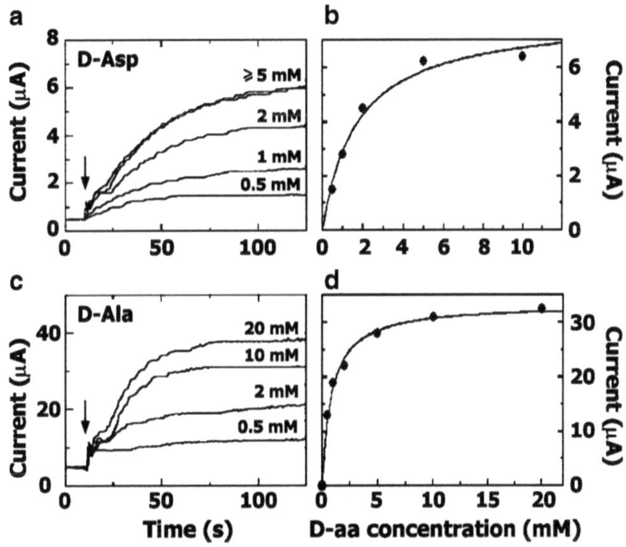

Fig. 2. Amperometric traces obtained with the Midaspro biosensor and M213R RgDAAO as biocatalyst and using standard solutions at different concentrations of D-Ala (**a**) and D-Asp (**c**). The corresponding calibration curves are depicted in the right panels (**b**, **d**).

3. Use standard solutions of substrate (D-Ala and D-Asp for wild-type and M213R RgDAAO, respectively) to generate a calibration curve: 5 mL of 0.5–20 mM D-Ala or D-Asp solutions are added in the electrochemical chamber.
4. When the recorded current is stable, inject 1.5 U of wild-type or 0.2 U of the M213R variant RgDAAOs.
5. Plot the current values detected after 3 min of recording with the different standard solutions (Fig. 2a, c) as a function of substrate concentration and fit the data points by a classic Michaelis–Menten equation (Fig. 2b, d). See Note 1.
6. At the end of the recording time, rinse the cell with 5 mL of washing solution (see Note 2).
7. For the analytical measurement, fill the electrochemical cell with 4 mL of working buffer solution and 1 mL of a biological

sample containing an unknown concentration of D-amino acid. When the recorded current is stable, inject 1.5 U of wild-type or 0.2 U of M213R variant RgDAAOs.

8. The amount of neutral or acidic D-amino acids is measured from the current intensity by using the calibration curves (Fig. 2b, d). See Notes 3 and 4.

3.2. Determination of the Total D-Amino Acid Content (Specialities Amperometric Biosensor)

The assay is performed by using several electrochemical cells, produced by a screen-printing process and which contain both a working electrode (a graphite disk, 28 mm in diameter) and a reference electrode (Ag/AgCl) (17). This device represents an improved version of the previous biosensor and relies on very close contact between the immobilized biocatalyst and the electrode surface. The response is calculated by the difference between signals from the working and the reference electrodes, the latter indicating the presence of the enzyme (see Notes 5 and 6). The engineered M213G and T60A/Q144R/K152E RgDAAO variants represent the best catalysts available (better if used simultaneously) to determine the total D-amino acid content in biological samples because they respond to all (neutral, acidic, and basic) D-amino acids (17). The T60A/Q144R/K152E RgDAAO variant exhibits the best correlation between theoretical activity values on various D-amino acid mixtures and experimental results (see Notes 7 and 8). Moreover, variability in response as a function of the D-amino acid composition is low for RgDAAO variants in the immobilized form (17), see Note 6. Due to the additive nature of the individual amino acid responses, this device can be used to analyze solutions containing both single species and complex D-amino acid mixtures.

The assay is performed as follows.

1. Fill both the electrochemical chambers with 3 mL of a 1-mM D-amino acid solution in working buffer solution.
2. Wait for a few minutes until the current trace is stable and add 0.5–4.0 U of M213G and/or T60A/Q144R/K152E RgDAAO variants in the working cell, but not in the reference one.
3. Record the amperometric response until it reaches a plateau value (~10 min): the intensity of the recorded current is directly related to the D-amino acid concentration.
4. Following each measurement, the screen-printed electrodes are discarded and replaced with new ones.
5. Use standard solutions of D-Ala (in the 0.25–5 mM concentration range) to generate a calibration curve.

Please note that this biosensor has been used to detect D-amino acids in complex samples such as cheese specimens (17). In this case, the procedure is as follows:

1. Suspend finely grated Grana Padano cheese in working buffer solution at a 0.1 mg/mL final concentration.

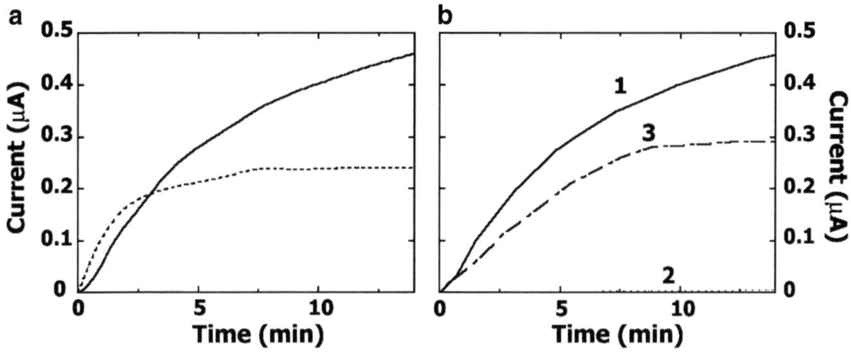

Fig. 3. Determination of D-amino acid content in cheese samples using the Specialities Amperometric Biosensor (18). (**a**) Response of the biosensor while measuring total D-amino acids content in Grana Padano cheese samples (*line*, 2 mL; *dotted line*, 1 mL). (**b**) The 2 mL cheese sample (*line*, 1) was treated for 20 min with 1.5 U of RgDAAO before measurement (*dotted line*, 2). Then, 1 mM D-Ala (*dashed line*, 3) was added to this sample, and after 10 min, it gave the same response as observed with 1 mM D-Ala alone (not shown). This confirms that the current response is effectively due to RgDAAO activity.

2. Incubate the suspension for 30 min at room temperature in a sonicator bath and centrifuge the sample at 3,200 × *g* for 30 min; recover the aqueous phase.
3. Centrifuge at 27,000 × *g* for 1 h; recover the resulting surnatant for the biosensor analyses. The best measurements are performed by adding an enzyme solution containing similar amounts (in terms of enzymatic units) of the M213G and T60A/Q144R/K152E RgDAAOs into the working electrochemical cell; when the difference in signal between the working and the reference electrode reaches a stable value, record the current value.
4. Calculate the total D-amino acid content in the cheese sample from the recorded amperometric current by using the calibration curve obtained using the D-Ala standard solutions and the enzyme mixture made up of M213G and T60A/Q144R/K152E RgDAAO variants (an example of the current traces obtained from a Grana Padano sample is depicted in Fig. 3), see Note 9.

3.3. Determination of D-Serine Content in the Central Nervous System by a DAAO-Based Microbiosensor

D-Serine is an endogenous ligand for *N*-methyl-D-aspartate (NMDA) receptors, and alterations in its concentration have been related to several brain disorders, especially schizophrenia (21). It represents therefore an important target neuromodulator for the pharmaceutical industry. To monitor D-serine levels *in vivo*, a specific microbiosensor has been developed (20). Here, a platinum microelectrode (25 μm × 150 μm) is covered with a membrane of poly-*m*-phenylenediamine (PPD); on the top of the membrane a

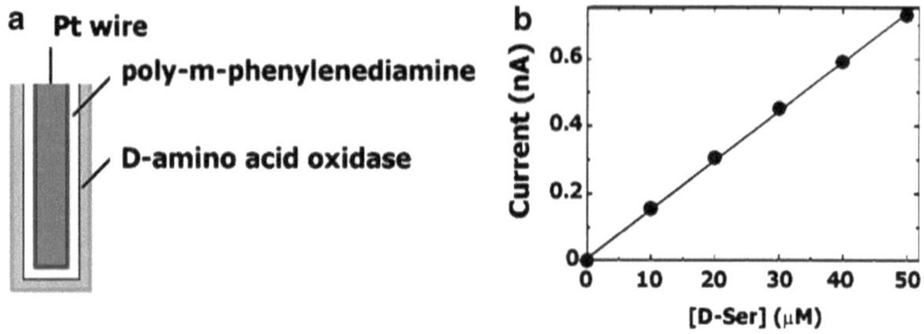

Fig. 4. Design of the RgDAAO microbiosensor for D-Ser detection *in vivo* (20). (**a**) Schematic representation of the microbiosensor: a platinum wire is covered with a layer of poly-*m*-phenylenediamine (PPD) and with the wild-type RgDAAO biocatalyst. (**b**) Calibration curve of a microbiosensor in the 0–50 μM D-Ser concentration range: the oxidation current shows a linear dependence on the D-Ser concentration.

layer of recombinant purified wild-type RgDAAO is deposited by dipping the Pt tip of the electrodes in the enzyme solution (Fig. 4a). The PPD membrane forms a steric barrier that allows H_2O_2 diffusion but blocks larger interfering molecules. The RgDAAO microbiosensors manufactured from a PPD-covered Pt wire, in fact, showed a 97–99% reduction in the interfering responses produced by endogenous oxidating molecules (20); by contrast D-Ser detection is only reduced by 12%. The RgDAAO microbiosensors and control biosensors used for *in vivo* experiments were covered with an additional Naflon membrane to protect the enzyme layer during penetration in the brain. This additional membrane did not change the sensitivity or the response time of the electrodes (see Note 10). D-Ser is detected by a two-step process: (1) RgDAAO catalyzes the oxidative deamination of the D-amino acids present in the proximity of the electrode (i.e., D-Ser is converted into hydroxypyruvate), generating equimolar amounts of H_2O_2 and (2) H_2O_2 is oxidized at the surface of a platinum wire connected to a patch-clamp amplifier. The resulting H_2O_2 oxidation current corresponds to the D-Ser concentration in the biosensor's microenvironment (Fig. 4b).

The *in vivo* experiments are performed as follows.

1. First calibrate the RgDAAO microbiosensor with D-Ser standard solution. The electrodes respond to changes in D-Ser concentration with an increase in the recorded oxidation current (see Note 11). The response time, defined as the duration of the signal increase by between 10 and 90%, is ~2 s.
2. Subsequently calibrate the microbiosensor in an artificial extracellular solution that closely resembles the ionic composition of rat cerebrospinal fluid. In this medium, the sensitivity of the electrodes are unchanged compared to PBS.
3. The RgDAAO microbiosensor is then implanted in the frontal cortex of anesthetized rats (81 mm lateral from the midline,

3 mm anterior from the Bregma, and 1.5 mm ventral from the Dura), side by side (0.5 mm) with a control biosensor (covered with BSA), see Note 12.

4. Wait for at least 30 min to allow currents to stabilize, then start recording them for ≥1 h. A difference in the stabilized background currents between the control and the RgDAAO microbiosensors should be observed (i.e., ~47 pA). This difference should be minimal in calibration tests made in PBS or in the artificial extracellular medium (i.e. 2–3 pA).
5. Provided that O_2 is not limiting, the relationship between the amplitude of the step in oxidation current detected by the microbiosensor and D-Ser concentration can be approximated by a Michaelis–Menten equation:

$$C = \frac{C_{\mathrm{max}}[D\text{-Ser}]}{K_{\mathrm{m,app}} + [D\text{-Ser}]}, \quad (1)$$

where C is the oxidation current and C_{max} and $K_{\mathrm{m,app}}$ are analogous to the classical V_{max} and K_{m} kinetic parameters as used for the free enzyme.

6. Calculate the D-Ser concentration using Eq. 1 and the calibration curve obtained with a different D-Ser standard solution (Fig. 4b). See Notes 13 and 14.

4. Notes

Midaspro Biosensor

1. With this calibration procedure, the amperometric signals from different electrochemical cells as well as those from different measurement sessions can be compared.
2. The main drawback of this system is represented by the operational stability. A good response reproducibility is obtained by washing the electrode with sodium hypochloride after each measurement. This entails adding a new aliquot of enzyme for each determination: however, only a very limited amount of RgDAAO is used for each single measurement.
3. The system was also used to estimate the overall content of neutral-basic D-amino acids in milk samples (maintained for 1–30 days at 4°C). The recovery of 1 mM D-Ala added to the crude milk samples is almost complete and no interference at an applied potential of +400 mV is detected (18). D-Amino acid concentration is below the detection limit in fresh milk, but increases with aging to reach a total concentration of 49 μM after 1 month at 4°C. These data are in good agreement

with the values reported in the literature and determined using chromatographic methods (~45 μM) (22).

4. An intense anodic signal is detected after adding the milk sample in the electrochemical cell; however, the amperometric trace returns to the initial value in ~15 min.

Specialities Amperometric Biosensor

5. The biosensor shows a well-defined and quick response after adding wild-type RgDAAO to a 1-mM D-Ala solution in the working cell. In the absence of the enzyme, no amperometric response is observed after adding up to 10 mM H_2O_2 and/or 1 mM D-Ala or pyruvate, indicating that the current response is not due to H_2O_2 or D-amino acid/α-keto acid detection (17). These results indicate that the electrons produced by the enzymatic reaction are directly transferred to the electrode: this device belongs to the third-generation group of biosensors (23).
6. The decrease in apparent K_m values for the D-amino acid offers an important advantage of using immobilized RgDAAO for biosensor-based measurements because of partitioning effects. This effect is especially evident for D-amino acids that bind weakly to DAAO, e.g., K_m for D-Pro is 21.5 and 5.6 mM and K_m for D-Val is 18.9 and 1.3 mM for free and immobilized RgDAAO, respectively (24).
7. The effect of the substrate composition on the amperometric response was evaluated using various immobilized RgDAAO variants and D-amino acid solutions (0.5 mM final concentration) containing different ratios of D-Ala (0–0.5 mM), D-Glu (0–0.4 mM), D-Lys (0–0.4 mM), D-Gln (0–0.16 mM), and D-Met (0–0.1 mM) (17).
8. The protocol for determining D-amino acids based on RgDAAO variants is simple, rapid (~10–15 min), and reliable. The easy, fast, and inexpensive production procedure (the overall cost is ~0.5 €/electrode) only requires a single screen-printed electrode for each amperometric analysis ("one-shot" disposable device), with no need to regenerate the electrode surface.
9. The RgDAAO-based device was also successfully used to measure the total D-amino acid content in cheese samples. The amount of D-amino acids determined in Grana Padano cheese samples (~6.1 mM corresponding to 6.7 mg D-amino acid per gram of cheese sample) (17) is in fairly good agreement with the value reported in the literature (10, 11). Interestingly, such a final D-amino acid concentration contains 54% D-Glu, 29% D-Asp, and 17% D-Ala, showing that the RgDAAO-based biosensor is a powerful analytical system even on complex matrices.

This electrochemical method compares favorably with standard methods for overall D-amino acid determination.

Microbiosensor

10. All microbiosensors were tested for the detection of serotonin (20 μM in PBS), D-Ser (1 μM in PBS), and H_2O_2 (1 μM in PBS) before use. Only the electrodes showing more than 7 pA/1 μM D-Ser and less than 4 pA/20 μM serotonin were used to determine D-Ser concentrations.
11. *In vitro* calibrations were performed in standard solutions prepared with PBS. The reference electrode was a chlorided silver wire placed directly into the recording chamber. Recordings were made under constant potential amperometry at +500 mV vs. Ag/AgCl (reference electrode).
12. Implanting the device in the CNS produced a small decrease in sensitivity (13.2%). This decrease is common to virtually all microbiosensors and generally occurs during the first 15 min following insertion into the brain.
13. The intraperitoneal injection of 1 g D-Ser/kg body weight induces a steady increase in oxidation current at the RgDAAO microbiosensor, but not at the control sensor, indicating that the electrochemical signal detected by the microbiosensor is specific for D-Ser (20). This increase in electrochemical signal probably arises from diffusion of D-Ser across the blood–brain barrier.
14. For detecting D-Ser *in vivo* increased selectivity, higher sensitivity, and miniaturization are required to allow implantation in order to implant the device in the CNS of small laboratory animals. These technical challenges have been overcome through: (1) the use of RgDAAO, a very active and selective enzyme; (2) the development of micrometric platinum wire electrodes for H_2O_2 monitoring; and (3) the use of a highly selective PPD layer to block the nonspecific oxidation of endogenous molecules.

Acknowledgments

The work reported in this paper was supported by grants from Fondo di Ateneo per la Ricerca (University of Insubria) to S. Sacchi and L. Pollegioni and from Fondazione Cariplo and Regione Lombardia (bando Cooperazioni Scientifiche Internazionali) to L. Pollegioni. We thank Carlo Rossetti, Jean-Pierre Mothet, and Stéphane Marinesco who over the years have actively contributed to the evolution of DAAO-based biosensors.

References

1. Leuchtenberger W, Huthmacher K, Drauz K (2005) Biotechnological production of amino acids and derivatives: current status and prospects. Appl Microbiol Biotechnol 69, 1–8.
2. Friedman M (1999) Chemistry, nutrition, and microbiology of D-amino acids. J Agric Food Chem 47, 3457–3479.
3. Friedman M (2010) Origin, microbiology, nutrition, and pharmacology of D-amino acids. Chem Biodivers 7, 1491–1530.
4. Gobbetti M, Simonetti M S, Rossi J et al. (1994) Free D- and L-amino acid evolution during sourdough fermentation and baking. J Food Sci 59, 881–884.
5. Casal S, Mendes E, Oliveira M B P P, Ferreira M A (2005) Roast effects on coffee amino acid enantiomers. Food Chem 89, 333–340.
6. Gandolfi I, Palla G, Marchelli R et al. (1994) D-alanine in fruit juices: a molecular marker of bacterial activity, heat treatment and shelf-life. J Food Sci 59, 152–154.
7. Pätzold R, Brückner H (2006) Gas chromatographic determination and mechanism of formation of D-amino acids occurring in fermented and roasted cocoa beans, cocoa powder, chocolate and cocoa shell. Amino Acids 31, 63–72.
8. Carlavilla D, Moreno-Arribas M V, Fanali S, Cifuentes A (2006) Chiral MEKC-LIF of amino acids in foods: analysis of vinegars. Electrophoresis 27, 2551–2557.
9. Ali H S, Pätzold R, Brückner H (2010) Gas chromatographic determination of amino acid enantiomers in bottled and aged wines. Amino Acids 38, 951–958.
10. Csapó J, Varga-Visi E, Lóki K, Albert C (2006) The influence of manufacture on the free D-amino acid content of Cheddar cheese. Amino Acids 32, 39–43.
11. Marchelli R, Galaverna G, Dossena A et al. (2006) D-Amino acids in food. In: Konno R, Brückner H, D'Aniello A, Fisher G, Fujii N, Homma H (eds), D-Amino Acids: A new frontier in amino acid and protein research. Nova Science Publishers, pp. 299–315.
12. Warnke M, Armstrong D W (2006) D-amino acid determination in foods, beverages, and biological samples. In: Konno R, Brückner H, D'Aniello A, Fisher G, Fujii N, Homma H (eds), D-Amino Acids: A new frontier in amino acid and protein research. Nova Science Publishers, pp. 317–336.
13. Pollegioni L, Piubelli L, Sacchi S et al. (2007) Physiological functions of D-amino acid oxidase: from yeast to human. Cell Mol Life Sci 64, 1373–1394.
14. Pollegioni L, Molla G, Sacchi S et al. (2008) Properties and application of microbial D-amino acid oxidase: current state and perspectives. Appl Microbiol Biotechnol 78, 1–16.
15. Sacchi S, Lorenzi S, Molla G et al. (2002) Engineering the substrate specificity of D-amino acid oxidase. J Biol Chem 30, 27510–27516.
16. Sacchi S, Rosini E, Molla G et al. (2004) Modulating D-amino acid oxidase substrate specificity: production of an enzyme for analytical determination of all D-amino acids by directed evolution. Protein Eng Des Sel 17, 517–525.
17. Rosini E, Molla G, Rossetti C et al. (2008) A biosensor for all D-amino acids using evolved D-amino acid oxidase. J Biotechnol 135, 377–384.
18. Sacchi S, Pollegioni L, Pilone MS, Rossetti C (1998) Determination of D-amino acids using a D-amino acid oxidase biosensor with spectrometric and potentiometric detection. Biotechnol Tech 12, 149–153.
19. Fantinato S, Pollegioni L, Pilone S M (2001) Engineering, expression and purification of a His-tagged chimeric D-amino acid oxidase from *Rhodotorula gracilis*. Enz Microb Technol 29, 407–412.
20. Pernot P, Mothet J P, Schuvailo O et al. (2008) Characterization of a yeast D-amino acid oxidase microbiosensor for D-serine detection in the central nervous system. Anal Chem 80, 1589–1597.
21. Pollegioni L, Sacchi S (2010) Metabolism of the neuromodulator D-serine. Cell Mol Life Sci 67, 2387–2404.
22. Gandolfi I, Palla G, Delprato L et al. (1992) D-amino acids in milk as related to heat treatments and bacterial activity. J Food Sci 57, 377–379.
23. Chaplin M, Bucke C (1990) Enzyme Technology. Cambridge University Press.
24. Pilone MS, Pollegioni L, Butò S (1992) Stability and kinetic properties of immobilized *Rhodotorula gracilis* D-amino acid oxidase. Biotechnol Appl Biochem 16, 252–262.

Chapter 22

Analysis of D-β-Aspartyl Isomers at Specific Sites in Proteins

Noriko Fujii and Norihiko Fujii

Abstract

Recent studies have shown that biologically uncommon D-β-aspartic acid residues accumulate in specific proteins during the aging process. However, aspartyl residues are not racemized uniformly because D-Asp appears to be confined to particular sites in these proteins. We here describe the method to identify the specific sites of D-β-aspartic acids inversion in proteins.

Key words: D-Amino acid, β-Aspartic acid, Aging, RP-HPLC, Mass spectrometry, Enantiomer, Isomer, Lens crystallin

1. Introduction

With the exception of glycine, all naturally occurring amino acids contain one or more asymmetric tetrahedral carbon atoms. Therefore, the asymmetric amino acid molecules comprise two non-superimposable mirror images, i.e., right-handed (D-enantiomer) and left-handed (L-enantiomer) structures. It is considered that equal amounts of D- and L-amino acids existed on the early earth before the emergence of life. However, during chemical evolution, only L-amino acids were selected for the formation of peptides and proteins from which life emerged. Homochirality is essential for the development and maintenance of life. Once the L-amino acid world was established, D-amino acids were excluded from living systems. Until recently, the homochirality of proteins composed of L-amino acids was believed to be maintained throughout the entire lifespan of an organism. However, D-aspartic acid (D-Asp) has been detected in various proteins from tissues such as teeth (1), bone (2, 3), aorta (4), ligament (5), brain (6–8), lens (9–11), retina (12), conjunctiva (13),

Loredano Pollegioni and Stefano Servi (eds.), *Unnatural Amino Acids: Methods and Protocols*,
Methods in Molecular Biology, vol. 794, DOI 10.1007/978-1-61779-331-8_22, © Springer Science+Business Media, LLC 2012

Table 1
D-β-Asp containing proteins are observed in various tissues of the living body

Tissue	Protein	Amino acid	Disease
Lens	αA-, αB-crystallin	D-Asp	Cataract
Retina	?	D-Asp	AMD
Conjunctiva	?	D-Asp	Pinguecula
Cornea	?	D-Asp	CDK
Skin	Elastin	D-Asp	Elastosis
Brain	β-Amyloid	D-Asp	Alzheimer
Brain	α-Synuclein	D-Asp	Parkinson
Aorta	Elastin	D-Asp	Arteriosclerosis
Teeth	Phosphophoryn	D-Asp	?
Bone	Osteocalcin	D-Asp	?
Ligament	Elastin	D-Asp	?

AMD age-related macular degeneration, *CDK* climatic droplet keratopathy, *?* unknown.

cornea (14), and skin (15) of elderly individuals (Table 1). The proteins in such tissues are metabolically inert. Hence, the presence of D-Asp in aged tissues of living organisms is thought to result from the racemization of aspartyl residues in these particular polypeptides. Of all the naturally occurring amino acids, aspartic acid is the most susceptible to racemization. The aspartyl residues are racemized non-uniformly presumably because of structural considerations, which make specific residues more susceptible to reaction than others. Previous studies identified the specific sites of D-Asp in αA- and αB-crystallins from lens Table 2 (10, 11), the β-amyloid protein in brain (8), histone of canine brain (16) and type I collagen telopeptide in urine (3).

Therefore, it is necessary to determine the nature of the aspartyl residues at specific sites within particular proteins. Here, we describe a convenient and robust biochemical method of identifying such sites in specific proteins. As an example, we analyzed lens crystallin. Previous study clearly showed that alpha A-crystallin containing large amounts of D-β-Asp may undergo abnormalaggregation to form massive and heterogeneous aggregates, leading to loss of its chaperone activity (Table 3).

Table 2
D-β-Aspartic acid residues in αA- and αB-crystallins from lenses of 80-year-old donors

Peptide	Site	D/L	Linkage
αA-T6	Asp-58	3.10	β
αA-T18	Asp-151	5.70	β
αB-T3	Asp-36	0.92	β
αB-T4	Asp-62	0.57	β

The D/L ratios of Asp-58 and -151 >1.0
D-Asp formation accompanied with isomerization of normal α-linkage to abnormal β-linkage

Table 3
Sedimentation coefficient and chaperon activity of α-crystallin from elderly and young donors

	1 Year	80 Year
$S_{20,W}$	17S	30–80S
Chaperon Act.	100%	40%

$S_{20,W}$ sedimentation coefficient (20°C in water)

2. Materials

All solutions are prepared with ultrapure water (e.g., Milli-Q grade water). Unless otherwise stated, reagents are special grade chemicals.

2.1. Separation of Peptides

1. Trypsin (Sequence grade, Promega, USA).
2. Acetonitrile (HPLC grade).
3. Trifluoroacetic acid (TFA) (HPLC grade).
4. Mobile phase: solution A: 0.1% TFA in water; solution B: 0.1% TFA in acetonitrile.
5. C18 column (e.g., TSK gel-ODS-80 TM, 4.6×250 mm, Tosoh, Japan).

2.2. Mass Spectrometry

1. α-Cyano-4-hydroxy-cinnamic acid (CHCA) (e.g., SIGMA, France).
2. Acetone (HPLC grade).
3. Neurotensin (e.g., SIGMA, France).

2.3. Determination of D/L Ratio of Amino Acids

1. 6 M HCl (Sequential grade, Thermo, USA).
2. 0.1 M Borate buffer is prepared by the dilution of 1 M potassium borate buffer, pH 10.4 (e.g., PIERCE, USA).
3. Acetonitrile and tetrahydrofuran are used for HPLC grade.
4. Mobile phase: solution A: 5% acetonitrile and 3% tetrahydrofuran in 0.1 M sodium acetate buffer, pH 6.0; solution B: 47% acetonitrile, 3% tetrahydrofuran in 0.1 M sodium acetate buffer, pH 6.0.
5. C18 column (Nova-Pak ODS, 3.9 × 300 mm, Waters, Japan).

2.4. Synthesis of Peptides Containing Asp Isomers

The peptides were synthesized by Fmoc (9-fluorenylmethoxycarbonyl) solid phase chemistry.

1. L-α-Asp: Fmoc-L-Asp (OtBu)-OH (e.g., Watanabe Chemical, Japan).
2. D-α-Asp: Fmoc-D-Asp (OtBu)-OH (e.g., Watanabe Chemical, Japan).
3. L-β-Asp: Fmoc-L-Asp-OtBu (e.g., Watanabe Chemical, Japan).
4. D-β-Asp: Fmoc-D-Asp-OtBu (e.g., Watanabe Chemical, Japan).

3. Methods

3.1. Purification of the Target Protein(s)

1. Extract the protein of interest from the tissue material and then purify by using various column chromatographic methods such as size-exclusion, ion-exchange, and reversed-phase high performance chromatography (RP-HPLC).
2. To facilitate analysis, several hundred micrograms of the purified protein must be prepared.
3. Dissolve an aliquot of the purified protein in a sample buffer 2% sodium dodecyl sulfate (SDS), 5% 2-mercaptoethanol, 10% glycerol, 0.05 M Tris (hydroxyl methyl) aminomethane–HCl (Tris–HCl), pH 6.8 and analyze by SDS–polyacrylamide gel electrophoresis (SDS–PAGE) according to the method of Laemmli (17).
4. Stain the gel with Coomassie brilliant blue and verify the homogeneity of the sample.

3.2. Digestion of the Purified Protein

Digest the protein obtained in Subheading 3.1 with trypsin using an enzyme-to-substrate ratio of 1:50 (mol/mol) in 0.1 M Tris–HCl buffer, pH 7.6 for 20 h at 37°C.

3.3. Separation of the Tryptic Peptides

1. Separate the resulting peptides obtained in Subheading 3.2 by RP-HPLC (JASCO, Japan) using a C18 column (TSK gel-ODS-80 TM, 4.6 × 250 mm, Tosoh, Japan) with a linear gradient

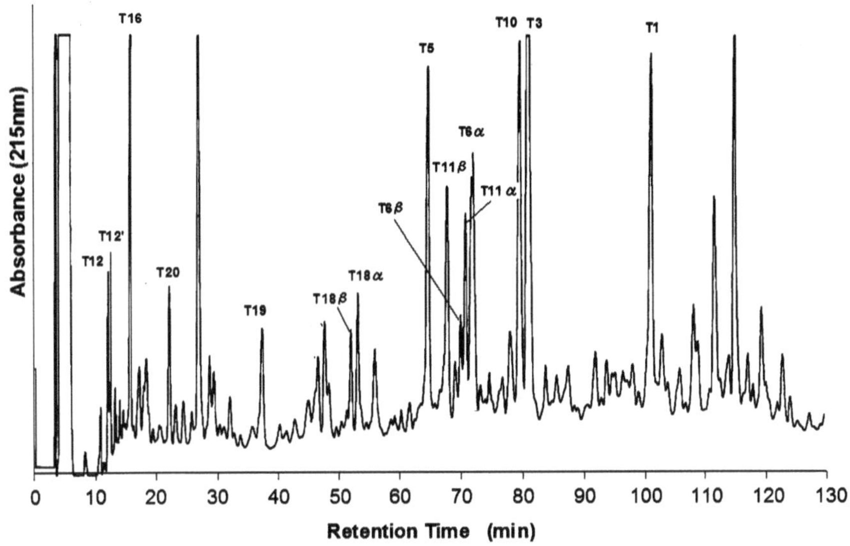

Fig. 1. Elution profiles of tryptic peptides of aged human αA-crystallin. Elution: solvent A: 0.1% trifluoroacetic acid/water; solvent B: 0.1% trifluoroacetic acid/acetonitrile. Column: TSK gel-ODS-80 TM, 4.6 × 250 mm, Tosoh, Japan. Detection: absorbance at 215 nm. Gradient: 0–40% B in 120 min. Flow rate: 0.8 mL/min. Peptides were detected by measuring their absorbance at 215 nm. αA-Crystallin-derived peaks were identified on the basis of amino acid sequence analysis, and TOF-MS analysis.

of 0–40% acetonitrile in the presence of 0.1% trifluoroacetic acid at a flow rate of 0.8 mL/min.

2. Monitor the eluate at 215 nm (Fig. 1).
3. Collect the eluate using a fraction collector (FRC10 Shimadzu, Japan).

3.4. Identification of the Tryptic Peptides by Sequence Analysis

1. Determine the amino acid sequences of the tryptic peptides by the Edman degradation method on a pulsed-liquid protein sequencer equipped with an on-line phenylthiohydantoin (PTH) amino acid analyzer (Applied Biosystems 477A, USA).
2. Apply the peptide sample (50 pmol) to be sequenced onto a solid surface. One common substrate is a glass fiber coated with polybrene as a cationic polymer.
3. Add the Edman reagent, phenylisothiocyanate (PTC) to the adsorbed peptide, together with a mildly basic buffer solution of 12% trimethylamine. This reacts with the amine group of the N-terminal amino acid.
4. Selectively hydrolyze the terminal amino acid with anhydrous acid and identify using chromatography by comparison with standards.
5. Repeat the cycle to determine more sequence.

3.5. Identification of the Tryptic Peptides by Mass Spectrometry

1. The other direct method by which the sequence of a protein can be determined is mass spectrometry.
2. Analyze the samples via seamless post source decay (PSD) on a curved field reflectron (CFR) by MALDI-TOFMS (AXIMA-TOF^2; Shimadzu, Japan), in positive ion mode.
3. Apply 1 μL of 20 mg/mL CHCA in acetone onto a stainless MALDI plate and allow to air dry before adding 1 μL of the peptide solution.
4. Carry out the mass analysis once the sample is dry.
5. Set the instrument to operate at an acceleration voltage of 20 keV in the *m/z* range from 10 to 5,000.
6. To improve mass resolution set the post-source pulsed extraction functioning to *m/z* 2,500.
7. Calibrate the TOF analyzer using the following external markers: a dimer of CHCA ($[2M+H]^+$; 379.09), neurotensin ($[M+H]^+$; 1,672.92).
8. Set the resolution for isolation of precursors to 150 for the PSD.
9. Perform a series of PSD experiments using a nitrogen laser set to an identical power setting.
10. Acquire and analyze the MS data with Kompact software (Kratos/Shimadzu-Biotech).
11. Carefully select each set of peak processing parameters, including peak smoothing and baseline subtraction, to give an adequate PSD spectrum for quantitative analysis.
12. Treat all PSD spectra with the identical set of parameters.

The procedures outlined in Subheadings 3.4 and 3.5 allow identification of the peptides.

3.6. Determination of the D/L Ratio of Amino Acids

1. Bake all glassware used in this procedure at 500°C for 3 h (see Note 1).
2. Add about 50 μL of peptide solution obtained from Subheading 3.3 to a glass surface and allow it to evaporate.
3. Hydrolyze the sample with gas-phase 6 M HCl at 108°C for 7 h (see Note 2) using a PicoTag Work Station (Waters, Japan).
4. Evaporate the hydrolysates under a reduced pressure environment and then dissolve in 50 μL of water.
5. Mix about 25 μL of the hydrolyzed sample solution with 70 μL of 0.1 M borate buffer, pH 10.4.
6. Add 5 μL of 5 mg/mL of *o*-phthalaldehyde (OPA) in MeOH solution and 5 μL of 11 mg/mL of *N-tert*-butyloxycarbonyl-L-cysteine (Boc-L-cys) in MeOH solution and incubate briefly to form diastereoisomers.
7. Apply about 40 μL of solution on the RP-HPLC (LC10A, Shimadzu, Japan) fitted with a C18 column (Nova-Pak ODS,

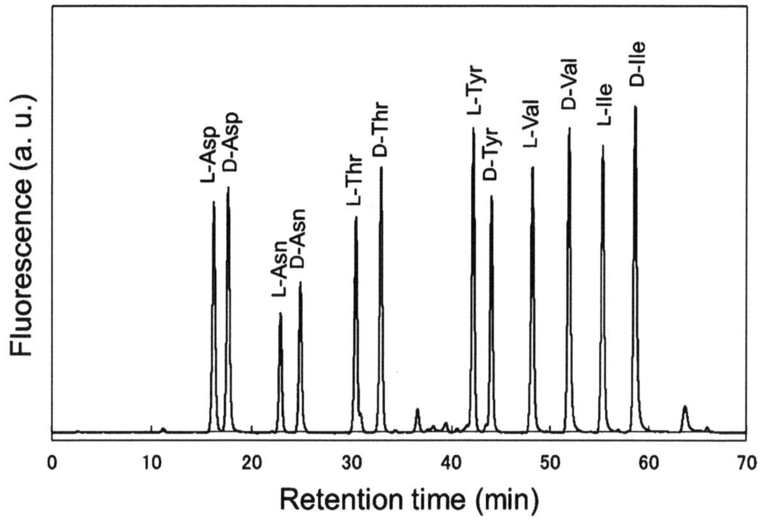

Fig. 2. RP-HPLC for enantiomeric separation of Boc-L-Cys/OPA amino acid derivatives. Elution buffer: solvent A: 5% acetonitrile, 3% tetrahydrofuran/0.1 M acetate buffer, pH 6.0 and solvent B: 47% acetonitrile, 3% tetrahydrofuran/0.1 M acetate buffer, pH 6.0. Column: Nova-pak ODS, 3.9 × 300 mm, Waters. Detection: Ex = 344 nm, Em = 433 nm. Gradient: solvent A–B in 120 min. Flow rate: 0.8 mL/min. Temperature: 30°C.

3.9 × 300 mm, Waters, Japan), using fluorescence detection (344 nm excitation wavelength and 433 nm emission wavelength).

8. Elute the bound material with a linear gradient of 5–47% acetonitrile plus 3% tetrahydrofuran in 0.1 M sodium acetate buffer, pH 6.0 in 120 min at a flow rate of 0.8 mL/min at 30°C (see Note 3).

9. Determine the D/L ratio of amino acids from the ratio of peak area in the chromatogram (Fig. 2).

3.7. Synthesis of Peptides Containing Asp Isomers (see Note 4)

1. The coupling reaction was carried out using single Fmoc amino acid (5 equiv), benzotriazole-1-yl-oxy-Tris–pyrrolidino-phosphonium hexafluorophosphate (PyBOP, 5 equiv), 1-hydroxybenzotriazole (HOBt) (5 equiv), and *N*-methylmorpholine (7.5 equiv) in dimethylformamide (DMF). The N-terminus Fmoc group was deblocked with 20% piperidine in DMF.

2. The cleavage of the peptide from resin and protective groups was achieved using a solution containing 90% trifluoroacetic acid (TFA), 5% 1,2-ethanedithiol and 5% thioanisole for 2 h.

3. The cleavage of the peptide containing Arg from resin and protective groups was carried out with 82.5% TFA, 5% water, 5% thioanisole, 3% ethylmethylsulfide, 2.5% 1, 2-ethanedithiol and 2% thiophenol for 6 h.

4. The cleavage of the peptide containing Trp from resin and protective groups was carried out with above reagent plus 2-methylindole (final concentration 1 mg/mL).

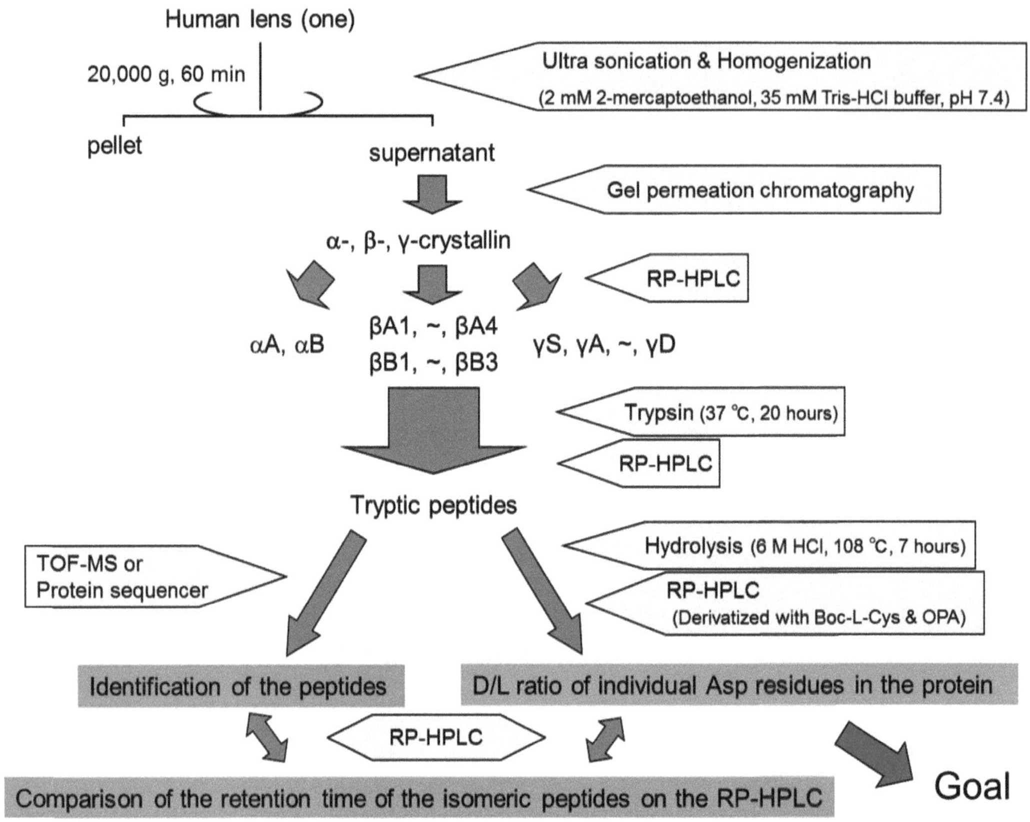

Scheme 1. The flow chart illustrates how specific sites of D-β-aspartyl residues in lens crystallins were identified.

5. The peptides were analyzed by HPLC to confirm their purity (in all cases higher than 95%).

3.8. Determination of β-Aspartyl Residues in the Peptide

The racemization of aspartyl residue accompanies isomerization to form β-aspartyl residues in proteins (see Note 5). The detection of β-aspartyl residue in peptides is possible by protein sequencing and mass spectrometry (see Note 6).

Subheadings 3.1–3.8 are summarized in Scheme 1. Lens crystallin is used as a model protein for analysis.

4. Notes

1. In order to avoid amino acid contamination, all glassware is baked at 500°C for 3 h.
2. Usually the protein is hydrolyzed for 20–24 h, but the racemization of amino acids proceeds under this condition during hydrolysis. Therefore, we examined the time to minimize the racemization of amino acids during hydrolysis. The most suitable time for hydrolysis was found to be 7 h. Using this protocol,

the yield of aspartic acid was more than 95% while racemization was reduced to less than 2%.

3. This method enables accurate analysis of amino acid enantiomers at the picomole level. We determined the D/L ratios of the individual Asp residues in αA- and αB-crystallin from only one lens.
4. β-Asp and α-Asp containing peptides are eluted at different times on the RP-HPLC. A β-Asp containing peptide tends to elute earlier than an α-Asp containing peptide on the RP-HPLC. Hence, the authentic peptides containing L-α-Asp, D-αAsp, L-β-Asp, and D-β-Asp were synthesized by Fmoc (9-fluorenyl-methoxycarbonyl) solid phase chemistry. The retention time for each isomeric peptide on the RP-HPLC was compared with that of the tryptic peptides from the samples, and then Asp isomer in the tryptic peptides was determined.
5. As shown in Fig. 3, the simultaneous formation of β- and D-Asp residues in the protein could be explained as follows: (a) when the carbonyl group of the side chain of the L-α-aspartyl residue is attacked by the nitrogen of the amino acid residue following the Asp residue, L-succinimide is formed by intramolecular

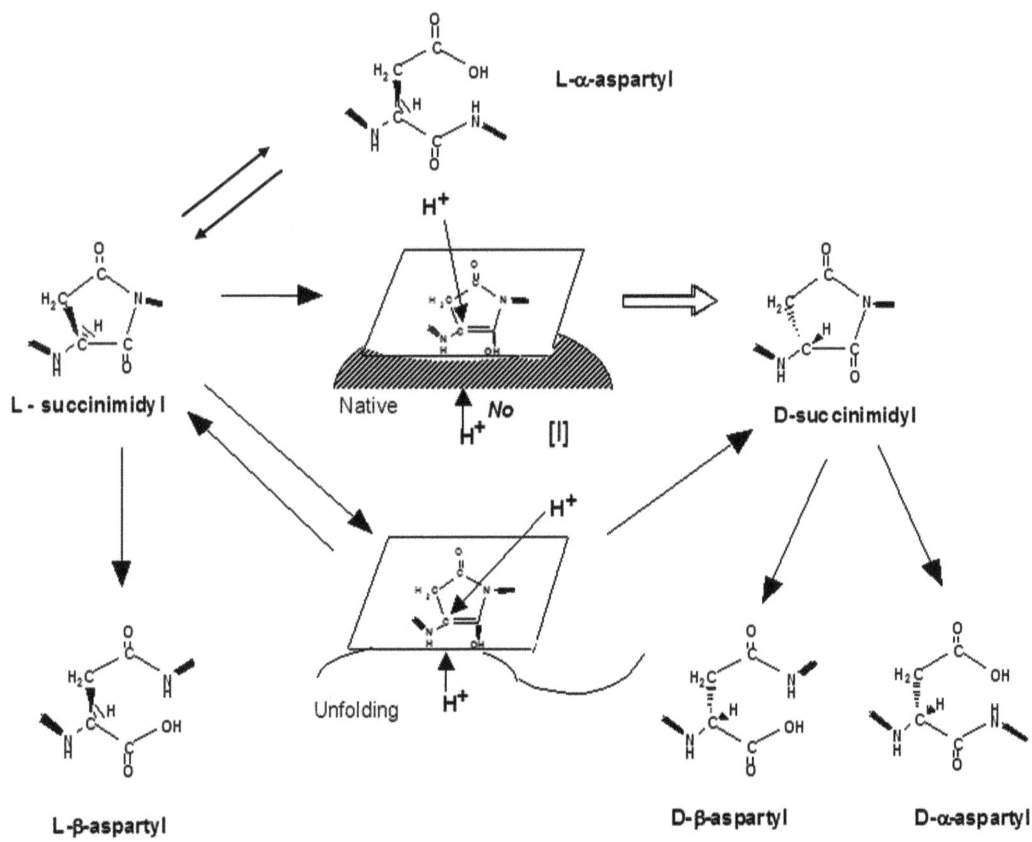

Fig. 3. Reaction pathway for spontaneous inversion and isomerization of aspartyl residues in proteins.

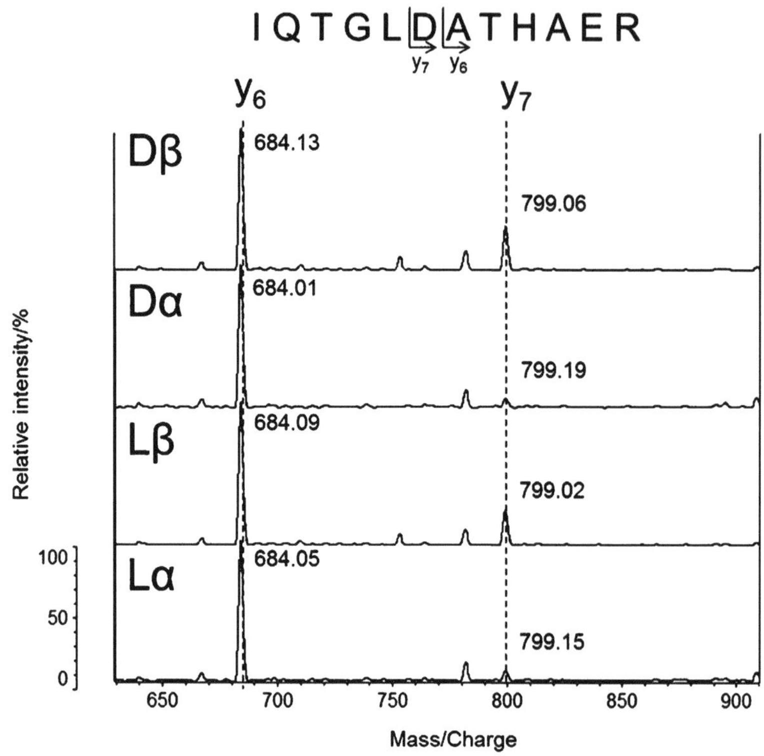

Fig. 4. Post source decay (PSD) spectra of T6 peptides obtained from αA-crystallin on a curved filed reflectron MALDI-TOF MS (AXIMA TOF[2], Shimadzu, Japan).

cyclization; (b) L-succinimide may be converted to D-succinimide through an intermediate [I] that has the prochiral α-carbon in the plane of the ring; (c) protonation of the intermediate [I] would occur from the upper or lower side of the plane in an ordinary peptide or protein; and (d) D- and L-succinimide are hydrolyzed at either side of their two carbonyl groups, yielding both β- and α-Asp residues, respectively. The rate of succinimide formation is expected to depend on the neighboring residue of the Asp residue. When the neighboring amino acid of the Asp residue has a small side chain, such as glycine, alanine, or serine, the formation of succinimide and racemization occur easily because there is no steric hindrance.

6. Subsequently, these peptides are applied to the protein sequencer. Because the β-Asp containing peptide is resistant to Edman degradation, the site of β-Asp in the peptide can be determined. However, it takes much more time to analyze.

   Recently, we found that MS/MS analysis using post source decay (PSD) with a curved field reflectron could distinguish between the β-Asp and α-Asp containing peptides. The relative content of β Asp in a peptide was successfully estimated from a unique ratio, y_n:y_{n+1}, derived from tryptic peptides of a protein (18) (Fig. 4).

Acknowledgment

This work was supported by a grant from the Ministry of Education, Culture, Sports, Science and Technology of Japan.

References

1. Helfman P M, Bada J L (1975) Aspartic acid racemization in tooth enamel from living humans. Proc Natl Acad Sci USA 72, 2891–2894.
2. Ohtani S et al. (1998) Changes in the amount of D-aspartic acid in the human femur with age. Growth Dev Aging 62, 141–148.
3. Cloos P A, Fledelius C (2000) Collagen fragments in urine derived from bone resorption are highly racemized and isomerized: a biological clock of protein aging with clinical potential. Biochem J 345, 473–480.
4. Powell J T, Vine N, Crossman M (1992) On the accumulation of D-aspartate in elastin and other proteins of the ageing aorta. Atherosclerosis 97, 201–208.
5. Ritz-Timme S, Laumeier I, Collins M (2003) Age estimation based on aspartic acid racemization in elastin from the yellow ligaments. Int J Legal Med 117, 96–101.
6. Fisher G H et al. (1992) Quantification of D-aspartate in normal and Alzheimer brains. Neurosci Lett 143, 215–218.
7. Shapira R, Chou C H (1987) Differential racemization of aspartate and serine in human myelin basic protein. Biochem Biophys Res Commun 146, 1342–1349.
8. Roher A E et al. (1993) Structural alterations in the peptide backbone of beta-amyloid core protein may account for its deposition and stability in Alzheimer's disease. J Biol Chem 268, 3072–3083.
9. Masters P M, Bada J L, Zigler J S Jr (1977) Aspartic acid racemisation in the human lens during ageing and in cataract formation. Nature 268, 71–73.
10. Fujii N et al. (1994) Simultaneous stereoinversion and isomerization at specific aspartic acid residues in alpha A-crystallin from aged human lens. J. Biochem 116, 663–669.
11. Fujii N et al. (1994) Simultaneous racemization and isomerization at specific aspartic acid residues in alpha B-crystallin from the aged human lens. Biochim Biophys Acta 1204, 157–163.
12. Kaji Y et al. (2007) Localization of D-beta-aspartic acid-containing proteins in human eyes. Invest Ophthalmol Vis Sci 48, 3923–3927.
13. Kaji Y et al. (2009) Immunohistochemical localisation of D-beta-aspartic acid in pingueculae. Br J Ophthalmol 93, 974–976.
14. Kaji Y et al. (2009) Immunohistochemical localisation of D-beta-aspartic acid-containing proteins in climatic droplet keratopathy. Br J Ophthalmol 93, 977–979.
15. Fujii N et al. (2002) The presence of D-beta-aspartic acid-containing peptides in elastic fibers of sun-damaged skin: a potent marker for ultraviolet-induced skin aging. Biochem Biophys Res Commun 294, 1047–1051.
16. Young G W et al. (2005) Protein L-isoaspartyl methyltransferase catalyzes *in vivo* racemization of Aspartate-25 in mammalian histone H2B. J Biol Chem 280, 26094–26098.
17. Laemmli U K (1970) Cleavage of structural proteins during the assembly of the head of bacteriophage T4. Nature 227, 680–685.
18. Yamazaki Y et al. (2010) Differentiation and semiquantitative analysis of an isoaspartic acid in human alpha-Crystallin by postsource decay in a curved field reflectron. Anal Chem 82, 6384–6394.

Chapter 23

Nutritional Value of D-Amino Acids, D-Peptides, and Amino Acid Derivatives in Mice

Mendel Friedman and Carol E. Levin

Abstract

This paper describes a method for determining the nutritional value of D-amino acids, D-peptides, and amino acid derivatives using a growth assay in mice fed a synthetic all-amino acid diet. A large number of experiments were carried out in which a molar equivalent of the test compound replaced a nutritionally essential amino acid such as L-lysine (L-Lys), L-methionine (L-Met), L-phenylalanine (L-Phe), and L-tryptophan (L-Trp) as well as the semi-essential amino acids L-cysteine (L-Cys) and L-tyrosine (L-Tyr). The results show wide-ranging variations in the biological utilization of test substances. The method is generally applicable to the determination of the biological utilization and safety of any amino acid derivative as a potential nutritional source of the corresponding L-amino acid. Because the organism is forced to use the D-amino acid or amino acid derivative as the sole source of the essential or semi-essential amino acid being replaced, and because a free amino acid diet allows better control of composition, the use of all-amino acid for such determinations may be preferable to protein-based diets.

Key words: D-Amino acids, D-Peptides, Amino acid derivatives, Nutritional evaluation, Bioavailability, Toxicity

1. Introduction

Most amino acids of importance in nutrition exist as L-isomers. During food processing, the L-amino acids may be racemized to their mirror image configuration, the D-isomers. D-Amino acids can also be synthesized by microorganisms (1–3). Racemization of L-amino acid residues to their D-isomers in food and other proteins is pH-, time-, and temperature-dependent. Although racemization rates of the 18 different L-amino acid residues in a protein vary, the relative rates in different proteins are similar. Racemization of amino acids and formation of D-peptide bonds and the crosslinked amino acid, lysinoalanine (LAL) can impair digestibility and nutritional quality (2, 4).

Loredano Pollegioni and Stefano Servi (eds.), *Unnatural Amino Acids: Methods and Protocols*,
Methods in Molecular Biology, vol. 794, DOI 10.1007/978-1-61779-331-8_23, © Springer Science+Business Media, LLC 2012

Two pathways are available for the biological utilization of D-amino acids: (a) racemases or epimerases may convert D-amino acids directly to L-isomers or to (D,L) mixtures; or (b) D-amino acid oxidases may catalyze oxidative deamination of the α-amino group to form α-keto acids, which can then be specifically reaminated to the L-form (5). Although both pathways may operate in microorganisms, only the second has been demonstrated in mammals. The amounts and specificities of D-amino acid oxidases vary in different animal species. In some, the oxidase system may be rate limiting in the utilization of a D-amino acid as a source of the L-isomer. Consequently, the kinetics of transamination of D-enantiomers would be too slow to support optimal growth. In addition, growth depression could result from nutritionally antagonistic or toxic manifestations of D-enantiomers.

The nutritional effectiveness of protein-bound essential D-amino acids depends on the amino acid composition, digestibility, and physiological utilization of released amino acids. Since an amino acid must be liberated by digestion before nutritional assimilation can occur, the decreased susceptibility to proteolytic enzymes of D–D, D–L, and L–D peptide bonds in D-amino acid-containing proteins to digestion is a major factor adversely affecting the bioavailability of protein-bound D-amino acids; reviewed in refs. 6–9. Since this protocol only tests free amino acid diets, it does not address the issue of impairment of digestibility caused by protein-bound D-amino acids.

As part of a program to evaluate the chemistry and the nutritional and toxicological potential of novel amino acids, including D-amino acids, crosslinked amino acids, and amino acid derivatives formed during food processing, we compared the weight gain in mice fed free amino acid diets in which the test compound was substituted for the relevant L-isomer. The results obtained reflect the ability of mice to utilize unnatural amino acids in the complete absence of the L-form. In the case of essential (L-His, L-Lys, L-Met, L-Phe, L-Trp, L-Val) or semi-essential (L-Cys, L-Tyr) amino acids (10), the mice must meet the entire metabolic demand from the D-isomeric forms or derivatives.

2. Materials

2.1. Animals

1. Weanling male mice, Swiss Webster strain (Simonsen Laboratories, Gilroy, CA). For convenience, Table 1 shows reported bioavailability studies in other animals and humans.
2. House mice in polycarbonate cages with stainless steel wire tops and pine shavings for litter.

Table 1
Summary of reported utilization of D-amino acids by different animal species and humans from multiple studies. Adapted from ref. 8

	D-Utilization (% relative to L-isomer)						
D-Amino acid	Rats	Mice	Poultry	Humans	Pigs	Dogs	Cats
Lys	0	0	0	0			
Trp	75 100	0	<10 17–40 20	<10	60 70	35	
Thr	0	0	0 100	0			
Met	100	100	100 100	0 36	50 100	100	100
Cys–Cys	0		0				
Arg	100 0		0				
His	>90	0	<10 20				
Phe	68 0	100	20 100 0	50			
Tyr	100			0			
Leu	100	0	100 Partial	0			
Val	0 50	0	Partial 100	0			
Ile	0	0	Partial 0 100	0			

2.2. Diets

Formulate standard diets from the ingredients listed in Table 2. A balanced array of amino acids replaces the protein normally in the diet. The diet is similar to the commercially available amino acid diet available from MP Biomedicals (Solon, OH), but lacks L-Cys–Cys and L-Tyr. This diet predates the current AIN 93 standard.

Table 2
Composition of amino acid diet

Ingredient	%
L-Ala	0.35
L-Arg · HCl	1.35
L-Asn	0.60
L-Asp	0.35
L-Glu	3.50
Gly	2.33
L-His · HCl	0.41
L-Ile	0.82
L-Leu	1.11
L-Lys · HCl	1.35
L-Met	1.17
L-Phe	1.51
L-Pro	0.35
L-Ser	0.35
L-Thr	0.82
L-Trp	0.174
L-Val	0.82
Cellulose	3.00
Corn oil	8.00
Cornstarch	20.00
Dextrose	38.33
Salts USP XIV	5.00
Zinc	0.0125
Cobalt	0.00057
Sodium acetate	1.31
Water (added)	5.00
Complete vitamin mixture	2.00
Total	100.00

AIN 93 adopted changes in vitamin and minerals, in the ratio of complex versus simple carbohydrates, and now includes high linolenic acid oil. We do not know if any of the differences between our diet and the AIN 93 standard will change the animal response to the D-amino acid evaluation diets. Amino acids should be 97% pure or better (Sigma Chemical Co., St. Louis, MO).

1. Alphacel™ from MP Biomedicals was used as the source for cellulose.
2. The salt mixture, U.S.P. XIV, is available commercially (MP Biomedicals: Solon, OH). It contains: calcium carbonate (6.86%), calcium citrate (30.83%), calcium biphosphate monobasic (11.28%), manganese carbonate (3.52%), magnesium sulfate·$7H_2O$ (3.83%), potassium chloride (12.47%), dipotassium phosphate (21.88%), sodium chloride (7.71%), copper sulfate·$5H_2O$ (0.00777%), ferric citrate (16–17% Fe) (1.52815%), manganese sulfate·H_2O (0.02008%), potassium aluminum sulfate (0.00923%), potassium iodide (0.00405%), sodium fluoride (0.0507%). AIN 93 mineral mixes contain the added minerals: zinc, chromium, selenium, molybdenum, silicon, nickel, boron, lithium, and vanadium, as well as other changes in composition.
3. Supplement the diet with zinc, 125 mg, and cobalt, 5.7 mg/kg diet. However, because these two minerals may interact negatively with sulfur amino acids, for diets with these amino acids, reduce zinc and cobalt to 6.25 and 0.29 mg/kg diet, respectively.
4. In our studies, we used a standard vitamin mixture which contained (per kg vitamin mix): 900,000 IU vitamin A acetate, 100,000 IU cholecalciferol, 5,500 IU DL-α-tocopherol, 100 g choline chloride, 1 g menadione, 4.5 g nicotinic acid, 1 g riboflavin, 1 g pyridoxine HCl, 1 g thiamin HCl, 3 g calcium pantothenate, 0.02 g D-biotin, 0.2 g folic acid, 5 g inositol, 45 g ascorbic acid, and 1.35 mg vitamin B-12 (crystalline). The currently commercially available vitamin mixes vary slightly from this mix. MP Biomedicals has a very similar mix with the following differences: levels of α-tocopherol are about four times higher, levels of menadione are about two times higher, levels of folic acid are about one-half as high, and *p*-aminobenzoic acid is now included in the mix.

3. Methods

3.1. Animal Care

1. Place one or two animals per cage.
2. Standardize the growth assay with six mice per group, and estimate potency based on response of 2–7 groups (growth data from 12 to 42 mice).
3. Assign mice to each treatment based on initial body weight blocks. Group mice into blocks according to weight, then an animal from each block is assigned to each treatment, such that

the end result is that each group will have nearly the same initial average body weight.

4. Maintain the environment at 22.2 ± 1.1°C and 50 ± 10% humidity. The light–dark cycle of 06.00–18.00 light and 18.00–06.00 dark may be regulated by automatic timer.
5. Provide feed and water *ad libitum*.
6. Weigh the mice after 14 days.

3.2. Design of Test Diet

1. To test the efficacy of a D-amino acid, replace the comparable or interchangeable L-amino acid in the diet in full and/or in part. Some amino acids may require special consideration (see Notes 1–5). Each amino acid (natural and unnatural) should be added back to the diet at graded concentrations, typically 0, 12.5, 25, 50, and 100% of ideal concentrations required for maximum growth.
2. A more complex design would include both the L- and the D-isomers in the same diet. For example, the L-isomer may be held constant at 25% of optimal, while varying the D-isomer to 0, 25, 50, and 100% of an optimal amount of the L-isomer.
3. When studying interchangeable amino acids, such as L-Met (see Note 2) and L-Cys (see Note 3), one must be careful to control the amount of both in the diet. For example, to test the availability of D-Cys, L-Met must be eliminated or reduced in the diet to prevent or minimize conversion of L-Met into L-Cys.
4. To test the efficacy of amino acid derivatives and dipeptides, follow the above procedure and use molar equivalents of the test compound to the L-amino acid.

3.3. Calculations

Efficacy of the test compound may be determined by the following four techniques: relative response, percent maximum growth response, replacement value, and relative potency. No one method is ideal for all cases. Multiple methods are often useful in a single study. Understanding the behavior of compounds in the growth assay is often complicated by other factors such as sparing effects of one amino acid for another, competition, and toxicity. Additional studies are needed to further define interactions of the unnatural amino acids *in vivo* (see Note 6).

Determine the mean body weight gain for each experimental group of six animals. Mean body weight gains may be compared by Duncan's multiple range test using individual values (11).

3.3.1. Relative Response (Efficiency)

This is the growth response of the test amino acid relative to the equivalent amount of L-amino acid. At any given level of feeding, the average net weight gain (relative to 0% L-amino acid in the diet) of mice on the test compound diet is divided by the average net weight gain of animals on the comparable L-amino acid diet,

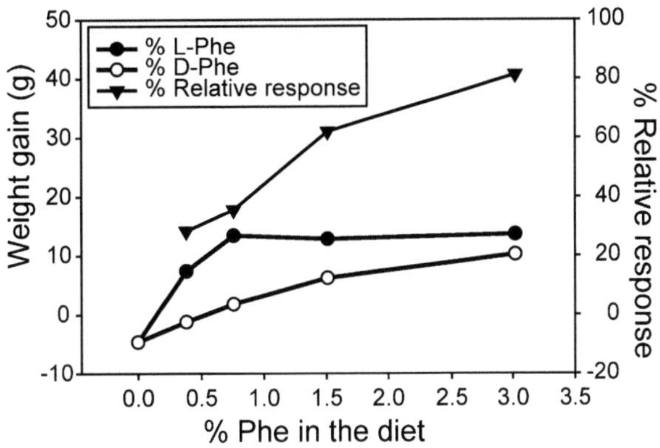

Fig. 1. Relationship of weight gain to % L- or D-Phe in amino acid diets fed to mice. The *triangles* represent the calculated relative response of D-Phe to L-Phe (average net weight gain of mice on the test compound diet divided by the average net weight gain of animals on the comparable L-amino acid diet, and the result multiplied by 100).

and the result multiplied by 100 to obtain percent relative response. This method can be used to calculate the relative response at any given feeding level. Note, however, that the response is likely to change as percent amino acid in the diet increases. This is due to the fact that the control diet reaches maximum growth before the test diet (Fig. 1). In this example, L-Phe reaches a maximum response at a faster rate than D-Phe, so at low dietary levels there is a larger difference between the two isomers. As dietary levels of D-Phe increase and weight gain improves, the relative response of D-Phe to L-Phe increases as well. At 100% of the required dose of L-Phe in the diet (1.51%), the relative response of D-Phe is 62%. Figure 1 is an example of a simple growth model where D-Phe is being converted *in vivo* to L-Phe with no complicating dynamics such as toxicity.

3.3.2. Percent Maximum Growth Response

This is the growth response of a test diet relative to the maximum growth from the relevant L-amino acid diet. The test diet may be a combination of compounds, and may be fed at varying concentrations. This method entails dividing the average net weight gain of the mice on the test diet by the average net maximum weight gain on the control (100% of the L-amino acid) diet, then multiplying the ratio by 100. Molar concentration is not a component of this equation. Figure 2 plots the percent maximum growth response when varying two dependent nutrients in the diet, L-Phe and L-Tyr. The response is complicated by the sparing effect of L-Tyr on L-Phe. The data show that L-Tyr exhibits only a partially sparing effect for L-Phe, and may even be inhibitory to mice at high L-Tyr to L-Phe ratios. This method is useful when studying the effect of more than one compound on growth or components that are mutually dependent, as it compares the growth induced by any given diet with the maximal growth by the optimal control diet.

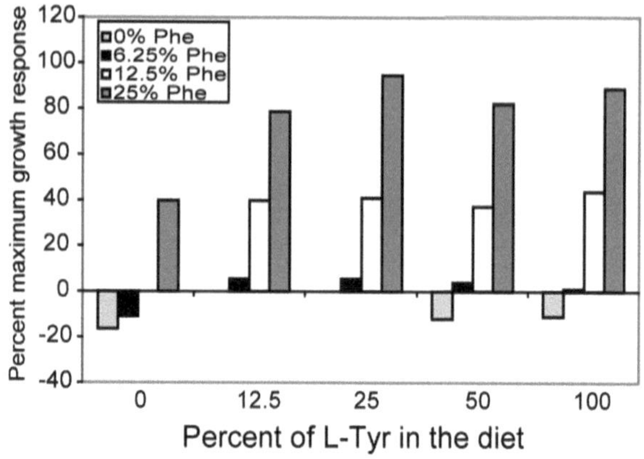

Fig. 2. Bar graph of the calculated Percent Maximum Growth Response of mouse diets with variable amounts of L-Tyr and L-Phe.

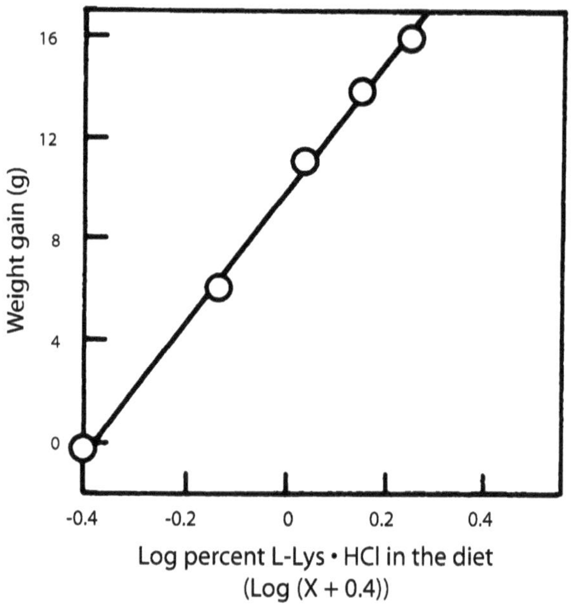

Fig. 3. Linear plot of mean weight gain versus log of L-Lys in the diet of mice (%). The empirical constant, 0.4, gives an optimum regression coefficient of 0.999. See Table 3 for results from corresponding growth studies with D-Lys and four lysine derivatives.

3.3.3. Replacement Value (Bioavailability)

Replacement value is defined as the relative concentration of the L-amino acid (obtained from a standard curve) that would produce the same weight gain observed with the molar equivalent of the test compound. It is especially useful when evaluating amino acid derivatives or peptides, as it is calculated on a molar basis. The equation for the standard curve of the control may be determined using regression analysis with a semi-logarithmic model with an empirical constant (Fig. 3). This plot is used to calculate the L-amino acid equivalent, and the result is then divided by the molar equivalent concentration of the L-amino acid to obtain the percent replacement value.

3.3.4. Relative Potency

Relative potency of the test compound is calculated as the ratio of the best fitting linear slope of the test compound to the best fitting linear slope of the standard control L-amino acid. This calculation is useful, as it provides a concentration-independent value that compares the nutritional value of the test compound to the L-amino acid control. Slope ratio analyses of growth data were performed as described by Finney (12). Growth curves were constructed from the average weight gains per group versus the molar concentrations of the test or control compound. Figure 4 is an example of such growth curves. The most linear part of the curve is situated between 1 and 5 mmol/100 g diet. The slopes were estimated from the best fitting linear part of the growth curve with the aid of the general linear model (GLM) using SAS software (13). Confidence intervals of potencies relative to the corresponding L-amino acid were estimated using Fieller's theorem (11, 14).

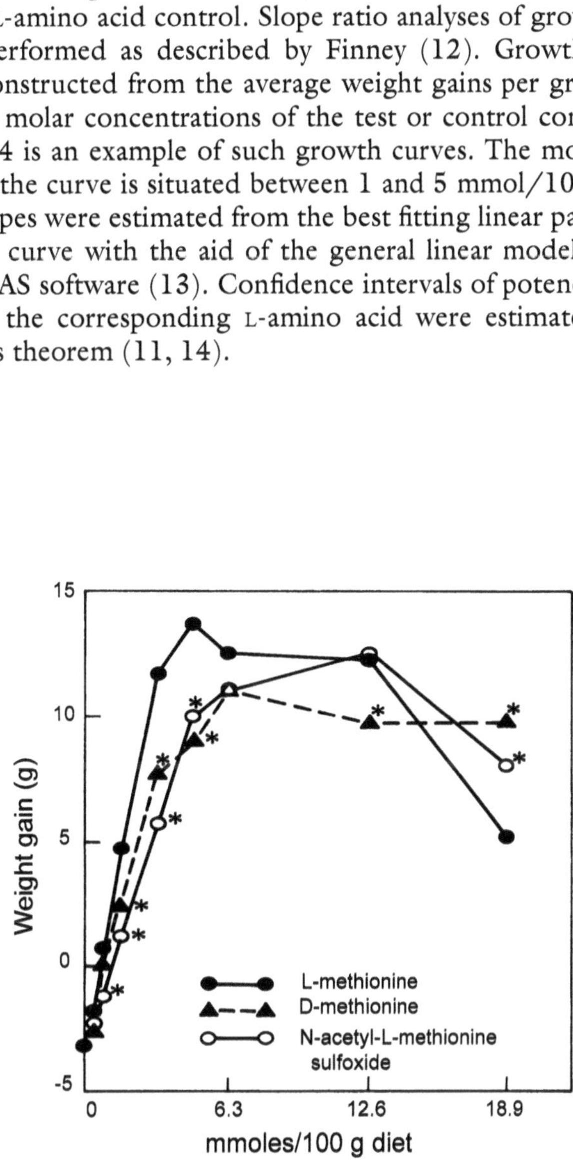

Fig. 4. Weight gain in mice fed increasing dietary levels of L- and D-Met and *N*-acetyl-L-Met sulfoxide for 14 days. Plotted are mean values ($n=6$), SEM = ±0.6 g; *asterisk* indicates significant difference from L-Met at the same dietary concentration, $P<0.05$.

4. Notes

Mice provide a good *in vivo* model to study the nutritional utilization and biological effects of unnatural amino acids. A major advantage of mouse bioassays is that they require about one fifth of the test substance needed for rats and can be completed in 14 days (15–20). Tables 3–5 list the nutritional values of specific D-amino acids, D-peptides, and amino acid derivatives as sources of the corresponding L-isomers. Here, we briefly discuss findings with specific unnatural amino acids.

1. Nutritional utilization of lysine derivatives. L-Lys is a limiting amino acid in cereal grains such maize, barley, rice, and wheat. The nutritional value of these grains can be improved by fortification or genetic modification to increase the lysine content (21, 22). The ε-NH_2 group of L-Lys residues in proteins is subject to chemical modification both *in vivo* and *in vitro*. Modifications *in vivo* include post-translational methylation to mono-, di-, and tri-methyl-L-Lys and interaction with glucose, especially in diabetics (23). Modifications *in vitro* include acylation, methylation, and lysinoalanine formation. Chemical modifications of lysine have been extensively studied in order to prevent Maillard browning reactions and lysinoalanine formation, as well as to protect proteins against reduction in nutritional quality and degradation by ruminant microorganisms. The replacement values of a number of Lys derivatives, determined with the aid of the standard curve in Fig. 3, as well as the calculated relative response, are listed in Table 3.

Table 3
Relative growth response of mice to D-Lys and L-Lys derivative diets by two measures: % replacement value and relative response. Adapted from ref. 16[a]

Test compound	Relative response	% Replacement value
D-Lys · HCl	−11.8	−4.4
ε-*N*-methyl-L-Lys · HCl	12.3	7.5
ε-*N*-dimethyl-L-Lys · HCl	9.4	4.6
ε-*N*-trimethyl-L-Lys · HCl dioxalate hemihydrate	8.2	3.8
LL-, DL-Lysinoalanine	8.3	3.8
α-*N*-acetyl-L-Lys	0.8	0.6
ε-*N*-acetyl-L-Lys	6.6	3.3

[a]Relative response = net growth of mice on test compound/net growth of mice on control compound. % Replacement value = percent amino acid equivalent/% molar equivalent of control

Table 4
Relative L-Met potency of unnatural Met, Met derivtives, and Met peptides. Adapted from ref. 20[a]

Test compound	Relative potency
D-Met	73.1
L-Met sulfoxide	85.4
D-Met sulfoxide	28.7
DL-Met sulfone	0
N-Acetyl-L-Met	89.1
N-Acetyl-D-Met	23.7
N-Acetyl-L-Met sulfoxide	58.7
N-Acetyl-D-Met sulfoxide	1.7
N-Formyl-L-Met	86.8
L-Met hydroxy analogue	55.4
D-Met hydroxy analogue	85.7
L-Met–L-Met	99.2
L-Met–D-Met	102.9
D-Met–L-Met	82.1
D-Met–D-Met	41.5

[a]Relative potency calculated as a ratio of the slope of the line best representing the test compound to the slope of the line best representing the control compound

The derivatives seem to be poorly utilized, with replacement values ranging between 0.6 and 7.5%. Additionally, D-Lys appears to not be utilized as source of L-Lys. Previous studies showed that D-amino acid oxidase did not convert D-Lys to the keto acid *in vivo*, and that D-Lys was largely excreted unchanged in rats (24). In contrast, we found that ε-NH_2-acetylated protein-bound L-Lys was utilized by rats as a nutritional source of L-Lys (25). Evidently, deacylase enzymes remove the acetyl group more rapidly from protein-bound acetylated L-Lys residues than from free L-Lys.

2. Nutritional utilization of methionine derivatives and isomeric dipeptides. L-Met is the limiting amino acid in legumes, but is abundant in cereals and nuts. Cereals (low in L-Lys and high in L-Met and L-Cys) and legumes (low in L-Met and high in L-Lys) consumed together are known to complement each other to provide high quality protein in the diet. The availability of both these amino acids is important in vegetarian diets,

Table 5
Growth response of mice as affected by dietary sulfur amino acids and derivatives in suboptimal L-Met (25%) diets. Test substances were added at a 100% molar equivalent of optimal L-Met

		Weight gain (g)			% Max growth response	
Test substance	Molar % of optimal L-Met	0% L-Met	25% L-Met	100% L-Met	Relative to 100% L-Met[a]	Change from 25% L-Met[b]
None	–	-3.4	7.7	16.0	58	–
L-Cys	100	-3.6	13.2		86	28
D-Cys	100	-4.2	5.6		47	-11
L-Cys–Cys	100	-1.8	11.8		79	21
	80	–	14.5		90	32
	40	–	16.0		98	40
	20	–	13.0		82	24
D-Cys–Cys	100	-2.5	5.0		43	-14
	25	–	10		71	13
DL- + meso-Lanthionine	100	-3.2	9.4		66	9
N-Acetyl-L-Cys	100	-3.0	15.8		99	42
L-Cysteic acid	100	-3.6	11.4		76	19
L-Cys sulfinic acid	100	-3.0	9.0		64	7
S-Methyl-L-Cys	100	-4.0	1.0		23	-35

[a]Net weight gain from diet containing the test compound divided by net weight gain from diet containing only 100% L-Met, e.g. (13.2–(-3.6))/(16.0–(-3.4)) = 86%
[b]Subtract % Max growth response of test compound from % Max growth response with no added test compound, both at the same supplemented L-Met level

especially in developing nations. During food processing L-Met can be chemically modified, including oxidation to L-Met sulfoxide and L-Met sulfone, racemization to D-Met, and degradation to sulfur compounds with undesirable flavors. Additionally, we found that excessive amounts of L-Met can inhibit growth. Therefore there is an interest in bioavailability of L- and D-Met substitutes to overcome these problems.

Table 4 lists the relative potency of Met, Met derivatives, and peptides we tested. Supplementation with L-Met of a diet devoid of other sulfur amino acids markedly stimulated growth of mice to a maximum between 3.15 and 6.3 mmol/100 g, after which additional L-Met inhibited growth. Derivatization of L-Met generally lowered potency. However, D-Met, the three isomeric dipeptides LL-, LD-, and DL-Met–Met, *N*-acetyl-L-Met, L-Met sulfoxide, *N*-formyl-L-Met, and D-Met hydroxy analogue were all well utilized as nutritional sources of L-Met. By contrast, the relative potency of the derivative L-Met hydroxy analogue, and the double derivative *N*-acetyl-L-Met sulfoxide was 55–59%. The relative potencies of D-Met sulfoxide, *N*-acetyl-D-Met, and DD-Met-Met ranged from 24 to 42%. *N*-Acetyl-D-Met sulfoxide and DL-Met sulfone were essentially not utilized. Several of the analogs were less growth-inhibiting at high concentrations in the diet than was L-Met. The results imply that some Met derivative may be better candidates for fortifying foods than L-Met because they are not as toxic at elevated concentrations.

3. Nutritional utilization of cysteine derivatives. L-Cys is a non-essential amino acid in the diet because it can be synthesized *in vivo* from the essential amino acid L-Met. L-Cys can supplement or spare dietary L-Met when that amino acid is limiting (26). Study designs can be complicated by this sparing effect. We found that in a diet containing no L-Met, addition of Cys or related sulfur-containing derivatives did not promote growth (Table 5). However, in the presence of suboptimal levels (25%), many of these compounds stimulated growth. Thus, *N*-acetyl-L-Cys, L-Cys, and LL-Cys–Cys were particularly effective growth promoters. D-Cys, DD-Cys–Cys, and *S*-methyl-L-Cys were growth depressing at 100% equivalent molar concentration. However, at 25%, DD-Cys–Cys was growth promoting. LL-Cys–Cys was also more effective at somewhat lower concentrations than the molar equivalent of L-Cys. Concentration responses to D-Cys and *S*-methyl-L-Cys were not measured. These growth-depressing effects, at equivalent molar concentrations to optimal L-Met, imply that these sulfur-containing amino acids and peptides may be toxic. The results with these amino acids demonstrate the complexity that may be encountered when test compounds are partially sparing for essential amino acids and also exhibit some toxicity (growth depression).

4. Nutritional value of D-phenylalanine and D-tyrosine. L-Tyrosine is a non-essential amino acid in the diet because it can be synthesized *in vivo* from the essential amino acid L-Phe. L-Tyr can supplement or spare part of the dietary need for L-Phe in the diet. This aspect is important for the management of the metabolic disease phenylketonuria, in which L-Phe is unable to convert to L-Tyr, leading to physiologically damaging concentrations of L-Phe in the body.

   Protein bound L-Phe and L-Tyr racemize rapidly to D-isomers upon treatment with alkali and heat. In our studies, we examined the ability of the D-isomers to replace L-Tyr and L-Phe, both individually and in combination. Higher concentration of D-Phe can replace the need for L-Phe in the diet (Fig. 1), possibly due to a lag in the *in vivo* conversion (pharmacokinetics) of D- to L-isomers. D-Phe does not appear to be toxic at the concentrations tested.

L-Tyr can replace some of the requirement for L-Phe in the diet (Fig. 2). A small increase in the level of L-Tyr (12.5%) doubled the growth of mice on an L-Phe-poor diet. Additional supplementation had no effect. In fact, at the higher levels, L-Tyr appeared to negatively impact L-Phe deficient mice. D-Tyr had no similar sparing effect on L-Phe. Addition of D-Tyr to amino acid or casein diets greatly depressed weight gain in mice (Fig. 5). This inhibition of growth was significantly reduced by increasing the L-Phe content of the amino acid

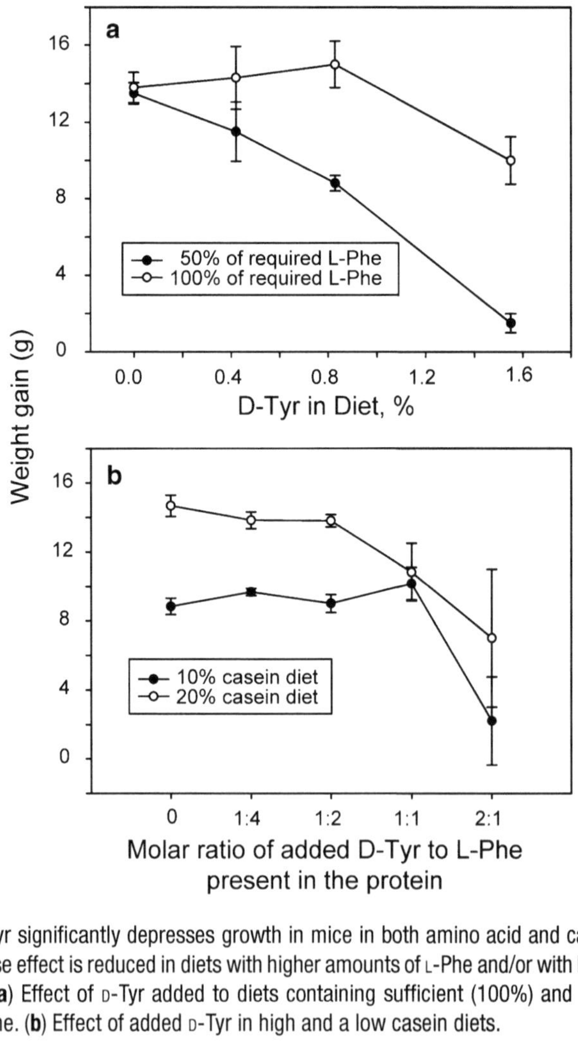

Fig. 5. D-Tyr significantly depresses growth in mice in both amino acid and casein diets. The adverse effect is reduced in diets with higher amounts of L-Phe and/or with high quality proteins. (**a**) Effect of D-Tyr added to diets containing sufficient (100%) and insufficient (50%) L-Phe. (**b**) Effect of added D-Tyr in high and a low casein diets.

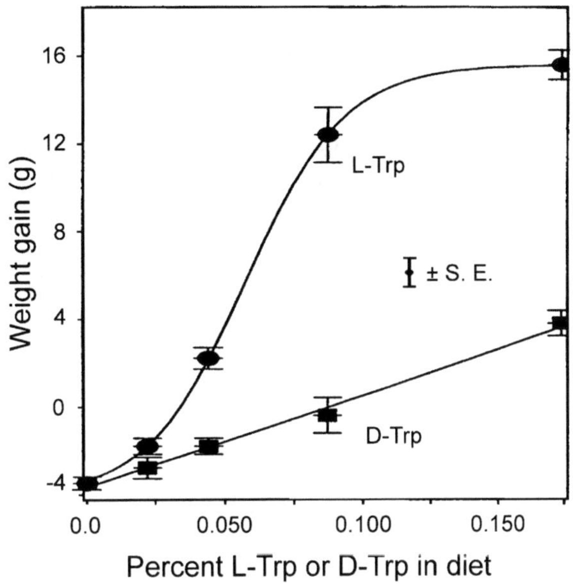

Fig. 6. Relationships of weight gains to percent of L- and D-Trp isomers in amino acid diets fed to mice.

diets or the protein content of casein diets. Our results demonstrate that D-Tyr cannot be utilized as a partial replacement for L-Phe. It creates a metabolic stress that becomes evident when D-Tyr and L-Phe are present in equimolar amounts. The potential of subchronic and chronic effects following exposure to lower levels of D-Tyr merits study.

5. Nutritional value of D-tryptophan. L-Trp is a nutritionally limiting amino acids in maize (27). Replacement of L-Trp with D-Trp in the diet followed a simple pattern, similar to the L- and D-Phe responses, not influenced by toxicity or by sparing effects (Fig. 6). Growth increased rapidly as L-Trp in the diet was increased and rapidly leveled off at a maximum. The growth curve for D-Trp relative to L-Trp had a shallow slope. Maximum growth was not achieved until the concentration of D-Trp was 2.5 times the corresponding level of L-Trp. It is also relevant to mention that reported studies using growth assays based on D-amino acid fortification of dietary proteins revealed large variations in the nutritive value of D-Trp among different animal species (Table 1).

6. Research needs. Largely unresolved are the following questions: (a) How do biological effects of D-amino acids vary, depending on whether they are consumed in the free state or as part of a food protein? (b) Do metabolic interactions, antagonisms, or synergisms among D-amino acids occur *in vivo*? (c) Do free and protein-bound D-amino acids interact (bind)

differently than L-amino acids in the digestive tract with active sites of the proteolytic enzymes, such as chymotrypsin, pepsin, and trypsin? (d) Does racemization of toxic bacterial, plant, and venom proteins alter protein conformations, charge distributions (isoelectric points), and affinities for cell membranes, resulting in protective effects against protein-induced *in vivo* toxicities (28–31)? (e) Will D-amino acid inhibit formation of biofilms produced by pathogenic bacteria such *Escherichia coli* and *Salmonella* in food (32). We are challenged to find answers to these questions.

References

1. Friedman M, Zahnley J C, Masters P M (1981) Relationship between *in vitro* digestibility of casein and its content of lysinoalanine and D-amino acids. J Food Sci 46, 127–131.
2. Friedman M (2010) Origin, microbiology, nutrition, and pharmacology of D-amino acids. Chem Biodiv 7, 1491–1530.
3. Arlorio M et al. (2009) D-amino acids and computer vision image analysis: A new tool to monitor hazelnuts roasting? Czech J Food Sci 27, S30.
4. Friedman M (2009) Dietary significance of processing-induced lysinoalanine in food. In: Stadler RH, Lineback DR (eds), Process-induced food toxicants: Occurrence, formation, mitigation, and health risks, John Wiley & Sons, Inc., Hoboken, New Jersey.
5. Brückner H, Fujii N (2010) Free and peptide-bound D-amino acids in chemistry and life sciences. Chem Biodiv 7, 1333–1336.
6. Friedman M (1991) Formation, nutritional value, and safety of D-amino acids. Adv Exp Med Biol 289, 447–481.
7. Friedman M (1999) Chemistry, nutrition, and microbiology of D-amino acids. J Agric Food Chem 47, 3457–3479.
8. Borg B S, Wahlstrom R C (1989) Species and isomeric variation in the utilization of amino acids. In: Friedman M (ed), Absorption and utilization of amino acids, CRC Press, Boca Raton.
9. Man E H, Bada J L (1987) Dietary D-amino acids. Annu Rev Nutr 7, 209–225.
10. Mercer L P, Dodds S J, Smith D L (1989) Dispensable, indispensable, and conditionally indispensable amino acids ratios in the diet. In: Friedman M (ed), Absorption and utilization of amino acids, CRC Press, Boca Raton, FL.
11. Duncan D B (1955) Multiple range and multiple F tests. Biometrics 11, 1–42.
12. Finney D J (1978) Statistical Method in Biological Assay, chapter 7, Macmillan, New York.
13. SAS Institute Inc. (1982) The GLM procedure. In: Ray A A (ed), SAS User's Guide: Statistics, SAS Institute, Inc., Cary, NC.
14. Zerbe G O (1978) On Fieller's theorem and the general linear model. Am Stat 32, 103–105.
15. Friedman M, Gumbmann M R (1979) Biological availability of ε-N-methyl-L-lysine, 1-N-methyl-L-histidine, and 3-N-methyl-L-histidine in mice. Nutr Rep Int 19, 437–443.
16. Friedman M, Gumbmann M R (1981) Bioavailability of some lysine derivatives in mice. J Nutr 111, 1362–1369.
17. Friedman M, Gumbmann M R (1984) The utilization and safety of isomeric sulfur-containing amino acids in mice. J Nutr 114, 2301–2310.
18. Friedman M, Gumbmann M R, Savoie L (1982) The nutritional value of lysinoalanine as a source of lysine for mice. Nutr Rep Int 26, 937–943.
19. Friedman M, Gumbmann M R (1984) The nutritive value and safety of D-phenylalanine and D-tyrosine in mice. J Nutr 114, 2089–2096.
20. Friedman M, Gumbmann M R (1988) Nutritional value and safety of methionine derivatives, isomeric dipeptides and hydroxy analogs in mice. J Nutr 118, 388–397.
21. Friedman M, Finot P A (1990) Nutritional improvement of bread with lysine and γ-glutamyl-L-lysine. J Agric Food Chem 38, 2011–2020.
22. Friedman M, Atsmon D (1988) Comparison of grain composition and nutritional quality in wild barley (*Hordeum spontaneum*) and in a standard cultivar. J Agric Food Chem 36, 1167–1172.
23. Friedman M (1982) Chemically reactive and unreactive lysine as an index of browning. Diabetes 31, 5–14.
24. Neuberger A, Sanger F (1944) The metabolism of lysine. Biochem J 38, 119–125.
25. Friedman M (1978) Inhibition of lysinoalanine synthesis by protein acylation. Adv Exp Med Biol 105, 613–648.

26. Gumbmann, MR, Friedman, M (1987) Effect of sulfur amino acid supplementation of raw soy flour on the growth and pancreatic weights of rats. J Nutr 117, 1018–1023.
27. Friedman M, Cuq J L (1988) Chemistry, analysis, nutritional value, and toxicology of tryptophan in food. A review. J Agric Food Chem 36, 1079–1093.
28. Friedman M (2001) Application of the S-pyridylethylation reaction to the elucidation of the structures and functions of proteins. J Protein Chem 20, 431–453.
29. Rasooly R, Do P M, Friedman M (2010) Inhibition of biological activity of *Staphylococcal enterotoxin* A (SEA) by apple juice and apple polyphenols. J Agric Food Chem 58, 5421–5426.
30. Rasooly, R et al. (2010) Inhibition of Shiga toxin 2 (Stx2) in apple juices and its resistance to pasteurization. J Food Sci 75, 296–301.
31. Rasooly R et al. (2010) Ingestion of Shiga toxin 2 (Stx2) causes histopathological changes in kidney, spleen and thymus tissues and mortality in mice. J Agric Food Chem 58, 9281–9286.
32. Kolodkin-Gal I et al. (2010) D-amino acids trigger biofilm disassembly. Science 328, 627–629.

Part V

Enzymes Active on D-Amino Acids

Chapter 24

Preparation and Assay of Recombinant Serine Racemase

Florian Baumgart, Clara Aicart-Ramos, and Ignacio Rodriguez-Crespo

Abstract

Serine racemase is a glial and neuronal enzyme that reversibly converts L-serine to D-serine, an endogenous co-agonist of *N*-methyl-D-aspartate receptor type glutamate receptors (NMDARs). Here we present methods to recombinantly express and purify serine racemase in bacteria and two complementary ways to determine D-serine levels in unknown samples. Furthermore, a detailed protocol of serine racemase activity assays is described that can be used to screen for activators and inhibitors in vitro.

Key words: Serine racemase, D-Serine, PLP enzymes, Racemization

1. Introduction

Serine racemase is a glial and neuronal enzyme that reversibly converts L-serine to D-serine, an endogenous co-agonist of *N*-methyl-D-aspartate receptor type glutamate receptors (NMDARs) (1–3). In addition, it has also been reported that serine racemase catalyzes the α,β-elimination of water from D-serine, thus regulating D-serine levels together with D-amino acid oxidase, an enzyme also expressed in the brain (4). Serine racemase has been purified by different groups, originally from brain tissue (2) or cultured eukaryotic cells (5) and finally in large quantities in bacterial expression systems (6). While studies of serine racemase in ex vivo tissue slices have contributed critically to the understanding of the expression patterns of this enzyme in the brain (1) and other tissues (7) and to the characterization of its functional interaction with NMDARs (1), it has been mostly in vitro studies of the purified enzyme that lead to the discovery of co-factors of serine racemase. Serine racemase directly binds to and is activated by Mn^{2+}, Mg^{2+}, and Ca^{2+} (6, 8) and

Loredano Pollegioni and Stefano Servi (eds.), *Unnatural Amino Acids: Methods and Protocols*,
Methods in Molecular Biology, vol. 794, DOI 10.1007/978-1-61779-331-8_24, © Springer Science+Business Media, LLC 2012

interacts with a specific region of the scaffolding protein GRIP (9, 10), which also increases racemase activity. Moreover, it has been shown that some nucleotides, most notably ATP, are potent activators of D-serine synthesis by serine racemase (8). The recently published crystal structure of recombinant human serine race mase (11) validated the importance of *in vitro* studies of purified enzyme preparations, in order to get precise mechanistic insight into the function of serine racemase and the possibility to rationally design small inhibitors for clinical use.

In this chapter, we present a detailed protocol to produce mammalian serine racemase in an *Escherichia coli* expression system (Subheading 3.1) and purify the recombinant protein by affinity chromatography (Subheading 3.2). Furthermore, we present a coupled colorimetric activity assay for serine racemase that permits the assessment of D-serine production in the presence of potential activators or inhibitors (Subheading 3.3). Finally, we also describe an HPLC method that allows for the detection of very small amounts of D-serine and that complements the colorimetric assay (Subheading 3.4).

2. Materials

Standard laboratory glassware, an orbital rocking incubator, a tabletop centrifuge and a Sorvall centrifuge are required in the respective steps.

2.1. Recombinant Expression of Serine Racemase

1. Serine racemase cDNA cloned in a bacterial expression vector (e.g. pCWori (6)) in frame with a hexa-His tag at the N-terminal end.
2. Antibiotics (e.g. ampicillin for pCWori) for expression plasmid selection and maintenance.
3. LB bacterial growth medium: 10 g/L Bacterial bacto tryptone, 5 g/L yeast extract, and 10 g/L NaCl in dH_2O, pH 7.0 (autoclaved); 15 g/L bacto-agar are added for soft agar plates (see Note 1).
4. 2× YT bacterial growth medium: 10 g/L Bacterial bacto tryptone, 16 g/L yeast extract, and 5 g/L NaCl in dH_2O, pH 7.0 (autoclaved). Generally, 3 L of medium are prepared per purification batch and distributed in four flasks (0.75 L/flask).
5. *E. coli* BL21 (DE3) cells (Novagen) for expression of serine racemase.
6. Incubators with swirling platform and thermostat for controlled temperatures (37 and 30°C).
7. Sorvall centrifuge for cell harvesting.

2.2. Affinity Purification of Serine Racemase

1. Resuspension buffer: 50 mM Tris–HCl, 150 mM NaCl, pH 7.0.
2. Protease inhibitors (1,000× stock solutions): 200 mM phenylmethylsulfonyl fluoride (PMSF), 1 M leupeptin, and 1 M aprotinin.
3. Lysozyme.
4. 1,000× DNAseI solution in 50 mM Tris–HCl, pH 7.0 (50 mg/mL).
5. Dithiotreitol (1 mM DTT).
6. Branson sonicator.
7. Sorvall centrifuge for sedimentation of cell debris.
8. Whatman filter paper for filtration of lysate.
9. Stock solution of 1 M imidazole, pH 7.0.
10. 5 mL Ni^{2+}-nitrilotriacetic column (Qiagen).
11. Wash buffer A: 50 mM Tris–HCl, 150 mM NaCl, 30 mM imidazole, pH 7.0.
12. Wash buffer B: 50 mM Tris–HCl, 150 mM NaCl, 40 mM imidazole, pH 7.0.
13. Elution buffer: 50 mM Tris–HCl, 150 mM NaCl, 200 mM imidazole, pH 7.0.
14. 7 M urea (dissolved in dH_2O) to regenerate the column.
15. 200 mM $NiSO_4$ dissolved in dH_2O.
16. Dialysis bags: Cut-off <10,000 Da.
17. 5 mM pyridoxal 5′-phosphate (PLP) (Sigma) (dissolved in dH_2O).

2.3. Colorimetric Activity Assay of Serine Racemase

Stock solutions are prepared as follows:

1. 200 mM MOPS buffer, pH 8.1.
2. 100 μM FAD (dissolved in dH_2O).
3. 1 mM DTT (dissolved in dH_2O).
4. 500 μM PLP (dissolved in dH_2O).
5. 5 mg/mL *o*-phenylenediamine (OPD, dissolved in dH_2O).
6. 1 mg/mL D-amino acid oxidase (DAAO, dissolved in 50 mM Tris–HCl, pH 7.0).
7. 1 mg/mL horseradish peroxidase (HRP, dissolved in 50 mM Tris–HCl, pH 7.0).
8. 1 M L-serine (ultrapure, NovaBiochem, Switzerland).

2.4. HPLC Assay to Determine Low D-Serine Levels

1. (1-Fluoro-2,4-dinitrophenyl)-5-L-alanine amide (Marfey reagent) 1% solution in acetone.
2. 1 M sodium acetate solution.
3. Block heater (for incubation at 40°C).

4. 2 M HCl.
5. C18 Kromasil 100 column (15 cm × 0.46 cm).
6. 20 mM Sodium acetate buffer, pH 4.0.
7. MeOH (HPLC grade).
8. Beckman detector module for detection at 340 nm.

3. Methods

3.1. Recombinant Expression of Serine Racemase

1. The plasmid containing the cDNA coding for serine racemase cloned in frame with a hexa-His affinity tag at the amino-terminal end of the protein is transformed into the *E. coli* strain BL21 (DE3). The cells are plated on LB/soft agar Petri dishes containing the appropriate antibiotic (100 μg/mL ampicillin in the case of pCWori) for selection of transformed bacteria. The plates are incubated 16–20 h at 37°C.
2. Next day, a fresh colony (from a plate not older than 2 weeks) of transformed *E. coli* BL21 (DE3) is transferred from the LB/soft agar plate into a flask of 20-mL liquid LB medium containing the appropriate antibiotic (100 μg/mL ampicillin in the case of pCWori). The flask is incubated 16–20 h at 37°C in an orbital incubator at 220 rpm.
3. Next day, four flasks each containing 750 mL (total 3 L) of autoclaved 2× YT medium are inoculated with 5 mL of the pre-culture/flask. The appropriate antibiotic (100 μg/mL ampicillin in the case of pCWori) is added to maintain the plasmid in the culture (see Note 2).
4. The flasks are incubated at 37°C in an orbital incubator at 220 rpm until the optical density at 600 nm (OD_{600}) reaches approximately 1.0 (typically ~5 h) and protein expression in the bacterial culture is induced by the addition of isopropyl-1-thio-β-D-galactopyranoside (IPTG) at a final concentration of 1 mM in the bacterial growth medium (see Note 3).
5. Continue the incubation of the induced bacterial culture in an orbital incubator at 220 rpm 16–20 h lowering the temperature to 30°C.
6. Next day, the cells are harvested by centrifugation in a Sorvall centrifuge in a F10S rotor at 5,140 × *g* for 20 min at 4°C. Pellets are transferred to plastic bags and the cell paste is extended to form a thin film in order to facilitate the later resuspension. The cells are frozen and stored at −80°C.

3.2. Affinity Purification of Serine Racemase

Serine racemase is expressed with a hexa-His tag which allows for its purification on a Ni^{2+}-nitrilotriacetic column.

All purification steps are performed at 4°C or on ice and centrifugations are carried out at 4°C.

1. The frozen bacterial pellets are broken up and transferred to a 100-mL glass beaker with a magnetic stirrer. Approximately 30-mL resuspension buffer per liter of culture medium is added and cell suspension is stirred on ice to homogeneity. Protease inhibitors (PMSF, aprotinin, leupeptin) are added. DNAseI is added to a final concentration 50 μg/mL, and 0.5 mg/mL lysozyme is added to facilitate cell lysis. The homogenate is stirred on ice for 30 min.
2. The cells are lysed by sonication: Perform four cycles of alternate sonication (50%, 2 min) and stirring (3 min on ice). Avoid heating.
3. In order to sediment unbroken cells and cell debris, centrifuge the lysate in a Sorvall centrifuge in a SS34 rotor at 23,500 × *g* for 30 min at 4°C.
4. Filtrate the lysate through Whatman filter paper. Add 10 mM imidazole and 5 μM PLP (final concentrations) to the filtrated lysate.
5. Load the clear filtrate on a Ni^{2+}-nitrilotriacetic column that has been equilibrated in resuspension buffer (see Note 4). After loading the lysate, wash the column briefly with 2 mL of resuspension buffer to remove remaining lysate on the column walls.
6. Add consecutively 100 mL wash buffer A and 100 mL wash buffer B to the column.
7. Elute serine racemase from the column. Start adding ~1.5 mL of elution buffer to the column, let it enter into the resin and close the column immediately. Wait 5 min and then add more elution buffer, open the column and start collecting fractions of 1 mL. Serine racemase should elute within the first ten fractions.
8. After the elution of the protein add 7 M urea to denature any remaining protein that binds unspecifically to the resin. Wash extensively with water and store the column in 20% EtOH (to avoid contamination).
9. Measure the absorbance of each eluted fraction at 280 nm (A_{280}) to determine the fractions that contain protein. Additionally, take small aliquots (5 μL) of each fraction and analyze them by SDS-PAGE and Coomassie staining to assess the purity of each fraction (Fig. 1). Pool the purest and most concentrated fractions (Fig. 2) and dialyze them three times for 2 h in 2 L 50 mM Tris–HCl, pH 7.0 to remove the imidazole.

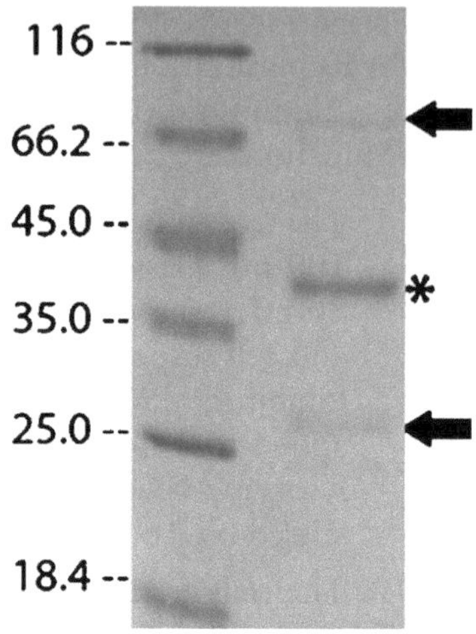

Fig. 1. Coomassie blue-stained SDS-PAGE of purified and dialyzed serine racemase (marked with an *asterisk*). The purity of the sample is ~95%. The large contaminating band that migrates at ~72 kDa (marked with an *arrow*) is an *E. coli* protein and the small contaminating protein that migrates at ~26 kDa (marked with an *arrow*) is a proteolytic fragment of serine racemase. Western blot using anti-His antibodies allowed us to conclude that only the mature serine racemase that migrates at ~37 kDa has the hexa-His tag (data not shown).

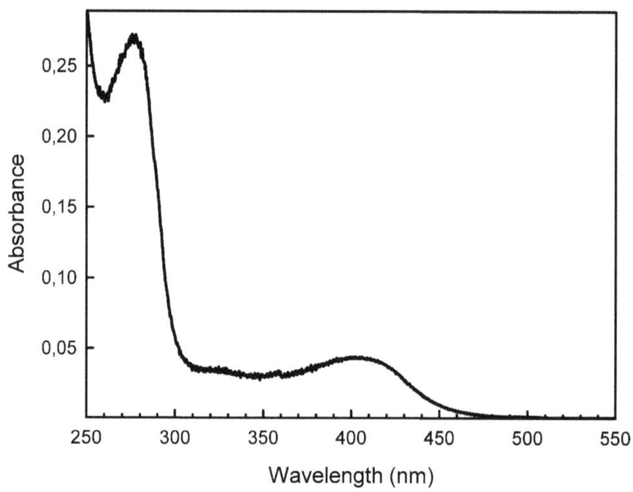

Fig. 2. Absorbance spectrum of purified recombinant SR in the 250–550 nm range. The ratio A_{275}/A_{415} is usually between 6 and 7.

Assess the protein quality and quantity by SDS-PAGE/Coomassie staining and by a protein spectrum in the range of 250–550 nm (Fig. 2)

10. Store serine racemase at 4°C for short-term use or shock-freeze it in liquid N_2 for storage at −80°C (see Note 5).

3.3. Colorimetric Activity Assay of Serine Racemase

We routinely perform a colorimetric activity assay for serine racemase to assess D-serine synthesis from L-serine. The assay couples the production of D-serine by serine racemase to the activities of

D-amino acid oxidase (DAAO) and horseradish peroxidase (HRP). D-serine is degraded by DAAO generating α-keto acid, NH_3, and hydrogen peroxide. The latter can be quantified using HRP together with the substrate *o*-phenylenediamine (OPD) that turns yellow (measured at 411 nm) upon oxidation. The activity assay is performed in a final volume of 200 μL mixing all the different components including proteins, cofactors, and L-serine in a single reaction cocktail. A negative control is included that does not contain serine racemase and that allows for the determination of unspecific oxidation of OPD throughout the course of the assay. A working table is included (Table 1).

1. The final volume per tube is 200 μL and reactions are performed in duplicates. The reaction cocktail is prepared on ice. Volumes of the reaction cocktail are given per tube and can be upscaled by multiplication with the number of the tubes. A 1× reaction cocktail (Table 1) contains 20 μL 200 mM MOPS, pH 8.1, 2 μL 100 μM FAD, 6 μL 1 mM DTT, 2 μL 500 μM PLP, 2 μL 5 mg/mL OPD, 20 μL 1 M L-serine 20 μL 1 mg/mL DAAO, and 2 μL 1 mg/mL HRP. All protein components (DAAO and HRP) should be added at the end. Mix gently.
2. For the negative control, pipette 74 μL of the cocktail and add 126 μL dH_2O to a final volume of 200 μL. Mix gently.
3. Add approximately 30 μg/tube of serine racemase to the remaining cocktail. Mix gently.
4. Pipette the adequate volume (74 μL negative control cocktail + volume of 30 μg serine racemase) of the complete cocktail into the respective tubes and add dH_2O to a final volume of 200 μL. Mix gently.

Table 1
Reaction cocktail for the Serine Racemase assay

Added/tube (μL)	Stock solution	Final concentration/tube
20	200 mM MOPS, pH 8.1	20 mM
2	100 μM FAD	1 μM
6	1 mM DTT	30 μM
2	500 μM PLP	5 μM
2	5 mg/mL OPD	50 μg/mL
20	1 M L-Ser	100 mM
20	1 mg/mL DAAO	100 μg/mL
2	1 mg/mL HRP	10 μg/mL
126 μL dH_2O	–	–

5. Centrifuge all the tubes in a tabletop centrifuge at maximum at room temperature for 1 min, in order to collect the entire volume on the bottom of the tubes.
6. Incubate the reactions at 37°C for 3–4 h.
7. Centrifuge all the tubes in a tabletop centrifuge at maximum at room temperature for 1 min, in order to collect the entire volume on the bottom of the tubes.
8. Measure all the samples either using a spectrophotometer at 411 nm or, alternatively, transferring all the reaction volumes to a 96-well plate and measuring the samples in a plate reader equipped with appropriate filters. The data can be either represented as relative activity or indicating the amount of D-serine produced in a given sample (see Note 6).

3.4. HPLC Assay to Determine Low D-Serine Levels

In the case of very low D-serine concentrations (<100 μM range), an alternative quantification assay can be used in order to readily detect D-serine produced by serine racemase (12).

1. 50 μL of a problem sample are mixed with 20 μL 1 M sodium bicarbonate solution. Vortex.
2. 200 μL Marfey reagent (1% solution in acetone) are added to derivatize D-serine and L-serine in the sample. Vortex.
3. Incubate the sample at 40°C in a block heater.
4. Cool down the reactions and add 10 μL 2 M HCl. Vortex.
5. The separation of derivatized D- and L-serine is achieved on a C18 Kromasil 100 column (15 cm × 0.46 cm). The samples are loaded in 20-mM sodium acetate buffer, pH 4.0 and elution is performed isocratically with methanol in 20-mM sodium acetate buffer, pH 4.0 (80:20). The absorbance is monitored at 340 nm. The typical elution position of L-Ser and D-Ser is shown in Fig. 3.

4. Notes

1. Work under sterile conditions when preparing growth media and plates! Growth medium (typically LB) for plates for bacterial cell growth is prepared like liquid medium and 15 g/L bacto-agar is added prior to autoclaving (bacto-agar does not dissolve at room temperature!). Moreover, a magnetic stirrer is added to flask and is autoclaved in the medium. After the autoclaving process, the bacto-agar should have dissolved and the hot liquid growth medium is put on a magnetic stirrer until it has cooled down to approximately 50–60°C. An appropriate amount of an antibiotic solution (ampicillin in the case of pCWori) is added to the growth medium under sterile conditions

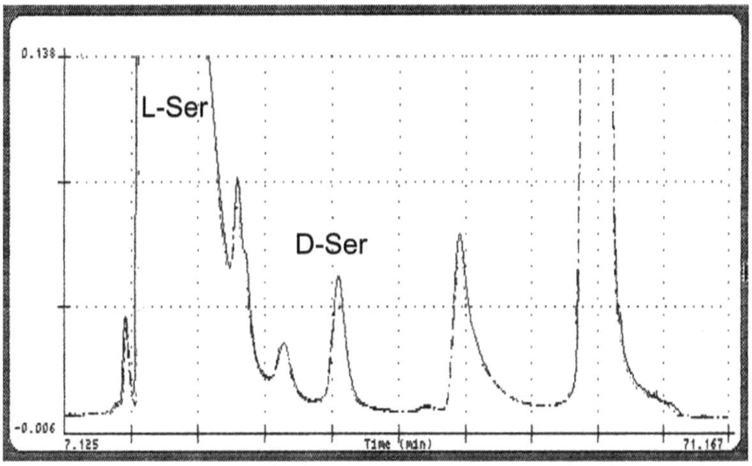

Fig. 3. Typical HPLC profile of derivatized a L- and D-serine mixture in DMEM using a C18 Kromasil 100 column. Elution was performed isocratically with methanol in 20 mM sodium acetate buffer, pH 4.0 (80:20).

(do not add the antibiotic earlier, since antibiotics are heat-sensitive and might be degraded otherwise) and the flask is left to stir again for about 15 s, in order to distribute the antibiotic homogeneously in the growth medium. Quickly pour about 5 mL/plate under sterile conditions and close the lid. Leave the plates without moving them for about 45 min until the agar has solidified. Store plates at 4°C for several months.

2. Add 100 μg/mL solid ampicillin (in the case of pCWori) to the 20 mL 16–20 h culture and dissolve it quickly in the flask. Then, distribute the 20 mL evenly between the flasks containing 750 mL 2× YT growth medium (5 mL/flask).
3. Weigh 0.238 g IPTG per liter of bacterial growth medium and dissolve it in a small volume of dH_2O (e.g. 4 mL). Then add equal volumes of the IPTG solution to the flask of incubating bacterial culture when OD_{600} reaches ~1.0, in order to induce recombinant protein expression.
4. Follow the supplier's instructions for operation and maintenance of the Ni^{2+}-nitrilotriacetic column. It is advisable to wash the column extensively with dH_2O prior to use and then saturate it with a solution of 200 mM $NiSO_4$ in dH_2O (~0.5 mL/mL resin) before equilibrating the column in resuspension buffer.
5. In our experience it is better to perform serine racemase activity assays with a fresh purification that has not been frozen previously. Freeze/thaw cycles seem to influence racemase activity negatively.
6. D-serine calibration curves are readily obtained in a range of 0–3 mM D-serine by mixing the negative control cocktail (no serine racemase) with different amounts of D-serine dilutions in water.

Acknowledgments

This work was supported by Grants BFU2006-05395 and BFU2009-10442 from the Spanish MICINN.

References

1. Wolosker H, Blackshaw S, Snyder S H (1999) Serine racemase: a glial enzyme synthesizing D-serine to regulate glutamate-N-methyl-D-aspartate neurotransmission. Proc Natl Acad Sci USA 96, 13409–13414.
2. Wolosker H, Sheth K N, Takahashi M et al. (1999) Purification of serine racemase: biosynthesis of the neuromodulator D-serine. Proc Natl Acad Sci USA 96, 721–725.
3. Baumgart F, Rodriguez-Crespo I (2008) D-amino acids in the brain: the biochemistry of brain serine racemase. FEBS J 275, 3538–3545.
4. Foltyn V N, Bendikov I, De Miranda J et al. (2005) Serine racemase modulates intracellular D-serine levels through an alpha,beta-elimination activity. J Biol Chem 280, 1754–1763.
5. De Miranda J, Santoro A, Engelender S, Wolosker H (2000)Human serine racemase: moleular cloning, genomic organization and functional analysis. Gene 256, 183–188.
6. Cook S P, Galve-Roperh I, Martinez del Pozo A, Rodriguez-Crespo I (2002) Direct calcium binding results in activation of brain serine racemase. J Biol Chem 277, 27782–27792.
7. Stevens E R, Esguerra M, Kim PM et al. (2003) D-serine and serine racemase are present in the vertebrate retina and contribute to the physiological activation of NMDA receptors. Proc Natl Acad Sci USA 100, 6789–6794.
8. De Miranda J, Panizzutti R, Foltyn V N, Wolosker H (2002) Cofactors of serine racemase that physiologically stimulate the synthesis of the N-methyl-D-aspartate (NMDA) receptor coagonist D-serine. Proc Natl Acad Sci USA 99, 14542–14547.
9. Baumgart F, Mancheno J M, Rodriguez-Crespo I (2007) Insights into the activation of brain serine racemase by the multi-PDZ domain glutamate receptor interacting protein, divalent cations and ATP. FEBS J 274, 4561–4571.
10. Kim P M, Aizawa H, Kim PS et al. (2005) Serine racemase: activation by glutamate neurotransmission via glutamate receptor interacting protein and mediation of neuronal migration. Proc Natl Acad Sci USA 102, 2105–2110.
11. Smith M A, Mack V, Ebneth A et al. (2010) The structure of mammalian serine racemase: evidence for conformational changes upon inhibitor binding. J Biol Chem 285, 12873–12881.
12. Martinez del Pozo A, Merola M, Ueno H et al. (1989) Stereospecificity of reactions catalyzed by bacterial D-amino acid transaminase. J Biol Chem 264, 17784–17789.

Chapter 25

Assay of Amino Acid Racemases

Masumi Katane, Masae Sekine, and Hiroshi Homma

Abstract

D-Amino acids play important physiological roles in the mammalian body. Recent investigations revealed that, in mammals, D-amino acids are synthesized from their corresponding L-enantiomers via amino acid racemase. This article describes a method used to measure amino acid racemase activity by high-performance liquid chromatography (HPLC). The assay involves fluorogenic chiral derivatization of amino acids with a newly developed reagent, and enantioseparation of D- and L-amino acid derivatives by HPLC. The method is accurate and reliable, and can be automated using a programmable autosampling injector.

Key words: D-Amino acid, Racemase, Enantioseparation, Chiral derivatization, High-performance liquid chromatography

1. Introduction

It was long believed that only the L-forms of amino acids were functional, and that D-amino acids were either unnatural or laboratorial artifacts. However, recent investigations have revealed a variety of D-amino acids that have important physiological functions in the body (1). Of the free D-amino acids identified in mammals, D-serine (D-Ser) and D-aspartate (D-Asp) have been the most intensively studied. Free D-Ser occurs in the mammalian forebrain at high levels and has been proposed to be a neuromodulator that binds to the glycine-binding site of the *N*-methyl-D-Asp (NMDA) receptor, a subtype of the L-glutamate (L-Glu) receptor, and potentiates glutamatergic neurotransmission in the central nervous system, for reviews see refs. 2, 3. Meanwhile, substantial amounts of free D-Asp are present in a wide variety of mammalian tissues, particularly the central nervous, neuroendocrine, and endocrine systems. Several lines of evidence suggest that D-Asp plays an important role in regulating developmental processes, hormone secretion, and steroidogenesis, for reviews see refs. 4, 5.

Loredano Pollegioni and Stefano Servi (eds.), *Unnatural Amino Acids: Methods and Protocols*, Methods in Molecular Biology, vol. 794, DOI 10.1007/978-1-61779-331-8_25, © Springer Science+Business Media, LLC 2012

Since the discovery of free D-Ser and D-Asp in mammals, much attention has been paid to the origins and synthetic pathways of these D-amino acids. In 1999, Wolosker et al. (6) first identified mammalian Ser racemase and purified it from rat brain. Since then, cDNAs of mammalian Ser racemases (EC 5.1.1.16) have been cloned from mouse (7), human (8), and rat (9), and the recombinant forms of mouse and human Ser racemases have been purified and functionally characterized (10–18). The three-dimensional structures of rat and human Ser racemases have also been determined by X-ray crystallography (19). Ser racemase protein and mRNA are present in several mammalian tissues, particularly the brain, liver, and kidney (7, 9, 20–22). In the mammalian brain, Ser racemase is predominantly detected in forebrain regions, such as the cerebral cortex and hippocampus (7, 21, 23), and it has been shown that Ser racemase is localized in neurons as well as in astrocytes and oligodendrocytes (7, 21, 23–28). Recently, three independent groups established Ser racemase knockout mice (23, 29, 30). In these mice, the D-Ser levels in the brain are reduced to 10–20% of those in wild-type control mice, indicating that D-Ser in the brain is predominantly synthesized by Ser racemase.

In contrast to D-Ser, understanding of the D-Asp biosynthetic pathway is much less complete. In cultured mammalian cells, biosynthesis of D-Asp was first established in rat pheochromocytoma PC12 cells (31) and subsequently demonstrated in a rat pituitary tumor cell line GH_3 (32), primary cultured rat embryonic neurons (33), and MPT1 cells (a subclone of PC12 cells) (34). The gene that encodes mouse Asp racemase was also identified recently and its cDNA was cloned (35). In mice, Asp racemase (EC 5.1.1.13) is abundant in the brain, heart, and testis, followed by the adrenal gland, and is expressed at negligible levels in the liver, lung, kidney, and spleen. Immunohistochemical analysis revealed coincident localization of Asp racemase and D-Asp in the mouse brain, pituitary gland, hippocampus, pineal gland, adrenal medulla, and testis (35). Interestingly, depletion of Asp racemase in the adult mouse hippocampus by retrovirus-mediated expression of short-hairpin RNA was shown to elicit profound defects in dendritic development and the survival of neurons (35). Thus, the current data suggest an essential role for Asp racemase in neuronal development, consistent with the previous proposal that D-Asp is important for the development and neurogenesis of the brain (33, 36).

The mammalian Ser and Asp racemases described above both require pyridoxal 5′-phosphate (PLP) for full activity. In addition to these racemases, PLP-independent racemases have also been identified. Since Wood and Gunsalus (37) discovered alanine (Ala) racemase in eubacterium *Streptococcus faecalis*, both PLP-independent and -dependent amino acid racemases have been identified in various eubacteria and archaebacteria, for reviews see refs. 38, 39. In eukaryotes, PLP-independent proline racemase has been identified in the eukaryotic protozoan, *Trypanosoma cruzi* (40). On the other hand, a variety of PLP-dependent eukaryotic racemases, including Ala, tryptophan (Trp), Ser, and Asp racemases, have been characterized. These PLP-dependent eukaryotic racemases are listed in Table 1 (6–8, 35, 41–61).

Table 1
PLP-dependent eukaryotic amino acid racemases and the assay methods used to determine their enzymatic activities

Enzyme	Species	Assay method[a]	References
Ala racemase	Diatom *Thalassiosira* sp.	(i) or (ii)	(41)
	Green alga *Chlamydomonas reinhardtii*	(iii) or (iv)	(42)
	Fungus *Tolypocladium niveum*	(ii), (v) or (vi)	(43)
	Fungus *Cochliobolus carbonum*	(v)	(44)
	Yeast *Schizosaccharomyces pombe*	(v) or (vi)	(45)
	Black tiger prawn *Penaeus monodon*	(v), (vi), (vii) or (viii)	(46–48)
	Crayfish *Procambarus clarkii*	(i) or (ii)	(49)
	Bivalve *Corbicula japonica*	(v) or (ix)	(50–52)
	Alfalfa *Medicago sativa* L.	(i) or (iv)	(53)
	Protozoan *Leishmania amazonensis*	(viii)	(54)
Trp racemase	Wheat *Triticum aestivum* L.	(ii) or (x)	(55)
Ser racemase	Silkworm *Bombyx mori*	(viii)	(56)
	Yeast *Schizosaccharomyces pombe*	(viii)	(57)
	Arabidopsis thaliana	(iv) or (xi)	(58)
	Rice *Oryza sativa* L.	(i)	(59)
	Barley *Hordeum vulgare* L.	(iv) or (xi)	(60)
	Rat	(viii) or (xii)	(6)
	Mouse	(viii) or (xii)	(7)
	Human	(viii) or (xii)	(8)
Asp racemase	Bivalve *Scapharca broughtonii*	(i) or (ii)	(61)
	Mouse	(viii)	(35)

[a]Assay methods:

(i) D- or L-amino acids formed enzymatically from the enantiomer were measured by reversed-phase HPLC using the OPA/NAC pre-column derivatization technique

(ii) D-Amino acids formed enzymatically from the enantiomer were determined with D-amino acid oxidase (DAAO) or D-Asp oxidase (DASPO). The 2-oxo acid produced in the DAAO or DASPO reaction was incubated with 2,4-dinitrophenylhydrazine, and the resultant hydrazone derivative was measured spectrophotometrically

(iii) D- and L-amino acids in the reaction mixture were first derivatized with 1-fluoro-2,4-dinitrophenyl-5-L-Ala amide, and then separated on a silica gel plate by two-dimensional thin-layer chromatography. Subsequently, the derivatives were recovered from the plate and quantitated by reversed-phase HPLC

(iv) D-Amino acids formed enzymatically from the enantiomer were measured with DAAO. Hydrogen peroxide produced in the DAAO reaction was measured spectrophotometrically using a chromogenic substrate(s) in the presence of peroxidase

(continued)

Table 1 (continued)

(v) D-Ala formed enzymatically from L-Ala was determined with DAAO coupled to L-lactate dehydrogenase (LDH). Pyruvic acid produced in the DAAO reaction was converted to L-lactate by LDH, together with oxidation of NADH, and consumption of NADH was measured spectrophotometrically

(vi) L-Ala formed enzymatically from D-Ala was converted to pyruvic acid by L-Ala dehydrogenase, together with the reduction of NAD^+, and the formation of NADH was measured spectrophotometrically

(vii) D- or L-amino acids formed enzymatically from the enantiomer were measured by HPLC with a chiral column. Copper sulfate was used as the mobile phase, and the amino acid/copper complex was detected with UV light

(viii) D- or L-amino acids formed enzymatically from the enantiomer were determined by reversed-phase HPLC using the OPA/Boc-L-Cys pre-column derivatization technique

(ix) L-Ala enzymatically formed from D-Ala was determined with Ala aminotransferase (ALT) coupled to LDH. Pyruvic acid produced in the ALT reaction was converted to L-lactate by LDH, together with the oxidation of NADH, and consumption of NADH was measured spectrophotometrically

(x) L-Amino acids formed enzymatically from the enantiomer were determined with L-amino acid oxidase (LAAO). The 2-oxo acid produced in the LAAO reaction was incubated with 2,4-dinitrophenylhydrazine and the resultant hydrazone derivative was measured spectrophotometrically

(xi) L-Amino acids formed enzymatically from the enantiomer were determined with LAAO. Hydrogen peroxide produced in the LAAO reaction was determined spectrophotometrically using a chromogenic substrate in the presence of peroxidase

(xii) D-Amino acids formed enzymatically from the enantiomer were measured with DAAO. Hydrogen peroxide produced in the DAAO reaction was determined by a chemiluminescent assay with peroxidase and luminol

The enzymatic activities of these racemases are assayed by measuring D- or L-amino acids formed from the corresponding enantiomer in the reaction mixture. There are two primary methodologies used for the enantiomeric determination of amino acids. One uses stereospecific enzymes and the other utilizes high-performance liquid chromatography (HPLC) (1). The assay methods for PLP-dependent eukaryotic racemases are summarized in Table 1. These assay methods use stereospecific enzymes such as D- or L-amino acid oxidase or L-amino acid aminotransferase. To measure D- and L-amino acids by HPLC, two methods have been primarily used: (1) diastereomeric derivatization of amino acids followed by separation on achiral stationary phases and (2) direct enantiomeric separation on chiral stationary phases. In the former method, amino acids in the reaction mixture can be determined by conventional reversed-phase HPLC following derivatization with *o*-phthalaldehyde (OPA) and an optically active thiol reagent, such as *N*-acetyl-L-cysteine (NAC) and *N*-*tert*-butyloxycarbonyl-L-cysteine (Boc-L-Cys) (62–64). In the report (65), we synthesized a novel chiral thiol compound, *N*-(*tert*-butylthiocarbamoyl)-L-cysteine ethyl ester (BTCC), which improved enantiomeric separation of amino acids by HPLC, particularly acidic amino acids. In this chapter, we describe an accurate and reliable assay method for the measurement of Asp racemase activity using the OPA/BTCC pre-column derivatization technique.

2. Materials

1. Instrument: a HPLC equipped with an octadecylsilyl silica gel (ODS) column and a fluorescence detector.
2. Dry acetonitrile: dry acetonitrile is prepared by addition of a molecular sieve.
3. Borate buffer (0.2 M, pH 10.2): dissolve 2.47 g of boric acid in 190 mL of H_2O, and adjust the pH to 10.2 by adding 5 M NaOH. Subsequently add H_2O to 200 mL.
4. HPLC-grade methanol and acetonitrile.
5. Phosphate buffer (50 mM, pH 6.8): dissolve 1.56 g of $NaH_2PO_4 \cdot 2H_2O$ in 190 mL of H_2O, adjust to pH 6.8 by adding 5 M NaOH. Subsequently add H_2O to 200 mL.
6. D- and L-Asp and D,L-Asp (racemic mixture): add 3 mL of H_2O to 66.6 mg of D- or L-Asp, or to 66.6 mg of D,L-Asp, and adjust to pH 6.8 by adding 5 M NaOH. Subsequently add H_2O to 5 mL (100 mM).
7. PLP: dissolve 12.36 mg of PLP in 1 mL of 10 mM phosphate buffer (pH 8.0) (50 mM).
8. Phosphate buffer (100 mM, pH 8.0): dissolve 3.12 g of $NaH_2PO_4 \cdot 2H_2O$ in 190 mL of H_2O and adjust to pH 8.0 by adding 5 M NaOH. Subsequently add H_2O to 200 mL.
9. Fluorogenic chiral derivatization reagents: *N*-(*tert*-butylthiocarbamoyl)-L-cysteine ethyl ester (BTCC) and *o*-phthalaldehyde (OPA).

2.1. Synthesis of BTCC

1. BTCC is synthesized from commercially available L-cysteine ethyl ester hydrochloride and *tert*-butyl isothiocyanate, as described in our previous report (65).
2. Mix L-cysteine ethyl ester hydrochloride (13.3 g), *tert*-butyl isothiocyanate (7.0 g), and triethylamine (7.2 g) and keep overnight at room temperature with gentle stirring in 300 mL of dry acetonitrile.
3. Subsequently, evaporate the acetonitrile and dissolve the resulting residue in ethyl acetate.
4. Wash this solution consecutively with 1 M hydrochloric acid, 5% sodium hydrogen carbonate, and distilled water.
5. Separate the organic layer and dry over anhydrous sodium sulfate.
6. After evaporation of ethyl acetate, a syrup of BTCC (approximately 9.93 g, reaction yield 65%) is obtained.
7. BTCC (158.4 mg) is mixed with commercially available OPA (80.5 mg) in 10 mL of acetonitrile and is used for the fluorogenic chiral derivatization of amino acids (Fig. 1).

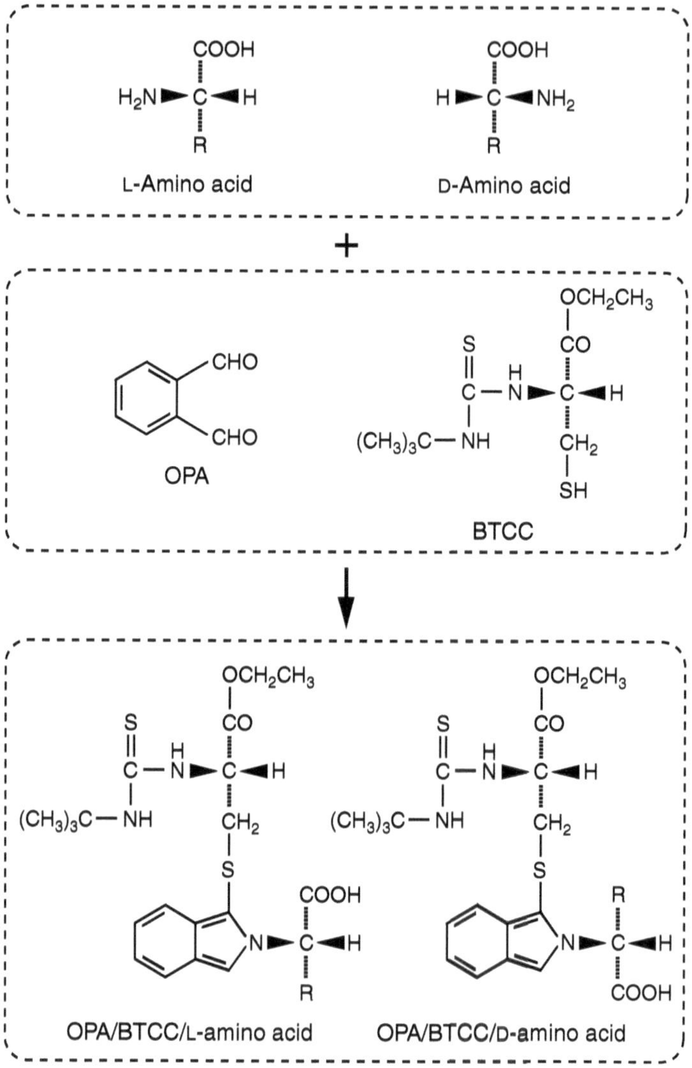

Fig. 1. Fluorogenic chiral derivatization of amino acids with BTCC and OPA. Reproduced from ref. 65 with permission from Elsevier.

8. This reagent solution is stable for approximately 3 months at 4°C. See Note 1.

3. Methods

3.1. Fluorogenic Chiral Derivatization

1. Fluorogenic chiral derivatization is carried out with an autosample injector (GILSON model 231) as follows.
2. Add a sample solution (10 μL) of assay mixture to a sample tube placed on the tray of the sample injector.

3. Seal the tube and add 30 μL of 0.2 M borate buffer and 20 μL of the derivatization reagent solution automatically by the autosample injector.
4. Mix automatically the solution in the tube by the injector by repeated suction and ejection.
5. During this process (approximately 1.5 min), almost complete fluorogenic chiral derivatization is achieved. See Note 2.
6. Inject automatically an aliquot of the mixture (5 μL) into the HPLC.

3.2. HPLC Separation

1. Column: mightysil RP-18 GP (75 × 3.0 mm, i.d. 5 μm, Kanto Chem. Co. Inc., Tokyo, Japan) maintained at 20°C.
2. Mobile phase: 50 mM phosphate buffer (pH 6.8):methanol:acetonitrile (70:10:20).
3. Flow rate: 0.5 mL/min.
4. Fluorescence detector: excitation = 340 nm; emission = 420 nm.
5. Time program: after injection of the sample, the mobile phase is delivered for 12 min and the column is washed with 80% aqueous acetonitrile for 4 min. The column is equilibrated again with the mobile phase for the next injection.
6. These steps are controlled by the HPLC machine.
7. Repeat this program every 23 min. See Notes 3–6.

3.3. Enzyme Assay

1. Asp racemase activity is assayed in a reaction mixture (200 μL) containing 100 mM sodium phosphate buffer (pH 8.0), 10 mM L-Asp as a substrate, and 50 μM PLP as coenzyme and the enzyme source.
2. For the assay of Asp racemase in *Thermoplasma acidophilum* HO-62 (66), incubate the reaction mixture at 70°C for 60 min.
3. Stop the reaction by cooling to 4°C and add 10 μL of 100% trichloroacetic acid.
4. Isolate the precipitated proteins by centrifugation at 15,000 rpm (20,600 × *g*) for 10 min and subject the supernatant to automated HPLC analysis.
5. Disrupt frozen *Thermoplasma acidophilum* HO-62 cells (67) at 4°C by sonication in ice-cold 10 mM sodium phosphate buffer (pH 7.4) containing 0.9% NaCl.
6. Centrifuge the homogenate at 3,000 rpm (900 × *g*) for 15 min to remove large cellular debris and then centrifuge again the supernatant at 33,000 rpm (100,000 × *g*) for 60 min.
7. Assay the resultant supernatant for Asp racemase activity. See Notes 7–9.

4. Notes

1. BTCC is readily prepared from commercially available reagents and the product can be used directly for fluorogenic chiral derivatization without any special refining.
2. The fluorogenic chiral derivatization reaction proceeds rapidly at room temperature under mild conditions.
3. The ethyl ester of BTCC increases the hydrophobicity of the resultant amino acid derivatives and presumably increases retention on a reverse-phase column with charged and polar amino acids. The rigid structure of the *tert*-butyl group of BTCC is presumed to improve the chiral separation of the derivatized amino acid enantiomers.
4. Standard D- and L-Asp were derivatized as described in the Subheading 3, and the derivatized enantiomers were separated as shown in Fig. 2. It is of note that derivatized D-Asp was eluted before the L-Asp derivative. Elution order is important for assaying racemase activity, since it is necessary for the accurate measurement of small amounts of one enantiomer (D-Asp in this case) produced from large amounts of the other enantiomer (L-Asp) added as substrate. D-Asp could be accurately measured when it was eluted before L-Asp, while it could not be measured when it was eluted after L-Asp because the large peak of L-Asp obscured its small peak. Linearities for D- and L-Asp were evaluated by measuring the correlations between peak areas and concentrations ranging from 0.03 to 3.0 mM, details are available in ref. 65. The correlations were satisfactory, evidenced by the regression coefficients (r^2) that were 0.9975 and 0.9977 for D- and L-Asp, respectively. The correlations of variance of the peak areas of 0.3 mM of D- and

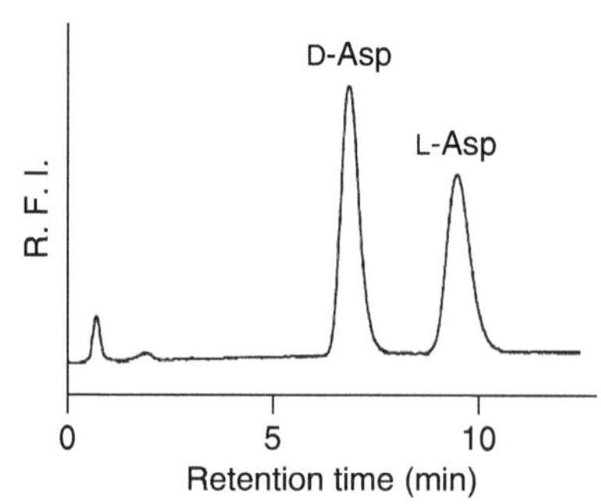

Fig. 2. Chromatograms of the derivatives formed from standard D- and L-Asp and OPA/BTCC. *RFI* relative fluorescence intensity. Reproduced from ref. 65, with permission from Elsevier.

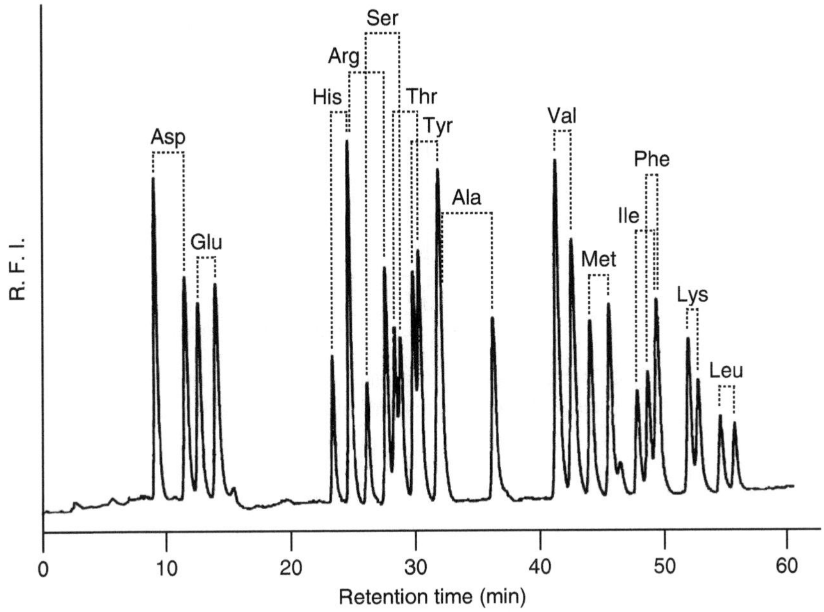

Fig. 3. Chromatogram of the derivatives formed from various standard D,L-amino acids. Mobile phase A, acetonitrile:20 mM phosphate (pH 7.0) (27.5:100). Mobile phase B, methanol:20 mM phosphate (pH 7.0) (90:100). Elution was carried out at flow rate of 1.0 mL/min for 12 min with mobile phase A, and then a linear gradient was applied for 15 min from 0 to 50% mobile phase B, followed by elution for 3 min with 50% B. Subsequently, a linear gradient from 50 to 75% B was applied for 37 min, and a final linear gradient was applied from 75 to 100% B for 5 min.

L-Asp were 3.3% and 4.5% ($n=10$), respectively, and those of the retention times for each isomer were 1.5% (D-Asp) and 1.1% (L-Asp) ($n=10$).

5. The enantioseparation of several amino acids (Glu, Ser, Ala and Phe) was successfully carried out by this method or methods with simple modifications (68), and the D-isomer was eluted before the corresponding L-isomer in most cases. Therefore, this racemase assay can be carried out with L-amino acids other than L-Asp as a substrate.
6. Figure 3 shows a chromatogram of various standard D,L-amino acids that were separated by the modified procedure described in the figure legend. The enantiomers of these amino acids were clearly separated, demonstrating that this procedure can be applied to assay these amino acids.
7. In Fig. 4, the reaction mixture obtained for Asp racemase from *Thermoplasma acidophilum* was separated following this method (66). A small amount of produced D-Asp was clearly detected even in the presence of a large amount of L-Asp. D-Asp could be quantified accurately with high sensitivity, even in the presence of excess amounts of L-Asp. The detection limit for D-Asp was approximately 1 pmol.

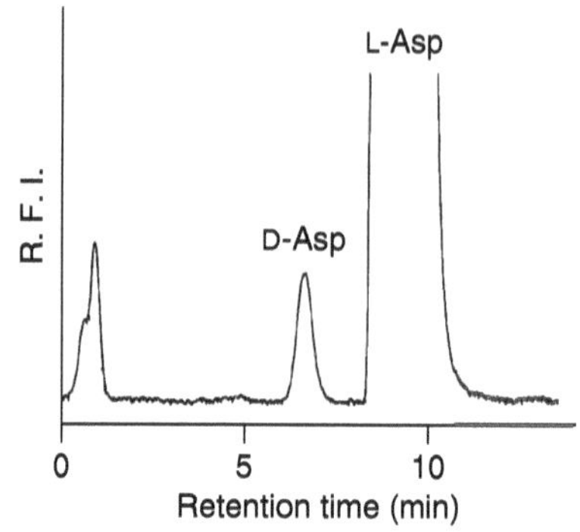

Fig. 4. Chromatogram of the reaction mixture for the Asp racemase assay in *Thermoplasma acidophilum*. Reproduced from ref. 65, with permission from Elsevier.

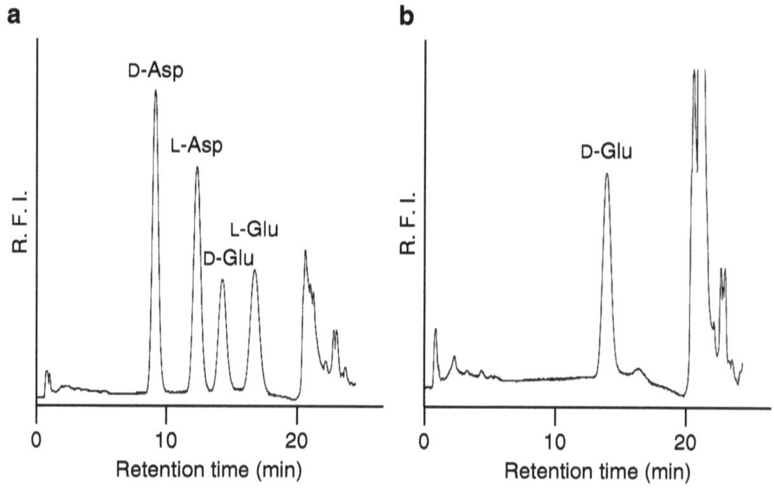

Fig. 5. Chromatogram of the reaction mixture obtained for D-amino acid aminotransferase from *Arabidopsis thaliana*. For fluorogenic chiral derivatization, a sample solution (10 μL) of assay mixture (68), 50 μL of 0.2 M borate buffer and 40 μL of the derivatization reagent solution were mixed. HPLC conditions were as follows: column, Inertsil ODS-3 (150 × 4.6 mm, i.d. 5 μm, GL Sciences, Inc., Tokyo, Japan) maintained at 25°C; mobile phase A, 20 mM phosphate buffer (pH 7.0):acetonitrile:tetrahydrofuran (75:22:3); and mobile phase B, 20 mM phosphate buffer (pH 7.0):acetonitrile:tetrahydrofuran (50:47:3). Elution was carried out at flow rate of 1 mL/min for 20 min with 100% mobile phase A, and then a linear gradient was applied for 30 min until 100% mobile phase B was reached. The column was washed with 80% aqueous acetonitrile for 10 min and equilibrated again with mobile phase A for the next injection. These steps were regulated by the system controller. The program was repeated every 80 min. Fluorescence detector: excitation = 340 nm; emission = 420 nm. (**a**) Chromatogram of the standard racemic mixture of Asp and Glu. Peaks of D- and L-Ala appear after 30 min. (**b**) With D-Ala and α-ketoglutarate as substrates, pyruvate and D-Glu were produced by D-amino acid aminotransferase from *Arabidopsis thaliana*.

8. This method is of practical use, since its efficiency is easily maintained by changing the guard column after approximately 500 injections. The fluorogenic chiral derivatization and injection into the HPLC can be carried out manually instead of through autosampling.
9. The method is also applicable to assays of other enzymes, such as stereospecific D-amino acid aminotransferase activity. Figure 5 shows a chromatogram of the reaction mixture obtained using the stereospecific D-amino acid aminotransferase from *Arabidopsis thaliana* (68). The method details are described in the figure legend.

References

1. Konno R, Brückner H, D'Aniello et al. (2007) D-Amino acids: a new frontier in amino acids and protein research, practical methods and protocols. Nova Science Publishers, Inc., New York.
2. Nishikawa T (2005) Metabolism and functional roles of endogenous D-serine in mammalian brains. Biol Pharm Bull 28, 1561–1565.
3. Scolari M J, Acosta G B (2007) D-Serine: a new word in the glutamatergic neuro-glial language. Amino Acids 33, 563–574.
4. D'Aniello A (2007) D-Aspartic acid: an endogenous amino acid with an important neuroendocrine role. Brain Res Brain Res Rev 53, 215–234.
5. Homma H (2007) Biochemistry of D-aspartate in mammalian cells. Amino Acids 32, 3–11.
6. Wolosker H, Sheth K N, Takahashi M et al. (1999) Purification of serine racemase: biosynthesis of the neuromodulator D-serine. Proc Natl Acad Sci USA 96, 721–725.
7. Wolosker H, Blackshaw S, Snyder S H (1999) Serine racemase: a glial enzyme synthesizing D-serine to regulate glutamate-N-methyl-D-aspartate neurotransmission. Proc Natl Acad Sci USA 96, 13409–13414.
8. De Miranda J, Santoro A, Engelender S, Wolosker H (2000) Human serine racemase: molecular cloning, genomic organization and functional analysis. Gene 256, 183–188.
9. Konno R (2003) Rat cerebral serine racemase: amino acid deletion and truncation at carboxy terminus. Neurosci Lett 349, 111–114.
10. Baumgart F, Mancheño J M, Rodríguez-Crespo I (2007) Insights into the activation of brain serine racemase by the multi-PDZ domain glutamate receptor interacting protein, divalent cations and ATP. FEBS J 274, 4561–4571.
11. Cook S P, Galve-Roperh I, Martinez del Pozo A, Rodríguez-Crespo I (2002) Direct calcium binding results in activation of brain serine racemase. J Biol Chem 277, 27782–27792.
12. De Miranda J, Panizzutti R, Foltyn V N, Wolosker H (2002) Cofactors of serine racemase that physiologically stimulate the synthesis of the N-methyl-D-aspartate (NMDA) receptor coagonist D-serine. Proc Natl Acad Sci USA 99, 14542–14547.
13. Foltyn V N, Bendikov I, De Miranda J et al. (2005) Serine racemase modulates intracellular D-serine levels through an α,β-elimination activity. J Biol Chem 280, 1754–1763.
14. Hoffman H E, Jirásková J, Ingr M et al. (2009) Recombinant human serine racemase: enzymologic characterization and comparison with its mouse ortholog. Protein Expr Purif 63, 62–67.
15. Nagayoshi C, Ishibashi M, Tokunaga M (2009) Purification and characterization of human brain serine racemase expressed in moderately halophilic bacteria. Protein Pept Lett 16, 201–206.
16. Panizzutti R, De Miranda J, Ribeiro C S et al. (2001) A new strategy to decrease N-methyl-D-aspartate (NMDA) receptor coactivation: inhibition of D-serine synthesis by converting serine racemase into an eliminase. Proc Natl Acad Sci USA 98, 5294–5299.
17. Strísovsky K, Jirásková J, Barinka C et al. (2003) Mouse brain serine racemase catalyzes specific elimination of L-serine to pyruvate. FEBS Lett 535, 44–48.
18. Strísovsky K, Jirásková J, Mikulová A et al. (2005) Dual substrate and reaction specificity in mouse serine racemase: identification of high-affinity dicarboxylate substrate and inhibitors and analysis of the β-eliminase activity. Biochemistry 44, 13091–13100.
19. Smith M A, Mack V, Ebneth A et al. (2010) The structure of mammalian serine racemase: evidence

for conformational changes upon inhibitor binding. J Biol Chem 285, 12873–12881.

20. Wang L-Z, Zhu X-Z (2003) Spatiotemporal relationships among D-serine, serine racemase, and D-amino acid oxidase during mouse postnatal development. Acta Pharmacol Sin 24, 965–974.
21. Xia M, Liu Y, Figueroa D J et al. (2004) Characterization and localization of a human serine racemase. Brain Res Mol Brain Res 125, 96–104.
22. Yoshikawa M, Kobayashi T, Oka T et al. (2004) Distribution and MK-801-induced expression of serine racemase mRNA in rat brain by real-time quantitative PCR. Brain Res Mol Brain Res 128, 90–94.
23. Miya K, Inoue R, Takata, Y et al. (2008) Serine racemase is predominantly localized in neurons in mouse brain. J Comp Neurol 510, 641–654.
24. Ding X, Ma N, Nagahama M et al. (2011) Localization of D-serine and serine racemase in neurons and neuroglias in mouse brain. Neuro Sci. 96, 157–163.
25. Kartvelishvily E, Shleper M, Balan L et al. (2006) Neuron-derived D-serine release provides a novel means to activate N-methyl-D-aspartate receptors. J Biol Chem 281, 14151–14162.
26. Verrall L, Walker M, Rawlings N et al. (2007) D-Amino acid oxidase and serine racemase in human brain: normal distribution and altered expression in schizophrenia. Eur J Neurosci 26, 1657–1669.
27. Yoshikawa M, Nakajima K, Takayasu N et al. (2006) Expression of the mRNA and protein of serine racemase in primary cultures of rat neurons. Eur J Pharmacol 548, 74–76.
28. Yoshikawa M, Takayasu N, Hashimoto T et al. (2007) The serine racemase mRNA is predominantly expressed in rat brain neurons. Arch Histol Cytol 70, 127–134.
29. Basu A C, Tsai G E, Ma C-L et al. (2009) Targeted disruption of serine racemase affects glutamatergic neurotransmission and behavior. Mol Psychiatry 14, 719–727.
30. Labrie V, Fukumura R, Rastogi A et al. (2009) Serine racemase is associated with schizophrenia susceptibility in humans and in a mouse model. Hum Mol Genet 18, 3227–3243.
31. Long Z, Homma H, Lee J-A et al. (1998) Biosynthesis of D-aspartate in mammalian cells. FEBS Lett 434, 231–235.
32. Long Z, Lee J-A, Okamoto T et al. (2000) D-Aspartate in a prolactin-secreting clonal strain of rat pituitary tumor cells (GH_3). Biochem Biophys Res Commun 276, 1143–1147.
33. Wolosker H, D'Aniello A, Snyder S H (2000) D-Aspartate disposition in neuronal and endocrine tissues: ontogeny, biosynthesis and release. Neuroscience 100, 183–189.
34. Long Z, Sekine M, Adachi M et al. (2002) Cell density inversely regulates D- and L-aspartate levels in rat pheochromocytoma MPT1 cells. Arch Biochem Biophys 404, 92–97.
35. Kim P M, Duan X, Huang A S et al. (2010) Aspartate racemase, generating neuronal D-aspartate, regulates adult neurogenesis. Proc Natl Acad Sci USA 107, 3175–3179.
36. Sakai K, Homma H, Lee J-A et al. (1998) Emergence of D-aspartic acid in the differentiating neurons of the rat central nervous system. Brain Res 808, 65–71.
37. Wood W A, Gunsalus I C (1951) D-Alanine formation; a racemase in *Streptococcus faecalis*. J Biol Chem 190, 403–416.
38. Yoshimura T, Esak N (2003) Amino acid racemases: functions and mechanisms. J Biosci Bioeng 96, 103–109.
39. Yoshimura T, Goto M (2008) D-Amino acids in the brain: structure and function of pyridoxal phosphate-dependent amino acid racemases. FEBS J 275, 3527–3537.
40. Reina-San-Martín B, Degrave W, Rougeot C et al. (2000) A B-cell mitogen from a pathogenic trypanosome is a eukaryotic proline racemase. Nat Med 6, 890–897.
41. Yokoyama T, Tanaka Y, Sato Y et al. (2005) Alanine racemase activity in the microalga *Thalassiosira* sp. Fish Sci 71, 924–930.
42. Nishimura K, Tomoda Y, Nakamoto Y et al. (2007) Alanine racemase from the green alga *Chlamydomonas reinhardtii*. Amino Acids 32, 59–62.
43. Hoffmann K, Schneider-Scherzer E, Kleinkauf H, Zocher R (1994) Purification and characterization of eucaryotic alanine racemase acting as key enzyme in cyclosporin biosynthesis. J Biol Chem 269, 12710–12714.
44. Cheng Y-Q, Walton J D (2000) A eukaryotic alanine racemase gene involved in cyclic peptide biosynthesis. J Biol Chem 275, 4906–4911.
45. Uo T, Yoshimura T, Tanaka N et al. (2001) Functional characterization of alanine racemase from *Schizosaccharomyces pombe*: a eucaryotic counterpart to bacterial alanine racemase. J Bacteriol 183, 2226–2233.
46. Fujita E, Okuma E, Abe H (1997) Partial purification and properties of alanine racemase from the muscle of black tiger prawn *Penaeus monodon*. Fish Sci 63, 440–445.

47. Yoshikawa N, Dhomae N, Takio K, Abe H (2002) Purification, properties, and partial amino acid sequences of alanine racemase from the muscle of the black tiger prawn *Penaeus monodon*. Comp Biochem Physiol B Biochem Mol Biol 133, 445–453.
48. Uo T, Ueda M, Nishiyama T et al. (2001) Purification and characterization of alanine racemase from hepatopancreas of black-tiger prawn, *Penaeus monodon*. J Mol Catal B Enzym 12, 137–144.
49. Shibata K, Shirasuna K, Motegi K et al. (2000) Purification and properties of alanine racemase from crayfish *Procambarus clarkii*. Comp Biochem Physiol B Biochem Mol Biol 126, 599–608.
50. Matsushima O, Katayama H, Yamada K, Kado Y (1984) Occurrence of free D-alanine and alanine racemase activity in bivalve molluscs with special reference to intracellular osmoregulation. Mar Biol Lett 5, 217–225.
51. Omura Y, Hayashi Y S, Matsushima O et al. (1985) Partial purification and characterization of alanine racemase from the brackish-water bivalve *Corbicula japonica*. J Exp Mar Biol Ecol 94, 281–289.
52. Nomura T, Yamamoto I, Morishita F et al. (2001) Purification and some properties of alanine racemase from a bivalve mollusc *Corbicula japonica*. J Exp Zool 289, 1–9.
53. Ono K, Yanagida K, Oikawa T et al. (2006) Alanine racemase of alfalfa seedlings (*Medicago sativa* L.): first evidence for the presence of an amino acid racemase in plants. Phytochemistry 67, 856–860.
54. Panizzutti R, de Souza Leite M, Pinheiro C M, Meyer-Fernandes J R (2006) The occurrence of free D-alanine and an alanine racemase activity in *Leishmania amazonensis*. FEMS Microbiol Lett 256, 16–21.
55. Rekoslavskaya N I, Yur'eva O V, Shibanova L A, Salyaev R K (1997) Synthesis and physiological function of D-tryptophan during wheat germination. Russ J Plant Physiol 44, 227–234.
56. Uo T, Yoshimura T, Shimizu S, Esaki N (1998) Occurrence of pyridoxal 5′-phosphate-dependent serine racemase in silkworm, *Bombyx mori*. Biochem Biophys Res Commun 246, 31–34.
57. Yamauchi T, Goto M, Wu H Y et al. (2009) Serine racemase with catalytically active lysinoalanyl residue. J Biochem 145, 421–424.
58. Fujitani Y, Nakajima N, Ishihara K et al. (2006) Molecular and biochemical characterization of a serine racemase from *Arabidopsis thaliana*. Phytochemistry 67, 668–674.
59. Gogami Y, Ito K, Kamitani Y et al. (2009) Occurrence of D-serine in rice and characterization of rice serine racemase. Phytochemistry 70, 380–387.
60. Fujitani Y, Horiuchi T, Ito K, Sugimoto M (2007) Serine racemases from barley, *Hordeum vulgare* L., and other plant species represent a distinct eukaryotic group: gene cloning and recombinant protein characterization. Phytochemistry 68, 1530–1536.
61. Shibata K, Watanabe T, Yoshikawa H et al. (2003) Purification and characterization of aspartate racemase from the bivalve mollusk *Scapharca broughtonii*. Comp Biochem Physiol B Biochem Mol Biol 134, 307–314.
62. Aswad D W (1984) Determination of D- and L-aspartate in amino acid mixtures by high-performance liquid chromatography after derivatization with a chiral adduct of *o*-phthaldialdehyde. Anal Biochem 137, 405–409.
63. Hashimoto A, Nishikawa T, Oka T et al. (1992) Determination of free amino acid enantiomers in rat brain and serum by high-performance liquid chromatography after derivatization with N-tert.-butyloxycarbonyl-L-cysteine and *o*-phthaldialdehyde. J Chromatogr 582, 41–48.
64. Nimura N, Kinoshita T (1986) *o*-Phthalaldehyde–*N*-acetyl-L-cysteine as a chiral derivatization reagent for liquid chromatographic optical resolution of amino acid enantiomers and its application to conventional amino acid analysis. J Chromatogr 352, 169–177.
65. Nimura N, Fujiwara T, Watanabe A et al. (2003) A novel chiral thiol reagent for automated precolumn derivatization and high-performance liquid chromatographic enantio separation of amino acids and its application to the aspartate racemase assay. Anal Biochem 315, 262–269.
66. Long Z, Lee J-A, Okamoto T. et al. (2001) Occurrence of D-amino acids and a pyridoxal 5′-phosphate-dependent aspartate racemase in the acidothermophilic archaeon, *Thermoplasma acidophilum*. Biochem Biophys Res Commun 281, 317–321.
67. Yasuda M, Oyaizu H, Yamagishi A, Oshima T (1995) Morphological variation of new *Thermoplasma acidophilum* isolates from Japanese hot springs. Appl Environ Microbiol 61, 3482–3485.
68. Funakoshi M, Sekine M, Katane M et al. (2008) Cloning and functional characterization of *Arabidopsis thaliana* D-amino acid aminotransferase-D-aspartate behavior during germination. FEBS J 275, 1188–1200.

Chapter 26

Assays of D-Amino Acid Oxidases

Gabriella Tedeschi, Loredano Pollegioni, and Armando Negri

Abstract

D-Amino acid oxidase and D-aspartate oxidase are two well-known FAD-containing flavooxidases that catalyze the same reaction (the oxidative deamination) on different D-amino acids. D-aspartate oxidase is specific for acidic D-amino acids (i.e., D-aspartate and D-glutamate) and D-amino acid oxidase is active on neutral and polar D-amino acids (a low activity is also detected on basic D-amino acids). The assay of these flavoenzymes is of utmost importance in different fields because D-amino acids are common constituents of bacterial cell walls, are present in foods and because free D-serine and D-aspartic acid were identified in brain and peripheral tissues of mammals. In this chapter, we report on the most used methods employed to assay the activity of D-amino acid oxidase and D-aspartate oxidase. Interestingly, their activity can be followed using different assays, namely D-amino acid or oxygen consumption, α-keto acid or ammonia production, or using artificial dyes as final indicator of the flavin redox reaction.

Key words: Enzymatic assay, D-Amino acid oxidase, D-Aspartate oxidase, Analytical detection

1. Introduction

D-Amino acids (D-aa, the D-isomer of amino acids) are common constituents of bacterial cell walls; D-glutamate and D-alanine are the most recurrent components. Their presence has been reported also in marine invertebrates as well as in animal tissues (1). D-aa are also present in foods: their occurrence is normally associated to food ingredients, processing, and ripening. Their presence is expected to yield a decrease in proteins digestibility, thus affecting the bio-availability of essential amino acids, and ultimately impairing the nutritional quality of food. As molecular marker of ripening and as index for the assessment of food quality both the amount of D-aa and the D/(D + L) ratio have been proposed (2).

In recent years, significant amounts of free D-serine and D-aspartic acid were identified in brain and peripheral tissues of

Loredano Pollegioni and Stefano Servi (eds.), *Unnatural Amino Acids: Methods and Protocols*,
Methods in Molecular Biology, vol. 794, DOI 10.1007/978-1-61779-331-8_26, © Springer Science+Business Media, LLC 2012

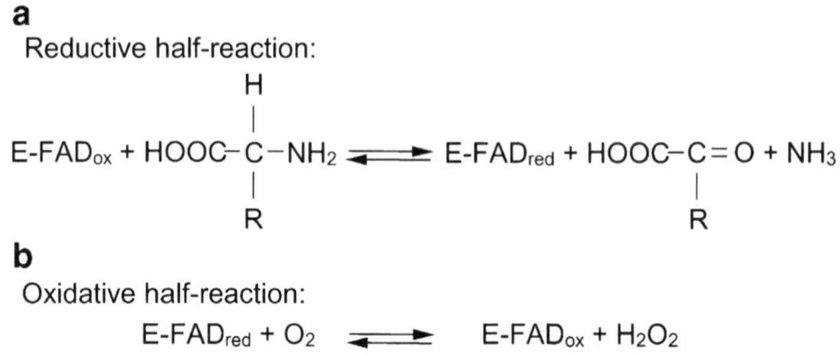

Fig. 1. Reductive and oxidative half-reaction catalysed by DAAO and DASPO. The R substituent is preferentially neutral or polar for DAAO and acidic for DASPO; D-proline is a substrate common to both flavooxidases.

mammals (3). D-Aspartic acid is mainly located in neurons of the neuroendocrine system and in neuroendocrine glands (pineal, adrenals, etc.); its presence in proteins increases with aging (4). D-Aspartic acid is specifically degraded by the FAD-containing flavoenzyme D-aspartate oxidase (DASPO, EC 1.4.3.1).

D-Serine has been recognized as a new neuromodulator present in the forebrain where it serves as ligand for the "glycine site" of *N*-methyl-D-aspartate subtype of glutamate receptors, thus modulating the glutamatergic neurotransmission (5). In humans, D-serine is produced by the isomerisation of L-serine catalyzed by the PLP-containing enzyme serine racemase and degraded by the α,β-elimination catalyzed by the same enzyme and/or by the deaminative oxidation due to D-amino acid oxidase (DAAO, EC 1.4.3.3), for a recent review see ref. 6. In order to unravel the functional consequences of D-aspartate and D-serine mediated signaling, it is essential to clarify the mechanisms controlling their metabolism and thus to assay the enzymatic activities related to their catabolism.

Both DAAO and DASPO catalyze the oxidative deamination of D-aa to the corresponding α-keto acid and ammonia. At this step, FAD is reduced and the catalytic cycle is completed through its reoxidation by O_2 to yield hydrogen peroxide (Fig. 1).

The reaction of DAAO and DASPO with D-aa, as reported in Fig. 1, can be followed using different assays, namely D-aa or oxygen consumption, α-keto acid or ammonia production, or using artificial dyes as final indicator of the flavin redox reaction (i.e., dyes that react with the product hydrogen peroxide).

2. Materials

2.1. Enzymes

1. Recombinant wild-type DAAO from the yeast *Rhodotorula gracilis* was expressed and purified from *Escherichia coli* cells (7). The enzyme concentration is determined using an extinction coefficient at 455 nm of 12.6/mM/cm.

Purified yeast DAAO shows a specific activity on D-alanine of 110 U/mg protein and a $K_{m,app}$ of 0.83 mM at pH 8.5 and 25°C (7); for a review see refs. 8, 9. During the years, a number of recombinant DAAO variants have been prepared showing an altered substrate specificity, see refs. 8–10.

2. Purified beef kidney DASPO is available either from natural source (11) or as recombinant protein overexpressed in *E. coli* (12). The specific activity of the purified enzyme on D-aspartate is 30 U/mg protein.
3. One DAAO/DASPO unit corresponds to the amount of enzyme that converts 1 μmol of D-aa (D-Ala or D-Asp, respectively) per minute at 25°C.

2.2. Oxygen Consumption Assay

1. Hansatech oxygen electrode (Hansatech Instruments Ltd, Pentney, UK).
2. 100 mM Disodium pyrophosphate, pH 8.5.
3. 100 mM D-Alanine, in 100 mM disodium pyrophosphate, pH 8.5.
4. 10 mM FAD, in water.

2.3. Spectrophotometric Assays

1. UV/Vis spectrophotometer and 1 mL semimicro-cuvettes.
2. Water bath with external circulation for temperature control.
3. 10 mM D-Phenylglycine, in 75 mM disodium pyrophosphate, pH 8.5.
4. 1 M D-Aspartic acid, in 200 mM triethanolammine-HCl, pH 7.6.
5. 1 mM 2,4-Dinitrophenylhydrazine (DNP), in 1 M HCl.
6. 10 mM *o*-Dianisidine dihydrochloride (o-DNS), in H_2O.
7. Peroxidase from horseradish (HRP, EC 1.11.1.7; grade I, Roche Diagnostics, Basel, Switzerland): prepare a 100 U/mL solution in 100 mM disodium pyrophosphate, pH 8.5.
8. 15 mM 4-Aminoantipyrine (4-AAP), in H_2O.
9. Phenol.
10. Glutamate dehydrogenase (GDH, EC 1.4.12; Roche Diagnostics, Basel, Switzerland): prepare a 1,500 U/mL solution in 75 mM disodium pyrophosphate, pH 8.5.
11. 15 mM NADH, in 75 mM disodium pyrophosphate, pH 8.5.
12. 100 mM α-Ketoglutaric acid, in H_2O.

2.4. Amplex® Red Assay (Molecular Probes, Eugene, OR, USA)

1. 50 mM Sodium phosphate, pH 7.4.
2. 0.5 M D-serine, in 50 mM sodium phosphate, pH 7.4 (stock solution).
3. Lysis buffer (daily prepared): 50 mM sodium phosphate, pH 7.4, 0.7 μg/mL pepstatine (1 μM), 1 μg/mL leupeptin, and 10 μM FAD.

4. 100 mM Sodium benzoate, in 50 mM sodium phosphate, pH 7.4.
5. 0.5 M Sodium azide, in 50 mM sodium phosphate, pH 7.4.
6. Cell samples: 3×10^5 cells (e.g., U87 glioblastoma cells or COS-7 fibroblasts) in 750 μL of lysis buffer.
7. HRP: prepare a 10 U/mL solution in 50 mM sodium phosphate, pH 7.4. This solution is stable at −20°C.
8. Amplex® Ultra Red Reagent: open the vial under nitrogen and add 166 μL of DMSO. The final concentration is 20 mM. Prepare 25 μL aliquots and store in the dark at −20°C.
9. 96-Well microplates.
10. Working Solution (1.5 mL): mix 1.3 mL of 50 mM sodium phosphate, pH 7.4, 150 μL of 500 mM D-serine stock solution, 30 μL of 500 mM NaN_3, 15 μL of 1 mM FAD, 30 μL of 10 U/mL HRP and 7.5 μL of 20 mM Amplex® Ultra Red reagent. This reagent is prepared fresh daily and it is conserved at 4°C.
11. Stop Solution: add 1.45 mL of ethanol in the Stop Solution Reagent vial. Mix 1.0 mL of this solution with 1.0 mL of milliQ-grade H_2O. This Stop Solution is stable for 1 month, at 4°C in the dark.
12. A fluorescent microplate reader (e.g., Infinite 200, TECAN).

2.5. Activity Assay of Immobilized D-Aspartate Oxidase

1. CNBr-activated Sepharose 4B (GE Healthcare, Piscataway, NJ, USA).
2. Buffer A: 100 mM potassium phosphate, pH 8.5, 0.5 M NaCl, 2 μM FAD.
3. Water bath with external circulation for external temperature control.
4. Spectrophotometer and 1 mL semimicro-cuvette.
5. Bench top centrifuge.
6. Rotating platform.

3. Methods

All assays were carried out at 25°C, using a temperature control system.

3.1. Determination Based on Oxygen Consumption

This is a direct enzymatic method based on a Hansatech Clark-type oxygen electrode (see Note 1). During DAAO/DASPO reaction, the pO_2 decreases in the oxymeter chamber and its value is recorded by the CB1D electrode control box connected to a chart recorder/PC.

In order to achieve a fast equilibrium after the addition of the enzyme, the assay solution in the oxymeter chamber is mixed by a magnetic stirrer. The assay is performed as follows:

1. Prepare the reaction mixture by mixing the solution containing the D-aa, disodium pyrophosphate buffer, pH 8.5 (for DAAO) or potassium phosphate, pH 7.4 (for DASPO) and water to a final volume of 1–3 mL. The final concentration of disodium pyrophosphate is 75 mM; the standard concentration for D-alanine is 28 mM; final concentration of potassium phosphate is 50 mM; the standard concentration for D-aspartate is 10 mM.
2. Start the chart recorder, and wait until the signal is stable (at 25°C the oxygen solubility in water is 0.25 mM; see Note 2).
3. Add 5–10 μL of DAAO (~0.1 μg of wild-type DAAO corresponding to 0.01 U) or of DASPO (~0.5 μg of wild-type DASPO corresponding to 0.015 U) and follow the time course of the reaction for 3–5 min.
4. Measure the initial oxygen consumption (ΔpO_2/min) and calculate the enzymatic activity using Eq. 1:

$$\text{enzymatic activity} = \frac{\Delta pO_2/\text{min} \times \text{oxygen solubility } (0.253\,\text{mM})}{\text{mg of added enzyme}} \times \text{assay volume } (1 \text{ or } 3\text{ mL}) = \text{U/mg protein.} \quad (1)$$

5. In order to determine the kinetic parameters of DAAO/DASPO on a specific substrate a Michaelis–Menten plot is built using the ΔpO_2/min values determined on a range of substrate concentrations (see Subheading 3.4).

3.2. Methods Based on Spectrophotometric Assays

All spectrophotometric assays are based on Lambert–Beer's law (see Note 3). The product concentration is determined using the following Eq. 2 and the known extinction coefficient of the compound of interest:

$$\frac{\mu\text{mol}}{\text{mL}} = \Delta\text{Abs}/(l \times \varepsilon), \quad (2)$$

where: $\Delta\text{Abs} = (\text{Abs}_{\text{final}} - \text{Abs}_{\text{initial}})$; l = cuvette path (1 cm); ε = extinction coefficient, expressed as/mM/cm.

3.2.1. Methods Based on Assay of α-Keto Acid Production

Direct Method for DAAO

DAAO activity is frequently assayed using a modification of the original method from Fonda and Anderson (13). It is based on the increase in absorbance at 252 nm that accompanies the oxidative deamination of D-phenylglycine into benzoylformic acid and ammonia (Fig. 2). See Note 4 for details related to the assay of different α-keto acids.

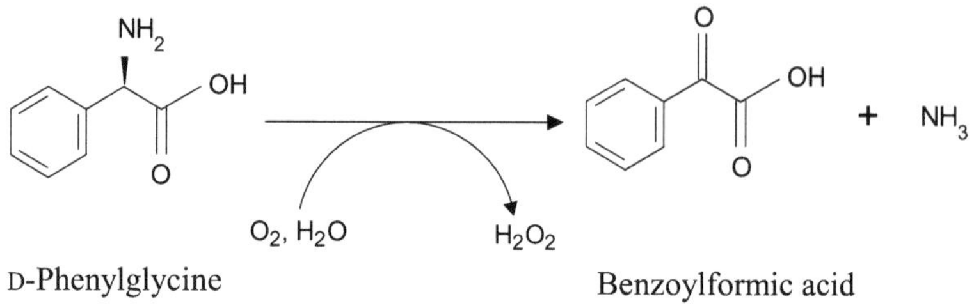

Fig. 2. Reaction of DAAO on D-phenylglycine.

The assay is set up as follows:

1. Prepare the reaction mixture directly in a quartz semimicro-cuvette, using 10 mM D-phenylglycine stock solution in 75 mM disodium pyrophosphate, pH 8.5.
2. Record the initial absorbance (Abs_i) at 252 nm of the mixture.
3. Add wild-type DAAO (~10 mU, 0.1 μg) to a final volume of 1 mL.
4. Record the absorbance change after the enzyme addition and measure the initial value ΔAbs/min.
5. Calculate the enzymatic activity using Eq. 3:

$$\text{Enzymatic activity} = \frac{\Delta \text{Abs}_{252\text{nm}}/\text{min}}{\varepsilon \times \text{mg added DAAO}} = \text{U/mg protein}, \quad (3)$$

and an extinction coefficient for benzoylformic acid $\varepsilon_{252\text{ nm}} = 14.3/\text{mM/cm}$.

Indirect Method with DNP

In this assay, the α-keto acids produced by the DAAO activity on its substrates is evaluated as 2,4-dinitrophenylhydrazone derivative (14), which is obtained by the reaction with 2,4-dinitrophenylhydrazine (DNP) (Fig. 3).

The 2,4-dinitrophenylhydrazone derivatives show a characteristic absorbance spectrum in the visible region, with an intensity maximum at ~445 nm (see Note 5). The assay is set up as follows:

1. The reaction mixture is prepared mixing 2 mM D-alanine, 75 mM disodium pyrophosphate, pH 8.5, and ~1 U of DAAO (all final concentrations) in a final volume of 300 μL.
2. Incubate at 25°C for 10 min.
3. Add 150 μL of 1 mM DNP, solubilized in 1 M HCl.
4. Stop the reaction with 1.05 mL of 0.6 M NaOH and incubate at room temperature for 5 min.
5. Read the absorbance value at 445 nm. In order to obtain the ΔAbs/min, divide the measured absorbance change for 10.
6. Prepare a blank solution using the same protocol, but replacing the D-aa with H_2O.

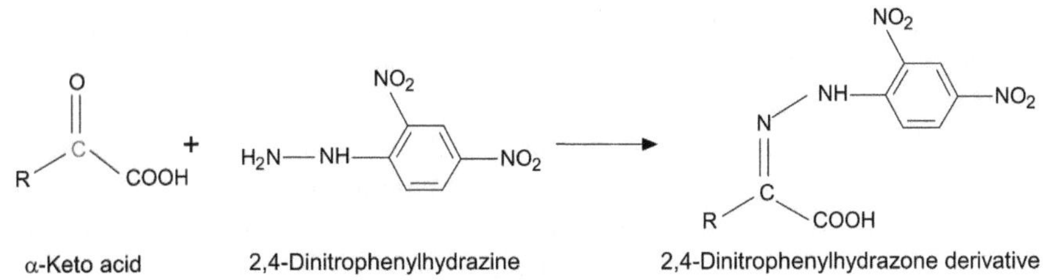

Fig. 3. Reaction of DAAO/DASPO-produced α-keto acids with DNP.

7. Determine the concentration of the α-keto acid derivative produced using Eq. 3 and an extinction coefficient $\varepsilon_{445\ \mathrm{nm}} = 3.5/\mathrm{mM/cm}$.
8. The same assay can be performed using a different D-aa. In this case, the extinction coefficient of the 2,4-dinitrophenylhydrazone derivative may be determined reacting with 2,4-dinitrophenylhydrazine with different concentrations of the α-keto acid corresponding to the employed D-aa.
9. The same assay has been optimized to assay DASPO activity (see Subheading 3.3).

3.2.2. Peroxidase Coupled Assays

The H_2O_2 produced by DAAO/DASPO activity is decomposed by peroxidase (HRP) while a colorless hydrogen donor is oxidized to colored compound (see Note 6).

Determination Based on Hydrogen Peroxide Production: Indirect Method with *o*-Dianisidine

In this assay, the compound *o*-dianisidine is used. It reacts with hydrogen peroxide to produce a red–brown dye with a broad absorption maximum centered at ~440 nm; the reaction is complete in 3 min and the color is stable for several hours. The assay is set up as follows:

1. Prepare the assay mixture in a disposable plastic semimicro-cuvette mixing 10 mM D-alanine/D-aspartate, 1 mM *o*-dianisidine, 20 μM FAD, and 1 U of HRP, in 75 mM disodium pyrophosphate, pH 8.5 for DAAO or in 200 mM triethanolamine-HCl, pH 7.6 for DASPO (all final concentrations, see Note 7); total volume 1 mL.
2. Incubate at 25°C for 5 min and record the initial absorbance of the mixture at 440 nm and then add the solution containing DAAO/DASPO (~0.01 U).
3. Monitor the absorbance change at 440 nm and calculate the $\Delta Abs_{440\ \mathrm{nm}}/\mathrm{min}$.
4. Determine the enzymatic activity using Eq. 3 and an extinction coefficient of 11.6/mM/cm.
5. The assay allows evaluation of DAAO/DASPO activities against any potential substrate simply by substituting D-alanine or D-aspartate with the compound to be tested (see Notes 8 and 9).

Applications: Detection of DASPO Activity Toward Neurotransmitter Agonists (15)

DASPO is strictly specific for D-aa since L-amino acids do not act as substrates or inhibitors. However, it is not strictly specific for D-aspartate, being active also on a variety of other D-aa, with a strong preference for dicarboxylic ones. The availability of large amounts of pure active enzyme allows to extend the characterization of the substrate specificity of DASPO to a series of other D-amino acids which are known to possess pharmacological activity as neurotransmitter agonists (15). The apparent kinetic parameters have been obtained by using the activity assay under conditions in which DASPO was shown to follow Michaelis–Menten kinetics with D-aspartate (from 2 to 30 mM) and variable amount of DASPO depending on the substrate tested.

Determination Based on Hydrogen Peroxide Production: Indirect Method with 4-Aminoantipyrine

This assay is based on the determination of H_2O_2 produced by the reaction catalyzed by DAAO/DASPO coupled to the reduction of 4-aminoantipyrine (4-AAP) catalyzed by peroxidase in presence of phenol (see Fig. 4).

The assay is set up as follows:

1. The reaction mixture is prepared directly in a disposable plastic semimicro-cuvette mixing 10 mM D-alanine/D-aspartate, 2 mM phenol, 1.5 mM 4-AAP, and 2.5 U of HRP, in 75 mM disodium pyrophosphate, pH 8.5 (all final concentrations). These conditions are the ones optimized for DAAO assay (see also Subheading 3.2.2).
2. Record the initial absorbance at 505 nm and add the enzyme preparation (<1 U) to a final volume of 1 mL and monitor the absorbance change.
3. The concentration of produced quinoneimine is determined using Eq. 3 and an extinction coefficient 6.58/mM/cm (16). Since two molecules of hydrogen peroxide react with 4-AAP to produce one molecule of quinoneimine (see Fig. 4), for the calculation of enzymatic activity (Eq. 3) a factor 2 must be added at the numerator.

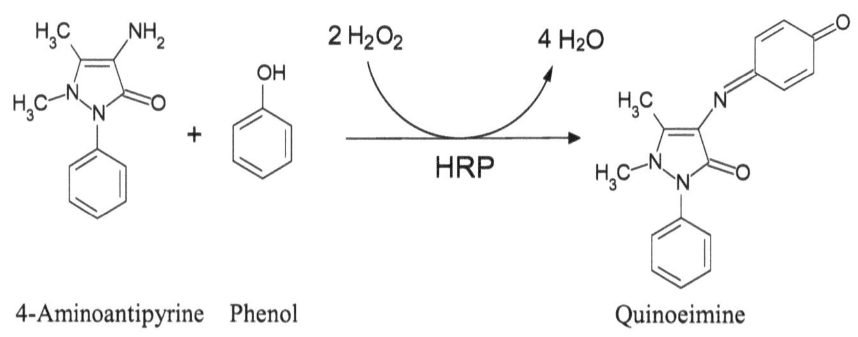

Fig. 4. Reaction of 4-AAP and phenol with H_2O_2 produced by DAAO/DASPO activity.

Amplex® Red/Ultra Red Assay

This assay is based on a fluorogenic substrate for HRP that reacts with hydrogen peroxide in a 1:1 stoichiometric ratio to produce a brightly fluorescent product. We set up a procedure to assay the DAAO activity on cell samples using a 96-well microplates (17). Procedure:

1. Prepare the cell sample by sonication (three cycles of 10 s plus 30 s in ice); centrifuge in an Eppendorf tube for 30 min at 13,000 rpm (16,000 × *g*) and 4°C. Recover the surnatant.
2. Add 50 μL of cell sample and 50 μL of Working Solution in each well maintaining the plate under nitrogen. Incubate on a shaker for 1 h, in the dark.
3. Add 20 μL of Stop Solution in each well and incubate for 5 min in the dark.
4. Read the fluorescence using a fluorimeter microplate reader: exc. = 535 nm (slit = 25 nm); em. = 590 nm (slit 20 nm).
5. The enzymatic activity is expressed as fluorescence change at 590 nm per 10,000 cells. The assay was performed on U87 glioblastoma control cells and on the same cells transfected with human DAAO expressing plasmid (17). See Notes 10–12.

3.2.3. Determination of Ammonia Production

The enzymatic activity of DAAO/DASPO can be also assayed measuring the ammonia released during the hydrolysis of the imino acid intermediate. In fact, the enzyme glutamate dehydrogenase (GDH) in presence of ammonia catalyzes the reductive amination of α-ketoglutaric acid to L-glutamic acid. During this reaction, the cofactor NADH is oxidized to NAD^+ (see Fig. 5).

The assay is set up as follows:

1. The reaction mixture is prepared in a quartz semimicro-cuvette mixing 10 mM D-alanine (for DAAO assay), 5 mM α-ketoglutaric acid, 0.25 mM NADH, and 20 U of GDH, in 75 mM disodium pyrophosphate, pH 8.5 (all final concentrations).
2. Record the initial absorbance of the mixture at 340 nm. Add the enzyme solution (~1 U of DAAO) to a final volume of 1 mL and monitor the absorbance at 340 nm. Calculate the $\Delta Abs_{340\ nm}/min$.
3. The concentration of NAD^+ enzymatically produced, corresponding to the amount of produced ammonia by DAAO reaction, is determined measuring the decrease of absorbance at 340 nm using Eq. 3 and an extinction coefficient of 6.3/mM/cm.
4. The same protocol can be used for the assay of DASPO by employing D-aspartate as substrate and 200 mM triethanolamine-HCl, pH 7.6, as buffer.

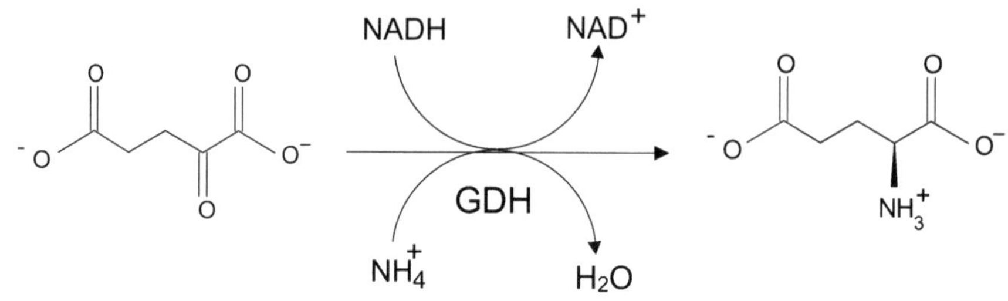

α-Ketoglutaric acid L-Glutamic acid

Fig. 5. Reaction of GDH on ammonia (produced by DAAO/DASPO reaction) and α-ketoglutaric acid.

3.3. Enzyme Activity Assay of Immobilized D-Aspartate Oxidase

DASPO can be immobilized by covalent binding to a resin such as Sepharose 4B (18). The enzyme kinetic parameters can be estimated by direct methods which involve reaction of the immobilized flavoenzyme with different reagents and the evaluation of the product freely released into solution so it can be separated from the support and assayed without interference. The oxidation of D-aspartate by DASPO produces iminoaspartate which, being very unstable, spontaneously forms oxaloacetate (through deamination). The method reported in the following paragraph is based on the measurement of this α-keto acid by reaction with 2,4-dinitrophenylhydrazine following the increase in absorbance at 445 nm (Fig. 3). See Notes 13 and 14.

1. Immobilize DASPO on CNBr-activated Sepharose 4B by incubating 17 mg of purified enzyme with 25 mL gel in buffer A for 2 h at room temperature under gentle rotation (18).
2. Add a tip of solid FAD and incubate the suspension at 4°C for 30 min (to stabilize the holoprotein of DASPO).
3. Add 20 μL of 1 M D-aspartic acid in buffer A to 2 mL of enzyme suspension to reach a final concentration of 10-mM D-aspartic acid and incubate the immobilized enzyme for 5 min at 37°C under gentle rotation following 1 min centrifugation at 1,000 × *g* at 4°C.
4. Acidify 1.5 mL of the surnatant by adding 200 μL of 25% trichloroacetic acid in water in order to precipitate proteins eventually present in solution.
5. After centrifugation at 12,000 × *g* at room temperature, add 0.1% DNP in 2 M HCl to the surnatant (100 μL of 0.1% DNP in 2 M HCl to 500 μL surnatant) and incubate at 37°C for 10 min.
6. Add 400 μL of 3.75 M NaOH to 600 μL of the solution at point 5. Incubate the sample at room temperature for 10 min before reading the absorbance at 444 nm.
7. Utilize a molar extinction coefficient of 4.72/mM/cm to calculate the enzyme activity.

8. Repeat the assay starting from step 3 but using different D-aspartate concentrations in order to calculate the apparent kinetic parameters V_{max} and K_m (see below).

3.3.1. Applications: Measurement of Free D-Aspartate in Biological Fluids by Means of Immobilized DASPO

The immobilized enzyme was used as a tool to quantify the amount of free D-aspartate in seminal plasma and in cerebrospinal fluid.

1. Add 1 mL of ethanol to 5 mL of biological fluid and incubate the sample at 0°C for 20 min in order to precipitate the proteins.
2. Upon centrifugation at 12,000 × g for 15 min and at 4°C, the surnatant was concentrated in speed vac for 20–30 min in an Eppendorf tube.
3. Add 1 mL of 50 mM Hepes, pH 8.0, and 10 μL of 20 μM FAD; the sample was solubilized by cycles of sonication and vigorous stirring (5 min each).
4. In order to measure the amount of free D-aspartate, the sample was split in three aliquots: two were treated as described in Subheading 3.2.3 starting from step 3 (without the addition of D-aspartate in the assay mixture), the third one was kept as negative control.

For both assay samples, and by using the calibration curve previously set up, a concentration of free D-aspartic acid of 344 and 154 μM in the seminal plasma and in the cerebrospinal fluid was calculated, respectively. See Note 15.

3.4. Determination of Steady-State Kinetic Parameters of DAAO/DASPO

The steady-state kinetic parameters of the reaction catalyzed by DAAO/DASPO can be determined by varying the concentration of the D-aa in the presence of constant concentrations of all other reagents and at fixed temperature (25°C). The experimental initial velocities (ΔpO_2/min for the oxygen consumption assay or ΔAbs/min for the spectrophotometric assays) are converted into enzyme units (U/mg protein, see Eq. 1) and fitted using the Michaelis–Menten equation (Eq. 4):

$$v = \frac{V_{max} \times [S]}{K_m + [S]}, \tag{4}$$

where v is the initial reaction velocity (μmol/min); V_{max} is the apparent maximal velocity; K_m is the apparent Michaelis–Menten constant (mM) for the assayed substrate (see Fig. 6). k_{cat} values (per second) are obtained from the V_{max} value divided by the enzyme concentration (1 mg of DAAO corresponds to 25 nmol) and 60 (to convert minutes in seconds).

DAAO isolated from different sources differ in substrate specificity as well as in kinetic properties (8, 9). A comparison of the apparent kinetic parameters (as determined at 25°C at fixed 0.253 mM oxygen concentration by means of the oxygen-consumption assay) of the most used DAAO forms is reported in Table 1.

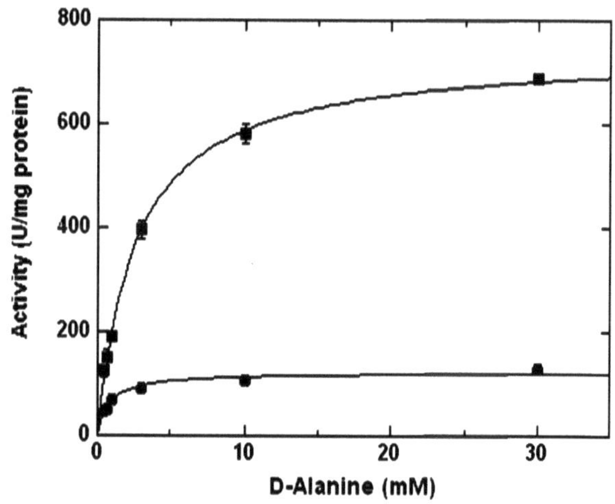

Fig. 6. Determination of steady-state apparent kinetic parameters V_{max} and K_m for the reaction of yeast DAAO with D-alanine as substrate by the oxygen consumption assay (20). Conditions: sodium pyrophosphate buffer, 100 mM, pH 8.5, at 25°C, and 0.253 mM (*filled circles*) or 1.2 mM (*filled squares*) oxygen concentration.

Table 1
Comparison of the kinetic properties of DAAO from different sources

	Human	Pig kidney	Yeast (*Rhodotorula gracilis*)	Yeast (*Trigonopsis variabilis*)
$k_{cat,app}$ (per second)	5.2	7.3	81	46
$K_{m,app}$ (mM)	1.3	1.7	1.0	7.0
$k_{cat,app}/K_{m,app}$ (/mM/s)	4.0	4.3	81	6.5

The apparent kinetic parameters were determined on D-alanine as substrate at 25°C and at air saturation (21% oxygen)

4. Notes

1. Calibration: the oxymeter is calibrated considering as 100% of oxygen content the amperometric value measured filling the oxymeter chamber with pure water (0.253 mM O_2 at 25°C), and as 0% the amperometric value measured following the addition of few crystals of dithionite.
2. During the assay the temperature may be strictly controlled since the oxygen solubility decreases with temperature increase (from 0.25 mM at 25°C to 0.214 mM at 37°C).

3. The DAAO and DASPO activity in biological samples as well on purified protein preparations can be determined using different spectrophotometric assays. These assays are based on both direct or indirect methods and on the simple Lambert–Beer's law.
4. The direct spectrophotometric method is the one based on the extinction coefficient of the α-keto acid produced by the DAAO/DASPO reaction. The concentration of the α-keto acid produced is determined from the absorbance change and by means of an appropriate calibration curve set up using different concentrations of the same α-keto acid (i.e., the calibration curve for pyruvate, the α-keto acid produced by D-alanine, gives a $\varepsilon_{340\ nm} = 0.017/mM/cm$). This method suffers of a low sensitivity.
5. Even in this case, the determination of the concentration of the 2,4-dinitrophenylhydrazone derived from the reaction is based on the set-up of a calibration curve obtained using different amounts of D-aa ($\varepsilon_{445nm} = 2.3/mM/cm$ for the 2,4-dinitrophenylhydrazone derivate of benzoylformic acid and of α-ketoglutaric acid, products originating from D-phenylglycine and D-glutamate, respectively).
6. The assay of DAAO/DASPO activity can be also based on the hydrogen peroxide produced during the reoxidation of the reduced flavin by dioxygen (Fig. 1b). These assays are indirect methods based on the reaction of H_2O_2 with an hydrogen donor to yield a colored product, with a known extinction coefficient (Figs. 3 and 4). In both cases, the coupling reaction is catalyzed by peroxidase (HRP).
7. For the determination of DASPO activity by the indirect method with DNP, *o*-dianisidine solution was prepared by adding 1 mg of *o*-dianisidine-2 HCl to 100 mL of 0.2 M triethanolamine-HCI, pH 7.6. The suspension was kept under gentle agitation for 30 min in the dark, then filtrated with a Whatman filter paper. If stored at 4°C in a dark glass bottle, the solution was stable for 1 week. In order to let oxygen concentration reach equilibrium, the solution was allowed to warm up to 25°C (while gently stirring) before use. The stability of the *o*-dianisidine stock solution can be increased to 1 month by adding 0.1% Triton X-100 without affecting DASPO activity toward D-aspartate.
8. In the same test, when using homogenates instead of purified enzyme preparations, long lag times (up to several min) before reaching maximum rate were observed. Care must be taken to avoid the presence of reducing agents (such as 2-mercaptoethanol) in the assay mixture.

9. When the activity of DASPO holoenzyme reconstituted from apoprotein and FAD was measured, the enzyme was incubated with FAD for 15 min in the reaction mixture without D-aspartate before starting the reaction by addition of the substrate.
10. On each plate different controls are also introduced:
    (a) A control without the cell sample: 50 μL of lysis buffer is added instead of the cell sample.
    (b) A control without the substrate D-serine: 150 μL of sodium phosphate buffer, pH 7.4, is used instead of the substrate. The fluorescence intensity values recorded on these wells are subtracted to the measurements performed on the assay wells.
    (c) A control blocking the DAAO activity: 3 mM benzoate (final concentration), a well-known DAAO inhibitor (8, 9, 19), is added before to start the reaction.
11. A calibration curve was obtained by adding known amounts of recombinant human DAAO to U87 crude extracts (0.02–0.4 mU range) (17, 19).
12. Optimal wavelength settings may vary between instruments: the excitation wavelength can be lowered (up to 490 nm) if saturation of excitation signal at 530 nm is observed.
13. To determine the extinction coefficient of oxalacetate, a calibration curve was set up following the procedure described above (Subheading 3.3, from step 3) but using oxaloacetate in buffer A at different final concentrations: 10 mM, 1 mM, 700 μM, 500 μM, 300 μM, 100 μM, 10 μM, 1 μM instead of D-aspartic acid. A sample containing buffer instead of oxaloacetate was kept as negative control and the corresponding absorbance was subtracted for the calculation of the calibration curve.
14. In order to evaluate the stability of the immobilized enzyme, an activity assay using a constant D-aspartate concentration (1 mM) was repeated at 1 day, 1 week, 1 and 8 months after the binding to Sepharose 4B. The enzyme was stored at 4°C and incubated at 15 or 37°C to assay the activity.
15. When the immobilized enzyme is used to evaluate the amount of free D-aspartate present in solution the incubation time must be carefully evaluated in order to obtain a result which is only related to the substrate concentration and independent from the incubation time. In the assay described above, this was achieved by measuring the absorbance at 445 nm after incubation of the enzyme with D-aspartic acid for different times: 2, 3, 4, 5, 10, 15, 30, and 60 min. A time of 30 min was required to reach the maximum absorbance value.

Acknowledgments

The works carried out in the laboratory of L. Pollegioni were supported by grants from Fondo di Ateneo per la Ricerca (University of Insubria) and from Fondazione Cariplo. LP thanks the support of Consorzio Interuniversitario per le Biotecnologie (CIB) and of Centro di Ricerca in Biotecnologie per la Salute Umana (University of Insubria). The work carried out in the laboratory of G. Tedeschi and A. Negri was supported by grants from Ministero dell'Università e della Ricerca Scientifica and from the Consiglio Nazionale delle Ricerche (Italy).

References

1. Friedman M (1999) Chemistry, nutrition, and microbiology of D-amino acids. J Agric Food Chem 47, 3457–3479.
2. Gandolfi I, Palla G, Delprato L et al. (1992) D-Amino acids in milk as related to heat treatments and bacterial activity. J Food Sci 57, 377–379.
3. Hashimoto A, Oka T (1997) Free D-aspartate and D-serine in the mammalian brain and periphery. Prog Neurobiol 52, 325–353.
4. Fujii N (2002). D-Amino acids in living higher organisms. Orig Life Evol Biosph 32, 103–127.
5. Snyder S H, Kim P M (2000). D-Amino acids as putative neurotransmitters: focus on D-serine. Neurochem Res 25, 553–560.
6. Pollegioni L, Sacchi S (2010) Metabolism of the neuromodulator D-serine. Cell Mol Life Sci 67, 2387–2404.
7. Fantinato S, Pollegioni L, Pilone M S (2001) Engineering, expression and purification of a His-tagged chimeric D-amino acid oxidase from *Rhodotorula gracilis*. Enz Microb Technol 29, 407–412.
8. Pollegioni L, Piubelli L, Sacchi S et al. (2007) Physiological functions of D-amino acid oxidases: from yeast to humans. Cell Mol Life Sci 64, 1373–1394.
9. Pollegioni L, Sacchi S, Caldinelli L et al. (2007) Engineering the properties of D-amino acid oxidases by a rational and a directed evolution approach. Curr Protein Pept Sci 8, 600–618.
10. Boselli A, Piubelli L, Molla G et al. (2007) Investigating the role of active site residues of *Rhodotorula gracilis* D-amino acid oxidase on its substrate specificity. Biochimie 89, 360–368.
11. Negri A, Massey V, Williams C H Jr (1987) D-aspartate oxidase from beef kidney. J Biol Chem 262, 10026–10034.
12. Simonic T, Duga S, Negri A et al. (1997) cDNA cloning and expression of the flavoprotein D-aspartate oxidase from bovine kidney cortex. Biochem J 322, 729–735.
13. Fonda M L, Anderson B M (1969) D-amino acid oxidase. IV. Inactivation by maleimides. J Biol Chem 244, 666–674.
14. Nagata Y, Shimojo T, Akino T (1988) Two spectrophotometric assays for D-amino acid oxidase: for the study of distribution patterns. Int J Biochem 20, 1235–1238.
15. Negri A, Tedeschi G, Ceciliani F, Ronchi S (1999) Purification of beef kidney D-aspartate oxidase overexpressed in *E. coli* and characterization of its redox potentials and oxidative activity towards agonists of excitatory amino acid receptors. Biochim Biophys Acta 143, 212–222.
16. Mori N, Sano M, Tani Y, Yamada H (1980) Purification and properties of sarcosine oxidase from *Cylindrocarpon didymum* M-1. Agric Biol Chem 44, 1391–1397.
17. Sacchi S, Bernasconi M, Martineau M et al. (2008) pLG72 modulates intracellular D-serine levels through its interaction with D-amino acid oxidase: effect on schizophrenia susceptibility. J Biol Chem 283, 22244–22256.
18. Tedeschi G, Negri A, Bernardini G et al. (1999) D-Aspartate oxidase is present in ovaries, eggs and embryos but not in testis of *Xenopus laevis*. Comp Biochem Physiol 124B, 489–494.
19. Molla G, Sacchi S, Bernasconi M G et al. (2006) Characterization of human D-amino acid oxidase. FEBS Letters 580, 2358–2364.
20. Pollegioni L, Falbo A, Pilone M S (1992) Specificity and kinetics of *Rhodotorula gracilis* D-amino acid oxidase. Biochim Biophys Acta 1120, 11–16.

Chapter 27

Enzymes Acting on D-Amino Acid Containing Peptides

Yasuhisa Asano

Abstract

Using a synthetic oligopeptide (D-Phe)$_4$, a microorganism *Bacillus cereus* DF4-B producing alkaline D-peptidase (ADP) was isolated. The enzymatic properties have been characterized; the enzyme showed D-stereospecific dipeptidyl aminopeptidase and endopeptidase activities. The enzyme was active toward (D-Phe)$_n$, Boc-(D-Phe)$_n$, (D-Phe)$_n$ methyl ester, D-Phe-NH$_2$, Boc-(D-Phe)$_n$ methyl ester, and Boc-(D-Phe)$_n$ *tert*-butyl ester, but not toward (D-Ala)$_n$ (n=2–4), (D-Val)$_3$, and (D-Leu)$_2$.

Key words: D-Amino acid, Alkaline D-peptidase, *Bacillus cereus*

1. Introduction

D-Amino acids are important starting materials for various pharmaceuticals, herbicides, and food additives: they act as intermediates for the preparation of β-lactam antibiotics such as semisynthetic cephalosporins and penicillins (1) or, like D-alanine, constitute an important part of a synthetic sweetener such as alitame (2). D-Amino acid-containing peptides are found in the bacterial peptidoglycan, D-alanine-containing dipeptides are present in rice plant, and D-amino acids are also components of the peptide antibiotics gramicidin S, tyrosidine, and bacitracin. Some oligopeptides are produced by fungi and bacteria via a non-ribosomal peptide synthesizing (NRPS) mechanism, in which the multidomain enzyme system itself acts as the template in the biosynthesis of the peptides, unlike the ribosome which uses mRNA as a template for protein synthesis (3). Skin secretions of the European frog *Bombina variegata* contain a family of hydrophobic peptides called bombinins H (4). Dermorphin is a hepta-peptide first isolated from the skin of South American frogs (5). These peptides contain a D-amino acid because of the action of "L-to-D-peptide isomerase", an epimerase.

Loredano Pollegioni and Stefano Servi (eds.), *Unnatural Amino Acids: Methods and Protocols*, Methods in Molecular Biology, vol. 794, DOI 10.1007/978-1-61779-331-8_27, © Springer Science+Business Media, LLC 2012

We are studying the properties of new D-stereospecific amino acid amidases to use them in the D-stereospecific peptide (amide) synthesis and D-stereospecific kinetic resolution of amino acid amides (1). We discovered three bacterial enzymes, D-aminopeptidase (EC 3.4.11.19), D-amino acid amidase (EC 3.5.1.-), and alkaline D-peptidase (ADP) (EC 3.4.15.-) from microorganisms. D-aminopeptidase can be used for the "dynamic kinetic resolution" of racemic amino acid amides (6), and the synthesis of D-amino acid *N*-alkyl amides in organic solvents. Most recently, D-stereospecific endopeptidase ADP has been used for L-peptide synthesis (7). Some peptidases act on peptides containing D-amino acids. Soluble *Streptomyces* carboxypeptidase DD catalyzes not only the transpeptidation reaction on the peptide intermediate in peptidoglycan biosynthesis, but also the hydrolysis of *N,N*-diacetyl-L-lysyl-$(\text{D-Ala})_2$ in water (8).

In this chapter, the chemical synthesis of D-Phe containing oligopeptides, which are not commercially available, is described. Using $(\text{D-Phe})_4$, *Bacillus cereus* Strain DF4-B, producing ADP was isolated from soil. The enzymatic properties are characterized.

2. Materials

1. DEAE-Toyopearl 650M, Butyl-Toyopearl 650M, HPLC columns G-3000 SW, and DEAE-5PW were purchased from Tosoh Corp. (Tokyo, Japan).
2. Sephacryl S-300 was purchased from Pharmacia (Sweden).
3. Marker proteins for molecular mass determination were from Oriental Yeast (Tokyo, Japan).
4. DNA manipulations were carried out under standard procedures (9).
5. $(\text{Boc})_2\text{O}$ (di-*t*-butyl dicarbonate) (10), Boc-ON (2-(*t*-butoxy-car-bonyloxyimino)-2-phenylacetonitrile) (11), and 4 N hydrogen chloride in ethylacetate were purchased from Kokusan Chemicals (Tokyo, Japan).
6. For microbial screening, a large amount of $(\text{D-Phe})_4$ was synthesized in-house as described in Subheading 3.
7. Thionyl chloride, carbodiimide, isobutyl chloroformate, absolute methanol, and other chemicals were purchased from Wako Pure Chemicals Co. (Osaka, Japan).
8. *N*-Nitrosomethylurea was from ICN K & K Laboratories, Inc. (Plainview NY, USA).
9. Palladium-carbon (10%) was purchased from Nippon Engelhard Inc. (Tokyo, Japan).

10. All other chemicals were from commercial sources and used without further purification.
11. LB Plates containing $(Phe)_4$ are prepared as follows: After autoclaving LB medium containing 1.5% agar, $(D\text{-}Phe)_4$ dissolved in 10% dimethylsulfoxide (w/v) is quickly added before solidification, to a final concentration of 5 mg per 20 mL (a typical volume required for a plate).

3. Methods

3.1. Synthesis of $(D\text{-}Phe)_4$ and Other Substrates for the Characterization of Alkaline D-Peptidase

1. Synthesize the methylesters of amino acids by slowly dropping 1.2–5 equivalents of thionylchloride into amino acids suspended in methanol at –20°C (12).
2. Synthesize Boc-amino acids from free amino acids by the use of $(Boc)_2O$ (10) or Boc-ON (11). Dissolve 33.3 g (0.20 mol) D-Phe and 16.8 g (0.20 mol) sodium bicarbonate in 60% 1,4-dioxane (200 mL); chill on ice. Slowly add to the solution 52.4 g (0.24 mol) Boc_2O. Then add 60 mL of trimethylamine (0.40 mmol) and 200 mL of 1,4-dioxane at 0°C, and stir for 24 h at room temperature. Wash the reaction mixture with chloroform, and acidify the water layer to pH 2–3with crystalline citric acid. Extract the solution by ethyl acetate and dry with $MgSO_4$, evaporate and crystallize from ethereal ethyl acetate (28.8 g, 98%).
3. Synthesize methylesters of the Boc-D-amino acids by methylation of the Boc-D-amino acids by ethereal diazomethane, which was evolved from *N*-nitrosomethylurea (12). Prepare diazomethane as follows: in a 500-mL round-bottomed flask, place 60 mL of 50% aqueous solution of potassium hydroxide and 200 mL of ether; cool the mixture to 5°C and add 20.6 g (0.20 mol) of *N*-nitrosomethylurea. Place the reaction flask in a water bath at 50°C. Distill the ether solution containing diazomethane and collect as ethereal diazomethane by cooling. Add to an appropriate amount of Boc-D-amino acid, dissolved in ethyl acetate, diazomethane in ether until the evolution of nitrogen gas stops and the yellow color of diazomethane does not modify anymore (meaning that the esterification reaction is completed) (see Note 1).
4. Prepare ammonia-saturated absolute methanol by bubbling ammonia gas into absolute methanol cooled at about –20°C by ice, methanol, and NaCl.
5. Synthesize amino acid amides by ammonialysis (13) of amino acid methylesters·HCl or Boc-amino acid methylesters, in absolute methanol saturated with dry ammonia gas at room

temperature. Suspend Boc-D-phe methylester (15.1 g, 0.05 mol) in 200 mL of methanol and chill to −15°C in a hood. Introduce a dry ammonia gas into the solution from a cylinder every day for 5 min under chilling at −20°C, and stir at room temperature. Monitor the disappearance of the substrate in the reaction mixture by TLC (1-butanol:acetic acid:water = 4:1:1, by vol.). Evaporate the reaction mixture and crystallize to obtain Boc-D-phe amide methyl ester (12.1 g, 80%).

6. Synthesized peptide substrates used to screen and test the substrate specificity from D- and L-Phe. NH_2 and COOH termini are protected by Boc (10) and methyl groups, respectively. Isobutyl chloroformate (14) and carbodiimide (15) condenses the monomer to a dimer and the dimer to a tetramer of Phe, respectively.

   Boc-D-Phe-D-Phe methyl ester synthesis is described as follows: dissolve Boc-D-Phe (53.5 g, 0.20 mol) in dry-THF (800 mL), in ice chilled with NaCl at −15°C, and under stirring. Add triethylamine (50.6 g, 0.5 mol), isobutyl chloroacetate (30.1 g, 0.222 mol), and then 47.9 g (0.222 mol) of D-Phe methyl ester·HCl, stir for 5 h at room temperature. Evaporate the reaction mixture and wash with saturated $NaHCO_3$ (700 mL), 10% citric acid (700 mL), and then with saturated NaCl solution (700 mL). Dry the solution with $MgSO_4$, evaporate and crystallize from ethyl acetate/hexane (67.2 g, 78%).

   Boc-$(\text{D-Phe})_4$ methyl ester was synthesized as follows: dissolve 20.6 g (0.05 mol) of Boc-D-$(\text{Phe})_2$, 21.4 g (0.06 mol) of D-$(\text{Phe})_2$ methyl ester, 7.10 g (0.053 mol) of 1-hydroxybenzotriazole and 6.1 g (0.06 mol) of triethylamine in dimethylformamide (400 mL), in a chilled ice with methanol and NaCl at −20°C, and under stirring. Add water-soluble carbodiimide·HCl (12.5 g, 0.65 mol) and stir overnight; the temperature is slowly brought to room temperature. Evaporate the reaction mixture and then add 500 mL of water. Filter the crystals formed and then wash with 10% citric acid (500 mL), water (500 mL), 5% $NaHCO_3$ solution (500 mL), and water (500 mL). Dry the crystals over P_2O_5, and recrystallize Boc-$(\text{D-Phe})_4$ methyl ester from ethyl acetate/hexane (30.4 g, 84%).

7. Carry out the deprotection of Boc (*tert*-butoxycarbonyl) and Z (benzyloxycarbonyl) moieties in 4 N hydrogen chloride in ethylacetate and by catalytic hydrogenation in the presence of 10% Palladium-carbon, respectively.

8. Synthesize the following peptide derivatives: Boc-D-Phe, D-Phe *tert*-butyl ester, $(\text{D-Phe})_2$·HCl, Boc-$(\text{D-Phe})_2$, $(\text{D-Phe})_2$ methyl ester·HCl, Boc-$(\text{D-Phe})_2$ methyl ester, $(\text{D-Phe})_3$·HCl,

Boc-(D-Phe)$_3$, Boc-(D-Phe)$_3$ methyl ester, Boc-(D-Phe)$_3$ *tert*-butyl ester, (D-Phe)$_4$·HCl, Boc-(D-Phe)$_4$, Boc-(D-Phe)$_6$, Boc-(D-Phe)$_4$ methyl ester, L-Phe methyl ester·HCl, (L-Phe)$_2$ methyl ester·HCl, (L-Phe)$_3$·HCl, (L-Phe)$_4$·HCl, Boc-(L-Phe)$_4$, Boc-(L-Phe)$_4$ methyl ester, (D-Phe)$_2$-D-Tyr·HCl, D-Tyr-(D-Phe)$_2$·HCl, D-Phe-(L-Phe)$_2$·HCl, L-Phe-(D-Phe)$_2$·HCl, (D-Phe)$_2$-L-Phe·HCl, (L-Phe)$_2$-D-Phe·HCl, D-Phe-L-Phe-D-Phe·HCl, L-Phe-D-Phe-L-Phe·HCl, D-Phe-L-Phe·HCl, and L-Phe-D-Phe·HCl.

3.2. Screening for (D-Phe)$_4$-Degrading Microorganisms

1. The ability of microorganisms to hydrolyze (D-Phe)$_4$ in LB medium (9) can be screened in enriched cultures at 30°C.
2. Dissolve (D-Phe)$_4$ in 10% Me$_2$SO (w/v) and then add to 2 mL of LB medium-containing soil samples.
3. Then aerobically shake the mixture for 2 days.
4. Transfer a loopful of the culture broth to the same medium and aerobically incubate under the same conditions.
5. Streak a small portion of the culture broth onto a plate of the same medium containing 1.5% agar and incubate at 30°C overnight.
6. Isolate strains forming clear zones around the colonies (Fig. 1), and monitor (D-Phe)$_4$ degradation in the liquid culture by

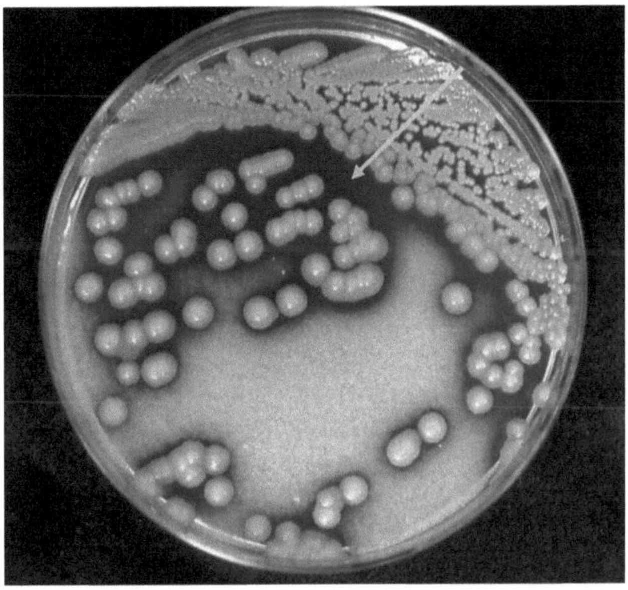

Fig. 1. Formation of halo on (D-Phe)$_4$ containing culture plates of *B. cereus* DF4-B. In the screening for (D-Phe)$_4$-degrading microorganisms, a microorganism was isolated based on its ability to give an halo around the colony. This microorganism was identified as *B. cereus* DF4-B, and it excreted alkaline D-peptidase outside the cells (16). The enzyme caused a halo (*black part* indicated by an arrow) by hydrolyzing insoluble (D-Phe)$_4$ (*white part*) contained in the medium.

TLC (chloroform/methanol/acetic acid = 8:2:1 or 10:2:1) and visualization with ninhydrin (see Note 2).

7. A bacterial strain isolated from a soil of Kanagawa Prefecture, Japan completely degraded the substrate by forming $(\text{D-Phe})_2$ (16) (see Note 3).

3.3. Activity Measurement of ADP

1. ADP activity is routinely assayed at 30°C by measuring the production of $(\text{D-Phe})_2$ from $(\text{D-Phe})_4$ by HPLC (15).
2. Prepare the reaction mixture by mixing 1 μmol of $(\text{D-Phe})_4$, 10 μL of Me_2SO, 50 μmol of Tris–HCl, pH 9.0, and 1 μmol of $MgSO_4$.
3. Start the assay by addition of the enzyme in a total volume of 500 μL.
4. Stop the reaction with 50 μL of 2 N $HClO_4$.
5. Estimate the amount of $(\text{D-Phe})_2$ formed with a Waters 600E HPLC apparatus equipped with a Cosmosil 5 C18-MS reverse-phase column (4.6 mm × 150 mm) at a flow rate of 1.0 mL/min, using 35% methanol in 5 mM KH_2PO_4/H_3PO_4 buffer, pH 2.9.
6. Monitor the absorbance of the eluate at 254 nm.
7. One unit of enzyme activity is defined as the amount of enzyme that catalyzes the formation of 2 μmol of $(\text{D-Phe})_2$ from 1 μmol of $(\text{D-Phe})_4$ per minute.

3.4. Substrate Specificity of ADP

1. Qualitatively examine the enzyme activity by thin-layer chromatography first and then quantitatively assay it by the following methods.
2. Measure the enzyme activity toward peptide substrates as described above with 2.5 U of enzyme.
3. The amounts of $(\text{D-Phe})_2$, D-Phe, L-Phe-D-Phe, Boc-D-Phe, and Boc-$(\text{D-Phe})_2$ are quantitatively assayed by HPLC on a Cosmosil 5 C18-MS reverse-phase column (4.6 mm × 150 mm) at a flow rate of 1.0 mL/min, using 5 mM KH_2PO_4/H_3PO_4 buffer, pH 2.9: methanol = 13:7 or 9:11 (v/v) as solvent system.
4. In order to determine the kinetic constants of the enzyme for the peptide substrate, prepare a reaction mixture containing 0.1–12.5 μmol of the substrate, 50 μmol of Tris–HCl, pH 9.0, 1 μmol of $MgSO_4$, 10 μL of Me_2SO, and 100 μL of the enzyme solution in a total volume of 500 μL (see Notes 4 and 5).

3.5. Properties of ADP

Measure the optimal pH, and temperature of ADP. Measure the molecular mass by HPLC. Judge the mode of action of the enzyme toward various synthetic substrates (see Notes 6 and 7).

3.6. Cloning of the adp Gene

1. Partially digest *B. cereus* genomic DNAs with *Mbo*I and fraction it by sucrose density gradient ultracentrifugation (5–25%; 100,000 × *g*, 16 h).
2. Purify DNA fragments of 3–6 kb and ligate into *Bam*HI-digested and dephosphorylated pUC118 by T4 ligase.
3. Select ampicillin-resistant transformants expressing ADP activity by monitoring halo formation from (D-Phe)$_4$: the host *E. coli* JM109 cells does not show ADP activity and thus only transformants will produce the halo.
4. A transformant exhibited ADP activity, harboring a plasmid designated pBDP2 with a 5-kb DNA insert was previously identified (16).
5. For subcloning, digest pBDP2 with *Eco*RI and *Sal*I, and then introduce the resulting 1.3-kb fragment into the *Eco*RI and *Sal*I sites of pUC118 to yield pBDP22 (15, 16) (see Note 7).

4. Notes

1. Diazomethane is toxic and explosive. Gloves and goggles should be worn and the solution be placed and handled in a hood door. It is also recommended that ground joints and sharp surfaces should be avoided and the connections be made with rubber stoppers. Diazomethane solutions should not be exposed to direct sunlight (12).
2. (D-Phe)*n* (*n* = 1–4) are well separated by TLC with the solvent employed here.
3. An enrichment culture in LB medium containing a synthetic substrate (D-Phe)$_4$ led to an isolation of a bacterial strain *Bacillus cereus* DF4-B. The extracellular enzyme hydrolyzing (D-Phe)$_4$ was purified and characterized (16).
4. Substrate specificity of ADP. The enzyme was active toward (D-Phe)$_4$ (relative activity: 100%, K_m value: 0.398 mM) and (D-Phe)$_3$ (90%, 0.127 mM), forming (D-Phe)$_2$ and D-Phe. The calculated V_{max}/K_m values against (D-Phe)$_3$, and (D-Phe)$_4$ were about 1,000 times higher than that for (D-Phe)$_2$. As shown in Table 1, the enzyme was active toward (D-Phe)*n*, Boc-(D-Phe)*n*, (D-Phe)*n* methyl ester, D-Phe-NH$_2$, Boc-(D-Phe)*n* methyl ester, Boc-(D-Phe)*n* *tert*-butyl ester, but not toward (D-Ala)*n* (*n* = 2–4), (D-Val)$_3$, and (D-Leu)$_2$. The enzyme is also active toward tripeptides L-Phe-(D-Phe)$_2$ (119%, 0.455 mM), L-Phe-D-Phe-L-Phe (28.1%, 1.63 mM), D-Tyr-(D-Phe)$_2$ (83.6%), (D-Phe)$_2$-D-Tyr (83.6%), and on Boc-(D-Phe)*n* (*n* = 2–4) forming Boc-D-Phe, (D-Phe)$_2$ and D-Phe. The enzyme digested (D-Phe)$_2$-D-Tyr into (D-Phe)$_2$ and D-Tyr, and D-Tyr-(D-Phe)$_2$ into

Table 1
Substrate specificity of ADP (16)

Substrate	Relative activity (%)	V_{max} (U/mg protein)	K_m (mM)	V_{max}/K_m
(D-Phe)$_6$	1.8			
(D-Phe)$_4$	100	199	0.4	500
(D-Phe)$_3$	90	130	0.1	1,020
(D-Phe)$_2$	0.2	13.7	50.1	0.3
D-Phe-L-Phe	<0.1			
(D-Phe)$_2$-L-Phe	14.9	30.6	0.5	59
L-Phe-(D-Phe)$_2$	119	154	0.5	346
L-Phe-D-Phe-L-Phe	28.1	66	1.6	41
D-Tyr-(D-Phe)$_2$	83.6			
(D-Phe)$_2$-D-Tyr	83.6			
D-Phe-OMe	15.0		1.8	
D-Phe-NH_2	0.1		0.1	
D-Phe-pNA	4.2			
Boc-(D-Phe)$_4$	1.8		0.8	
Boc-(D-Phe)$_3$	3.2		1.1	
Boc-(D-Phe)$_2$	7.0			
Boc-(D-Phe)$_3$-O^{t}Bu	1.2		0.3	
Boc-(D-Phe)$_4$-OMe	0.5		0.2	
Boc-(D-Phe)$_3$-OMe	0.7		0.3	
Boc(D-Phe)$_2$-OMe	1.4			
Ampicillin	8.9	262	73.1	3.6
Penicillin G	9.7	250	48.9	5.1

The following compounds were not used as a substrate: (L-Phe)$_4$, (L-Phe)$_3$, D-Phe-L-Phe-D-Phe, D-Phe-(L-Phe)$_2$, (L-Phe)$_2$-D-Phe, (L-Phe)$_2$, L-Phe-D-Phe, L-Phe methyl ester, L-Phe amide, Boc-(L-Phe)$_3$, Boc-(L-Phe)$_4$ methyl ester, D-Leu *p*-nitroanilide, D-Ala *p*-nitroanilide, D-phenylglycine amide, (D-Ala)$_5$, (D-Ala)$_4$, (D-Ala)$_3$, (D-Ala)$_2$, (D-Val)$_3$, (D-Leu)$_2$, and D,L-Ala-D,L-Phe

D-Tyr-D-Phe and D-Phe. The enzyme had esterase activity toward D-Phe methyl ester and (D-Phe)$_2$ methyl ester. The products from Boc-(D-Phe)$_3$ *tert*-butyl ester were Boc-D-Phe, D-Phe and D-Phe *tert*-butyl ester. A dimer was formed when D-Phe methyl ester and D-Phe-NH_2 were the substrates. The enzyme also showed poor β-lactamase activity toward ampicillin (8.9%, 73.1 mM) and penicillin G (9.7%, 48.9 mM).

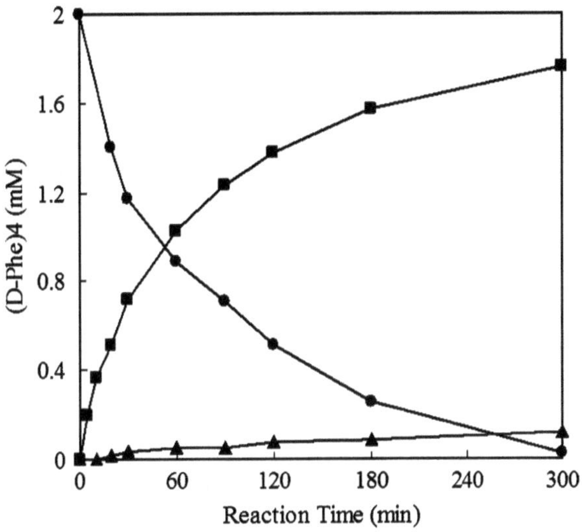

Fig. 2. The time course of (D-Phe)$_4$ hydrolysis by ADP. The reaction mixture consisted of 2 mM of (D-Phe)$_4$, 100 mM of Tris–HCl, pH 9.0, 2 mM $MgSO_4$, 2% (v/v) of DMSO, and 0.038 units of the enzyme in a total volume of 500 μL. The reaction was carried out at 30°C and stopped with $HClO_4$. The amount of (*filled circle*) (D-Phe)$_4$, (*filled square*) (D-Phe)$_2$, and (*filled triangle*) D-Phe in the supernatant was measured by a Waters HPLC equipped with a Cosmosil 5 C18-MS at a flow rate of 1.0 mL/min with a mixture of methanol (55%) and 5 mmol $KH_2PO_4 \cdot H_3PO_4$, pH 2.9 (45%) (16).

5. Time course of the (D-Phe)$_4$ degradation. (D-Phe)$_4$ was hydrolyzed to (D-Phe)$_2$ and D-Phe (Fig. 2), while no (D-Phe)$_3$ was detected. These results coincide with the kinetic properties of the enzyme described above. The mode of action of the enzyme was examined with the synthetic substrates D-Tyr-(D-Phe)$_2$ and (D-Phe)$_2$-D-Tyr. When D-Tyr-(D-Phe)$_2$ was the substrate, D-Phe was released first, then D-Tyr was slowly formed. When (D-Phe)$_2$-D-Tyr was used as a substrate, D-Tyr was released first, then D-Phe was slowly formed. In both the reactions, the second peptide bond from the N-terminus of the substrate was hydrolyzed first. These results showed that the enzyme acts as a D-stereospecific dipeptidylendopeptidase.
6. Properties of the enzyme. The enzyme is an endopeptidase that acts D-stereospecifically upon peptides composed of aromatic D-amino acids, recognizing the D-configuration of the amino acid, whose carboxy-terminal peptide bond is hydrolyzed (16). The enzyme had an optimum pH at around 10.3 (and it was stable in a relatively wide pH region). Thus, the enzyme was named "alkaline D-peptidase" (D-stereospecific peptide hydrolase, EC 3.4.15.-). Although the enzyme showed poor β-lactamase activity toward ampicillin and penicillin G, carboxypeptidase DD (17) and DAP (18) activities were not detected. The enzyme can be classified as a member of the

"penicillin-recognizing enzymes" family (8). The enzyme was inhibited by PMSF, indicating that a serine residue is responsible for exerting the enzyme activity.

7. Structure of ADP. The gene coding for ADP (*adp*) was cloned into plasmid pUC118, and a 1,164-bp open reading frame consisting of 388 codons – encoding for a polypeptide of *Mr* of 42,033 Da – was identified as the *adp* gene (19). ADP would contain a signal peptide: the *Mr* of the subunit was calculated about 36,000 Da by SDS-PAGE, 37,000 Da by HPLC, and 37,952 Da by mass spectroscopy.

References

1. Asano Y (2010) Tools for enzyme discovery. Industrial enzymes, biocatalysis and enzyme evolution. In: Baltz R H, Davies J E, Demain A (eds) Manual of Industrial Microbiology and Biotechnology, 3rd edn. American Society for Microbiology, pp 441–452.
2. Glowaky R C, Hendrick M E, Smiles R E, Torres A (1991) Development and uses of alitame. A novel dipeptide amide sweetener. ACS Symp. Ser. 450, Sweeteners, 57–67.
3. Montavon T J, Bruner S D (2010) Nonribosomal peptide synthetases. In: Vederas J C (ed) Comprehensive natural products II, Chemistry and Biology, Vol 5, pp 619–655.
4. Jilek A, Mollay C, Tippelt C et al. (2005) Biosynthesis of a D-amino acid in peptide linkage by an enzyme from frog skin secretions. Proc Natl Acad Sci USA 102, 4235–4239.
5. Krell G (1994) Conversion of L- to D-amino acids: posttranslational reaction. Science 266, 996–997.
6. Asano A, Yamaguchi S (2005) Dynamic kinetic resolution of amino acid amide catalyzed by D-aminopeptidase and α-amino-ε-caprolactam racemase. J Am Chem Soc 127, 7696–7697.
7. Wehofsky N, Pech A, Liebscher S et al. (2008) D-amino acid specific proteases and native all-L-proteins: a convenient combination for semisynthesis. Angew Chem Int Ed Engl 47, 5456–5460.
8. Frère J-M, Joris B, Dideberg O et al. (1988) Penicillin-recognizing enzymes. Biochem Soc Trans 16, 934–938.
9. Maniatis T, Fritsch E F, Sambrook J (1982) Molecular Cloning: A Laboratory Manual, Cold Spring Harbor Laboratory, Cold Spring Harbor, NY.
10. Tarbell D S, Yamamoto Y, Pope B M (1972) New method to prepare *N-t*-butoxycarbonyl derivatives and the corresponding sulfur analogs from di-*t*-butyl dicarbonate or di-*t*-butyl dithiol dicarbonates and amino acids. Proc Natl Acad Sci USA 69, 730–732.
11. Ito M, Hagiwara D, Kamiya T (1977) Peptides. VI. Some oxime carbonates as novel *t*-butoxycarbonylating reagents. Bull Chem Soc Jpn 50, 718–721.
12. Arndt F (1943) Organic Syntheses, Coll. Vol. 2, 165.
13. Altman J, Shoef N, Wilchek M, Warshawsky A (1983). Bifunctional chelating agents. Part 1. 1-(*p*-Aminophenethyl)-ethylenediaminetetraacetic acid. J. Chem. Soc. Perkin Trans I 365–368.
14. Vaughan J R Jr, Osato R L (1952) The preparation of peptides using mixed carbonic-carboxylic acid anhydrides. J Am Chem Soc 74, 676–678.
15. Sheehan J C, Preston J, Cruikshank P A (1965) A rapid synthesis of oligopeptide derivatives without isolation of intermediates. J Am Chem Soc 87, 2492–2493.
16. Asano Y, Ito H, Dairi T, Kato Y (1996) An alkaline D-stereospecific endopeptidase with beta-lactamase activity from *Bacillus cereus*. J Biol Chem 271, 30256–30262.
17. Duez C, Piron-Fraipont C, Joris B et al. (1987) Primary structure of the *Streptomyces* R61 extracellular DD-peptidase. Eur J Biochem 162, 509–518.
18. Asano Y, Kato Y, Nakazawa A, Kondo K (1992) Structural similarity of D-aminopeptidase to carboxypeptidase DD and β-lactamase. Biochemistry 31, 2316–2328.
19. Komeda H, Asano Y (2003) Genes for an alkaline D-stereospecific endopeptidase and its homologue are located in tandem on *Bacillus cereus* genome. FEMS Microbiol Lett 228, 1–9.

INDEX

Loredano Pollegioni and Stefano Servi (eds.), *Unnatural Amino Acids: Methods and Protocols*,
Methods in Molecular Biology, vol. 794, DOI 10.1007/978-1-61779-331-8, © Springer Science+Business Media, LLC 2012

P

R

S

T

U

MIX
Papier aus verantwortungsvollen Quellen
Paper from responsible sources
FSC® C105338

If you have any concerns about our products,
you can contact us on
ProductSafety@springernature.com

In case Publisher is established outside the EU,
the EU authorized representative is:
Springer Nature Customer Service Center GmbH
Europaplatz 3, 69115 Heidelberg, Germany

Printed by Libri Plureos GmbH
in Hamburg, Germany